Contemporary Ergonomics 2006

Contemporary Ergonomics 2006

Editor

Philip D. Bust

Loughborough University

© 2006 Taylor & Francis

Establishing Ergonomics Expertise
*Malcolm Jackson, David Hitchcock, Fari Haynes, Dona Teasdale,
Becky Hilton, Evelyn Russell, Frances Flynn, Moira Coates & Rachel Mortimer*
© Crown Copyright 2006/DWP

Management of Manual Handling Risk in Welsh Care Homes
A.D.J. Pinder, P. Marlow & V. Gould
© Crown Copyright 2006/HSL

A Method for Testing Pressures on Fingers in Entrapment Hazards
Edmund Milnes
© Crown Copyright 2006/HSL

Slips and Trips in the Prison Service
Anita Scott & Kevin Hallas
© Crown Copyright 2006/HSL

Falls from Vehicles
Kevin Hallas, Anita Scott & Mary Miller
© Crown Copyright 2006/HSL

Assessing the Impact of Organisational Factors on Safety in a High Hazard Industry
Johanna Beswick, Shuna Powell, Martin Anderson & Alan Jackson
© Crown Copyright 2006/HSL

Reproduced with permission of the Controller of her Britannic Majesty's Stationary Office. The views expressed are those of the Author and do not necessarily reflect the views or policy of the Controller or any government department. The Controller, any government and Taylor & Francis Ltd accept no responsibility for the accuracy of any recipe, formula instruction contained within this population.

Printed and bound in Great Britain.

All rights reserved. No part of this book may be reprinted or reproduced or utilised in any form or by any electronic, mechanical, or other means, now known or hereafter invented, including photocopying and recording, or in any information storage or retrieval system, without permission in writing from the publishers.

Every effort has been made to ensure that the advice and information in this book is true and accurate at the time of going to press. However, neither the publisher nor the authors can accept any legal responsibility or liability for any errors or omissions that may be made. In the case of drug administration, any medical procedure or the use of technical equipment mentioned within this book, you are strongly advised to consult the manufacturer's guidelines.

ISBN10   0 415 39818 5   ISBN13   978 0 415 39818 3

CONTENTS

Preface   xvii

DONALD BROADBENT LECTURE

An engineer's view of human error   3
T.A. Kletz

APPLICATIONS OF ERGONOMICS

On documentation for the design of school buildings   13
M.A. Tainsh

Is that a gun? The influence of features of bags and threat items on detection performance   17
X. Liu, A.G. Gale, K. Purdy & T. Song

The warship as a sailor's home: From HMS victory to the type 45 destroyer   23
S. Wells & P. Thorley

Towards the development of a low cost simulator to support NVG training   28
J. Wu, K.Y. Lim & C.W. Koh

Re-evaluating kao's model of educational ergonomics in light of current issues   33
A. Woodcock

The kids on the adults armchair, the ergonomy of the computers   38
N. Özdener, O.E. Özdener, Z. Sütoluk & M. Akbaba

COMPLEX SYSTEMS/TEAMS

Understanding bakery scheduling: Diverse methods for convergent constraints in user-centered design   45
P.C.-H. Cheng, R. Barone, N. Pappa, J.R. Wilson, S.P. Cauvain & L.S. Young

Work compatibility improvement framework: Defining and measuring the human-at-work system   50
A. Alhemood & A. Genaidy

Team training need analysis for safety critical industries   54
H. Livingstone, C. Deighton & R. Luther

Development for a tool testing team reliability   58
I.H. Smith, C.E. Siemieniuch & M.A. Sinclair

Ergonomics aspects of "good engineering governance" in the design process 62
J.V. Nendick, C.E. Siemieniuch & M.A. Sinclair

Knowledge-centred design as a generative basis for user-centred design 67
R. Barone & P.C.-H. Cheng

On the use of social network analysis to understand group decision-making 70
S. Cook, C.E. Siemieniuch & M.A. Sinclair

The influence of sharing displays on team situation awareness and performance 73
A.P. Banks & W.J. McKeran

CONTROL ROOMS SYMPOSIUM

En-route air traffic control rooms 1965–2005 79
H. David

Air Traffic Control consoles 1965–2005 83
H. David

Control room ergonomics: Observations derived from professional practice 88
K.Y. Lim

Media, signs and scales for the design space of instrument displays 93
M. May & J. Petersen

Development of a prototype overview display for Leningrad nuclear power station 98
A. Anokhin & E. Marshall

Operations room redesign for the Australian anzac class frigates 103
S. Sutherland, S. Hanna & S. Cockshell

CCTV in control rooms: Meeting the ergonomic challenges 109
J. Wood

DEFINING ERGONOMICS

Ergonomics advisors – a homogeneous group? 117
C. Williams & R.A. Haslam

Establishing ergonomics expertise 122
M. Jackson, D. Hitchcock, F. Haynes, D. Teasdale, B. Hilton, E. Russell, F. Flynn, M. Coates & R. Mortimer

Putting the "design engineering and specification" back into human factors 127
K.Y. Lim

What did you do in the war Hywel? Some foundations of ergonomics 132
R.B. Stammers

DESIGN

Maximising VMC: The effects of system quality and set-up on communicative success 139
O.I. Martino, C. Fullwood, S.J. Davis, N.M. Derrer & N. Morris

A cognitive study of knowledge processing in collaborative product design 144
Y.F. Qiu, Y.P. Chui & M.G. Helander

Comparisons between user expectations for products in physical and virtual domains 149
B. Sener, P. Gültekin & Ç. Erbug

DESIGN – ENGAGE PROJECT

Introduction to project engage: designing for emotion 157
M. Tito & T.H.C. Childs

Implementation of affective design tools in the clothing industry 159
M.-J. Such, C. Solves, J. Olaso, B. Mateo, J. Montero, C. Chirivella, P. Vera & R. Dejoz

Validation study of Kansei engineering methodology in footwear design 164
C. Solves, M.-J. Such, J.C. González, K. Pearce, C. Bouchard, J. Mª Gutierrez, J. Prat & A.C. García

Safety semantics: A study on the effect of product expression on user safety behaviour 169
I.C.M. Karlsson & L. Wikström

A semantic differential study of combined visual and tactile stimuli for package design 174
B. Henson, D. Choo, C. Barnes & T.H.C. Childs

Drivers' experiences of material qualities in present and future interiors of cars and trucks 179
L. Sperling & P. Eriksson

Of the interaction of cultural and emotional factors on industrial design 184
P.-H. Dejean, M. de Souza & C. Mourthe

Understanding user experience for scenario building: A case in public transport design G. Hasdoğan, N. Evyapan & F. Korkut	189
Subjective assessment of laminate flooring S. Schütte, A. Lindberg & J. Eklund	194
RealPeople S. Chhibber, S. Porter, J.M. Porter & L. Healey	199
HADRIAN R. Marshall, J.M. Porter, R. Sims, D.E. Gyi & K. Case	200

HCI SYMPOSIUM – KNOWING THE USER

Validating diagnostic design knowledge for air traffic management: A successful case-study B. Hill & J. Long	203
Human factors evaluation of systems that recognise and respond to user emotion K. Hone & L. Axelrod	208

HCI SYMPOSIUM – USABILITY AND BEYOND

Eye-centric ICT control F. Shi, A.G. Gale & K. Purdy	215
Applying the keystroke level model in a driving context M. Pettitt, G. Burnett & D. Karbassioun	219
The physical world as an abstract interface D. Edge, A. Blackwell & L. Dubuc	224

HCI SYMPOSIUM – ACCESS AND INCLUSIVITY

Accessibility vs. usability – Where is the dividing line? S. Chandrashekar & R. Benedyk	231
A technique for the client-centred evaluation of electronic assistive technology G. Baxter & A. Monk	236
Synergy of accessibility, usability and acceptance: Towards more effective and efficient evaluation C.M. Harrison & H.L. Petrie	241
A framework for evaluating barriers to accessibility, usability and fit for purpose S. Keith, G. Whitney & W. Wong	246

HCI – APPLICATIONS

The decay of malleable attentional resources theory 253
M.S. Young & N.A. Stanton

User responses to the learning demands of consumer products 258
T. Lewis, P.M. Langdon & P.J. Clarkson

E-learning support for postgraduate students 263
A. Woodcock, A.-M. McTavish, M. De, L. Slater & I. Beucheler

HCI – INTERFACES

A haptic Fish Tank virtual reality system for interaction with scientific data 271
W. Qi

Roles for the ergonomist in the development of human–computer interfaces 276
H. David

Fundamental examination of HCI guidelines 280
Y. Dadashi & S. Sharples

Virtual and real world 3D representations for task-based problem solving 285
I. Ashdown & S. Sharples

Here's looking at you: A review of the nonverbal limitations of VMC 290
S.J. Davis, C. Fullwood, O.I. Martino, N.M. Derrer & N.Morris

The effect of an icebreaker on collaborative performance across a video link 293
C. Fullwood, N.M. Derrer, O.I. Martino, S.J. Davis & N. Morris

An initial face-to-face meeting improves person perceptions of interviewees across VMC 296
N.M. Derrer, C. Fullwood, S.J. Davis, O.I. Martino & N. Morris

Mindspace 299
H. David

HOSPITAL ERGONOMICS

Hospital bed spaces: Patient experiences and expectations 303
K.Taylor & S. Hignett

Postural analysis of loading and unloading tasks for emergency ambulance stretcher-loading systems 308
A. Jones & S. Hignett

Management of manual handling risk in Welsh care homes 313
A.D.J. Pinder, P. Marlow & V. Gould

Keeping abreast of the times 318
H.J. Scott & A.G. Gale

Fatigue experienced by cytology screeners reading 321
conventional and liquid based slides
J. Cole & A.G. Gale

INCLUSIVE DESIGN – IN THE BUILT ENVIRONMENT SYMPOSIUM

Decent homes as standard, but are they inclusive? 329
M. Ormerod, R. Newton & P. Thomas

Accessible housing design for people with sight loss 333
C. Lewis, J. John & T. Hill

Inclusive product design: Industrial case studies from the 338
UK and Sweden
H. Dong, O. Bobjer, P. McBride & P.J. Clarkson

Smart home technology in municipal services; state of the art 343
in Norway
T. Laberg

Look in, turn on, drop out 348
A.G. Gale & F. Shi

Live for tomorrow – future-proof your home 353
A. Wright

Taking the tablets home: Designing communication software for 358
isolated older people at home
G. Dewsbury, P. Bagnall, I. Sommerville, V. Onditi &
M. Rouncefield

The benefits of adapting the homes of older people 363
P. Lansley

Industry's response to inclusive design: A survey of current 368
awareness and perceptions
J. Goodman, H. Dong, P.M. Langdon & P.J. Clarkson

Exploring user capabilities and health: A population perspective 373
U. Persad, P.M. Langdon & P.J. Clarkson

"Sleeping with the enemy!" – a survival kit for user and 378
provider collaborators
J. Mitchell, M. Turner, S. Ovenden, J. Smethurst &
T. Allatt

What's in it for me? Altruism or self interest as the "driver" for inclusion 382
J. Middleton, J. Mitchell & R. Chesters

"Choices not barriers" housing strategy – learning from disabled people 387
C. Wright

Barriers against people with different impairments in their homes and neighbourhoods 392
A. Beevor, P. Mortby, C. Townsend, J. Mitchell, M. Sanders & R. Waller

Innovation and collaboration between users and providers 395
L. Birchley, G. Davies, S. Gamage, C. Hodgkinson, J. Mitchell & J. Smethurst

Tactile communication in the home environment 399
E. Ball & C. Nicolle

Developing new heuristics for evaluating Universal Design standards and guidelines 404
C.M. Law, J.A. Jacko, J.S. Yi & Y.S. Choi

Design, usability and unsafe behaviour in the home 409
H.J. McDermott, R.A. Haslam & A.G.F. Gibb

Lifestyles and values of older users – a segmentation 414
P.W. Jordan

Going outside of the front door: Older people's experience of negotiating their neighbourhood 419
R. Newton, M. Ormerod & V. Garaj

An evaluation of community alarm systems 423
S. Brown, M. Clift & L. Pinnington

A strategic spatial planning approach to public toilet provision in Britain 426
C. Greed

The challenge of designing accessible city centre toilets 431
J. Hanson, J.-A. Bichard & C. Greed

Accessible housing? One man's battle to get a foot through the door 436
H.J. McDermott, R.A. Haslam & A.G.F. Gibb

INCLUSIVE DESIGN – IN SOCIETY

Guide dogs and escalators: A mismatch in urban design 443
D.E. Gyi & L. Simpson

COST 219ter: An evaluation for mobile phones 448
E. Chandler, E. Dixon, L.M. Pereira,
A. Kokkinaki & P. Roe

Designing from requirements: A case study of project spectrum 453
A. Woodcock, D. Georgiou, J. Jackson & A. Woolner

INCLUSIVE DESIGN – IN TRANSPORT

My camera never lies! 461
D. Hitchcock, J. Walsh & V. Haines

Collection of transport-related data to promote inclusive 465
design door-to-door
R. Sims, J.M. Porter, S. Summerskill, R. Marshall,
D.E. Gyi & K. Case

Developing the HADRIAN inclusive design tool to provide 470
a personal journey planner
J.M. Porter, K. Case, R. Marshall, D.E. Gyi & R. Sims

Can a small taxi be accessible? Notes on the development 475
of microcab
P. Atkinson

METHODS AND TOOLS

Knowledge representation for building multidimensional 483
advanced digital human models
N.C.C.M. Moes

Conducting research with the disabled and disadvantaged 489
D. Hitchcock, V. Haines & S. Swain

Agile user-centred design 494
M. McNeill

Manikin characters: User characters in human computer modelling 499
D. Högberg & K. Case

Effects of viewing angle on the estimation of joint agles in the 504
sagittal plane
I. Lee

Evidence-based ergonomics 509
A. Genaidy & J. Jarrell

Validity of dual-energy X-ray absorptiometry for body 513
composition analysis
J.A. Wallace, K. George & T. Reilly

Body composition in competitive male sports groups 516
J.A. Wallace, E. Egan, K. George & T. Reilly

OCCUPATIONAL HEALTH AND SAFETY

"Let's be careful out there" the Hong Kong Police OSH system 521
M. Dowie, P. Haley, N. Heaton, R. Mason &
D. Spencer

A method for testing pressures on fingers in entrapment hazards 526
E. Milnes

Slips and trips in the prison service 531
A. Scott & K. Hallas

Falls from vehicles 536
K. Hallas, A. Scott & M. Miller

Human variability and the use of construction equipment 541
P.D. Bust, A.G.F. Gibb, C.L. Pasquire & D.E. Gyi

The relationship between recorded hours of work and 546
fatigue in seafarers
P. Allen, E. Wadsworth & A. Smith

OIL, GAS AND CHEMICAL INDUSTRIES SYMPOSIUM

The impact of psychological ill-health on safety 551
C. Amati & R. Scaife

Checking failures in the chemical industry: How reliable are 555
people in checking critical steps?
J. Henderson & S. Cross

Early human factors interventions in the development of an FPSO 560
J. Fisher & W.I. Hamilton

Shiftwork on oil installations 565
A. Smith

Assessing the impact of organisational factors on safety in a 570
high hazard industry
J. Beswick, S. Powell, M. Anderson & A. Jackson

Integrating human factors in an oil platform control room 575
during organisational change
Z. Mack & L. Cullen

Why do people do what they do? 580
R. Lardner & G. Reeves

PHYSICAL ERGONOMICS

Physical ergonomic design aspects of computer workstations: 587
A Liechtenstein perspective
E. Kessler, S. Mills & S. Weinmann

Exposure assessment to musculoskeletal load of the upper 590
extremity in repetitive work tasks
U.M. Hoehne-Hückstädt, R.P. Ellegast & D.M. Ditchen

Data management system for analysis of occupational 593
physical workload
I. Hermanns & R.P. Ellegast

Effects of non-neutral posture on human response to 598
whole-body vertical vibration
G. Newell, S. Maeda & N. Mansfield

A systematic review and meta-analysis of lower back disorders 603
among heavy equipment operators
M. Makola & A. Genaidy

TRANSPORT

Encouraging co-operation in road construction zones 607
T. Wilson

Development of a method for ergonomic assessment of a 611
control layout in tractors
D. Drakopoulos & D.D. Mann

Workload associated with operation of an agricultural sprayer 616
A.K. Dey & D.D. Mann

Effects of the common cold on simulated driving 621
A. Smith

Top Gear on cars – experts' opinions and users' experiences 625
P.W. Jordan

BIONIC – "Eyes-Free" design of secondary driving controls 630
S. Summerskill, J.M. Porter, G. Burnett & K. Prynne

Mind the gap? – What gap! 635
G. Hayward & S. Bower

Designing a system for European road accident investigation 640
C.L. Brace, H. Jahi, L.K. Rackliff & M.E. Page

Challenges in the usability evaluation of agricultural mobile machinery 645
P. Nurkka

Research on the influence of design elements on driving posture in China 648
N. Zou, S.-Q. Sun, M.-X. Tang & C.-L. Chai

Developing systems to understand causal factors in road accidents 653
C.L. Brace

AUTHOR INDEX **657**

PREFACE

Contemporary Ergonomics 2006 is the proceedings of the Annual Conference of the Ergonomics Society, held in April 2006 at Robinson College Cambridge, UK. The conference is a major international event for Ergonomists and Human Factors Specialists, and attracts contributions from around the world.

Papers are chosen by a selection panel from abstracts submitted in the autumn of the previous year and the selected papers are published in *Contemporary Ergonomics*. Papers are submitted as camera ready copy prior to the conference. Each author is responsible for the presentation of their paper. Details of the submission procedure may be obtained from the Ergonomics Society.

The Ergonomics Society is the professional body for ergonomists and human factors specialists based in the United Kingdom. It also attracts members throughout the world and is affiliated to the International Ergonomics Association. It provides recognition of competence of its members through its Professional Register. For further details contact:

The Ergonomics Society
Elms Court
Elms Grove
Loughborough
Leicestershire
LE11 1RG
UK

Tel: (+44) 1509 234 904
Fax: (+44) 1509 235 666

Email: ergsoc@ergonomics.org.uk
Web page: http://www.ergonomics.org.uk

DONALD BROADBENT LECTURE

AN ENGINEER'S VIEW OF HUMAN ERROR

Trevor A Kletz

*Department of Chemical Engineering, Loughborough University,
LE11 3TU*

Various sorts of human error are described and illustrated by examples. Better training or supervision can prevent some errors but the most effective action we can take is to reduce opportunities for error, or minimise their effects, by changing designs or methods of working.

Why is it that when someone bangs their head on a scaffold pole sticking through the ladder they are climbing the first question asked is, "Why weren't you wearing your safety helmet?" rather than, "How did the scaffold pole come to be sticking through the ladder in the first place?" – Ann Needham, HSE

Introduction

Much of the literature on human error does not make it clear that people make errors for different reasons and that we need to take different actions to prevent or reduce the different sorts of error. Also, blaming human error diverts attention from what can be done by better engineering. This paper describes four sorts of human error and illustrates them by example:

- Those that occur because someone does not know what to do. The intention is wrong. They are usually called mistakes.
- Those that occur because someone knows what to do but decides not to do it. They are usually called violations though often the person concerned genuinely believes that a departure from the rules, or the usual practice, is justified. Non-compliance is therefore a better name.
- Those that occur because the task is beyond the physical or mental ability of the person asked to do it, often beyond anyone's ability. They are called mismatches.
- Errors due to a slip or a momentary lapse of attention. The intention is correct but it is not carried out.

More than one of these factors may be involved. I like this classification because it directs our thoughts towards methods of prevention.

Mistakes

Someone does not know what to do or, worse, thinks he or she knows but does not. (*It ain't so much the things we don't know that get us in trouble. It's the things we know that ain't so.* – Artemus Ward)

The obvious solution is to improve the training and/or instructions but before doing so we should first see if we can simplify the task or remove opportunities for error by changing the design or method of working (see Section 7.4).

Some of these errors are due to a lack of the most elementary knowledge of the properties of the materials or equipment handled, some to a lack of sophisticated knowledge and others to following the rules when flexibility was needed. However many rules we write we can never foresee every situation that might arise and people should therefore be trained to diagnose and handle unforeseen problems. This is true in every industry but particularly true in the process industries.

Sometimes people are given contradictory instructions. Those that are obviously contradictory are unusual, though not unknown. For example, one incident occurred because operators were asked to add a reactant at 50°C over 40 minutes. The heater was not powerful enough (or so they believed) so without telling anyone they added it at a lower temperature. The result was an unwanted reaction and a spoilt batch.

More common are instructions with implied contradictions. For example, operators, foremen or junior managers may be urged to achieve a certain output, or complete a repair, by a certain time. It may be difficult to do this without relaxing one of the normal safety instructions. What do they do? Perhaps the manager prefers not to know. In cases like this the unfortunate subordinates are in a "heads I win, tails you lose" situation. If there is an accident, they are in trouble for breaking the safety rules. If they stick to the rules and the output or repair is not completed in time they are in trouble for that reason.

A manager should never put his staff in this position. If he believes that a relaxation of the usual safety procedures is justified – sometimes it is – then he should say so clearly, preferably in writing. If he believes that the usual safety procedures should be followed, then he should remind people, when asking for experiments or urgent repairs or extra output, that they are not to be obtained at the cost of relaxing the normal safety procedures. **What we don't say is as important as what we do say.** If we talk a lot about output or repairs and never mention safety then people assume that output or repairs are what we want and all that we want, and they try to give us what we want.

Violations (non-compliance)

Many accidents have occurred because operators, maintenance workers or supervisors did not carry out procedures that they considered troublesome or unnecessary. For example, they did not wear the correct protective clothing or follow the full permit-to-work procedure. To prevent such accidents we need to convince people that the procedures are necessary as we do not live in a society in which we can expect people to obey uncritically. A good way of doing this is to describe (or, better, discuss) accidents that occurred because the procedures were not followed. In addition, we should keep our eyes open and check from time to time to see that the procedures are being followed. It is bad management to say, after an accident, "I didn't know it was being done that way. If I had known I would have stopped it". It is a manager's job to know.

Discussions are usually more effective than lectures and written reports, as more is remembered and those taking part are more committed to the conclusions if they have developed them and not simply been told what they are.

Before trying to persuade people to follow the rules we should first see if we can simplify the task or remove opportunities for error. If the wrong method is easier than the right method, people will be tempted to use the wrong method. To quote from the report on a fatal accident, "On previous occasions men have entered the vessel without complete isolation. It seems that this first occurred in an emergency when it was thought essential to get the vessel back on line with the minimum of delay.... Since everything went satisfactorily, the same procedure has apparently been adopted on other occasions, even when, as in this case, there was no particular hurry. Gradually, therefore, the importance of the correct procedure seems to have been forgotten...."

It is not just operators and supervisors who cause accidents by failing to follow the rules. Accidents also occur because managers ignore a rule in order to maintain output. Often the rules they break are not written down but are merely "accepted good practice". Like operators they do not suspend the rules because they are indifferent to injury but because they do not see the need for the rules and want to get the job done. For example, a procedure is introduced after an accident. Ten years later the reasons for it have been forgotten and someone in a hurry, keen to increase output or efficiency, both very desirable things to do, says, "Why are we carrying out these time-consuming procedures?" No one knows and the procedures are scrapped.

Note that if instructions are wrong violations can prevent an accident. Instructions may be wrong as the result of a slip (for example, someone writes "increase" when they meant "decrease"), ambiguity, or ignorance on the part of the writer.

Violations are the only sort of human error for which blame is sometimes justified but before blaming anyone, manager, designer or operator, we should ask:

- Were the instructions known and understood?
- Was it possible to follow them?
- Did they cover the problem?
- Were the reasons for them known?
- Were earlier failures to follow the rules overlooked? Turning a blind eye tells everyone that the rules are unimportant.
- Was he or she trying to help? Many violations are committed with the best of motives. We need to protect ourselves from those who try to help as well as those with other motives.
- If there had not been an accident, would he or she have been praised for his or her initiative? There is a fuzzy barrier between showing initiative and breaking the rules.

Mismatches

A few accidents occur because individuals are unsuited to the job. More occur because people are asked to carry out tasks which are difficult or impossible for anyone, physically or, more often, mentally. For example:

- People are overloaded. Computers make it too easy to supply people with more information than they can handle and they switch off (themselves, not the computer).
- They are underloaded, do not stay alert and then do not notice when something requires attention – the night watchman syndrome. With increasing automation some people are concerned that process operators will be placed in this position and have suggested that some tasks should be left on manual control in order to give operators something to do. However, rather than ask men to do something that machines can do more efficiently it would be better to look for useful but not essential tasks which will keep them alert but which can be set aside if there is trouble. Suitable tasks are calculating and plotting efficiencies, catalyst life or energy usage or studying training materials. This is the process equivalent of leaving the ironing for the baby-sitter.
- They are asked to go against established habits. We expect that turning a knob clockwise will increase the response and errors will occur if designers ask people to break this habit.
- They are expected not to develop mind-sets. Unfortunately we all do. We have a problem; we think of a solution; then we fail to see the snags in our idea. It is difficult to avoid mind-sets. Designers should not assume that operators will logically consider all the evidence in a dispassionate way. They should assume that they will behave as they have behaved in the past and should try to avoid the opportunities for wrong decisions. There are some examples in Kletz 2001 (of which this paper is a summary).

Errors due to slips or lapses of attention

The errors described so far can be prevented by:
- Providing better training or instructions.
- Motivating people better, by means of training and supervision.
- Designing plants and systems of work so that people are not asked to carry out tasks that are physically or mentally difficult or impossible.

Even if everyone is well-trained and well-motivated, physically and mentally capable of doing all that we ask and they want to do it, they will still make occasional errors. They will forget to close or open a valve, will close (or open) the wrong valve, will close (or open) the valve at the wrong time, will press the wrong button or make a slip in calculation. These errors are similar to those of everyday life, though their consequences are greater. Reason and Mycielska (1982) have described the psychological mechanism.

Training will not prevent errors of this type. Slips and lapses do not occur because people are badly trained but because they are well-trained. Routine tasks are then delegated to the lower levels of the brain and are not continually monitored by the conscious mind. We would never get through the day if every action required our full attention, so we put ourselves on autopilot. Our conscious minds check from time to time that all is well but not when we are stressed or distracted. Errors are also liable to occur when the smooth running of the program is interrupted for any reason.

Since we cannot prevent these slips we should accept that they will occur from time to time and design accordingly. We should design our plants and methods of working so as to remove opportunities for error (or provide opportunities for recovery or guard against the consequences). For example, rising spindle valves whose position is obvious at a glance are better than valves with non-rising spindles.

In short, do not try to change people but accept them as we find them and change the work situation. I use the phrase "work situation" rather than "design" because it is often impracticable to change the design. Safety by design should always be our aim, but often it is impracticable and we have to settle for a change in procedures. However, some people change procedures when a simple change in design is possible (see Sections 7.2–7.4).

Even if errors come into one of the first two categories, mistakes and violations, we should try to remove opportunities for error. This may be a better solution than trying to motivate or train people to carry out difficult or unwelcome tasks.

We do not say how equipment ought to behave. We find out how it actually behaves and design accordingly. In the same way we need an engineering approach to human error: Find out how people actually behave and design plants and procedures accordingly. The problem is not how to prevent bad people hurting others but how to prevent good people hurting others. Figure 1 summarises the message of this paper.

ERROR TYPE	ACTION TO PREVENT
MISTAKES – Does not know what to do	Better training & instructions/CHAOS
VIOLATIONS – Decides not to do it	Persuasion/CHAOS
MISMATCHES – Unable to do it	CHAOS
SLIPS & LAPSES OF ATTENTION	CHAOS

CHAOS = Change Hardware And/Or Software

Figure 1. Types of human error.

Note that estimates of the probability of human error are estimates of the probability of this fourth type of error. We can estimate the probability that someone will forget to close a valve, or close the wrong valve but not the probability that he (or she) will not have been trained to close the valve, or will decide not to close it, or will be unable to do so (for example, because the valve is too stiff or out of reach). Each of these probabilities can vary from 0 to 1. At the best we can assume that errors of these types will continue in a plant at much the same rate as in the past, unless there is evidence of change.

Consider the nitrogen blanketing of storage tanks. It is easy to estimate the reliability of the instruments, the availability of the nitrogen supply and the probability that the operator will forget to carry out any manual operations. It is more difficult to estimate the probability that someone will deliberately neglect or isolate the system because he is not convinced of its importance. This is the most likely reason for failure[1].

Management errors

These are not a fifth type of error. They occur because senior managers do not realise that they could do more to prevent accidents. They are thus mainly due to lack of training, but some may be due to lack of ability and a few to a deliberate decision to give safety a low priority. They are sometimes called organisational failures but organisations have no minds of their own. Someone has to change the culture of the organisation and this needs the involvement, or at least the support, of senior people.

Most senior managers genuinely want fewer accidents. They urge their staff to do better and give them the resources they need but they do not see the need to get involved in the detail. In contrast, if output, costs or product quality are a problem they expect to know in detail what is wrong, agree the actions taken and monitor progress. Kletz (2001) describes several accidents in which the underlying cause was management failure. If nitrogen blanketing is isolated or neglected, as described above, then the managers have not realised that instructions alone are not sufficient and that all protective equipment should be checked regularly and the results reported. An official report said, "... having identified the problem and arrived at solutions, he turned his attention to other things and made the dangerous assumption that the solution would work and the problem would go away. In fact it did not" (Hidden, 1989).

The reason why safety problems often get less senior management attention than costs, output and efficiency is that safety problems are latent (that is, hidden) until an accident occurs, while data on costs, outputs and efficiencies are continuously available.

An example

Reference 1 discusses the four sorts of human error in greater detail with many examples. Here is chosen to show how accidents can be prevented by better design. The operator is the last line of defence against poor design and poor management. It is poor strategy to rely on the last line of defence.

Figure 2 shows the simple apparatus devised in 1867, in the early days of anaesthetics, to mix chloroform vapour with air and deliver it to the patient. If it was connected up the wrong way round liquid chloroform was blown into the patient with results that were usually fatal. Redesigning the apparatus so that the two pipes could not be interchanged was easy; all that was needed were different types of connection and/or different sizes of pipe. Persuading doctors to use the new design was more difficult and the old design was still killing people in 1928. Doctors believed that highly skilled professional men would not make such a simple error but as we have seen slips occur only when we are well-trained.

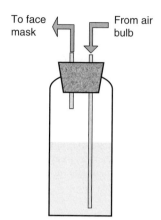

Figure 2. Early chloroform dispenser. If it is connected up the wrong way round liquid chloroform is blown into the patient.

Do not assume that chemical engineers would not make similar errors. In 1989, in a polyethylene plant in Texas, a leak of ethylene exploded, killing 23 people. The leak occurred because a vessel was opened for repair while the air-operated valve isolating it from the rest of the plant was open. It was open because the two compressed air hoses, one to open the valve and one to close it, had been disconnected and then replaced wrongly. The accident, some might say, was due to an error by the person who re-connected the hoses. This is an error waiting to happen, a trap for the operator, a trap easily avoided by using different types or sizes of coupling for the two connections. This would have cost no more than the error-prone design (Figure 3) (Anon., 1990).

But do not blame the designer instead of the operator. Why did he or she produce such a poor design? What was lacking in his or her training and the company standards? Was a safety engineer involved? Was the design Hazoped? Reports do not always look for these underlying causes.

As well as the poor design and the slip by the operator there was also a violation, a decision (authorised at a senior level) not to follow the normal company rules and industry practice which required a blind (slip-plate) or double isolation valves and a lock on the isolation valve(s).

Many other accidents are described in the complete version of this paper which can be obtained by emailing T.Kletz@Lboro.ac.uk.

Conclusions

To prevent accidents we can take the following actions, the first when possible, then the second, and so on:

- Avoid the hazards – the inherently safer solution.
- Add on passive protective equipment (that is, equipment that does not contain moving parts or has to be commissioned).
- Add on active protective equipment (which needs regular testing and maintenance).
- Rely on procedures.
- Use behavioural science techniques to increase the extent to which people follow correct procedures. These techniques can greatly reduce everyday accidents but are less effective in preventing process accidents. For all accidents they should be the last line of defence; whenever possible we should remove opportunities for errors.

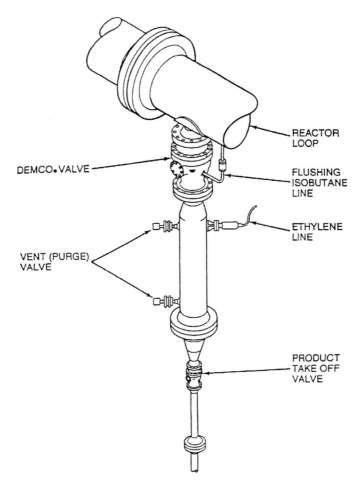

Figure 3. Typical arrangement of settling leg on polyethylene plant.

Unfortunately, as some of my examples show, the "default" action in some companies is to start at the bottom of the bullet list above and work upwards. Some programmes for conferences on human error are devoted entirely to methods of changing behaviour of operators and do not mention the need, whenever possible, to change the conditions under which people work.

To get better designs we have to change the behaviour of designers and those who accept designs. Many accidents are said to be due to organisational weaknesses. But organisations have no minds of their own. To change them we have to change the actions of senior managers. How can we change the behaviour of designers and managers? Could the techniques of behavioural science be adapted?

Errors by those at the bottom of the organisation tree result in relatively minor accidents. At the worst they may injure or even kill someone or wreck a machine or a batch of product. Errors by senior people – those at or near the top of the tree, people who choose processes and sites, determine organisation, manning and design and training policies – can kill tens or hundreds, even thousands as at Bhopal, and wreck a whole factory or company. To take an everyday example, children wreck their toys. When they reach 18 they start wrecking their parents' cars. Senior people look at the safety record, realise it is not good enough, tell their juniors to do better and set targets for improvement. They do not ask themselves, "What should I do better?"

Acknowledgement

Figure 2 is included by kind permission of the Institution of Chemical Engineers.

References

Anon., 1990, *The Phillips 66 Company Houston Chemical Complex Explosion and Fire*, (US Dept. of Labor, Washington DC).
Hidden, A., 1989, *Investigation into the Clapham Junction Railway Accident*, (HMSO, London), paragraph 16.66.
Kletz, T.A., 2001, *An Engineer's View of Human Error*, 3rd edition, (Institution of Chemical Engineers, Rugby).
Reason, J. and Mycielska, C., 1982, *Absent Minded? The Psychology of Mental Lapses and Everyday Errors*, (Prentice-Hall, Englewood Cliffs, New Jersey).

APPLICATIONS OF ERGONOMICS

ON DOCUMENTATION FOR THE DESIGN OF SCHOOL BUILDINGS

M.A. Tainsh

Krome Ltd, UK

The UK Government has polices and initiatives on the inclusion of pupils with Special Educational Needs (SENs) within mainstream education. Consequently, the Department for Education and Skills (DfES) has produced guidelines for the design and use of school buildings. Building Bulletin 77 was published in draft early in 2005. The ergonomics4schools Special Interest Group (SIG) has debated this document. It has commented on it, and discussed these comments with the DfES. The Bulletin is highly structured in accordance with an architectural approach which includes consideration of well defined categories of SEN. This paper considers the Bulletin and similar documents issued recently by the Ministry of Defence. Some key issues are discussed including the need for and character of "Use" studies.

Introduction

It is current UK Government policy to provide education for pupils with Special Educational Needs (SENs) in the same buildings and classes as others. As a result of this policy, the Department for Education and Skills (DfES) has published a consultative document with guidance on the design of school buildings, and some consideration given to equipment (Ref 1). The guidance is for architects and engineers when "planning, briefing and designing school accommodation across all educational settings where there are pupils who have SEN and disabilities". As there is a policy for integration, this guidance will need to be used when considering the design of all schools.

The scale of the design problem is substantial. It is not simply a question of catering for "the average student" or even clearly described sets of students whose characteristics have known means and standard deviations. There is a requirement for architects and engineers to meet the needs of the 15–20% of all pupils who have some form of SEN or disability. Further, this number of SEN pupils appears to be increasing, and the design problem needs to be addressed frequently, now. During 2004, a school was built every 4 days (Ref 1).

Aim of paper

This paper describes Building Bulletin 77 briefly and considers some of the high level systems engineering/ergonomics issues that it highlights. It compares it with the MoD documentation for Human Factors Integration (HFI) (Refs 2 and 3).

Description of Building Bulletin 77

This is a highly structured document in seven parts, as described below.

Part 1 Key Issues: The following issues are addressed: safety and security, health and well being, communication and interaction, sensory stimulus and information, mobility and access, behavioural development, activity and expression, social awareness and participation, and spiritual support. These issues are put into the context of teaching approaches, the learning environment, community use and design quality. It contains sets of high level goals which might well be included in the high level portion of HFI statements of requirements.

Part 2 Categories of Special Need: These are defined in terms of four categories:

(a) Disability characterised in terms of physical or mental impairment with substantial and long term effects on day-to-day activities
(b) SENs which call for special educational provision
(c) Medical needs
(d) Mental-health needs.

Both categories (a) and (b) are described in detail. This provides the Target Audience Description (TAD) from a medical, and/or psychological point of view.

Part 3 LEA's Strategic Planning: Here, there is a discussion on the importance of liaison and joint working with local authorities and other agencies. These strategic management lie outside the scope of this paper.

Part 4 Whole School Approach: Attention is drawn to the need to design the school as a whole i.e. a single entity rather than separate sets of areas or volumes. It includes design issues and requirements.

Part 5 Technical Advice on Best Value: There is technical guidance to ensure that best practice is followed in the areas described within the whole school approach, and associated equipment. It describes how the requirements and issues might be addressed and contains some technical details and these are supported by Part 6.

Part 6 Project Planning: This links directly from Part 5, and provides guidance on technical issues which may need to be considered during the planning phase of project.

Part 7 Case Studies: This contains appendices covering Acts of Parliament and other important topics.

Review of Building Bulletin 77 by SIG members

In May 2005 the Bulletin was reviewed by members of the SIG and then was discussed in three virtual conferences.

It was generally concluded that Building Bulletin 77 addressed issues of spaces and volumes rather than tasks, and these were generally characterised with mean values. Standards and guidance tended to be given in spatial measures, or as appropriate, for example, with noise and light. The main conclusion, of the SIG, was that there was a lack of concepts associated with the use of the school.

Experience in other applications

Military organisations within the UK have long been interested in the best means of specifying concepts of use along with requirements for purposes of acquisition. This has been particularly important because of the critical nature of so many of their systems and the high costs involved. There is a current emphasis on capability linked to operational effectiveness rather than physical entities and their characteristics (which may be left to the supplier organisations to propose and specify). Within these UK military organisations, ergonomics issues are considered in as much as they impact on capability and operational effectiveness, or factors that influence them such as costs and manpower. This is reflected in the HFI policy (Ref 4).

Table 1. Comparison of presentation of ergonomics material.

Subjects	Building Bulletin 77	STGP documents
Concept of capability	Not addressed	Addressed
Concept of operations and use	Not addressed	Addressed
Design goals	Addressed fully	Addressed fully
Design issues – across all six HFI domains (as defined in Ref 4)	Mainly covers human engineering, health and safety issues	Well addressed in Ref 3
TAD	Addressed fully	Addressed as part of HFI domains
Standards	Addressed to a limited degree	Not addressed. They are seen as lying outside this set of volumes

Comparison of documentation

It is interesting to compare the treatment of important subjects within Building Bulletin 77 with that in the UK MoD's Sea Technology Group Publications, (which have been developed over a period of nearly fifteen years from predecessor documents). The list of subjects in the left-hand column of Table 1 is based on an understanding of UK government equipment acquisition practices, while the assessments are derived from the beliefs expressed in the SIG review and conferences. It is believed that that in the case of Building Bulletin 77, the treatment of the topics associated with capability and operations, both related to concepts of use, should be assessed as poor – equally the TAD and standards have been treated well.

Relationships between capability, operations, use, TAD and standards

These relationships have been developed for over three decades within military organisations, and so the concepts are generally familiar and accepted by those using them. Capability refers to the "fitness to perform operations", whereas "the operation" is defined in terms of tasks with goals to be achieved. "Use" refers to the set of operations that might be performed. The TAD is the user characteristics, and the standards here are equipment features or task-based measures.

The concept of use applies at a number of levels including whole school, department and individual/group. One needs a statement for the school as a whole which could be linked with a concept of capability, individual departments with a concept of function, and then groups and individuals with concepts of jobs and tasks.

The concepts of "capability", "operations" and "use" are essential otherwise it is difficult to refer easily to fitness. It is particularly important for schools where the TAD is well defined and broadly based, as one would wish to know how all the various constituent members of the school's user population fits to the design.

Likely future issues within the ergonomics/HFI domains

It is feared that Building Bulletin 77, as currently drafted, is a missed opportunity for describing the contribution of ergonomics. Further, it is not just the needs of the pupils who will be misunderstood but also the staff and members of the external community.

Potential for use studies

There is a need for studies emphasising use. Much of the current work on the topic of "use" has come from military applications which are "equipment-intensive systems" not "user-intensive" systems such as schools. By "equipment-intensive", it is meant that high proportions of critical tasks are undertaken by equipment and the overall goals can only be understood in terms of users working with equipment. By contrast "use-intensive systems" will have goals that are expressed in terms of individual or group achievement and the equipment available has a lesser importance.

In equipment-intensive systems, the acquisition process, of which design is a part, will include use studies that involve simulations and task descriptions but these will be closely tied to the demands of the equipment characteristics. In user-intensive systems during the design phase, there will be a greater emphasis on the needs of the users and the means of satisfying them.

Conclusions and way ahead

The results of "use" studies need to be expressed in a form where they can form part of a statement of user requirements. These results need to be brought to the attention of the DfES and local authorities. However, there is now some work in UK at Coventry, and in Australia, New Zealand and Colombia. There is much to done to ensure that the needs of the pupils, staff and local community are met.

Acknowledgements

The contributions of Eddy Elton and Steve Bayer at ESRI, Loughborough (who wrote a substantial portion of the submission to the DfES) are acknowledged.

References

Building Bulletin No 77, 2005 (Department for Education and Skills).
Human Factors Integration Management Guide. Sea Technology Group Publication 10. Issue 3, 2003.
Human Factors Integration Technical Guide. Sea Technology Group Publication 11. 2004.
Human Factors Integration. www.ams.mod.uk/ams/content/docs/hfiweb/data/current/intro.pdf

IS THAT A GUN? THE INFLUENCE OF FEATURES OF BAGS AND THREAT ITEMS ON DETECTION PERFORMANCE

Xi Liu, Alastair Gale, Kevin Purdy & Tao Song

Applied Vision Research Centre, University of Loughborough, Loughborough, LE11 3UZ, UK

> An experiment is reported where naïve observers searched 50 X-ray images of air passenger luggage for potential terrorist threat items. For each image their eye movements were recorded remotely and they had to rate their confidence in whether or not a potential threat item was present. The images were separately rated by other naïve observers in terms of; visual complexity of bags, the familiarity and visual conspicuity of threat items. The visual angle subtended by guns and the familiarity of threat items influenced the detection rate. Eye movement data revealed that the complexity of bags and the conspicuity of threat items also influenced visual search and attention.

Introduction

X-ray screening of passengers' luggage at airports is an important way of protecting against terrorism on aeroplanes. However, the effectiveness of detecting and recognizing potential terrorist threat items is determined by the level of human performance that can be achieved. Inevitably such performance is variable and it is important to find methods to both improve and maintain such performance to be consistently high. The key interest lies in maintaining a low false negative error rate so that potential threat items are not missed. What are the visual and cognitive factors of this visual inspection task and what features of luggage bags and threat items influence detection and visual search?

A strong foundation for understanding the airport screeners' task can be found in radiology. Here, visual search studies over the past 30 years have provided a theoretical model to classify search error types based on visual dwell time (Kundel et al, 1978) and models of understanding the detection process (e.g. Swensson, 1980). The first step of visual search of such images is a fast, parallel global scanning stage. Then spatial attention is allocated serially to potential target areas. "Pop-out" happens if a target is obvious because of simple local properties or other reasons, e.g., size, brightness and contrast. Every selected area is scrutinized by the visuo-cognitive system. Based on eye movement data from various experiments, failure to find a target could result from search, detection or recognition if the target falls within the useful field of view which is defined as the area around the fixation point from which the information is being processed (Mackworth, 1974).

Visual search and detection performance are influenced by familiarity with target features (Kundel & La Follette, 1972) and the size and conspicuity of a target (nodule) – with larger and more conspicuous nodules receiving less visual attention than smaller and less conspicuous nodules (Krupinski et al, 2003). Visual complexity also influences the cognitive processing of pictorial stimuli; more complex stimuli are more difficult to process (Alario & Ferrand, 1999), and detection performance (Schwaninger et al, 2004). Moreover, for successful detection a target requires observers not only scrutinize the appropriate area but also to recognize the target when they look at it even when it is camouflaged.

Two studies are reported here. The first uses eye position recording of naïve observers as they searched X-ray luggage images for threat items such as; guns, knives and improvised explosive devices (IEDs). In the second, other observers, again naïve to the images, evaluated luggage images which contained a threat item using the following variables: visual complexity of bags; familiarity, and visual conspicuity of the threat items. The purpose was to understand the factors underlying the inspection of these complex images by investigating whether or not there was a relationship between the threat and luggage features that people attended to and how such images were perceived by other observers.

Method

Participants

Ten naïve participants (4 female and 6 male) took part in the visual search study. Ten other naïve participants (4 female and 6 male) took part in the evaluation study. They all had no prior knowledge of examining such X-ray images.

Stimuli and apparatus

Ten participants examined fifty X-ray images of which twenty-five were normal luggage items. The remaining images contained potential threat items, including; guns, knives and IEDs. Five images contained multiple threat items and for all others only one threat item was present – only data for the single threat items are considered here. Participants' eye movements were recorded by a Tobii eye-tracker (X50). They viewed images on a 53 cm monitor at a distance of 70 cm.

Procedures

Visual search

Participants were first familiarised with the type of images and the procedure. They then examined 50 luggage images in a random order, for an unlimited viewing time. For each image the participant had to rate their confidence in whether or not threat items were present and indicate such threat location in the luggage image (Liu et al, in press).

AOIs

The eye movement data needed to be related to the threat items. The approach taken was to generate an area of interest (AOI) around each threat item and then to consider eye fixation data which fell within each AOI. Given the irregular form of threat items the dimensions of the area of interest (AOI) were determined by: the form of each threat item; the visual angle subtended by the fovea, and the accuracy of the eye-tracker (0.5~0.7°). The foveola is the area of the retina that provides the best visual resolution and subtends a visual angle of about 1.2° (Schwartz, 1994). The AOI was estimated to be an area which had a similar form to the threat item profile, but was 1 cm wider than the edge of the item, which at a viewing distance of 70 cm subtended 1° visual angle.

Raw eye movement data were then grouped into meaningful fixations on the basis of fixation radius and the minimum fixation duration. Fixation radius was set at 2.5° – a typical useful field of view size in medical imaging (e.g. Nodine et al, 1992) and it has been previously shown reasonable to apply this to luggage images (Gale et al, 2000). The dwell time in an AOI was obtained from the fixation clusters on it. These eye position parameters were analysed with the subjective variables detailed in the next section. Other than the relationship between the threat item features and visual search, just how these features influence detection rate was also explored in this study.

Image evaluation

Ten participants evaluated twenty luggage images, which each contained a single target threat item, on a five-point rating scale. For each image they had to rate the overall visual complexity of the luggage item, then the threat item was clearly demarcated and they had to rate their familiarity with the threat depiction and its visual conspicuity within the image. The visual complexity of the bag was explained to participants as meaning "the degree of difficulty in providing a verbal description of an image" (Heaps & Handel, 1999). This is a multi-dimensional concept which relates to the quantity of objects, clutter, symmetry, organization and variety of colours in an image. The familiarity of a threat item, in this study, meant the degree of similarity between a participant's concept of such a threat (e.g. the X-ray image of the side view of a hand gun) and the shape of a threat item on the screen (e.g. the X-ray image of a gun as seen "barrel–on"). The conspicuity of an object relates to object properties and its surroundings, such as brightness, colour, contrast, etc. The visual conspicuity of a threat item refers to the discrepancy between the appearance of a threat item and its local background – an obvious threat item should stand out from the immediate background.

Results

Data were analysed using Receiver Operating Characteristics methodology and the mean overall performance of the naïve participants had an accuracy measure (A_z) value of 0.74. The overall results are presented elsewhere (Liu et al, in press).

For TP decisions, the time to first enter the AOI (1033.76 ms) was shorter than that of FN decisions (1393.82 ms) and the dwell time for FN decisions (5918.19 ms) was longer than that for TP decisions (5804.53 ms). However, the differences were not significant ($F (1, 188) = 1.419, p = .235$) and ($F (1, 198) = .020, p = .888$) respectively. The number of data points considered for dwell time is more than for the "time to enter AOI" analysis as some target threats were not fixated.

Four guns were used as potential threat items in this study, two of them were presented in front (end–on) view and the other two were presented in side view. Only one decision was an error for the side view gun images compared to eight incorrect decisions for the end–on view guns. The visual angle subtended by the gun influenced the detection rate ($\chi^2 = 13.789, df = 1, p < .001$).

The subjective variable "bag visual complexity" affected the time to first enter the AOI ($F (1, 18) = 5.723, p < .05$), this being significantly longer for a complex than a simple bag. The total dwell time on a complex bag (14.67 s) was longer than that on a simple bag (12.50 s), however the difference was not significant ($F (1, 18) = 1.430, p = .247$). The visual complexity of bags did not influence the detection rate significantly ($F (1, 18) = .022, p = .882$).

The familiarity of the threat item did not influence the time to first enter the AOI ($F (1, 18) = .008, p = .931$) or the dwell time on the AOI ($F (1, 18) = 4.359, p = .051$). Familiarity significantly influenced detection rate ($F (1, 18) = 5.432, p < .05$). The threat items detected by eight or more observers were rated with significantly higher familiarity scores (more familiar) than those detected by seven or fewer observers (Figure 1).

The conspicuity of a threat item influenced the time to first enter its AOI ($F (1, 18) = 6.055, p < .05$) and the rate of dwell time on AOI of total dwell time on bag

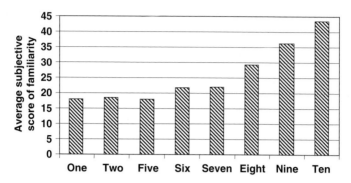

Figure 1. Number of participants detecting a threat item (e.g. two = two out of ten) as a function of the familiarity score of the threat item (a higher score represents a more familiar threat item).

image ($F (1, 18) = 5.319$, $p < .05$). As the degree of conspicuity of the threat item increased, the time to first enter the AOI decreased and the rate of dwell time on AOI of total dwell time on image increased significantly. Threat items detected by seven or more observers were scored higher conspicuity (more conspicuous) than those detected by six or fewer observers, but the difference was not significant ($F (1, 18) = .480$, $p = .497$).

Overall, the first time to enter AOI of guns and knives was significantly longer ($F (1, 188) = 11.401$, $p < .01$) than that of IEDs (1730.73 ms vs. 778.48 ms, respectively). Total dwell time on AOI of guns and knives was shorter (3895.86 ms) than that of IEDs (7140.43 ms), and the difference was significant ($F (1, 198) = 19.165$, $p < .01$).

Discussion

The results demonstrated that naïve individuals, after a very short familiarisation of the X-ray appearance of threats, can perform better than chance in examining luggage images for other threat items. Three subjective evaluation variables were used to explore luggage and threat item features and their influence on detection rate and visual search by naïve observers. The familiarity of threat items influenced detection rate significantly although both the visual complexity of bags and the conspicuity of threat items did not. This implies that object knowledge of threat items is very important for detection and recognition. A clear definition of a target is an effective way for improving detection performance (Kundel & La Follette, 1972) and training target materials should be maximally heterogeneous to ensure skill generalization (McCarley et al, 2004).

Schwaninger et al (2004) found that the viewpoint, the bag complexity and superposition influenced detection performance. The present study also showed that the apparent visual size of a gun influenced detection rate significantly. Target materials employed during training should include targets presented at difficult visual angles, even when the targets can be assumed to be known by most people. Naïve observers knew more about gun and knife appearances a shown by the shorter times ("first enter AOI" and dwell) than with IEDs.

Visual complexity of the bag influenced the time to enter AOI and the conspicuity of threat items significantly influenced the time to enter AOI and the rate of dwell time on AOI of total dwell time on the bag image. This indicated that a complex background and an obscure target increased the difficulty of cognitive processing, therefore increased the time to enter target AOI and those two variables were related to each other. Threat conspicuity is related not only to targets themselves but also the background. Bag complexity contributes to threat item conspicuity and would influence the dwell time distribution on an image. Conspicuous threat items attracted attention earlier in search and had more dwell time compared to total dwell time on the image. Missed targets did not appear to attract attention for approximately 360 ms later in the time to first enter the AOI, compared to detected targets. According to most models of visual search, search begins with a global scanning stage and scrutiny is allocated to selected features. One of the possible reasons that naïve participants allocated more attention to a conspicuous target was that they lacked the object knowledge of a threat item.

Conclusions

Naïve participants were able to identify X-ray images of potential threat items, although they took a long time to do so. Not surprisingly, they were more familiar with the appearances of guns and knives than IEDs. The visual angle subtended by a gun and the familiarity of threat items influenced the detection rate significantly. The complexity of bags and the conspicuity of threat items influenced the initial attention on target threats.

Acknowledgement

This research is supported by the EPSRC "Technologies for Crime Prevention and Detection" programme and is carried out in collaboration with QinetiQ and Tobii.

References

Alario, E-X. and Ferrand, L. 1999, A set of 400 pictures standardized for French: Norms for name agreement, image agreement, familiarity, visual complexity, image variability, and age of acquisition, *Behavior Research Methods, Instruments, & Computers*, **31**(3), 531–552.

Gale, A.G., Mugglestone, M., Purdy, K.J. and McClumpha, A. 2000, Is airport baggage inspection just another medical image? In E.A. Krupinski (ed.) Medical Imaging: Image Perception and Performance. Progress in Biomedical Optics and Imaging, 1(26), 184–192.

Heaps, C. and Handel, C.H. 1999, Similarity and features of natural textures, *Journal of Experimental Psychology: Human Perception and Performance*, **25**, 299–320.

Kundel, H.L. and La Follette, P.S. 1972, Visual search patterns and experience with radiological images, *Radiology*, **103**, 523–528.

Kundel, H.L., Nodine, C.F. and Carmody, D. 1978, Visual scanning, pattern recognition and decision-making in pulmonary nodule detection, *Investigative Radiology*, **13**, 175–181.

Krupinski, E.A., Berger, W.G., Dallas, W.J. and Roehrig, H. 2003, Searching for nodules: what features attract attention and influence detection? Academic Radiology, **10**, 861–868.

Liu, X., Gale, A.G. and Purdy, K. in press, Development of Expertise in Detecting Terrorist Threats at Airports. In D. de Waard et al (Eds). Proceedings of the HFES Europe Conference, 2005.

Mackworth, N.H. 1974, Stimulus density limits the useful field of view. In: R.A. Monty and J.W. Senders (eds.), Eye movement and psychological processes, (Lawrence Erlbaum, New Jersey), 307–321.

McCarley, J.S., Kramer, A.F., Wickens, C.D., Vidoni, E.D. and Boot, W.R. 2004, Visual skills in airport-security screening, *Psychological Science*, **15**, 302–306.

Nodine, C.F., Kundel, H.L., Toto, L.C. and Krupinski, E.A. 1992, Recording and analyzing eye-position data using a microcomputer workstation, *Behavior Research Methods, Instruments & Computers*, **24**(3), 475–485.

Schwaninger, A., Hardmeier, D. and Hofer, F. 2004, Measuring visual abilities and visual knowledge of aviation security screeners, IEEE ICCST Proceedings, 258–264.

Schwartz, S.H. 1994, Visual Perception: A Clinical Orientation, (East Norwalk, Connecticut: Appleton and Lange), 3–21.

Swensson, R.G. 1980, A two-stage detection model applied to skilled visual search by radiologists, *Perception & Psychophysics*, **27**(1), 11–16.

~~Wom 35 Jan~~

NHS Jobs Phond

Wom 35 Jan jan
Verified by the password
Wom 34 an jan
Password registry
Personal credit
Sharon Sheen

Priority
Username: Sarahsh36@notion.com
8/w ukexma kl3sa2x1

THE WARSHIP AS A SAILOR'S HOME: FROM HMS VICTORY TO THE TYPE 45 DESTROYER

Stephen Wells[1] & Paula Thorley[2]

[1] *BAE Systems, Type 45 Project, Filton, Bristol, BS34 7QW*
[2] *BAE Systems, Advanced Technology Centre, Filton, Bristol, BS34 7QW*

In Nelson's day, a seaman's home was a small wooden chest for his personal belongings, and space to sling a hammock on the gun deck. Today conditions are far better, driven not only by the expectations of seamen but also by an economic need to provide better accommodation. This paper examines the background to the changes in accommodation standards, surveys the expectations and current view of existing warship crew, looks at the accommodation in the Royal Navy's latest vessel – the Type 45 Destroyer, and considers what further changes might be expected in the future.

Introduction

For centuries seamen in the Navies of the world have been seen as little more than an inconvenience necessary for a ship to fire guns and launch missiles. This is changing. This change is coming about not through altruism but through hard economics. If trained personnel leave they take with them the investment in their training and their years of experience.

In the UK, the largest part of the Defence Budget is personnel. When personnel leave they represent a loss to the Defence Budget: a budget, which is itself, decreasing. To survive, therefore, the Royal Navy (RN) must make the job of being at sea as attractive as possible. Press Gangs are no longer an option. Potential recruits see the RN as simply another career option – and an inconvenient one at that.

Habitability is a key area where change is coming about. The most junior rating expects personal space and privacy. Slinging a hammock on the gun deck is no longer an option. For the time that a ship is at sea, the accommodation on board is the sailor's home. If that home is comfortable, they are more likely to stay in the RN. If that home is not to their liking, they will leave.

This paper looks first at the history: where have we come from. It then looks at what has been achieved and the driving forces behind that change, reporting on the current view from an existing warship's crew. Finally, it speculates on what further changes might take place.

The history

We are all used to the stories of life in the Royal Navy a couple of hundred years ago: hammocks on the gun deck; accommodation spaces with a headroom of four feet; no privacy; poor food; no communication with home; and violence.

We listen to this and wonder how the sailors of those days put up with it. In asking this question we are making a false comparison. We should not compare conditions today with conditions now. Rather, we should look at conditions afloat in the past and compare them with conditions ashore at the same time.

Lambert makes the point that at the time, for most people, the food on board a RN ship would have been of a better quality, and more regular, than food ashore, "… our predecessors would have considered it superior to anything available on shore. For them such regular, hot, protein-rich meals, together with a nearly limitless supply of beer, would have been a luxury." (Lambert, 2002).

The point is that conditions ashore have improved significantly in the last few hundred years. While conditions at sea were once better than on land, at some point in the Twentieth Century the position reversed. Conditions afloat had not kept up with conditions on land.

What is driving the change?

Two hundred years ago the RN was attractive because it offered better conditions. Today, people are still attracted by better conditions. The difference is that by the beginning of the Twentieth Century, the RN was no longer offering better conditions. Historically a RN ship had a large number of crewmembers living in a small space. There was no privacy. There was little space for personal belongings. Today, a naval recruit expects both privacy and personal space.

During the early days of the development of the Type 45 Destroyer, the Human Factors team visited HMS Kent, a Type 23 Frigate, to talk to the crew about accommodation. The Junior Ratings were prepared to accept bunks three high instead of two if that meant more personal locker space.

In Nelson's day communication between a seaman and home was effectively non-existent. A sailor would leave to go to sea and, if he survived, would reappear months or years later. A common theme amongst English folk songs is the returning sailor who finds that he has been given up for dead or who has changed so much that his fiancée no longer recognises him. Today we have the Internet, email and text messaging. Today's recruits expect to have moment-to-moment contact with friends and family back home.

For the RN, the consequence of these changes in expectation is a loss of personnel. Everyone who leaves the RN takes with them their experience and the investment in their training. It takes many years to train a specialist role such as Principal Warfare Officer (PWO) or Anti Air Warfare Officers (AAWO). If a candidate for one of these roles leaves the RN, they effectively write-off many years of investment in training.

To counter this, the RN is forced to take more account of people in the design of ships. For the RN, this is not simply a case of humanitarian concern, it is economic necessity. Ultimately, if changes are not made, no one will join. There will be no navy.

Where are we today?

A warship is sometimes compared to a village or small town. It is imagined to be self-contained providing, on board, all the services needed by the crew. As well as the basics requirements of working, sleeping and eating, it must also provide recreation, somewhere to keep fit, perhaps a library, a laundry and a hairdresser. It has a shop (NAAFI), a bank and a police force.

This idea that a warship is an isolated unit is changing. The outside world is affecting the assumptions. The RN is not operating in isolation – it must take account of conditions ashore simply in order to retain personnel. The first stage is to preserve

Figure 1. Illustration from Jack modelling of a Type 45 bunk with back rest.

Figure 2. Illustration from the INM Mock up of the Type 45 Bunk.

the warship as a self-contained unit, but improve the living conditions on board and improve contact with home. Eventually, the idea of the ship as an autonomous unit may have to be abandoned.

To understand more about off-duty life on a current warship, a survey was undertaken with a sample of 12 crew on a Royal Navy Frigate. The questionnaire focused on three key areas: accommodation, entertainment and recreation facilities, and communications with home. There was a general theme running through each set of questions, asking crew to rate on a 5-point scale: expectations prior to joining; the current view; and would it be an improvement if ... (based on known changes for a future warship). The 5-point response scale ranged from very poor to very good.

Analysis of the response data revealed that prior to joining the RN, the mean expectation rating for all off-duty living facilities was expected to be fair. However, when asked to rate their current view of facilities, the mean response rate decreased consistently, with the exception of communications with home, which increased slightly. The largest difference was noted between the expectation, and current view, of the washing and exercise facilities provided. It was also these two areas that were given the highest improvement rating when crew were asked how they would rate private washing facilities and a dedicated fitness suite.

The proposed changes that were rated the lowest included sharing a larger mess area with more crewmembers and, therefore, having fewer televisions. The crew were concerned that the camaraderie, which comes about from the current mess system, would be lost if they were required to share a larger recreational area with more people. Although they suggested that it would be likely that existing crew would still socialise within their original mess groups.

One of the more moot points amongst the crew was communications with home. The younger crew members who have come to use email and mobile telephones as the normal means of communication when on shore, welcome, indeed expect, the proposed increase in the number of email terminals and private telephone facilities on the future ships to maintain regular contact with friends and family at home. However, the older crewmembers, many of which have been used to long periods away from home without regular contact, are now finding it difficult to adopt these modern modes of communication.

The change in the design of warships is reflected in the specifications for those ships. The contract for building the Type 45 Destroyer, for example, explicitly specifies the anthropometric range for which equipment is expected to be operable, the numbers of personnel and the standard of the accommodation. Gone are the Junior Rates dormitories of earlier ships – the Type 45 are required to accommodate Junior Rates in cabins with six berths. Minimum floor areas are specified for the accommodation for each crew member. For example, 1.9 square metres per Junior Rates was specified for cabins (In the event 1.78 square metres was achieved). Minimum numbers of washing and toilet facilities are specified and, reflecting changes in the make up of crews, all washing and toilet facilities are to be gender neutral.

The berths themselves were modelled in the Jack Anthropometric modelling tool before being physically tested at the Institute of Naval Medicine (Green and Bridger, 2005).

Change in the future

Earlier we compared a ship to a village with all services locally provided and no one travelling very much. In the English countryside, self-contained villages like this ceased to exist sometime in the nineteenth century when the development of the railways enabled workplace and living space to be separated. The commuter was born.

Outside the Navy, many workers already have two homes. The "real" home and the hotel room during the week. The hotel room is inconvenient but provides comfort, entertainment, instant communication with home and the availability of food nearby. This inconvenience is set against the ability to get back to the "real" home at the weekend. Could this model apply to the RN?

In fact, that is the way things seem to be going. In the RN an initiative called "Topmast" (Tomorrow's Personnel Management System) proposes to go someway down this track. There is no proposal yet to fly sailors home for the weekend from anywhere in the world. However, it is proposed that someone who has booked a family holiday would be flown home despite the ship being thousands of miles away.

Perhaps this then is the future. Rather than being a self-contained village with a fixed team working together for weeks or months on end, think of a workplace with rather more fluid manning. The on-board accommodation could be seen not as an attempt to recreate "home", but as a hotel where people stay for a few days at a time. The sailor as a commuter? Perhaps, eventually, we could see the accommodation on board ship outsourced to Hilton, Travelodge or Holiday Inn.

References

Green A and Bridger RS, 2005, "Evaluation of the Posture of Tall Males in the T45 Bunk". In "Contemporary Ergonomics 2005", (Taylor and Francis, London).

Lambert, Professor A., 2002, "Life at Sea in the Royal Navy of the 18th Century". BBC Website accessed July 28th 2005. "http://www.bbc.co.uk/history/war/trafalgar_waterloo/life_at_sea_01.shtml". Professor Lambert is "Laughton Professor of Naval History" at Kings College, London.

"Royal Navy Command and Organisation: Fleet Command and Organisation". RN Website accessed August 2nd, 2005, "http://www.armedforces.co.uk/navy/listings/l0006.html".

TOWARDS THE DEVELOPMENT OF A LOW COST SIMULATOR TO SUPPORT NVG TRAINING

Jiajin Wu, Kee Yong Lim & Chai Wah Koh

Centre for Human Factors and Ergonomics, School of Mechanical & Aerospace Engineering, Nanyang Technological University, Nanyang Avenue, Singapore 639798

This paper presents the second stage of work concerned with the development of a low cost simulator for Night Vision Goggle (NVG) training. It extends previous work on the digital creation of simulated NVG images of the resolution chart targeted at NVG eye piece adjustment training. The present work is concerned with the development of a similar digital manipulation technique aimed at creating simulated NVG imagery to support object recognition training. Tests to assess the physical and performance fidelity of and the efficacy (transfer of training) afforded by the digitally created simulated NVG images have been completed. The results reported in this paper, reveal that the low cost training simulator under development, is a promising alternative to current methods of NVG training.

Background

For over thirty years, fighter pilots rely on NVGs to see, maneuver and fly at night. Although NVGs improve perception at night, their use (and misuse) have led to accidents and loss of life. To ensure effective operations and to minimize the risk of flying with an NVG, adequate training and exposure to different viewing conditions are necessary. However, NVG equipment costs are high, and their availability and lifespan are limited. Therefore, it is desirable to conduct NVG training without the need for an actual NVG. This project addresses this desire by developing a low cost NVG training simulator.

As the scope of NVG training is very broad, the development of the training simulator has been divided into several modules. One of the fundamental skills a pilot using an NVG would require is the ability to focus the NVG. Failure to achieve appropriate NVG focal adjustment might lead to poor visual acuity and visual fatigue. To this end, the first phase of the project is concerned with the development of a prototype simulator to support NVG focal adjustment training. A requirement of this work is to develop a method to create a simulated NVG image of the resolution chart and its transitions during adjustment of the objective and diopter lens. Tests to ascertain the fidelity of the simulated image and transitions during eye piece adjustment, have been completed and benchmarked against focal adjustment outcomes obtained with an actual NVG (AN/AVS 6 Gen III Aviator NVG). The results show that NVG images, NVG photo-images and digitally created simulated NVG images of the resolution chart are equivalent in fidelity, when used for focal adjustment training (Lim and Quek 2003).

The next phase of the project is concerned with the extension of the capability of the NVG simulator to include training of visual scanning techniques to enhance object recognition. To this end, a NVG photo-image database is collated and the characteristics of NVG images are examined to establish a basis for creating and assessing simulated

Figure 1.   Set up for capturing NVG photo-images.

NVG images of scenery (as opposed to an NVG resolution chart). The results of tests to ascertain the fidelity of the simulated images are reported in this paper.

Method

To digitally create simulated NVG images, a database of NVG photo-images is first collated for subsequent reference and comparison. For this purpose, the NVG and camera are secured together and affixed to a metal plate so that shaking can be minimized during photo-taking to avoid blurred images (see Figure 1). NVG photo-images are then taken using a Sony DSC F828 digital camera with 8 mega pixel resolution.

Next NVG images and photo-images are examined to uncover and document unique characteristics of NVG images that should be replicated in the simulated images to be created. The following characteristics of NVG images are noted:

1. NVG images are monochrome green.
2. NVG images are "noisy" with a snowy effect under low luminance.
3. NVG images are blur with a scintillating effect.
4. Trees appear light green when seen through an NVG.
5. Water surfaces appear black with graduated changes in color and ripples.
6. The appearance of the sky is similar to a water surface but without ripples.
7. Lights have a surrounding ring termed the "halo" effect.
8. Reflections are different on solid and water surfaces as the latter might include ripples.

To create simulated NVG images, the above characteristics and effects are replicated digitally using the functions and filters of Adobe Photoshop 7 (see Figure 2).

Assessment of the fidelity and efficacy of the simulated NVG images

The experiment setup used is consistent with the viewing conditions of an NVG eye test lane with a luminance level of $1.3 \, cd/m^2$. Tubes are used to restrict the subject's field of view to 40°. Using PowerPoint, simulated NVG images are presented to subjects by projection onto a screen (1.6 by 0.8 m).
Subjects are selected according to the following criteria:

- Myopia less than 300°
- Visual acuity of 6/6 with or without visual aid (tested using a Snellen Chart)

Photo-image taken through an actual NVG   Digitally created simulated NVG image

Figure 2.   A comparison between a NVG photo-image and a simulated NVG image.

- No sight related illness
- No consumption of alcohol and/or medicine within 24 hours of the experiment.

Three sets of tests are conducted to verify the fidelity of the digitally created simulated NVG images, namely physical fidelity test, performance fidelity test and transfer of training test. Each of these tests will now be reviewed in turn.

To assess physical fidelity, NVG photo-images and simulated NVG images are displayed side by side to subjects who are then asked to identify the image that they feel is the NVG photo-image. An actual NVG need not be used in this case as it has already been demonstrated by Lim and Quek (2003) that NVG photo-images are equivalent in fidelity. For this test, subjects are divided into 4 groups. The first group of subjects comprises naive participants, while the remaining three groups are subjects who have gone through the experiment to verify transfer of training (see later). This experiment design allows for a comparison between naive subjects and subjects who have been exposed to an actual NVG view of a terrain model board.

The performance fidelity test aims to verify the fidelity of visual cues provided by digitally created simulated NVG images for object recognition in a scene. Pre-test training is given to assist subjects and to reduce variation due to individual differences. For this test, subjects are shown two sets of 8 slides of an image. One set of images comprises NVG photo-images, while the other set of images are simulated NVG images. Subjects are required to identify the same objects in each image set, with the presentation sequence of the image set balanced across subject groups. To reduce learning effects, the slides in each image set are randomized in presentation sequence.

To verify the efficacy of the digitally created simulated NVG images and the appropriateness of its application in a training simulator, a positive transfer of training must be demonstrated. To enable a cross comparison in this test involving object recognition, the subjects are divided into 3 groups, namely:

- Group 1: training with NVG photo-images of outdoor scenery. This method has been used in NVG object recognition training by the military.
- Group 2: training with an actual NVG looking at a terrain model board ($\varnothing$ 1.2 m). This method has also been used in NVG object recognition training by the military
- Group 3: training using the digitally created simulated NVG images.

Following their training session, all three groups of subjects are tested on object recognition when they look through an actual NVG at various views of a terrain model

Table 1. Mean performance of subjects across 10 slides.

N = 24	S1	S2	S3	S4	S5	S6	S7	S8	S9	S10
Mean	0.25	0.21	0.46	0.59	0.63	0.71	0.58	0.42	0.67	0.63

Table 2. ANOVA of performance fidelity test results (accuracy & time).

Source	Measure	Sum Sq	df	Mean Sq	F	F_{crit} 0.05	F_{crit} 0.01	Sig.
Expt	Object ID	0.01	1	0.0	0.1	4.3		P > 0.05
	Time	68.7	1	68.7	9.9	4.3	5.8	0.00
Error (Expt)	Object ID	3.1	23	0.1				
	Time	159	23	6.9				
Question	Object ID	36.5	7	5.2	23.7	2.0	2.4	0.00
	Time	46.5	7	6.6	2.37	2.0	2.4	0.03
Error (Question)	Object ID	35.4	161	0.2				
	Time	452	161	2.8				
Expt Question	Object ID	0.2	7	0.0	0.3	2.0		P > 0.05
	Time	19.7	7	2.8	1.9	2.0		P > 0.05
Error (Expt Question)	Object ID	13.7	161	0.1				
	Time	238	161	1.5				

board. The subjects are asked to identify lit street lamps, water surfaces with ripples and other objects located on the terrain model board. It may be expected that in view of the test conditions and set up, Group 2 would perform best. However, if the training efficacy afforded by the simulator and digitally created NVG images is equivalent, the performance of all three groups would not be significantly different (note: in retrospect, to be conclusive a new experiment would be conducted with a fourth group that would not be provided any training prior to the test).

Results

The results of the test concerning physical fidelity (Table 1) show that with exceptions of slides 1, 2 and 6, subjects generally could not differentiate between NVG photo-images and digitally created simulated NVG images. Further examination of the slides reveal that for the exceptions, the digitally created simulated NVG images have not been done well, and subjects could see differences in color, brightness, sharpness and uniformity of the images. Thus, greater care needs to be taken when creating the simulated NVG images. The results of an analysis that includes subjects who have been given prior exposure to actual NVG imagery performance, show that such a prior exposure does not make a difference. It may thus be inferred that the physical fidelity of simulated NVG images do not have to exactly the same as actual NVG imagery.

A repeated measures ANOVA is used to analyze the results of the performance fidelity test (see Table 2). The results show that subject performance in object recognition tests in terms of correct identification, is similar across test conditions. The results imply that the visual cues provided to subjects to enable object recognition are equivalent between NVG photo-images and digitally created NVG images. However, a significant difference in performance time is observed. Further examination of the data reveals that subjects generally perform slower with simulated NVG images. Three possible

Table 3. ANOVA of the results for transfer of training assessment.

		Sum Sq	df	Mean Sq	F	$F_{crit\ 0.05}$	Sig.
Accuracy	Between Groups	1.4E−02	2	7.1E−03	0.10	3.29	$P > 0.05$
	Within Groups	33.0	33	0.9			
	Total	33.0	35				
Time	Between Groups	111	2	55.5	0.29	3.29	$P > 0.05$
	Within Groups	2040	33	61.8			
	Total	2151	35				

reasons may account for this observation. First, there might be differences in Powerpoint file size of the image sets and thus the slide transition time may be affected, i.e. slower loading time of simulated NVG images. Subsequent checks show that this is not the case, as there is little difference in loading time of slides across image sets. Second, as the performance time is manually recorded using a stop watch, the reaction time of the experimenter may have caused a difference. However, such errors would be uniformly distributed across test conditions and not result in a systematic bias. Third, the visual cues although present may have been more difficult to discern in digitally created simulated NVG images than in NVG photo-images. Thus, subject performance in terms of correct identification of objects may not be affected, but their performance time might be. As there are differences in difficult of object recognition across the slides in a test set, the ANOVA results reveal a significant difference across questions as expected.

A one way ANOVA is used to analyze the results of the test to assess the transfer of training (Table 3). The results show that there is no significant difference between the performances across the three groups of subjects. It is thus concluded that all the different routes of training enabled the same level of performance. Consequently, it may be concluded that a training simulator that employs the digitally created simulated NVG images would provide a promising alternative.

References

Lim, K.Y. & Quek, S.M. (2003). Development & assessment of a method for developing a low cost simulator for NVG focal adjustment training. Internal research report.

Miller II, R.E. & Tredici, T.J. (1992). Night vision manual for flight surgeons. Ophthalmology Branch, Armstrong Laboratory, Human Systems Centre (AFMC), Brooks Air Force Base (AL-SR-1992-0002).

Ruffner, J.W., Antonio, J.C., Joralmon, D.Q. & Martin, E. (2001). NVG training technologies and situational awareness. Proc. Advanced Technology Electronic Defense System Conference/Tactical Situational Awareness Symposium. Patuxent River, MD: Naval Aviation Systems Team (PMA-272).

US Army Aviation Training Brigade (1998). NVG training course. Fort Rucker, Al: U.S. Army Aviation Centre.

RE-EVALUATING KAO'S MODEL OF EDUCATIONAL ERGONOMICS IN LIGHT OF CURRENT ISSUES

Andree Woodcock

Coventry School of Art and Design, Coventry University,
Priory Street, Coventry, CV1 5FB, UK

Kao's (1976) paper was one of the first to outline the field of educational ergonomics. This was written when little was known about the detrimental effects of furniture on children and before the introduction of computers. This paper reviews the models that have been forward to sketch out this area of research and assesses the applicability of Kao's model in light of current educational experiences.

Introduction

With a few notable exceptions, ergonomists have ignored the field of educational ergonomics, or have chosen to focus on one particular aspect, such as furniture design or the use of IT, rather than focussing on wider issues and developing a systems approach to the field. My motivation in writing this paper, is to draw attention to the fact that educational ergonomics is one of continuing importance in contemporary ergonomics, and to highlight this by presenting three examples which show that education is in crisis. Had such issues been raised in military or manufacturing contexts, ergonomists would have developed interventionist strategies and pressed for organisational change. However, education seems to be almost a taboo area. I will then outline three models that have been presented as a framework for educational ergonomics, and consider the extent to which one of the earliest, that proposed by Kao, is still applicable.

Education in crisis?

In the UK we are bombarded with reports that the government is successfully tackling the falling standards in education. Despite showing increases in literacy, numeracy and measurements of student achievement, the public remains sceptical of new initiatives, research findings and the measurement instruments themselves. Additionally those most affected by education (teachers and students) are struggling with the present system, for example:

Teachers

- 38% of all absences of headteachers are attributed to work related stress
- 49% of those who retire owing to ill-health cite psychiatric reasons
- Jarvis' (2002) analysis of stress in the profession, found contributory factors relating to workload, long working hours, poor status and pay, role overload, evaluation apprehension and systemic factors such as lack of government support, lack of information about change, constant change and the demands of the National Curriculum.

MacSween, Associate Director of the Audit Commission revealed that it is "push" not "pull" factors that cause teachers to leave their jobs, and that the most effective solution for retaining staff is to target people's whole experience of work. Thorp (Director of Worklife Support) emphasized the importance of a school's climate and culture on the health of both staff and students. A positive climate and culture comprised of effective leadership, a good working team, open dialogue and a safe environment. It is not just staff that are effected by the negative climate that is found in schools.

Young people

- 700,000 secondary school pupils skip lessons each year
- In 2003, truancy varied in secondary schools between 4 and 48% (between 4 and 47% for primary schools), (Eason, 2003)
- 50,000 families in the UK educate their children at home. This is expected to rise to half a million children by 2010
- 19,000 children attempt to commit suicide every year (one every half an hour)
- 2 million children see their GP each year for emotional and psychological problems.

Obviously schools may only sometimes be a contributory factor. However Marr and Field (2001) commented that in school bullying "a child is harmed, or may attempt suicide, as a result of the deliberate actions of another in an environment where the responsible adults have failed to provide a mechanism for reporting, intervening, and dealing with physical and psychological violence". The social, organisational and indeed physical structures of schools are failing to support the most vulnerable of its users. One of the basic tenets of ergonomics is that if you design for the most vulnerable, all will benefit. Places of learning are no longer respected, as witnessed by the attacks on schools themselves.

School buildings

- 1 in 8 schools suffer from arson attacks each year, with an average of 20 schools in England and Wales damaged or destroyed each week
- Two-thirds of the offenders are aged 10 to 14.

Such facts are openly acknowledged and may be viewed as symptoms of a deep malaise in education. The financial implications of failing to address core problems is considerable, for example:

- Teacher absence through stress – £19 million a year
- Staff turnover – £40,000 a year
- Truancy prevention programme – £1 billion
- Loss attributable to arson attacks – £100 million a year. This could be cut by 90% if sprinklers were installed (Sinott, 2003).

It would seem that schools and the education they deliver are no longer fit for purpose. Woodhead (2002) based on his experiences as Chief Inspector of Schools criticises much of the current educational system. Is the answer to these and similar problems a more integrated, systems approach to the consideration of education? The following section provides an overview of previous models and how current research maps onto these.

The history of educational ergonomics

Robson (1874) considered a systems approach to the design of school buildings, when he advised architects to consider the users, the tasks they would have to undertaken and

Table 1. Overview of Components in Educational Ergonomics.

Kao (1976)	Smith (2003)	Mokdad (2005)
Ergonomics of learning	Community and family factors	Design of academic curricula
Instructional Ergonomics	Personal factors	Design of teaching
Ergonomics of Educational Facilities	Class design	Evaluation of academic performance
Ergonomics of Educational Equipment	Academic programme	Development of human resources
Ergonomics of Educational Environment	Teaching	Design of educational environment
	Organisational design and management	Design of educational rules and regulations
	Classroom/building ergonomics	

the effects of the quality of the building on the occupants and their sense of well being. The need to consider the design of furniture and its effects on those required to sit still for long periods of time was also echoed by Leibreich (1872). The adoption of poor posture in schools was even then considered responsible for life long effects on both gait and vision. Moving forward, Roth (1966) in his consideration of school architecture and urban planning, was influenced by Pestalozzi, with the school providing an extension of the education provided by the family. For example, nursery schools should be close to the parental home, small in scale and have a homely, intimate environment – to provide the child with much needed security as they moved into a strange new world.

Three models have delineated the field of educational ergonomics, as summarized in Table 1. However, only the one proposed by Smith (2003, and based on earlier work by Smith and Smith e.g. 1966) could potentially address the wider and complex problems found in education (as noted above) by including community, family and personal factors as well as the academic programme itself.

In his 1976 paper, Kao outlined an interdisciplinary field of educational ergonomics that would lead to a systems approach to educational effectiveness. He saw educational institutions as essentially "work systems where the objectives include effective and successful dissemination of knowledge and cultivation of intellectual sophistication". In his model, educational efficiency was related to 5 components. In this the success of the learning by the students and the delivery of instruction by the teachers also were related to the ergonomic design and development of educational facilities, educational equipment as well as the educational environments. If this model is updated to include recent technological developments, then it can be made applicable to current issues (such as e-learning). The interrelationships become more specific in the cybernetic model proposed by Smith that can explain some behavioural problems associated with peer pressure and substance abuse.

In this educational ergonomics is defined as the field of human factors concerned with the interaction of educational performance and educational design. Smith specified seven sets of factors that could influence educational performance. He noted that although there has been a debate about failing standards and underachievement in the US (similar to the UK) there has been little transfer of ergonomics principles and techniques into the education arena, despite proven success in other fields. Whilst the cybernetic model proposed characterised the influences on the student a similar set of factors could affect the teacher.

Lastly, from a review of previous papers in the educational domain, Mokdad (2005) recently concluded that the "dissemination of ergonomics knowledge and practice to improve the quality of educational work is both urgent and important" and that educational ergonomics could legitimately focus on six areas (as outline in Table 1). These overlap with the areas proposed by Kao, with a reduced focus on generic skill acquisition (as this model was developed for higher institutions).

Relationship of the models to current research

Using the papers included on the Ergonomics for Children and Educational Resources web site (http://education.umn.edu/kls/ecee/default.html accessed in December 2005) as a database, a marked increase in the number of submissions has occurred since 2000. Research has been submitted from Sweden, UK, USA, Japan and New Zealand (personal correspondence has also been received from Latvia, Colombia and Taiwan). The studies have included preschool, primary and secondary school students and their teachers; products ranging from computers and computer games, child car seats, barriers, meal toys, backpacks and furniture; with a predominance of papers on back pain in relation to furniture (especially chairs), laptops, backpacks and workstations; physiology, anthropometry and discomfort in lower back, vision and RSI. Backpain and seating have been topics of concern throughout the period, with the first mention of computers in 1996.

Using an adaptation of Kao's model, most of the papers can be mapped on to "Instructional Ergonomics" (textbook and printed material design, teaching machines and self instructional devices (in which I have also included some of the IT studies), AV presentation research, research in classroom instructions, lecture and class preparation, educational tv research (again this has been expanded to include some of the studies of IT) and the "Design of Educational Equipment" in which issues relating to computers and the relationship of chair design to posture and back pain have predominated. In relation to the current status of education and the way schools are used/way in which material is taught, Kao's model is still applicable and predates much of the avenues of investigation in elearning, distance/remote working, group and individual activities. Where Kao has emphasised penmanship, perhaps we could now also include keyboard skills (a legitimate though again under-researched areas).

However, whole areas of education have been omitted from study – such as the effective design of the academic year, lesson scheduling, teacher skills and recruitment, environmental design as identified by Kao. This is worrying, as a picture is emerging where buildings, curricular and tasks are being designed and researched by others, without any input from ergonomists. There are obvious links between design and educational performance as indicated by Smith's model. We can predict factors likely to cause stress in teachers, and have developed interventions to successfully address this; likewise we can define the optimum acoustics, lighting, colour for improved educational performance. Such interventions will lead to a better overall environment, which may in turn reduce the level of disengagement felt by school users and provide more financially viable alternatives.

Conclusion

Ergonomists still have an opportunity to make a difference in the design of schools and education, and of improving the immediate and long term health of all stakeholders, by applying their knowledge and methods to the design of future school environments and curricular, in consultation exercises and in making sure that the voices of all stakeholders are heard. Such interventions are now necessary, timely and financially imperative.

References

Eason, G. 2003, Up to half of pupils play truant, Accessed Dec 1st 2005, on url: http://news.bbc.co.uk/1/hi/education/3116760.stm

Ergonomics for Children and Educational Resources web site Accessed Dec 1st 2005, on url: http://education.umn.edu/kls/ecee/default.html

Jarvis, M. 2002, Teacher Stress, *Stress News*, Accessed Dec 1st 2005, on url: http://www.isma.org.uk/stressnw/teachstress1.htm

Kao, H.S.R. 1976, On educational ergonomics, *Ergonomics*, 16, 6, 667–681

Leibreich, R. 1872, *School life in its influence on sight*, in Robson (*ibid*)

Marr, N. and Field, T. 2001, *Bullycide: Death at Playtime*, Success Unlimited, ISBN 0952912120

Mokdad, M. 2005, Educational Ergonomics: Applying ergonomics to higher education institutions, *Fourth International Cyberspace Conference on Ergonomics*, 15th Sept – 15th Oct. Accessed Dec 1st 2005, on url: http://cyberg.wits.ac.za/cb2005/educat2.htm

Robson, E.R. 1874, *School Architecture*, Victorian Library Edition, Leicester University Press, republished in 1973

Roth, A. 1966, *New School Building*, Thames and Hudson: London

Smith, K.U. and Smith, M.F. 1966, *Cybernetic Principles of Learning and Educational Design*, New York; Holt, Rhinehart and Winston

Smith, T.J. 2003, Educational ergonomics; educational design and educational performance, *Proceedings of the XVth Triennial Congress of the International Ergonomics Association*, Aug 24–29, Seoul, Korea (CD ROM)

Sinott, S. 2003, *Arson in Schools*, Accessed Dec 1st 2005, on url: http://www.epolitix.com/EN/ForumBriefs/200410/914f0048-d450-4cf2-84c9-bb6bc896867f.htm

Woodhead, C. 2002, *The Standards of Today*, Adam Smith Institute, downloadable from http://www.adamsmith.org/pdf/the-standards-of-today.pdf

THE KIDS ON THE ADULTS ARMCHAIR, THE ERGONOMY OF THE COMPUTERS

Nureddin Özdener, Olga Eker Özdener, Zeynel Sütoluk & Muhsin Akbaba

Çukurova University, Faculty of Medicine, Department of Public Health, Adana, Turkey

The kids are using the school's or internet cafe's general computers in Turkey. The statistics that have published in the developed countries show that every year rather much kids see damages tied to accumulating injures on their head, neck and hand.

Searching the computer stations that the kid's use at the internet cafes. At Turkey, in Adana, 5 internet cafe's computer stations that are used by the children aged 15, are searched. The computer environment working conditions lists of the Cornell University is carried out on this study.

The computer users in our search are, 18 people. The age average is 12.5 year. This users have been using the computers since average 2.6 years. The time that is entered in front of the computer is average 1.8 hour/day. If the ergonomically findings are appraised; 16.6% of the kid's upper arms and elbows were near the body. 22.2% of the people's wrists, hands were flat; 38.8% of the chairs are supporting the waist and arm and these are adjustable. 44.4% of the mouses were near the keyboard and 33.3% were suitable to the users hands. 44.4% of the computers weren't shining and reflecting. There were no computers that had foot support.

The Computer Work Stations at the internet cafes aren't suitable for kids.

Introduction

The period that environment skill accordance makes most difficulties for people is bachelorhood. The changes in sizes and physiological functions arise with the difficulties related with posture. Firstly, rapid increase in foot and leg length is observed in the acceleration of growth in bachelorhood and transverse growth of hips, then the increase in the front-back radius of breast, expansion of shoulders and increase of body length follows it (Güler, 2004). From clothes to tolls, many instruments become important in posture contemplation. To make creations that ease all these consistencies, it is not possible to find a standard boy or girl (Güler, 2004). Today, in most countries children are growing up and maturing more rapidly. Environmental caution has important effect on this as health factors' development. The slip of acceleration and maturation of growing to the earlier period is designated as century's inclination. These changes; differs according to countries, geography, social and cultural properties.

The growing curve of Turkish children according to their age till 13–14 is nearly same as the curves given for West Europe and USA white race children. The statistics published in developed countries show that every year lots of children come to harm according to head, neck, clasping of heads. Especially in growing up phase of children, head sphere, breast depth and hand clerens radius are the most important variables that must be considered in instrument creation (Godfrey, 1986) Being healthy adult possibilities increase by prevention of accidents in childhood and growing in secure media. However the adults' health is affected by high calorie, cigarettes and other bad habits, and the effects of environment such as air pollution, sound pollution, infections and stress, the positive and negative effects that people face with in early life is the important determinants for sensitivity of health in post-ages to harmful effects.

In the school age, the most active period of people life in bodily, psychological and social side, for the problems come up with the change and development of body structure and take pains over forming strong physiological structure.

The ergonomically deficiencies in children's school age interest children's health intimately (Dirican and Bilgel, 1993). In ergonomics, that is the most disciplined area in the life and activities of the children in this age, perceptual ergonomics, anthropometrics, work physiology and biomechanics are in relation.

Today, the rapid dispersion of computers makes the ergonomics problems related with computers important in school and daily life of the children. It is noticed that an average American child spend 3 hours in front of computer (Purvis and Hirsch, 2003).

In developing countries, children use general computers such as school and internet cafe more than individual computers. The statistics that published in developed countries show that every year many children come to harm because of carpal tunnel syndrome, accrued bruises in heads, necks and hands. In literature, it is noticed that chronic pains in hands, wrists, neck and back are seen in children related with computer usage. The increase in the number of children using computer bring in its train that children working in the computer working stations that are designed for adults influated from dissonant situations.

How much is it true to leave children with the equipments that are designed for adults?

Bad working conditions and the computer stations that are not designed ergonomically may have negative effects on developing children's body.

Goal

Searching the ergonomically compatibility of computer stations that children use in internet cafes.

Materials And Methods

In this descriptive study, in the five internet cafes in Turkey, Adana Turgut Özal Bulvari, the computer stations that children below the age of 15 are using, is examined. The control list of working conditions in computer media that was developed by Cornell University is applied.

Findings

There are 18 computer users that 17 of them are boys and 1 of them is girl, in our study. The average age is 12.5 (min. 9–max. 15). They use computers nearly 2.6 years. The average time passed in front of computer is 1.8 hours/day (min. 1 hour–max. 5 hours). While 77.7% of people have computers at their school, the rate of people having computers at their houses is 22.2%. When positive results are evaluated ergonomically, the 16.6% of children's upper arm and elbows were near to the body. In 22.2% people wrists and hands are straight and not folded to right or left. 38.8% of chairs support the waist region and they were adjustable type and arm supported. In 44.4%, mouse was near to the keyboard. 33.3% of them best fit to users' hand. In 44.4% of computers, there is no shining and reflection in screen. In none of computer stations, the wrist supports are padded enough and sharp sides are prevented. No one has foot support. None of the screen has shining and reflection filters.

Discussion

In the first study of Cornell University, 40% of primary education children are under postural risk related to working at computer work stations. When the time in front of the computer increases, the risk of muscle-skeleton system deformation increases. When this is considered, in America, a research is done to study the muscle-skeleton system complaining and using computer in 152 of high school students. A meaningful relation is founded between the time that is passed in front of the computer and muscle-skeleton system complaining. When to muscle-skeleton system is studied in people that using furniture designed compatible and not, however odds ratio is in boundaries they have meaningful value. Children using computers on incompatible designed furniture's have higher complaining possibilities (Barrero and Hedge, 2002).

Special lightening and no reflection in screens is especially important in working ergonomics in computers. In the study, which is about two types of school chairs and tables, it shows that in standard furniture's pose and way of sitting gives more negative results than ergonomically designed (Schroder, 1997).

In school age children, chronic muscle-skeleton system pains are widespread problems in all over the world. In one of the study of World Health Organization, 30% of age of 11 children has back-ache in the last week in America. This rate increases with the increase in age. In most of the regions in Europe, there is an increase in children's back-ache complaining in the last 15 years (Jacops, 2002).

Benefit from faulty type keyboards, fault in writing techniques results in back, neck and shoulder pains and problems related with hands, wrists and eyes (Williams, 2002).

Important advantages are provided with simple ergonomically arrangements in computer usage of primary school children. The problems that can be formed after the usage of computers at homes, schools and game saloons are specified in this study (Marshal et al, 1995).

In a study that is prepared in America, a meaningful relation between time passed in front of computer and muscle-skeleton system complaining is determined.

In internet cafes different size of chairs and tables must be used because of the differences in the length of children (Knight, 1999).

Putting a small pillow or rolled towel to the back is very useful while sitting working long time in a bad pose causes cumulative bruises in backbones and digestion system sicknesses. 60% of school age children are complaining about neck, shoulder pains and the most important factor is insufficient furniture's (Güler, 2004).

Providing special lightening is appropriate, because visual works deformity posture.

When to muscle-skeleton system is studied in people that using furniture designed compatible and not, however odds ratio is in boundaries they have meaningful value. Children using computers on incompatible designed furniture's have higher complaining possibilities. To decrease the possibilities of painless and mutilating piques in children, American orthopaedist association and professional health council suggest these:

- If adults and children at home, school and internet cafes use the same computers, than the computer hardware must be adjustable/changeable according to every child
- Computer screen must be adjustable according to eye level

- Computer hardware must be complete and correct for the children and bask support must be satisfied. Chair's arm must be in the position to satisfy the angle 70°–135° between elbows and keyboard
- Elbows must be in nature position while writing
- Mouse must be close to the keyboard to prevent extending arms to reach.

Results

Computer working stations in internet cafes are not in comfortable posture for children. There is no interference for the adaptation of adults computer media for children. The computer work stations in internet cafes must be modified for children that they can use more comfortable. Forming behaviour change by transferring ergonomics with education is important in developing countries. Parents must investigate the compatibility of bureau ergonomics in internet cafes that their children go. When neck, back or elbow pains occur, cumulative trauma must be investigated in children that deal with computers.

References

Barrero M. and Hedge A. (2002) Computer Environments for Children: a Review of Design Issues, Work, **18**, 227–237

Dirican R, Bilgel N, 1993, Public Health, Second Edition, Uludag University, Bursa, 405–428

Godfrey S et al, 1986, Scenario Analiz of Children's Ingestions analysis; Proceeding of the Human Factors Society, 30. Annual Meeting. Santa Monica Ca:, Human Factors Society, 566–569

Güler C, 2004, *Ergonomy for Doctors and Engineers with Dimension of the Health*, First Edition, (Palme, Ankara)

Jacops K, 2002, Are Backpacks Making Our Children Beasts of Burden? The Proceeding of the XVI Annual International Occupational Ergonomics and Safety Conference, http://education.umn.edu/kls/ecee/pdfs/BackpacksMakgChldrnBeastsofBurden_Jacobs.pdf, 2 December 2005, visiting on web

Knight G., Noyes J, 1999, Children's Behavior and the Design of The School Furniture, Ergonomics **42**, 747–760

Marshal M. et al, 1995, Effect of Work Station Design on Sitting postrur in Young Children, Ergonomics, **38**, 1932–1940

Purvis JM, Hirsch SA, 2003, Playground Injury Prevention. Clinical Orthopaedics, **409**, 11–19

Schroder I, 1997, Variational of Sitting Posture and Physical Activity in Different Types of School Furniture Coll Anthropology, **21(2)**, 397–403

Williams IM, 2002, Students' Musculoskeletal and Visual Concern, The Proceeding of the XVI Annual International Occupational Ergonomics and Safety Conference, http://education.umn.edu/kls/ecee/pdfs/StudentsMuscAndVisualConcerns_Williams.pdf, 2 December 2005, visiting on web

COMPLEX SYSTEMS/TEAMS

UNDERSTANDING BAKERY SCHEDULING: DIVERSE METHODS FOR CONVERGENT CONSTRAINTS IN USER-CENTERED DESIGN

P.C.-H. Cheng[1], R. Barone[1], N. Pappa[1], J.R. Wilson[2],
S.P. Cauvain[3] & L.S. Young[3]

[1]*Department of Informatics, University of Sussex, Brighton, BN1 9QH, UK*
[2]*Institute of Occupational Ergonomics, University of Nottingham, NG7 2RD, UK*
[3]*BakeTran, 97 Guinions Rd., High Wycombe, HP13 7NU; and Campden and Chorleywood Food Research Association, Chipping Campden, Gloucestershire, GL55 6LD, UK*

The overall aim of the ROLLOUT project is to evaluate an approach to the design of user-centered information technology to support complex problem solving. Production planning and scheduling in bakeries has been chosen, because it presents many substantial and diverse challenges. This paper interprets them as requirements that tools to support bakery planning and scheduling should address. A wide variety of knowledge acquisition techniques have been used gain a comprehensive understanding of the problem at multiple levels.

Introduction

The ROLLOUT project's overall aim is to evaluate a *Representational Epistemological* (REEP) approach to the design of user-centered IT systems to support complex problems and conceptual learning (Cheng, 2002). The REEP approach advocates understanding the underlying conceptual structure of a domain and then using that understanding to invent new graphical representations to serve as tools or computer interfaces to support problem solving. The approach has been applied to instructional domains and for event and personnel scheduling (e.g., Cheng, 2002). In the current work, bakery planning and scheduling provides a challenging test case for the generality of the approach. It is a stringent test because of the inherent complexities of the domain, the problem solving demands that are imposed and the diversity of problem contexts that must be covered, not least with respect to both in-store and plant bakeries. ROLLOUT is the particular graphical software tool that the project is developing and evaluating using the REEP approach (Barone & Cheng, 2005).

The problem of production planning and scheduling mars the efficiency and productivity of the baking industry. It is a longstanding problem whose importance has attracted a consortium of ten commercial partners to the project, spanning plant (factory-scale) producers, supermarkets and equipment manufacturers. On the academic research side of the consortium there are teams from two universities and a division of the Campden and Chorleywood Food Research Association. The CCFRA is a membership based research organization and its Cereal Processing Division brings expertise in bakery science and technology to the project. The project is on-going and approaching its final quarter, at the time of writing.

The aim of this paper is to provide a summary of our understanding of the nature of this problem and to describe challenges facing the development of an effective user-centered IT system to support it. The paper will first outline the methods used to gain this knowledge.

Knowledge acquisition and studies

Bakery planning and scheduling are complex activities, involving multiples levels, over a range of time scales and multiple points of view. Hence, a variety of methods have been used to examine the nature of the problem from different perspectives. The perspectives and the methods are ones often advocated for studies of complex task environments (e.g., Endsley, Bolté & Jones, 2003; Preece, Rogers, Sharp, Benyon, Holland & Carey, 1994).

Seven perspectives were covered. (1) The *bakery system analysis and modelling perspective* concerns the setup and capabilities of particular bakery configurations. This includes the specification of the products made in the bakery. (2) The *analysis of extant representations and tools* concerns the physical materials and IT systems that are currently used for the tasks. (3) The *cognitive task analysis* perspective examines individual's psychological process of thinking about component sub-tasks, especially from an information processing perspective. (4) The *studies of problem solving* specifically examines how the tasks, overall, are done. (5) The *ecological or ethnographic analyses* focus on social interactions and collaborations in the task environment. (6) The *analysis of goals and belief systems* address what normative views and goals exist. (7) The *comprehensions and situational awareness perspective* concerns how well bakers understand the current and future operational states of their bakery.

Twelve methods were used. (A) In preliminary bakery site visits the project team members toured plant and in-store bakeries accompanied by the CCFRA team who provided contextualizing explanations. (B) Initial questionnaires were sent to selected planners and schedulers and included items on the methods, tools and practices of bakery scheduling. Although asked for concrete examples, the respondents typically gave normative descriptions. (C) Materials and digital tools used in the bakeries were examined. Demonstrations of existing software applications were also given to the project team. (D) Demonstrations of ROLLOUT to the members of the project consortium provided a venue for the elicitation information about the generic contexts within which planning/scheduling occurs in bakeries. (E) Job shadowing occurred as part of an extended field study, in which bakery managers were observed throughout a shift of production. (F) Interviews were conducted with bakery managers, on site at the bakeries and at the CCFRA, using free form, semi-structured and formal decisions probes on different occasions. (G) Bakery managers were given scheduling problems based on a simplified in-store bakery, which they attempted to solve in a conventional manner. Their performance was recorded, and verbal and action protocols obtained for selected solutions. (H) In the goal rating task bakers rated a list of common goals from different perspectives. (I) The domain specific temporal reasoning task involved the bakers making judgments about temporal relations among different production stages of runs. It was found that reasoning forwards in time was easier than reasoning backwards in time. (J) To obtain detailed information about the configuration of bakeries, structured elicitation templates were designed for the bakers to complete. (K) An experiment was conducted comparing ROLLOUT and an equivalent spreadsheet for managing the production in a working bakery. The training bakery at the CCFRA was used to mimic two simplified but realistic in-store bakeries. Experienced bakers from in-store bakeries used ROLLOUT or the spreadsheet, separately, and at regular intervals they were questioned about their awareness of current and future bakery operating status.

Each perspective is informed by several methods. Although the interrelations among them will not be elaborated here, the outlines indicate the scope of the research on which the claims in the next section are grounded.

Challenges of bakery planning and scheduling

The different requirements for a tool to support bakery planning and scheduling that have been identified are presented grouped by theme.

Inherent complexity and variability of baking

An effective tool should address the inherent complexities of the process of baking. Baking involves many and varied process stages. Stage durations range from a few seconds to about two hours. Bakeries typically have a wide variety of product types, which may involve different processing stages and durations. All bakeries involve a combination of batch and sequential processing stages; i.e., whole product *en masse* versus sequential conveyor-based unit by unit processing. Further, some processes that are hybrids; e.g., starting as bulk process and ending as a sequential process. Products that have differing process durations poses a particular problem for the scheduling of conveyor-based processes as the conveyor must run at different speeds, which requires special strategies when such transitions occur.

An effective tool must be able to cope with variability of bakery setups, which of differ along a number of dimensions. Different types, quantities and capacities of equipment are available. The equipment will have different performance capabilities, such as the thermal lag of an oven, when different product types require different temperatures. Such characteristics are a concern for scheduling as they mean equipment may not be used even if empty and because they imply particular preferred orderings of product types.

Many operatives are needed to run bakeries. A shift or plant manager, who directs the operatives, typically manages the coordination of activities across time and space. A tool should be compatible with such patterns of working but ideally should also allow users to comprehend the plan/schedules for themselves and support collaborative approaches.

Difficulties of the planning and scheduling tasks

The actual planning and scheduling tasks present many challenges for a support tool. The time critical nature of the fermentation process means pausing production is impossible without compromising product quality. In plant bakeries the situation is worse as they are designed for continuous production. Further, the relatively long times of some of the process stages exacerbates the problem by making it difficult to predict the state of production into future and hence to prevent problems such as clashes or inefficient production.

In plant bakeries planning and scheduling cannot be treated as separate tasks, because similar products from different orders must be aggregated in to sequences runs, but the orders will have different deadlines. This has been less of a problem with in-store bakeries historically. However, the current trend toward "just in time" production for freshness means that planning and scheduling must be considered together. Hence, it is necessary for a tool to support bakers to successfully integrate both tasks.

A support tool must also be compatible with the different strategies used in bakeries. Production is usually based on historical records for production on similar days, with specific adjustments made for the day. Such adjustments are typically done working forwards in time, if the production does not have many tight deadlines, using baker's knowledge of time and capacity constraints to judge whether the schedule is feasible. In special situations, such as plant breakdowns, or for exceptional orders, the bakers attempt to reason backwards from the deadlines, but this approach tends to be avoided. This is consistent with the experimental findings that such inferences are hard and the lack of evidence of any attempts to improve production by directly examining novel permutations of orders from scratch. This is an area where a support tool could be of some benefit.

Bakery planning/scheduling goals considered are diverse and may be complementary or contradictory. For example, deadlines are viewed as particularly important, but quality and efficiency are also highly valued. A support tools should be able to deal with the range of goals and the trade-offs among them that bakers make.

Heterogeneity of tasks

This aspect warrants its own subsection and concerns the diversity and scope of activities that the bakery schedulers would ideally like a tool to support. Firstly, the granularity at which different bakers would like to examine the schedule will differ. For some, where the production has homogenizing processing times, then all the process stages can be considered as a singe aggregated process. For others, there is a need to understand the potential interactions at each process stage for each production run.

Second, the character of the tasks differs between plant and in-store bakeries. In plant bakeries production runs are typically comprised of tens mixes, each producing hundreds of units per mix. In contrast, each mix in an in-store bakery produces tens of units and successive mixes are usually different product types. The problem for the in-store bakers is more or less started from scratch each day, as production stops overnight. Plant bakeries run 24 hours per day and typically six days per week. Thus, experienced managers have strategies to deal the propagation of problems over successive days.

Third, bakers do not draw a clear conceptual boundary between planning and scheduling tasks and other related problems. They see merit in a planning and scheduling tool also serving as a communication aid to others, such as supermarket store managers, who need to comprehend the operation of the bakery.

Expertise

Bakery planning and production requires expertise in many forms, in addition to an underlying knowledge of baking. First, generic knowledge is needed about the nature of plans and schedules involving reasoning about discrete and continuous temporal relations. Second, there are various forms of bakery specific expertise, including: knowledge of the specific capabilities and operation of the bakery equipment; knowledge of the process specifications of types of product; understanding and skill in the use of the paper and IT systems for monitoring and managing production; understanding of the particular planning and scheduling rules and heuristics that have developed for a bakery. It is not possible capture all this knowledge in a tool, rather a successful tool to support bakers should allow them to be able use their knowledge of planning and scheduling effectively.

Conclusion

A variety of methods have been used to gain an understanding of the nature of bakery planning and scheduling from many perspectives. Bakery scheduling is a major challenge for many reasons, such as: numerous and often conflicting goals and constraints to be satisfied; the functional and structural complexity of the production system; the variability and uncertainties of process attributes for many different products and potential schedules. An effective tool to support bakery planning and scheduling should address these perspectives. Developing such a tool is also a challenge because support for knowledge and reasoning is needed at different levels, ranging from low level basic cognition about events, through generic problem solving methods and heuristics, to overarching organizational issues and commercial constraints. These challenges provide a stringent set of requirements for the ROLLOUT system that the project is designing. The evaluation of ROLLOUT will show whether it meets these requirements and hence will provide evidence for the utility of the underlying REEP approach to user-centered design.

Acknowledgements

This work was supported by the PACCIT research programme, funded by the ESRC, EPSRC and DTI (RES-328-25-001). We thank the members of the ROLLOUT consortium for their substantial contributions to this research and for their continued support.

References

Barone, R., & Cheng, P.C.-H. (2005). Structure determines assignment strategies in diagrammatic production scheduling. In A. Butz, B. Fisher, A. Kruger & P. Olivier (Eds.), *Smart graphics 2005* (pp. 77–89). Berlin: Springer-Verlag.

Cheng, P.C.-H. (2002). Electrifying diagrams for learning: Principles for effective representational systems. *Cognitive Science, 26*(6), 685–736.

Cheng, P.C.-H., Barone, R., Cowling, P. I., & Ahmadi, S. (2002). Opening the information bottleneck in complex scheduling problems with a novel representation: Stark diagrams. In M. Hegarty, B. Meyer & N. H. Narayanan (Eds.), *Diagrammatic representations and inference: Second international conference, diagrams 2002* (pp. 264–278). Berlin: Springer-Verlag.

Endsley, M. R., Bolté, B., & Jones, D. G. (2003). *Designing for situation awareness: An approach to user-centered design.* London: Taylor & Francis.

Preece, J., Rogers, Y., Sharp, H., Benyon, D., Holland, S., & Carey, T. (1994). *Human-computer interaction.* Wokingham, UK: Addison-Wesley.

WORK COMPATIBILITY IMPROVEMENT FRAMEWORK: DEFINING AND MEASURING THE HUMAN-AT-WORK SYSTEM

Ali Alhemood[1] & Ash Genaidy[2]

[1]*Kuwait Institute for Scientific Research, P.O. Box 24885, 13109 Safat, Kuwait*
[2]*University of Cincinnati, Department of Mechanical Industrial and Nuclear Engineering, University of Cincinnati, Cincinnati, OH 45221-0072, USA*

> This paper introduces the first two phases of the Work Compatibility (WC) Improvement Framework based on the Six-Sigma cycle. The objective is to identify the elements that constitute the human-at-work system and explain how the WC model characterizes the system in order to obtain the compatibility indices (i.e., performance measures of the system). A cross-sectional study was conducted in three small manufacturers that measured the work environment characteristics and the health status of the respondents. The trend shown by musculoskeletal disorders and compatibility indices is consistent with the assertion that higher compatibility level will decrease the likelihood of musculoskeletal and stress disorders in the workplace.

Introduction

Genaidy and Karwowski (2003) introduced the Work Compatibility Model (WCM) as a comprehensive approach to improve human performance at work, embedding previous models including motivation-hygiene theory, job characteristics theory, balance theory, person-environment fit, and demand-control. While previous efforts have concentrated on establishing the fundamentals of the WCM, the implementation of the model as a decision-support framework is yet to be developed.

Imai (1986) presents two contrasting approaches to process improvement from the Western and Japanese perspectives. The Western approach, reengineering, emphasizes innovation or discontinuous changes while the Japanese approach, *Kaizen*, emphasizes continuous improvement. While continuous improvement is achieved through small lasting steps, innovation requires abrupt changes that will have an impact in the short-term.

Because organizations have both long-term and short-term goals, they require a combination of continuous and discontinuous changes. Six-sigma is a modern approach that interprets both practices. This approach seeks to drive out variability, reduce waste in processes and ultimately drive culture transformation. Six-sigma projects have been implemented in American industries such as General Electric, AlliedSignals, Polaroid, Asea Brown Boveri and Texas Instruments. The DMAIC (Define-Measure-Analyze-Improve-Control) model has been implemented as part of the problem-solving methodology of six-sigma strategy to characterize and optimize processes.

The Work Compatibility Improvement Framework (WCIF) can be defined as the identification, improvement and maintenance of the well-being characteristics of the workforce through the application of engineering, medicine, management, and human sciences methodologies, technologies, and best practices. The WCIF would be introduced in phases: (1) define and measure, (2) analyze, (3) improve and (4) control, following the steps of DMAIC approach. This paper introduces the initial steps, i.e., define and measure, to be taken to implement the WCIF in business organizations.

Define phase

Problem definition

The Work Compatibility Model is a bottom-up approach that seeks to improve the worker's well-being and consequently improve the well-being of the organization as a whole. According to the WCM, the well-being of the organization can only be explained by the interaction between workers and the work environment. The work environment consists of many elements involving physical and non-physical elements that affect the human performance.

Musculoskeletal disorders have been identified as a major cause of sickness absence and disability. However, the composition of the risk factors is not well understood to date. While physical load has been traditionally associated with back disorders, there have been studies showing some association with non-physical factors. Efforts in musculoskeletal epidemiology seek to develop a set of etiologic variables that comprises the most relevant risk factors of the work environment.

Project goal

The purpose of the *Work Compatibility Model* is to identify elements in the work environment that do not contribute to better performance and require intervention. The current challenge for human resource management is how to maximize human performance, moving beyond what people should do to what people potentially can do. To explore the full potential of workers, a broader view of the well-being of workforce is required, not limited to economic incentives.

The scope of the project could be further explained by using another Sig-Sigma tool: the SIPOC (Supplier – Input – Process – Output – Customer). As a systematic tool, the SIPOC shows the suppliers, inputs, process, outputs and customers of a business process. SIPOC establishes the boundaries by showing the start and end points and divides the entire scope of the project into manageable partitions. For the process improvement efforts addressed in this article, different parts of the process are defined as follows:

(1) Suppliers: The organizational setting or work environment.
(2) Inputs: Physical and non-physical components that replenish or deplete the energy of workers.
(3) Process: The physiological process that results in a state of distress or arousal.
(4) Outputs: The well-being of workers.
(5) Customers: Workers and management.

Study design

A cross-sectional study was conducted in three small manufacturers located in the Midwest United States. The companies chosen for this feasibility study work with a variety of technologies oriented toward industrial clients. The employees were asked to complete two surveys administered by researchers. The first survey, the *demand-energizer instrument* assesses the work environment, and the *work outcome survey*, identifies musculoskeletal and stress symptoms the workers experienced. In addition, demographic information was gathered including age, gender, body weight, height and time in the company. Surveys were submitted to the study participants around the same time and were collected within a six-month period, spanning from the fall of 2003 to the winter of 2004.

Measure phase

Operational definition

Work Compatibility is defined as the balance between energy expenditure and energy replenishment. The work environment interacts with the worker taking or providing energy simultaneously. Work compatibility not only seeks a balance between energy expenditure and replenishment but also an increase in the energy level, resulting in more productive work. A work characteristic is a physical or non-physical element of interest in the work environment that permits classification and inferences. The WCM includes 166 characteristics based on a systematic comparison of different taxonomies including the Position Analysis Questionnaire (McCormick et al., 1969), the Ergonomics Task Analysis (Rohmert and Landau, 1983), and the NASA Task Load Index (Hart and Staveland, 1988).

Measurement of variables

The level of the effect of a work characteristic on the subject was measured using five linguistic levels, that is, (1) "Not at all", (2) "A little", (3) "Moderately", (4) "A lot", (5) "Entirely". "Not at all" means that the characteristic is not present in the work environment or has no effect on the subject. "Entirely" means that the characteristic has a complete incidence in the performance of the subject, either consuming (demand) or providing (energizer) resources for the human-at-work system. The demand and energizer instrument measures the extent to which a work-environment characteristic creates a demand or energizer on the respondent. The survey is divided into sections for each of the twelve work-environment domains.

Results and discussion

Work Compatibility is a multi-dimensional entity that seeks to meet the following specifications:

(1) Optimum conditions. The compatibility level is a measure of how close work characteristics are to the optimum conditions: high energizer and demand levels. Values that depart from the optimum value will result in lower productivity and higher health risks.
(2) Consistency among workers. The compatibility level depends on the distribution of the respondents. A significant proportion of respondents with low scores will result in low compatibility for the organization or as a whole.
(3) Spread of work characteristics. The compatibility level depends on how close the work characteristics are of each other. A high separation among work factors results in strain conditions that cannot be sustained.

Several indices were developed to address the complexity of work compatibility taking into account the aforementioned specifications.

The results show that each measure captures some of those three attributes with emphasis. The discrete model identifies optimum conditions and consistency across respondents. It does not fully account for the spread of work characteristics. The continuous transformation also identifies optimum compatibility and accounts for the spread of work characteristics. However, this transformation does not identify consistency across respondents.

A detailed analysis of the differences between the two calculations methods (i.e., discrete and continuous transformation) shows that each method is able to predict some of the outcomes of the model. While both models overlap, there is still a portion that cannot be explained by either model. This issue is further explored in subsequent research.

Both the physical and non-physical environment is intertwined in the worker's health status. Based on trend of cases reported with musculoskeletal and stress disorders for different body parts, it is possible to state that the higher work compatibility levels present, the fewer musculoskeletal and stress disorder cases. Evidence of the interaction between physical and psychosocial factors can be found in studies in athletics. The effects of the work environment are best understood by the combined effect of multiple variables both physical and non-physical.

Concluding remarks

The Work Compatibility Improvement Framework is an implementation of the Work Compatibility Model as part of an improvement effort that leads change in organizations. The DMAIC framework has been used to represent the Work Compatibility Improvement Framework in four phases. This article introduces the first two phases.

References

Genaidy, A. and Karwowski, W., 2003, Human performance in lean production environments, *Human Factors and Ergonomics in Manufacturing*, **13**, 317–330.

Hart, S.G. and Staveland, L.E., Development of NASA-TLX (task load index): results of empirical and theoretical research. In P.A. Hancock and N. Meshkati (eds.) *Human Mental Workload 1988*, (Amsterdam, The Netherlands: North-Holland).

Imai, M., 1986, *Kaizen: The Key to Japan's Competitive Success 1986* (New York, NY: Random House Inc).

McCormick, E.J., Jeanneret, P.R. and Mecham, R.C., 1969, *The Development and Background of the Position Analysis Questionnaire*, Report Nonr-1000 No. 5 (Washington, DC: Department of Navy).

Rohmert, W. and Landau, K., 1983, *A New Technique for Job Analysis* (New York, NY: Taylor & Francis).

TEAM TRAINING NEED ANALYSIS FOR SAFETY CRITICAL INDUSTRIES

Harriet Livingstone, Carole Deighton & Rebecca Luther

Air Affairs (UK) Ltd, 6 Old Flour Mill, Queen Street, Emsworth PO10 7BT

This paper is part of the background reading for a workshop which will be run by Air Affairs (UK) Ltd at the Ergonomics Society Conference on Thursday 6th April. The workshop is suitable for Training Needs Analysis Practitioners and Managers, especially Human Factor Consultants, Occupational Psychologist and Training/ Human Resources Personnel working within safety critical industries including military, health, oil, gas and nuclear. Further information on workshops that Air Affairs (UK) Ltd offer can be found at **www.airaffairs.co.uk** or by contacting Harriet Livingstone (tel: 01243 389124 email **harrietl@airaffairs.co.uk**).

Introduction

Salas et al (1992) have a commonly cited definition of teams; they define a team as "a distinguishable set of two or more people; who interact dynamically, interdependently, and adaptively; towards a common and valued goal; who have been assigned specific roles or functions to perform; and who have a limited life span membership". One valuable development in team research over the last few years is the development of comprehensive theories of what it means to be effectively engaged in team working. This research is important as it highlights and defines the two distinguishable tracks that co-develop over the maturation of a team – *task work + team work*. The task work track involves the operations-related activities of the team and focuses on the technical accomplishment of individual and team tasks. For example task work would involve an individual's knowledge of the procedures to correctly respond to the sound of an alarm in an operations room, or the teams' knowledge of joint planning techniques. Team work includes those activities that serve to strengthen the quality of functional interactions, relationships, co-operation, communication and co-ordination of team members. For example team work would include team members providing regular feedback to appropriate team members to maintain a shared awareness of current and future work requirements and to effectively manage workload within the team.

When conducting a team training needs analysis (TNA) both the task work and the team work requirements of the team would have to be elicited. The focus of this paper is on considering what a **team work** TNA would need to elicit to ensure that the training and development requirements of a team are understood.

Team Competencies

A trend in team training is to move towards an examination of team training requirements by defining team competencies. This is a process which requires "starting at the end and working backwards" (Pascual et al 2001), due to the fact that the analyst begins by defining the required performance and works back from this to define the

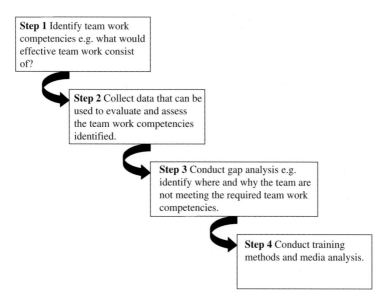

Figure 1. Four step team work training analysis.

requisite information. Figure 1 gives a simplified four step approach to the conduct of a team work training needs analysis using the competency approach.

The analyst aims to define what one would expect to find if the team is performing effectively (team work competency) and uses this to direct the training needs analysis. The starting point therefore is in establishing the team competencies and the requisite information e.g. the behaviours, skills and knowledge associated with each competency. For example the analyst may identify that a team work competency required by a team is that: "constructive feedback is provided to all team members". The team work TNA would then aim to understand what, how and who should provide the feedback. This data would be used to establish if the team was meeting the competency (training gap analysis) and if not then the most cost effective training methods and media would need to be identified.

There is a vast amount of literature on team effectiveness that could be used to conduct the team work analysis, but a recent study (Deighton et al 2005) found that a major source of data is from work that has been conducted for the U.S military, as summarised within Cannon-Bowers et al (1995) review of team work competencies. Cannon Bowers et al (1995) define team competencies as: 1) the requisite knowledge, principles and concepts underlying the team's effective task performance; 2) the required skills and behaviours necessary to perform the team tasks effectively; and 3) the appropriate attitudes on the part of team members (about themselves and the team) that foster effective team performance. The following three sections present a high level review of the potential output of the team work TNA required to conduct the gap and method/media analysis using Cannon Bowers (1995) team work competencies.

Team Work Knowledge

Cannon Bowers et al (1995) found that to be able to explore competency in team work knowledge the analyst would need to identify whether the team members have: accurate shared mental models (e.g. do the team members have compatible knowledge structures?); an understanding of the nature of teamwork and teamwork skills; knowledge of the teams overall goals objectives and missions; knowledge about fellow members' roles, responsibilities, strengths and weaknesses. Therefore the TNA would need to make explicit:

- The team's goals, objectives and missions.
- The team's relationship with other teams and within the organisation.
- Team members' roles and responsibilities.
- The procedures, sequences, and timing for team goals.
- Appropriate communication structures and style for each role.
- The team work skills required by each role on team tasks.
- Team members' skills, abilities, preferences, and experiences.
- Strengths and weaknesses of each of the team members.

Team Work Skills

Cannon-Bowers et al (1995) identified eight team work skills that an effective team would be expected to demonstrate. The eight skills are adaptability, shared situation awareness, performance monitoring and feedback, leadership/team management, interpersonal relations, co-ordination, communication and decision making. A team work TNA would aim to identify the behaviours expected from the team to demonstrate that they were competent in the eight team work skills by gaining information on how:

- A team should use information gathered from the task environment to adjust strategies through the use of compensatory behaviour and re-allocation of intra-team resources (***Adaptability***).
- Team members should develop compatible models of the team's internal and external environment, including how they should arrive at a common understanding of the situation and apply appropriate task strategies (***Shared situation awareness***).
- Team members should give, seek and receive task clarifying feedback, including how they should accurately monitor the performance of team-mates, provide constructive feedback regarding errors, and offer advice for improving performance (***Performance monitoring and feedback***).
- Leaders/managers of the team should direct and co-ordinate the activities of other team members, assess team performance, assign tasks, motivate team members, plan and organise, and establish a positive atmosphere (***Leadership/team management***).
- The quality of team members' interactions should be optimised through either resolution of dissent, utilisation of co-operative behaviours, or use of motivational re-enforcing statements (***Interpersonal relations***).
- Team resources, activities and responses should be organised to ensure that tasks are integrated, synchronised, and completed within established temporal constraints (***Co-ordination***).
- Information should be exchanged, clarified and acknowledged between two or more team members. What the prescribed manner and technology is required for certain communication and specific roles (***Communication***).
- Information should be gathered and integrated, alternatives identified, best solution is selected and consequences evaluated (in team context, emphasises skill in pooling information and resources in support of a response choice) (***Decision making***).

Team Work Attitudes

Team work attitudes are those that have been shown to have a direct bearing on the team's interaction processes and on the ability of an individual to flourish in a team context (Cannon-Bowers et al, 1995). Cannon Bowers et al (1995) identified six team work attitudes that impact on a team's functioning:

- Positive attitude to teamwork – a positive belief in the importance of team work skills (e.g. leadership, co-ordination and communication).
- Well developed team concept – a belief in prioritising the team goals above those of its individual members. It involves the capacity to take others' behaviour into account during group interaction, as well as the belief that a team approach is superior to an individual one.
- Collective efficacy – a belief in the ability of the team to perform effectively as a unit, given a set of specific demands.
- Cohesion – a belief that the members must remain as a group.
- Mutual trust – an attitude held by team members regarding the aura or mood of the team's internal environment.
- Shared vision – a commonly held attitude regarding the direction, goals and mission of an organisation, or team.

In conducting a team work TNA the analysts would need to identify the behaviours or beliefs **specific** to the team that would be associated with each of these attitudes. For example, within a business development team to know if a team had shared vision, all team members would have to know what the long term and short term goals of the business development team were.

Summary

An approach to team work training needs analysis is to define the competencies expected of a team engaged in effective team working and use the competencies to direct and guide the training needs analysis process. For example if there is a requirement for team members to have compatible knowledge of roles and responsibilities, then the analyst needs to make explicit what the team members roles and responsibilities are. A recent study conducted by Air Affairs (UK) ltd found that Cannon-Bowers (1995) team work KSA provide a useful starting point for the conduct of a team work TNA from a competency perspective (Deighton et al, 2005).

References

Cannon-Bowers, J.A., Tannenbaum, S.I., Salas, E. & Volpe, C.E. (1995). Defining Competencies and Establishing Team Training Requirements. In R. Guzzo & E. Salas and Associates (Eds.), Team Effectiveness and Decision-Making in Organizations (333–380). San Francisco, CA: Jossey-Bass.
Deighton, C., Livingstone, H., & Luther, R. (2005). Information Exploitation Competencies Study. V 1.0. issued to DCBM/Defence J6. Unclassified.
Pascual, R., Mills, M., & Henderson, S. (2001). Training and Technology for Teams. Chpt 9 in Noyes, J. and Bransby, M. (Eds). People in Control. Human factors in control room design. The Institute of Electrical Engineers.

DEVELOPMENT FOR A TOOL TESTING TEAM RELIABILITY

I.H. Smith[1], C.E. Siemieniuch[2] & M.A. Sinclair[1]

[1]Department of Human Sciences, Loughborough University, Leicestershire, LE11 3TU

[2]Department of Electrical and Electronic Engineering, Loughborough University, Leicestershire, LE11 3TU

Human reliability assessment techniques (HRAs) have been in existence since the 1960's. Following Dougherty's (1990) comments, a second generation of more complex HRAs were created, all measuring the reliability of individuals. However, often a team of people interact with a system, not an individual. This can increase or decrease the individual's reliability. A new generation of HRAs is needed to assess the effects of teamwork on reliability. During the development of a new tool, the model of team reliability needs to be validated. This is to be partially accomplish at a workshop at the Annual Conference of the Ergonomics Society, where HRA experts and other interested parties can critique the model. These opinions will then be utilised to enhance the model.

Introduction

A human interacting with a system can be analysed to determine the errors that could occur, what factors could help mitigate these errors, and the probability of these errors occurring. This is done by using human reliability assessments (HRA). HRAs are qualitative and quantitative measurements of the risks and errors that can occur in a system because of human actions, not by a fault of the system. HRA have been developed for designers and users to understand the technical difficulties of using a product or system. As, no matter how good the product is, it is impossible to make the product error proof: humans are inevitably fallible.

As a field of research HRA has been around since 1960s. Predicting the probability of error can be a controversial topic because probabilities are based on random behaviour and humans are not random; some factors that can affect them that are consistent (Redmill, 2002). HRA techniques have accounted for factors that influence the error probability in the form of "performance shaping factors" (PSFs). The task, the individual and the environment define the performance shaping factors. There are three main approaches to HRA (Kirwan, 2002):

1. Human error identification – what can go wrong?
2. Human error quantification – how often will a human error occur?
3. Human error reduction – how can human error be prevented from occurring or its impact on the system reduced?

The first generation of HRA techniques began in the 1970s, e.g. THERP (Swain & Guttmann, 1980) and HEART (Williams, 1986). Criticism by Dougherty (1990) triggered a new generation of techniques that included the most recent knowledge of error and human behaviour (Redmill, 2002). Second-generation techniques such as, CREAM (Hollnagel, 1998) have improved the reliability of the HRA for individuals.

However, often a team of people interacts with a system, not just an individual. There is now a call for a new generation of HRA techniques to account for team interaction with a product or system. The interactions between the team members can increase and decrease the reliability of each individual and hence the overall reliability of the team. Some of the PSFs of team reliability are communication, trust and resource management (Sasou & Reason, 1999).

CHLOE (Miguel, Wright, & Harrison, 2002) has been developed to take into account the effect of teams on reliability. Miguel et al (2002), wrote "collaborative errors may be caused by factors such as a lack of [situational awareness (SA)], misunderstandings between participants, conflicts and failures of co-ordination" (p. 4). CHLOE is a qualitative method, and in a time when corporate manslaughter is becoming more prominent and system reliability is measured in probabilities, human reliability also needs to be quantitative.

Aim

Therefore the aim of the workshop is to validate, using expert judgements, a model of team reliability and tool that will quantitatively measure team reliability.

Definition of a team

The tool being validated will be looking at a team of people that are either interacting with the same system or piece of equipment, and whose procedures are all critical in achieving the same overall goal. The tool does not look into how an organisation works, and so it will not go into details of the command structure. However, the communication structure, and decision structure are important to team reliability. Several group topologies that should be considered are shown in Figure 1.

Error classifications

Before team reliability can be measured, how and why humans perform errors must be understood. There are three main aspects that can cause error; the task, the environment and the individual. There are several types of errors that humans can perform, such as execution errors, or errors of cognition. There are also, many causes of error, e.g. bad design, or cognitive overload. Below are some of the main classification of errors.

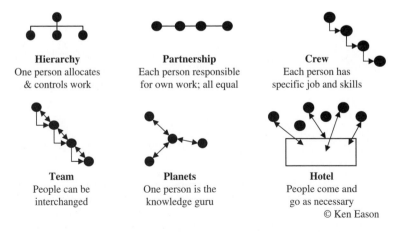

© Ken Eason

Figure 1. Group Topologies need to switch planets and teams.

Kletz (1999) presented four classifications of errors.

1. *Mistakes* are errors that are made because the correct procedure is not known and the intention of the action is wrong.
2. *Violations* are actions that are known to be wrong but are thought of as being the most suitable action at the time given the information known; again the correct procedure is not followed.
3. A *mismatch* is where the task and the cognition of the operator are not compatible, for example the operator could be overloaded or may have established a habit and cannot change their viewpoint when new information is offered.
4. A *slip* is where the intention is correct but that action is wrong, for example, pressing the wrong knob on a control panel. A lapse is where an action is missed.

Rasmussen (1982) presented error classifications based on the cognitive functions,

1. *Skill based errors* are errors related to variability of force, space or time.
2. *Rule based errors* are errors that are related to cognitive mechanisms, such as classification, recognition or recall.
3. *Knowledge based errors* are errors in planning, prediction and evaluation.

These definitions of errors are for individuals, not teams. Errors of execution, such as slips, or skill based errors, are affected less by team reliability. Cognitive errors are affected by team reliability, as cognitive overload, can be augmented by the presence of other team members. Taxonomies of team error should also be used, such as Sasou & Reason's (1999) individual and shared errors.

Model of team reliability

Before the tool can be developed a model of how the individuals in a team interact should be produced. One viewpoint of team reliability is represented in Figure 2. A procedure is performed by a team of people. Each person has an individual reliability score (R_n), as measured using a 1st or 2nd generation HRA technique. The interactions between team members, such as communication and trust create the PSFs for the team. But as each individual, their task and their environment is unique, effects of the interaction should be calculated separately, creating new "interaction reliability" scores R_{ni}. These are then combined together to produce the overall team reliability score, R_E.

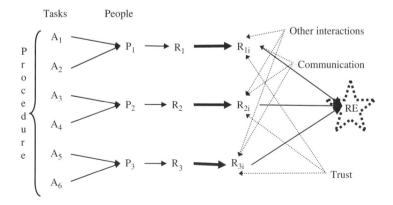

Figure 2. Model of Team Reliability.

Tool vision

Following interviews with potential users of the tool (HRA and design experts in a defence systems integration company) a list of tool requirements was formed. These illustrated that it was necessary to develop a tool that can:

- quantitatively assess the reliability of a team.
- qualitatively predict areas of high risk.
- be usable and accessible to non-human factors experts.
- educate designers in the importance of the human factors to produce high usability and reliability.

Structure of the workshop

The workshop will consist of a brief introduction into the reasoning for collaborative HRA. The model that is used to represent team interactions will be described and explained. There will then be a discussion on some of the issues that may effect interactions within teams, and what errors may be produced from these interactions.

Subsequent work

Following the workshop the opinions expressed will be considered and further adaptations to the model and tool will be made.

References

Dougherty, E. M. 1990, Human reliability analysis – where shouldst thou turn? *Reliability Engineering and System Safety,* **29,** 283–299

Hollnagel, E. 1998, *Cognitive Reliability and Error Analysis Method.* (Elsevier, London)

Kirwan, B. 2002, Human reliability assessment. In J. R. Wilson & E. N. Corlett (eds.) *Evaluation of Human Work,* Second Edition, (Taylor and Francis, London), 921–968

Kletz, T. A. 1999, *An Engineer's View of Human Error,* Third Edition, (Institute of Chemical Engineers, London)

Miguel, A., & Wright, P. 2002, *Chloe Version 1: An Application to Nimrod Case Study Scenarios,* (DCSC, York)

Miguel, A., Wright, P., & Harrison, M. 2002, *Chloe: A Reference Guide*, TR/2002/33

Rasmussen, J. 1982, Human errors. A taxonomy for describing human malfunction in industrial installations. *Journal of Occupational Accidents,* **4,** 311–333

Redmill, F. 2002, Human factors in risk analysis. *Engineering Management Journal,* 171–176

Sasou, K., & Reason, J. 1999, Team errors: Definition and taxonomy. *Reliability Engineering and System Safety,* **65,** 1–9

Swain, A. D., & Guttmann, H. F. 1983, *Handbook of Human Reliability Analysis with the Emphasis on Nuclear Power Plant Applications,* (No. NRC NUREG-CR-1278), (Sandia N. L., Washington, DC)

Williams, J. C. 1986, *A Proposed Method for Assessing and Reducing Human Error (HEART),* Paper presented at the 9th Advances in Reliability Technology Symposium

ERGONOMICS ASPECTS OF "GOOD ENGINEERING GOVERNANCE" IN THE DESIGN PROCESS

J.V. Nendick[1], C.E. Siemieniuch[1] & M.A. Sinclair[2]

[1]*Dept of Electrical & Electronic Engineering,*
Loughborough University, LE11 3TU
[2]*Dept of Human Sciences, Loughborough University, LE11 3TU*

At least 75% of design projects, including capability acquisition projects, do not meet all their goals. In part, this is due to emergent, unwanted and unexpected behaviour. In turn, this behaviour is usually due either to the intrinsic complexity, of the product, or induced complexity because of the organisation and operation of the project. Engineering governance provides a way of containing the effects of complexity, and after discussing complexity, this paper outlines an approach to governance in industry.

Introduction

Engineering governance is focussed on the control that is present through the hierarchy of the organisation with respect to the engineering function. This control is an important lever for the executive management responsible for corporate governance who want to assure customers, stakeholders, shareholders and the law that projects will meet the requirements (both customers' and business). In essence, engineering governance addresses the twin questions, "Are we doing the right things?" and "Are we doing those things right?". If the answers to these two questions are affirmative, then the organization is heading towards a "no nasty surprises" state of operations.

However, this rather simplistic viewpoint ignores the contribution of complexity to the problems of control. In this paper we discuss some of the ways in which complexity can manifest itself behaviourally, and then discuss ways and means by which these effects can be sequestered, ameliorated, and, in certain situations, quenched. These ways and means can be seen to be a part of engineering governance, albeit at a somewhat deeper level than implied in the questions above.

What is complexity, and how does it manifest itself?

There is a considerable literature on complexity; one good text for explaining it is Rycroft & Kash (1999). The main findings are that complexity usually manifests its presence by emergent, almost always undesirable, behaviour, not envisaged by the system designers nor expected by the systems operators. Secondly, it has diffuse origin; decomposition of the system to discover its causes may show that individually the system components are reliable, but not when grouped. Thirdly, complex problems require at least equal complexity in the knowledge and approaches to solve them (a version of Ashby's Law, 1956), though the solution might be simple. Finally, it is said that for innovation to happen within the organisation, at least some parts must operate "at the edge of chaos". The implication is that complexity will always be a factor in the organisation's behaviour; channelling its effects is a challenge for management.

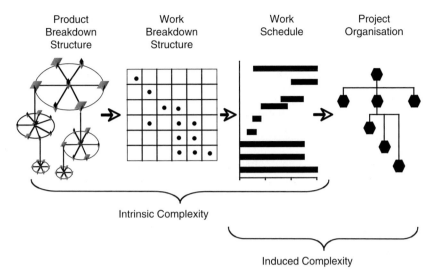

Figure 1. Diagram outlining the organising and resourcing of a design project. This depicts simply the relationships between intrinsic and induced complexity.

Complexity can be decomposed into two overlapping classes; firstly, intrinsic complexity which arises because the problem we are facing is by its nature a complex one (e.g. wide area traffic management), and secondly, induced complexity because we have organised ourselves inappropriately to address it (e.g. project teams appointed on availability grounds, not on expertise). Our concern, from a governance aspect, is mainly with the latter, though because of the overlap we discuss both. Fig. 1 shows this, for the design process.

The symptoms of complexity within a design project include:

- the slow, gradually-spreading realization that the project is much more difficult than originally thought, due to unexpected interactions and feedbacks;
- project management characterized by near-continuous fire-fighting, due at least in part to commitment to inflexible work schedules;
- considerable rework of supposedly completed components, due in part to out-of-phase development of components in a concurrent engineering environment;
- self-evidently dysfunctional teams, because of organizational problems not recognized early enough;
- failures of organisational learning, because nobody has the time to attend to this, since they are dealing with all the issues above;
- failures in the delivery of service, leading to contract penalties.

Some of these symptoms arise from intrinsic complexity, some from induced complexity, and others from both. The common characteristic is that they are all in-house; however, the complexities behind them may have external origins, and Fig. 2 below explains this in part.

The increase of complexity on the provider's side has now to be accommodated, and its associated emergent behaviour, additional to that present before, needs to be contained. The sources of complexity in this diagram, additional to those listed before, include:

- Many agents, of different kinds
- An evolving, uncertain, environment (client, suppliers, weather, etc.)

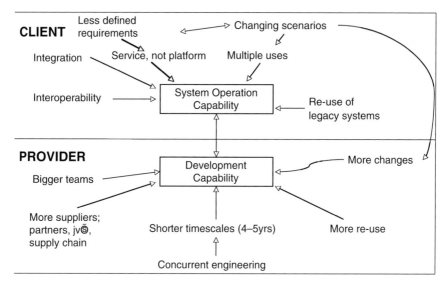

Figure 2. Client-provider relationships, as they are changing over this decade. On the client side, the drivers that are altering the nature of the need are shown. The result is that the client specifies a desired capability to be delivered, not a product or system. The provider has now to decide this, and the implications are shown. In effect, much of the complexity the client had to manage before is exported to the provider.

- Lots of connections between agents, who are
- Communicating in parallel
- Some degree of behavioural autonomy for agents, with
- Multiple steady states for agents, and
- Evolution of the agents
- Interactions between agents across system boundary
- Interactions between different goals within an agent
- Interactions between agents with different goals
- Language/culture differences
- Restricted time (deadlines, interruptions, etc.).

If only a few of these characteristics are present, it is likely that emergent behaviour will arise, and that it will not be to the advantage of the project.

Containing complexity

It should be noticed that we talk about "containing" complexity, not its elimination. Hence, there is a need for continuous attention to the control measures, emphasising the need for engineering governance. We discuss firstly intrinsic complexity, and then induced complexity. Organising for intrinsic complexity involves the following:

- Modularity in design, to enable containment of complexity
- Maturity of system components is vital – i.e. the state of knowledge, and quality of knowledge management is critical. These are long-term issues

- Points to the need for an architecture for core components of the system, and rigid adherence to standards
- Requires a good prior understanding of the problem context; especially of interactions and non-linearities
- Requires stability of project environment – budget, timescales, client consistency and coherence, partners, etc. (this is best addressed within induced complexity).

It will be noted that all of these are concerns for engineering governance; unless there are policies, procedures and practices for these, and they are maintained, intrinsic complexity can spiral out of control. In addition to these are the demands for induced complexity:

- Need to consider containment measures at project strategy level, workgroup level, and individual level.
- Make use of the important role of humans as "Complexity Absorbers" – situation may be complex, but a simple plan may suffice. This capability depends on:
  - Trust in other system components (especially the human ones)
  - Situation awareness, and shared situation awareness
  - Excellent communications
  - Knowledge & experience
- These all depend on job design, the organisation of work, responsibilities and authority over resources, culture, values, and many other organizational aspects.

These all fall within the ambit of engineering governance; the inclusion of social considerations such as culture and policies, becomes evident; to draw an analogy, if people start from the same place, and want to march in the same direction, and they all march in the same fashion, it becomes much easier to control the march.

An engineering governance framework

We report on nearly-completed work within an aerospace company. Several case studies have been executed in different business units, and corroborative work has commenced into other classes of organisation in other domains, to generality of the findings.

For a commercial business, governance must address four aspects;

- Meeting legal requirements for health & safety, probity, and so on
- Ensuring the development, at acceptable risk, of competitive offerings for its customers
- Ensuring the offerings are to specification
- Delivering the offering to the customer to the business benefit of the enterprise.

From these, a given project will establish its own business objectives. For each of these, it will be necessary to develop a governance process, covering the business objective stakeholders, with appropriate metrics and with a process owner responsible for good governance for that objective. Typically, in the company concerned, this individual turns out to be a Chief Engineer. Chief Engineers usually have responsibilities for all four of the aspects above, so most of the governance processes will be owned by such an individual within the project, and streamlined processes can therefore be adopted.

Some governance mechanisms will already be in existence; design reviews, for example. Others may need to be extended, or developed; appraisals of individuals and their contributions, for example. There are three key issues that must be borne in mind in developing governance:

- Essentially, governance involves humans. Therefore, the processes and the metrics must be human-sensitive (i.e. as unobtrusive as possible), else false data or no data will accrue.

- Governance should measure only that which is necessary to achieve the business objective. It is a mistake to try to measure everything. With a little subtlety, it should be possible to adopt metrics which will indicate emergent behaviour and its likely source to enable containment of the complexity to occur.
- There must be clear responsibilities to act on the basis of the measures, and procedures and resources available for actions to take place.

Finally, a UML class diagram, together with a process for using it, are nearly finalised to enable managers and others to develop appropriate governance structures for their projects.

References

Ashby, W.R. 1956 *An introduction to Cybernetics*. (Chapman & Hall, London)
Rycroft, R.W. & Kash, D.E. 1999 *The complexity challenge*. (Pinter, London)

KNOWLEDGE-CENTRED DESIGN AS A GENERATIVE BASIS FOR USER-CENTRED DESIGN

Rossano Barone & Peter C-H. Cheng

Representation & Cognition Group
Department of Informatics, University of Sussex, Falmer, BN1 9QH, UK

The article considers the merits of a heuristic for the design of problem solving interfaces that focuses on capturing the relational structure of the system underlying the problem. We argue that this is a powerful and economical method that will tend to satisfy unspecified information requirements for a broad range of interpretive tasks. We describe how these information effects depend on capturing certain subsumption relations over different levels of a system in the structure of the graphical representation. Concrete examples are given with respect to a prototype graphical interface for the domain of bakery scheduling.

Introduction

Numerous advances have been made in identifying the kinds of knowledge relevant to informing the design of graphical interfaces for complex tasks that inquire into the nature of the user, task, problem and representation (e.g. Vicente 1996, Zhang et al 2002, Cheng et al in press). The approach we have termed *representational epistemology* whilst being sympathetic with the importance of multi-factor considerations, depart somewhat from other approaches in that emphasis is placed on acquiring the elementary structure of the problem system and encoding this in the structure of the external representation. We propose that this approach is a powerful and economical design heuristic because it will tend to satisfy unspecified information requirements for a broad range of interpretive tasks that result as a side effect of diagrammatising the elementary structure.

One exemplar of this approach is a prototype interface called ROLLOUT currently being developed for bakery planning and scheduling. A screen shot of one of its views is shown below. This display shows details of how individual mixes are assigned to specific equipment over a series of processing stages. Each processing stage comprises of one or more sub-bars that represent individual instances of processing equipment. Each mix unit is indicated by strips of coloured rectangles that form a diagonal structure over a sequence of processing stages. See Barone and Cheng (2005) for a detailed account.

Structure preservation

Four levels of information underpin the relational structure of a system: (1) the global reference structures; (2) attributes; (3) relations and (4) interrelations. A **Global reference structure** (GRS) is dimensions that may have an ordinal, interval or ratio structure (e.g., the temporal GRS in bakery scheduling). An **Attribute level** comprises the attributes types of objects that vary along their respective GRS (e.g., start time and end time of particular process steps on the temporal GRS). We classify the GRS and

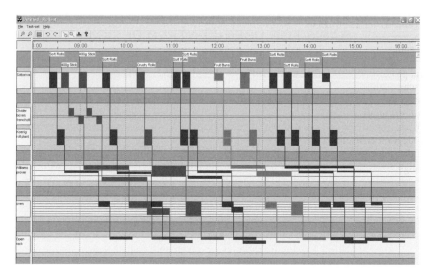

Figure 1.   Screenshot of the ROLLOUT interface.

attribute levels as elementary structure of the system. Higher levels include the **Relational level** that consists of the relations between attributes of objects indexed on their GRS (e.g. duration between the start-time of consecutive processes on the temporal GRS). The **Inter-relational level** involves the global patterns made over elements of a model that interleave background knowledge in interpretations (e.g., the temporal profile of the use of equipment over a period of time).

For any model of a system, types of relational and interrelational structures subsume types of the elementary structure. This is because interrelational and relational structures are composed of the elementary constituents of the model. We propose that the pattern of subsumption relations that exists in a model of a system holds in a graphical representation if it encodes the elementary structure of the system. Encoding the elementary structure means employing graphical GRS that instantiate the constraints of the system's GRS. This is a perspective of the notion of intrinsic structure described by Palmer (1976) that underlies the behaviour of representations we call *Law Encoding Diagrams* (e.g., Cheng 2002). Under such circumstances the space of relations and interrelations bound or fixed by the elementary structure, will result as side effect. These expressions will be derivable from the representation in a manner commonly attributed to diagrams (Shimojima 1999). Table 1 provides several examples of relational and interrelational expressions that are side-effects of the elementary structure diagrammatised in ROLLOUT.

Conclusion

Designing representations that intrinsically encode the elementary structure of a system is particularly useful when it is problematic to specify all the kinds of information required for a task. As much of the task relevant information is derivable from the representation, the representation is as complicated as the user wants it to be. This entails that different kinds of users with different kinds of goals and background knowledge can share the same representation, better supporting collaboration. Although the representation is not guaranteed to be the best for every task it will typically have greater

Table 1. Examples of relational and interrelational expressions in ROLLOUT

Meaning	Graphical expression
Mix process	A diagonal strip
Machine capacity	Overlap relation between a mix rectangle and equipment bar
Assignment clash	Overlap between rectangles of neighbouring diagonal strips
Equipment use	Profile of filling of an equipment bar by mix rectangles
Process continuity	Horizontal alignment of concurrent rectangles in diagonal strip
Production efficiency	Magnitude of spaces between neighbouring diagonal strips
Production continuity	Homogeneity of spaces between neighbouring diagonal strips
Dough changes	Changes in the colour of neighbouring diagonal strips

task generality than more abstract information displays that were designed to match specific tasks.

Acknowledgements

This work was supported by the PACCIT research programme, funded by the ESRC, EPSRC and DTI (RES-328-25-001). We thank the members of the ROLLOUT consortium, especially Stan Cauvain and Linda Young, for their substantial contributions to this research and for their continued support.

References

Barone, R., Cheng, P. C.-H. (2005). Semantic Traits of Graphical Interfaces To Support Human Scheduling. In *Proc. IV05, IEEE: Los Alamitos, California*, 607–612.

Cheng, P. C-H., Barone, R., Pappa, N., Wilson, J. R., Cauvain, S., Young, L. (In press). Understanding bakery scheduling: diverse methods for convergent constraints in user-centred design. *This volume*.

Cheng, P. C-H. (2002). Electrifying diagrams for learning: principles of complex representational systems. *Cognitive Science*, **26**, 685–736.

Palmer, S. E. (1978). Fundamental aspects of cognitive representation. In: Rosch, E., & B. B. Lloyd, B. B. (Eds.), *Cognition and Categorization*. Hillsdale, N.J.: Lawrence Erlbaum, 259–303.

Shimojima, A. (1999). Derivative Meaning in Graphical Representations. In *Proc. 1999 IEEE Symposium on Visual Languages*, 212–219.

Vincente, K. J. (1996) Improving dynamic decision making in complex systems through ecological interface design: A research overview. *Systems Dynamics Review*, **12**(4) 251–279.

Zhang, J., Patel, V. L., Johnson, K. A., Malin, J., & Smith, J. W. (2002). Designing human-centered distributed information systems. *IEEE Intelligent Systems*, **17**(5), 2–47.

ON THE USE OF SOCIAL NETWORK ANALYSIS TO UNDERSTAND GROUP DECISION-MAKING

Shan Cook[1], Carys Siemieniuch[2] & Murray Sinclair[1]

[1]*Department of Human Sciences, Loughborough University, Leicestershire, LE11 3TU*
[2]*Department of Electrical and Electronic Engineering, Loughborough University, Leicestershire, LE11 3TU*

Appropriate group decision-making is considerably influenced by the structure of the group. To understand the group arrangement you must look beyond formal organizational charts using Social Network Analysis (SNA) to reveal the informal, often hidden, networks. Once the patterns of interaction within the group are understood, it is possible to intervene to reinforce/alter existing networks or create new patterns. The subsequent group structure may be better able to make decisions through this shaping.

Group decision-making

Group decision-making has many advantages over making decisions as individuals. The sharing of skills and knowledge makes the group a powerful decision-making entity. Whilst there are liabilities to making decisions as a group, such as pressure to conform and competitiveness; Maier (1967) suggested many assets including pooled resources; the greater ability to avoid ruts in thinking; a better acceptance & understanding of decisions.

The list of factors in Table 1 is not exhaustive but it does illustrate the complexity of interrelated factors that can play a part in the effectiveness of decision-making in groups. It should be noted that excess in any one area may not be advantageous, balance is the key. Also, there are subtle interactions between the factors that influence their impact upon the group and its decision-making capability.

Table 1. Factors influencing the effectiveness of groups and decision-making.

Groups	Decision-making
Goals and function	Prior experience of similar problems
Rules and norms	Information availability and timeliness
Leadership style and cohesion	Expertise
Power, influence and control	Training as decision maker
Group typology	Communication flow and media
Size of group	Time requirements of the decision
Decision and communication structures	Knowledge including knowledge management
Cultural influences	
Group structure including roles, hierarchies, formality, centralization	

To summarise, the group is better equipped to make decisions if it is:

- Agile and adaptive: able to respond quickly to changing circumstances;
- Collaborative: uses resources effectively and efficiently;
- Innovative: able to see problems from new perspectives; and
- Has a strong understanding of the composition of the group and the roles of its members e.g. the trust between individuals.

Social network analysis

SNA is a statistical and graphic methodology for understanding patterns of interconnectedness and interactions within and without a group. Many of the measures examine aspects of groups that are central to leadership and thus influential to group decision-making. Such measures include: centrality, distance, power, influence and cohesion, which may highlight group attributes (e.g. bottlenecks in information flow) and individual attributes (e.g. trusted sources of advice). Collectively, SNA is considered to be a clean method for comparison between layers within a group, subgroups or between groups. For a comprehensive text, see Wasserman and Faust (1994).

Tool for seeing reality not formality

Informal relationships among individuals are typically more representative of work processes in an organization than relationships established by position within the formal structure. Often, success in organizations may depend less on reporting structure and more on an unofficial web of contacts. Unfortunately, informal networks often compete with and are fragmented by, such aspects of organizations as formal structure, work processes, geographic dispersion, human resource practices, leadership style, and culture.

Social network analysis can be an invaluable tool for systematically assessing and then intervening at critical points within an informal network. At its most basic level, redrawing hierarchies as ego nets enables better understanding of why some set ups work and others don't. Within an organization, for example, before undertaking costly or radical changes to organizational structures, SNA examines who talks to whom and about what. Thus making it possible to understand how the organization currently works. This enables the enhancement of:

- Information flow. Optimizing information flow through restructuring formal hierarchies to correspond the actual flow paths. Thereby reducing path lengths, avoiding bottlenecks and improving interlinking between individuals.
- Adaptability and agility. By knowing it's reality, the organization can better meet demands planned or unexpected as it is already conscious of its strengths through existing links and can be more agile to meets it's weaknesses.
- Innovation. It is only possible to plug gaps in knowledge if an organization is aware of their existence. Leading to increased innovation, productivity, and responsiveness.
- Collaboration. An awareness of relatedness in existing projects, knowledge and skills facilitates future collaborative projects (of particular note for time-critical projects).
- Knowledge and skills. SNA notes individuals from whom information/skills are sought and thus suggest structures to better enable access to these key personnel.
- The individual. SNA highlights vital knowledge and connections essential for retention/promotion/reassignment and individuals to reward for "invisible" work.
- Communities of practice. SNA provides a suitable method for monitoring the ramifications of Communities of Practice (CoP). In addition SNA may illustrate how

CoPs (formal and informal) are best used, identify areas where they may need to be created and subgroups and to ascertain the degree to which informal are CoPs are sympathetic to formal group goals.

Applications of SNA

SNA has been applied to fields as diverse as anthropology, ecology, epidemiology, linguistics, organization studies, political science, psychology and sociology; with influence in statistics. Topics of interest may include: terrorist networks, emergency planning, political campaigns; networks of innovators/entrepreneurs/professionals; project team efficiency; location of experts; succession planning; knowledge management; tacit knowledge retention and organizational restructuring.

Recommendations following SNA

The uses of SNA are numerous, however, some examples are:

- Encouraging inter-group connections for reintegration and re-engagement of peripheral individuals. For example, systems to help new employees or events to bring people together from different parts of the company (e.g. Special Interest Groups, SIG).
- A subgroup that has proportionally more ties within, as opposed to ties with other groups (betweenness) may result in poor knowledge exchange, which inhibits innovation. Thus the encouragement of links between subgroups is important to the feeding of information and ideas.
- Pivotal roles within the network may enable an individual to unduly impact upon a group by controlling information or decision-making. Recognition of such detrimental positions would acquaint the organization to the need for reassignment/reallocation of personnel/information/decision-making rights and so forth to ensure a suitably balanced group.

Conclusion

The crux of Social Network Analysis is the reporting of (a) the current situation within the group; (b) what needs to be retained for continued success; and (c) what could be altered for improvement. It can also be used for comparison following any changes. SNA cannot tell a group what decisions to make or how to make them. The methodology does, however, enable informed decisions regarding the structuring of the group to ensure it has the optimal decision making capability.

References

Maier, N.R.F. 1967, Assets and liabilities in group problem solving: The need for an integrative function, *Psychological Review*, **74**(4), 239–249.
Wasserman, S., and Faust, K. 1994, *Social Network Analysis: Methods and Applications*, (Cambridge University Press, Cambridge).

THE INFLUENCE OF SHARING DISPLAYS ON TEAM SITUATION AWARENESS AND PERFORMANCE

Adrian P. Banks & William J. McKeran

Department of Psychology, University of Surrey, Guildford, Surrey, GU2 7XH

A simulated process plant was used to investigate the use of displays to enhance team situation awareness and hence performance. Sixteen teams operated the simulation with either shared or non-shared displays. Unexpected faults were introduced in some trials. There was no overall difference in performance with shared and non-shared displays. However, in trials when faults occurred teams using shared displays performed better than teams with non-shared displays. Team interactions were more complex with non-shared displays and became less effective when coping with the demands of fault finding. Teams with shared displays performed better if they did not need to maintain situation awareness with questioning. Thus the benefits of shared displays lay in maintaining performance when faults arose by reducing the need for complex team interactions to develop situation awareness and negotiate action.

Introduction

The influence of shared displays on team performance has been studied for a number of years. This study develops earlier work by investigating the effect on the team process and situation awareness in teams under routine task performance and when unexpected errors arise.

Situation awareness is "the perception of the elements in the environment within a volume of time and space, the comprehension of their meaning and the projection of their status in the near future" (Endsley, 1995). Lack of situation awareness is often cited as a cause of accidents (Rodgers, Mogford & Strauch, 2000) and so developing situation awareness in teams is of practical importance. Endlsey (1995) suggests that optimising interface design is one route to achieving this. This study will investigate the effect of shared displays on improving situation awareness and, given the effect of situation awareness on accidents, test both routine operation and effect of unexpected faults. It is predicted that teams with shared displays will perform better as they will have superior situation awareness, and that this will be accentuated when unexpected faults arise.

Participants will work in pairs to operate a process plant with either shared or non-shared displays. Their communication will be coded in order to investigate their interactions. The relevant codes will be "Question" – when a question is asked about the system state; "Prompt" – when a suggestion for action is made; and "Rebuttal" – when a suggestion is refused. It is expected that questions will be used to develop team situation awareness and prompts and rebuttals to negotiate the team's actions.

Method

Participants

A total of 32 students at the University of Surrey volunteered to participate in this study.

Apparatus

A simulated process plant was used. The purpose of the process was to pasteurise beer by heating it to a specific level. Beer entered the plant, passed through a heater and if heated to the correct temperature was stored as pasteurised beer. If overheated it was dumped as waste. If underheated it was recycled to be heated again. There were two controls, one for each team member: the heat control and the rate of flow of beer. To heat the beer to the correct level these two controls needed to be balanced – a greater flow of beer requires a higher level to heat it to the correct temperature and vice versa. Therefore the two team members were required to work interdependently to achieve this balance. The task was complicated by the variable rate of flow of beer into the system and power into the heater. There were displays for the beer flowing in, the power flowing in to the heater, the rate of beer recycling, the amount of pasteurised beer and the amount of wasted beer.

Procedure

Participants were randomly allocated to conditions and teams of two. After an introduction and practice they completed four ten minute trials. In the two "No Fault" trials the process continued as had been explained. In the two "Fault" trials a fault was introduced without warning after two minutes. In one trial the heater became fixed at the highest level. In the other trial the recycle display became fixed at zero. The order of trials was counterbalanced. In the "Shared" condition both participants could see all of the displays. In the "Non-shared" condition the participant operating the heater could not see the rate of flow of beer into the plant and the participant operating the flow rate could not see the rate of power flowing into the heater. Display type was a between subjects factor.

Results and Discussion

The task required teams to both pasteurise as much beer as possible whilst minimising the amount of waste. Therefore the total pasteurised beer minus the total wasted beer was used as a metric of overall performance.

There was no significant main effect of display type ($F(1,14) = 0.34$, $p > 0.05$). Performance was worse in trials in which a fault was introduced ($F(1,14) = 33.25$, $p < 0.0001$). There was a significant interaction of display type and faults ($F(1,14) = 4.48$, $p = 0.05$). This interaction was investigated further using post hoc t tests with Bonferroni adjustment. There was no significant difference between Shared and Non-shared display types in trials with No faults ($t(14) = -0.29$, $p > 0.05$, $d = 0.15$). There was also no significant difference between Shared and Non-shared display types in trials with Faults ($t(14) = 1.57$, $p < 0.01$, $d = 0.78$). Therefore the source of the interaction cannot be clearly specified; however the effect size for trials with Faults is greater than that for trials without Faults, suggesting that the decrease in performance as a result of Faults is greater with Non-shared display types.

The relations between communication codes and performance provide an indication of the process used to complete the task. When there were no faults, teams with non-shared displays performed best when there were fewer prompts and more rebuttals. This suggests a large number of unhelpful suggestions were being made and performance was best when these were accurately rejected. Hence high performance arose from a complex dialectic process. In contrast, with shared displays high performance was associated with fewer rebuttals (this correlation approached significance). This suggests a different process in which accurate suggestions are made first time. This process changes when faults are introduced. With shared displays greater number of questions leads to worse performance. Questioning is indicative of teams seeking information

Table 1. Team performance as a function of condition

	Shared		Non-Shared	
	Mean	S.D.	Mean	S.D.
No fault	2798.75	506.06	2895.50	786.63
Fault	2356.44	684.93	1939.94	310.98

Table 2. Correlation of team communication and performance as a function of condition

Communication code	Shared		Non-Shared	
	No fault	Fault	No fault	Fault
Question	−0.19	−0.74*	−0.19	0.13
Prompt	−0.11	0.19	−0.69*	−0.42
Rebuttal	−0.52	−0.17	0.73*	0.19

* = correlation is significant at 0.05 level.

to develop situation awareness, suggesting higher performance arose when teams did not need to rely on this. In contrast, no overall pattern of communication arose with non-shared display types, possibly indicative of unclear communication patterns as a result of demands of the fault.

In conclusion, these results suggest that shared and non-shared displays are equally effective when there are no faults in the system; however there is more complex team interaction with non-shared displays. The difference in performance between shared and non-shared displays was greater when unexpected faults arose. These findings suggest that this is due to the complexity of team interactions with non-shared displays becoming excessive, whereas shared displays reduce the need for complex team interactions to develop situation awareness and negotiate action. Therefore distributed displays can qualitatively change how team tasks are performed, and should be tested fully in fault finding situations as well as during routine performance.

References

Endsley, M.R. 1995, Toward a theory of situation awareness in dynamic systems. *Human Factors*, **37**, 32–64

Rodgers, M.D., Mogford, R.H. and Strauch, B. 2000, Post hoc assessment of situation awareness in air traffic control incidents and major aircraft accidents. In M.R. Endsley and D.J. Garland (eds.) *Situation Awareness Analysis and Measurement, 2000*, (LEA, New Jersey), 73–112

CONTROL ROOMS SYMPOSIUM

EN-ROUTE AIR TRAFFIC CONTROL ROOMS 1965–2005

Hugh David

R+D Hastings, 7 High Wickham, Hastings, TN35 5PB, UK

En-Route Air Traffic Control (ATC) rooms contain mainly sector control suites. A sector may have up to six geographically adjacent sectors in three-dimensional airspace. Control room floors are two-dimensional and subject to practical constraints. Although executive and planning controllers were grouped together in early control rooms, it is now usual to place the executive and planner for each sector together. Sector suites are usually placed in lines, with associated sectors placed together. In recent years, the traditional darkened room with control units facing the walls has been replaced by day lighted rooms using furniture similar to office furniture.

Introduction

The specific control room layouts referred to in this paper are shown in plan and photograph in the PowerPoint presentation accompanying this paper on the CD record of this conference.

En-route ATC is of particular interest to the ergonomist because the design of control rooms is, or should be, determined by the requirements of the operators, rather than by physical constraints, such as the need to see where aircraft are on the surface of an airfield. (The earliest ATC facilities were concerned with the control of aircraft on and around airfields. The familiar and iconic towers are the direct descendants of these early facilities, and their design is constrained and complicated by their direct view of the airfield.) En-route control (away from the airports) requires only radio contact with aircraft, and, more realistically, radar coverage. Although the earliest "en-route" control centres were located on airfields, this was more for administrative than technical reasons.

It quickly became necessary to divide the airspace into geographical "sectors", so that each sector generated approximately equal workload – or at least, so that none were consistently overloaded. Traffic loads vary considerably within and between days. Controllers controlling adjacent sectors needed to be able to coordinate their activities. Each sector usually requires at least a planning and a radar controller, and it is common to "band-box" (combine) adjacent sectors at times of low traffic. Other physical constraints (such as the need for access for maintenance) further complicate the layout of control rooms.

Real-time simulation has been a successful tool for the assessment of sectorisation and control room layout. Real-time simulators themselves, particularly the flexible digital simulators of which Eurocontrol was a pioneer, have additional problems with the addition of simulator supervisory positions, "feed" sectors, measurement staff with their equipment and simulator pilots' consoles.

Background

An "en-route" ATC centre is, physically, a large room, in which teams of controllers maintain a round-the-clock traffic control service. Within the room sector teams control sectors of airspace, divided horizontally and vertically, seated at suites of working positions. A sector may have up to four adjacent sectors plus two underlying and two overhead sectors. A sector suite in a row of suites will have only one or two neighbours, and cannot therefore be adjacent to all the neighbours of its sector. The detailed equipment of suites has varied considerably over the years, although the development and generalisation of digital display technology has done much to reduce the physical constraints. The detailed design of sector working positions is addressed in a companion paper (David, 2006).

Here it should be mentioned that en-route control was initially exercised by "planning controllers" equipped with "strip-boards" which contained rows and columns of "strips" in strip holders. A strip was a slip of stiff paper, in a strip-holder that could be stacked with other strip-holders in accordance with the particular requirements of the position. The controller could annotate the strip to record flight changes as they were given to the aircraft.

Radar was initially introduced as an auxiliary tool, and the radar controller was considered to be an assistant to the planner. Using the classic "rotating scan" primary radars, with the primitive facilities originally available, radar controllers could handle few aircraft even if they followed the pre-defined networks of routes. Primary radar, familiar from Second World War documentaries, showed only "paints" or "blips" as the radar scan rotated. The large, heavy, radar display units were sometimes placed vertically, so that the screen image was horizontal.

Development

By 1971, (McCluskey and Zoellner, 1971) simulations of planned systems did not normally use horizontal displays, although they were still used for the "dummy" sectors, probably due to a shortage of vertical displays. "Dummy" sectors simulated sectors adjacent to those under investigation, to provide realistic entry and exit coordination. (Over the years, the name has changed from "dummy" or "ghost" to "feed" sectors, but the function has not changed.)

In initial studies for the projected Maastricht Upper Airspace Centre (McCluskey and Zoellner, 1971) Executive and Planning positions were situated back to back at some metres distance. Radar displays were provided at the Planning positions, in addition to the strip boards. Equally, "mini-strips" were provided for the Radar Executive, as he was then called.

A similar back-to-back arrangement was used in a 1972 (McCluskey et al 1973) simulation for the Shannon area. In this system (simplifying for clarity) the strips were initially passed to the Planning controller, who examined groups of strips, then passed them to the Radar controller, via an Input operator, who entered the Planner's modifications into the system, using a keyboard. In practice, the Planner retained groups of strips until he was completely satisfied with his solution, and then passed them to the Input Operator, whose work pattern involved long periods of inactivity, followed by intense effort, made more stressful by the urgent requests from the Radar controller for the strips for aircraft, which were entering his sector. This produced some high heart-rate readings. When the keyboard developed an intermittent malfunction of the most important key (which the engineering staff could not identify) heart rate rose to extreme levels.

An exploratory simulation for the Karlsruhe centre in 1975 (Clarke, 1976a), which simulated only one sector, employed an in-line organisation, involving nine working positions. These were, in order from left to right, a Assistant with three strip bays charged with strip distribution, a two-man planning section with one standard radar and twelve bays of strips, an "on-route" executive section with two radar controllers and a coordinator with five bays of strips between them, and finally, a similarly laid-out "off-route" executive section with a coordinator between two radar controllers. The main innovation in this simulation was the extensive use of small EDDs to transfer information between the various controllers, supplementing the physical movement of strips.

A research simulation in 1976 (Clarke, 1976b) was remarkable for the provision of colour-coded simulated radar, and for the provision of colour-coded EDDs (Electronic Data Displays) for the executive positions. Although the back-to-back positioning of Planner and Executive was maintained, the Planners and Executives for the two sectors simulated were situated diagonally opposite each other, not directly back to back, so that a certain amount of direct coordination was possible.

An operational simulation in 1980 (McCluskey 1980) of potential subdivisions of the area of Belgian airspace covered by the Maastricht UAC maintained the diagonal back-to-back organisation, but had EDDs and TIDs (Touch Input Displays) instead of strip boards at the Executive positions. A companion study, concerned with the details of inter-sector and external coordination at Maastricht UAC (Upper Air Traffic Control Centre), (Zoellner, 1980) showed that the simulated sector layouts corresponded to the actual Maastricht centre organisation.

However, a simulation the next year (Fuehrer, 1981) was carried out specifically to explore the arrangement of working positions. It showed a strong preference for the in-line arrangement, although where two sectors were side-by-side, there was no consensus about whether the Executives or Planners should be placed at the common boundary. This conclusion was thought to reflect the particular circumstances of the two areas involved.

A simulation of a proposed new layout for the Düsseldorf FIR (Anon, 1980) involved three sectors each divided into two sub sectors. It was assumed that Executives and Planners for each sub sector would sit together, with the two radar controllers next to each other in the middle and the Planners outside them.

A few years later a similar philosophy was applied to Maastricht UAC sectors, (Braun, 1988) although one sector had a Planner between two Executives, and other sectors had Assistants between the Executive and the Planner.

Although the ODID simulations were not intended to represent particular control rooms, ODIDII (Prosser and David 1989) simulated two single sectors in different UACs. It was interesting in that the two working positions essentially shared three large displays, and no strips were employed. In practice, it was observed that the EC and PLC in each sector shared tasks on an informal basis, depending on their relative workload.

The 1990 Shannon Simulation (Kroll et al 1991) involved lower sectors in addition to several middle-level sectors. All sectors had Executive and Procedural controllers together. The two Procedural controllers for the lower sectors were sited side-by-side. The middle-level sectors had all working positions in a row, with Procedural controllers to the right of Executives. Adherents to national stereotypes may care to know that the North sector was, in some places, south of the South sector.

Conclusion

The improvements in the technology of communication, particularly in the facilitation of communication between controllers, in the same centre or outside, might appear to make it possible for controllers to operate the same sector while physically separated. In fact, the coming together of Planning and Executive display facilities, discussed in David (2006) at this conference has led to sharing of facilities, and cemented the two more firmly. The distinction between Planner and Executive has become less important, particularly where they act in concert. Future systems may make different allocations of work, corresponding to a more efficient use of airspace – contemporary planning facilities imply extensive reliance on route structures, which may not be available in the near future.

The location of sector teams and sector suites is now less important than it used to be, but occasional advantage can be taken where the bulk of traffic flows from one sector to another. Whether, using the current work allocation, Planners or Executives should be placed together depends primarily on the nature of the traffic concerned.

In general, it seems probable that most centre layouts will consist of rows of working positions, simply for ease of access and maintenance. Developments in displays will make it much easier to switch the functioning of a working position.

References

Anon, 1984, *Simulation of a proposed new ATC Organisation in the Düsseldorf FIR*, EEC Interim Report No. 175. (Eurocontrol Experimental Centre, Bretigny-sur-Orge, France)

Braun, R., 1988, *Real Time Simulation of new sectorisation plans concerning the airspace presently controlled by the EAST and RUHR Sectors of MAASTRICHT UAC*, EEC Report No. 213. (Eurocontrol Experimental Centre, Bretigny-sur-Orge, France)

Clarke, L.G., 1976a, *Simulation for the Operational Evaluation of Colour Displays in Air Traffic Control*, EEC Report No. 102. (Eurocontrol Experimental Centre, Bretigny-sur-Orge, France)

Clarke, L.G., 1976b, *Simulation for the Operational Concept defined in the Karlsruhe 1 Operational Plan*, EEC Report No. 93. (Eurocontrol Experimental Centre, Bretigny-sur-Orge, France)

David, H., 2006, Air Traffic Control Consoles 1965–2005. In P. Burst (ed.) *Contemporary Ergonomics 2006*, (Taylor and Francis, London)

Fuehrer, B., 1981, *Simulation of Back-to-back and In-line arrangements of Working Positions at Maastricht UAC*, EEC Report No. 135. (Eurocontrol Experimental Centre, Bretigny-sur-Orge, France)

Kroll, L., David H., and Clarke, L., 1991, *Real-Time Simulation of Shannon UAC*, EEC Report No. 242. (Eurocontrol Experimental Centre, Bretigny-sur-Orge, France)

McCluskey, E. and Zoellner, R., 1971, *Simulation of the operational plan MADAP A*, EEC Report No. 70. (Eurocontrol Experimental Centre, Bretigny-sur-Orge, France)

McCluskey, E., 1980, *Real Time Simulation of Split of Brussels West Sector – Maastricht UAC*, EEC Report No. 136. (Eurocontrol Experimental Centre, Bretigny-sur-Orge, France)

McCluskey, E., Zoellner, R. and David, H., 1971, *Simulation Relative to the implementation of Automatic Data Processing at Shannon*, EEC Report No. 64. (Eurocontrol Experimental Centre, Bretigny-sur-Orge, France)

Prosser, M., and David H., 1989, *ODID II Real-Time Simulation*, EEC Report No. 226. (Eurocontrol Experimental Centre, Bretigny-sur-Orge, France)

Prosser, M., David H., and Clarke, L., 1991, *ODID III Real-Time Simulation*, EEC Report No. 242. (Eurocontrol Experimental Centre, Bretigny-sur-Orge, France)

Zoellner, R., 1980, *Study of the Inter-centre and Inter-sector Coordination Mechanisms in the Eurocontrol Region*, EEC Technical Note No. 11/80. (Eurocontrol Experimental Centre, Bretigny-sur-Orge, France)

AIR TRAFFIC CONTROL CONSOLES 1965–2005

Hugh David

R+D Hastings, 7 High Wickham, Hastings, TN35 5PB, UK

Air Traffic Control (ATC) consoles were originally designed by engineers to accommodate technical equipment, giving priority to mechanical and electronic requirements. Traditionally Executive (Radar) controllers work with radar displays, while Planning Controllers work with racks of "strips" in "strip holders". The increasing use of digital data handling has provided Executive controllers with Synthetic Dynamic Displays (SDDs) while Planners are provided with a variety of Electronic Data Displays (EDDs), which are increasingly rendering the traditional strip redundant. Future ATC consoles will increasingly resemble other computer working positions.

Introduction

(The consoles referred to in this paper are shown in diagrams and photographs in the PowerPoint presentation accompanying this paper on the CD record of this conference.)

En-route ATC is of particular interest to the ergonomist because the design of consoles (at least for en-route centres) rooms is, or should be, determined by the requirements of the operators, rather than by physical constraints, such as the need to see where aircraft are on the surface of an airfield. (The earliest ATC facilities were concerned with the control of aircraft on and around airfields. The familiar and iconic towers are the direct descendants of these early facilities, and their design is constrained and complicated by their direct view of the airfield.) En-route control (away from the airports) requires only radio contact with aircraft, and, more realistically, radar coverage. Although the earliest "en-route" control centres were located on airfields, this was more for administrative than technical reasons.

It quickly became necessary to divide the airspace into geographical "sectors", so that each sector generated approximately equal workload – or at least, so that none were consistently overloaded. Traffic loads vary considerably within and between days. Controllers controlling adjacent sectors needed to be able to coordinate their activities. Each sector usually requires at least a planning and a radar controller, and it is common to "band-box" (combine) adjacent sectors at times of low traffic. Other physical constraints (such as the need for access for maintenance) further complicate the layout of control rooms.

Real-time simulation has been a successful tool for the assessment of sectorisation and control room layout. The detailed design of working positions has rarely been the direct concern of the users of Real-time simulators. A good deal may, however, be learned from the equipment used in real-time simulators, particularly if viewed over an extended period of years, combined with the results of such control position design studies as have been made over the years.

Background

An "en-route" ATC centre is, physically, a large room, in which teams of controllers maintain a round-the-clock traffic control service. Within the room sector teams control sectors of airspace, divided horizontally and vertically, seated at suites of working positions. A sector suite usually consists of a Procedural Controller (Planner) position and an Executive (Radar) position. Other positions, such as coordinators or assistants are occasionally provided, although these have become less frequent in recent years. The detailed equipment of suites has varied considerably over the years, although the development and generalisation of digital display technology has done much to reduce the physical constraints. The arrangement of sector suites within the control room is addressed in a companion paper (David, 2006).

En-route control was initially exercised by "planning controllers" equipped with "strip-boards" which contained rows and columns of "strips" in strip holders. A strip was a slip of stiff paper, in a strip-holder that could be stacked with other strip-holders in accordance with the particular requirements of the position. The controller could annotate the strip to record flight changes as they were given to the aircraft.

Radar was initially introduced as an auxiliary tool, and the radar controller was considered to be an assistant to the planner. Using the classic "rotating scan" primary radars, with the primitive facilities originally available, radar controllers could handle few aircraft even if they followed the pre-defined networks of routes. Primary radar, familiar from Second World War documentaries, showed only "paints" or "blips" as the radar scan rotated. The long duration phosphor used took several revolutions to fade, providing a trail for the aircraft. The large, heavy, radar display units were sometimes placed vertically, so that the screen image was horizontal. In early centres, aircraft were identified by asking them to change heading, then to resume their track. The radar controller or an assistant observed which blip moved then placed a marker on the surface of the tube, identifying the aircraft. These markers were sometimes known as "shrimp boats", from the appearance of the display.

EUROCONTROL studies

In 1970, in a simulation for the Rome area (Martin and Zoellner, 1971) Secondary Surveillance radar, which provided a numerical identifier and an indication of height, made the "shrimp boat" obsolete, but horizontal displays were still used occasionally. In this simulation, radar controllers were provided with strip boards, but did not in fact use them. They were abandoned after a few trials.

By 1971, (McCluskey and Zoellner, 1971) simulations of planned systems did not use horizontal displays, although they were still used for the "dummy" sectors, probably due to a shortage of vertical displays. In this study, radar displays were provided at the Planning positions, in addition to the strip boards. Equally, "mini-strips" were provided for the Radar Executive, as he was then called. Strip marking rules required changes in flight plans to be marked on the strip. Different colours were used by the Planner, Executive or Assistant.

An exploratory simulation for the Karlsruhe centre in 1975 (Clarke, 1976a), which simulated only one sector, employed an in-line organisation, involving nine working positions. These were, in order from left to right;

- an Assistant with three strip bays charged with strip distribution,
- a two-man planning section with one standard radar and twelve bays of strips,
- an "on-route" executive section with two radar controllers and a coordinator with five bays of strips between them,
- and finally, a similarly laid-out "off-route" executive section with a coordinator between two radar controllers.

The main innovation in this simulation was the extensive use of small EDDs to transfer information between the various controllers, supplementing the physical movement of strips.

A research simulation in 1976 (Clarke, 1976b) was remarkable for the provision of colour-coded "penetron" displays for both radar and EDDs. These "penetron" displays were "calligraphic" displays, which "wrote" by deflecting an electron beam onto a tube face with two layers of phosphor. They could show the equivalent of a radar image, but could show a restricted range of foreground colours only (not including blue or cyan), and had problems in displaying red in sufficient intensity. They could not vary the background colour. The colour-coded EDDS (Electronic Data Displays) for the executive positions were designed to replace the strips used at those positions. In spite of the technical deficiencies of the displays, they were favourably received by controllers, although they had reservations about how colour could best be employed.

An operational simulation in 1980 (McCluskey, 1980) of potential subdivisions of the area of Belgian airspace covered by the Maastricht UAC had EDDs and TIDs (Touch Input Displays) instead of strip boards at the Executive positions. By this time, it was accepted that executive controllers would draw their information from electronic displays, without using strips.

The Operational Displays and Input Development (ODID) simulations were intended originally as "quick and dirty" experiments to resolve problems in the introduction of electronic display systems. (Experience, regrettably, showed that they were not quick.)

ODID I (Prosser and David, 1987) tested colour raster-scan tabular displays, based on a UK concept. During this simulation the participating controllers developed conventions for colour use, using direction of flight coding for the lines replacing strips, with colour highlights for outstanding tasks.

ODID II (Prosser and David, 1989) simulated two single sectors in different UACs. It was interesting in that the two working positions essentially shared three large displays, and no strips were employed. In practice, it was observed that the EC and PLC in each sector shared tasks on an informal basis, depending on their relative workload.

ODID III (Prosser et al, 1991) represented a radical shift in ATC display technology. All information, for either controller, was displayed on a single 20-inch $2K \times 2K$ pixel raster-scan display. The display was controlled by a two-button mouse, which was used to select and set up the displays. The dynamic radar image, with auxiliary input and conflict warning displays and aircraft labels which could be "opened" to provide complete data and used to modify flight plans was found highly satisfactory. The Planning controllers' main problem was that they did not have a "radar" picture. Physically, the display was vertical, with a map displayed above it, and standard communications panels between the screen and the mouse area.

ODID IV (Graham, 1994) simulated Copenhagen airspace. Although it was primarily concerned with exploring further the displays initiated in ODID III (Prosser et al, 1991), it showed an equally radical approach to control room layout. The working positions simulated were constructed from ordinary office furniture, with only the high-precision displays, which weighed about 75 Kilograms, supported by special trolleys. The Copenhagen control room now has natural lighting, since the displays are bright enough to read under normal office lighting.

Other studies

Although EUROCONTROL appears to be the only body that has made a series of studies over the years in which console design has played a part, other organisations have made special purpose studies.

The Federal Aviation Authority (FAA), in the United States, undertook a project to produce a common console for the Advanced ATC System. This involved a meticulous analysis and definition of all ATC activities, producing a specification twelve thousand of pages long. (Ammerman et al (1984) is part of this specification.) This was the basis of a design competition between a number of US corporations, from which two (IBM and Hughes) were selected to produce working prototypes. The prototype selected for evaluation was that produced by IBM. This involved a 20-inch square 2000 by 2000 pixel colour main display, with an auxiliary high resolution colour monitor above it, two small Voice Switching and Control System panels to control communications, a full-size moveable keyboard, and a moveable rolling-ball control. Unfortunately, by the time this process had been completed, control methods and expectations had changed. When the prototype was submitted to evaluation by working controllers, it was decisively rejected. The project was abandoned.

The Arnott Design Group carried out a design and evaluation of an ergonomically suitable ATC workstation (Arnott et al, 1988). They envisaged an essentially desk-like working position, for a pair of controllers, using a mouse to control and input data, together with standard high-quality computer monitor displays.

The UK Civil Aviation Authority (CAA) commissioned a study by the Royal College of Art (Reynolds and Metcalfe, 1992) on the best use of colours. This study recommended the use of subdued colours and unsaturated hues for area colours, with the minimum use of bright colours. This recommendation has been widely adopted, in the ODID studies among others.

Conclusion

The trend in control console design has been away from specialised units designed to accommodate special-purpose devices towards more generic units, designed to employ standard displays and control devices. Computer displays and mice have replaced Cathode Ray Tubes, strips and trackballs. In, for example, the ODID simulations, the Planning and Executive display systems were physically identical, although Planners and Executives might use different display windows. Simultaneously, the controllers themselves have been observed to share their working displays. The formal distinction between Planner and Executive has become less important, particularly where they act in concert. Future systems may make different allocations of work, corresponding to a more efficient use of airspace – contemporary planning facilities imply extensive reliance on route structures, which may not be available in the future.

Developments in displays will make it much easier to switch the functioning of a working position, and, probably, make consoles lower and more like generalised computer work stations.

References

Ammerman, H.A., Ardrey, R.S., Bergen, Bruce, Fligg, Jones, G.W., Kloster, G.V., Lenorovitz, D., Phillips, M.D., Reeves and Tischer, 1984, *Sector Suite Man-Machine Functional Capabilities and Performance Requirements*, (CTA Inc., Englewood Colorado)

Arnott, J., Nakashima, R. and Stager, P. 1988, *Development and Evaluation of an Ergonomically Sound Air Traffic Control Workstation*, (Arnott Design Group, Toronto, Canada)

Clarke, L.G. 1976a, *Simulation for the Operational Evaluation of Colour Displays in Air Traffic Control,* EEC Report No. 102. (Eurocontrol Experimental Centre, Bretigny-sur-Orge, France)

Clarke, L.G. 1976b, *Simulation for the Operational Concept defined in the Karlsruhe 1 Operational Plan*, EEC Report No. 93. (Eurocontrol Experimental Centre, Bretigny-sur-Orge, France)

David, H. 2006, En-route Air Traffic Control Rooms 1965–2005. In P. Burst (ed.) *Contemporary Ergonomics 2006*, (Taylor and Francis, London)

Graham, R.V. 1996, *ODID IV Real-Time Simulation*, EEC Report No. 256. (Eurocontrol Experimental Centre, Bretigny-sur-Orge, France)

Martin, R. and Zoellner, R. 1970, *Simulation of the Rome Terminal Area*, EEC Report No. 46. (Eurocontrol Experimental Centre, Bretigny-sur-Orge, France)

McCluskey, E. and Zoellner, R. 1971, *Simulation of the operational plan MADAP A*, EEC Report No. 70. (Eurocontrol Experimental Centre, Bretigny-sur-Orge, France)

McCluskey, E. 1980, *Real Time Simulation of Split of Brussels West Sector – Maastricht UAC*, EEC Report No. 136 (Eurocontrol Experimental Centre, Bretigny-sur-Orge, France)

Prosser, M. and David, H. 1987, *ODID I Real-Time Simulation*, EEC Report No. 217. (Eurocontrol Experimental Centre, Bretigny-sur-Orge, France)

Prosser, M. and David, H. 1989, *ODID II Real-Time Simulation*, EEC Report No. 226. (Eurocontrol Experimental Centre, Bretigny-sur-Orge, France)

Prosser, M. David, H. and Clarke, L. 1991, *ODID III Real-Time Simulation*, EEC Report No. 242. (Eurocontrol Experimental Centre, Bretigny-sur-Orge, France)

Reynolds, L. and Metcalfe, C. 1992, *The Interim NATS Standard For The Use Of Colour On ATC Displays*, CS Report 9213, (Civil Aviation Authority, London, U.K.)

CONTROL ROOM ERGONOMICS: OBSERVATIONS DERIVED FROM PROFESSIONAL PRACTICE

Kee Yong Lim

Centre for Human Factors and Ergonomics, School of Mechanical & Aerospace Engineering, Nanyang Technological University, Nanyang Avenue, Singapore 639798

This paper presents the author's observations derived from personal experience in the specification and evaluation of a number of control rooms. Examples are drawn from these cases and used as illustrations to highlight some human factors (HF) design considerations in control room design and the pitfalls that might be encountered typically in HF practice. From the above review, some observations are made relating to the adequacy of ISO 11064 for HF design of control rooms.

Background

Basic equipment in early control rooms comprises physical panel displays, some with controls incorporated, while others have separate control stations. Such control rooms may be found in refinery and nuclear power stations. Peripheral equipment may include some work tables, shelves for documentation and dot matrix printer(s). The physical display panels serve two main purposes:

1. The display panels update the operator on the current status of the plant to enable process monitoring and control of basic physical parameters such as pressure (P), temperature (T), volumetric rate (V), mass (M), length/distance (L/D) and time (T). Secondary parameters derived from the basic set, such as power and energy, may also be presented to inform the operator on the values of process variables of immediate interest, e.g. trend data, throughput, output, consumption. Using these data, the operator troubleshoots and controls the plant.
2. The schematic of the display panels provide the operator and supporting personnel a spatial map of the machinery and equipment layout, i.e. a pictorial representation of the layout of the plant in the real world. With this display, the operator is able to visualize plant process flows, trace lines against process and instrumentation diagrams, and designate the status of equipment with tags (operational/under maintenance). In doing so, the operator must remain aware that such information has to be checked against the actual layout of the plant in the real world, as the information extracted from panel displays and process and instrumentation diagrams, might not always be up to date. These shortcomings stem from oversight, laziness, version control and/or difficulties in modifying the physical displays and/or documentation.

With the advent of desktop computers, replacement of physical panel displays with computer consoles began in the early eighties. During this time, some mistakes were made with respect to the use of nested displays (due to limited screen estate) to convey information on the layout of the plant previously shown on the physical display panels. As a result, operators found themselves lost in cyber space when they tried to navigate through the computer displays. At this time, the author observed that operators refused to use the computer consoles and continued to rely on panel displays to operate the

Figure 1. Cinema size projection display panel with large adjacent CRT monitors.

Figure 2. A large electronic screen display to complement desktop monitors.

plant (a refinery). The nested digital displays not only imposed a higher memory load on the operator (the disparate displays failed to provide a coherent picture of the plant "at a glance"), but the interactions required for screen navigation also intruded into their main task of process control. In desperation and without understanding the problem confronting operators, the management of the company decided to board up the display panels to force the operators to use the computer consoles!

In recent years, the display technology has improved and a larger screen estate for displaying the plant layout has become available, including the following:

1. Rear projection screens (see Figure 1).
2. Composite screens of various configurations (see Figure 2). Composite screens, if implemented well, can complement desk workstations to minimize the problems of becoming lost in cyber-space (see earlier account).
3. Front projection screens (e.g. an Elumens immersive dome). The use of Elumens domes in control rooms is still limited due to cost, and will not be discussed further.

While the use of larger screens may help to solve some problems, they also introduced new ones. For instance, large rear projection screens suffer from a lower display resolution and luminance. Further, inappropriate implementation might make the specification of an optimal viewing distance from a seated control console, difficult if not seemingly impossible. In the case of the traffic management control room shown in Figure 1, specification of a viewing distance is complicated by the following interacting concerns:

- the sheer size of the cinema size projection screen
- what the operator needs to see in the cinema size projection display

- the time of day, internal and external lighting conditions, and type of camera, all of which affect the quality of the image fed to the projection screen
- the co-location of the large CRT monitors with the rear projection display, with respect to what the operator needs to be able to see in the monitors.

The rationale given for the use of a large projection display was that it was intended to support co-ordinated decision making when various emergency services were involved in the handling of an incident. This function, however, could not be realized in full as the resolution of a projection display was found to be rather low especially at night. As for the CRT monitors, they were used as check displays to inform the operator of the range of camera views that might be selected for projection onto the large screen. Note that depending on the time criticality and number of camera feeds, a composite multi-view or a scrolling display of various views may also be used. During routine work, the operators would sit at their individual control consoles and look at "usual" size computer monitors.

Standalone consoles to control a single type of equipment (as opposed to controlling and trouble-shooting various equipment and processes) may be simpler to specify. Such a system is found in the semi-automated remote crane control room in Singapore (see also Lim and Quek 2002). The task of the operator entailed by this system is as follows:

- Container pick up: involves landing a gripper (termed a spreader) onto a container, locking onto it and raising it to a certain height above the ground/prime mover, before handing it over to the computer to move it to a designated landing location.
- Container landing: involves landing a container onto the ground or prime mover, releasing the container, raising the spreader to a certain height, before handing it over to the computer to move it to the next pick-up location.

In this case, the clustering of consoles in a control room was decided by the company largely to enable convenient supervision. The project remit given to the author (acting as a consultant) was confined only to anthropometric design and the layout of the control consoles. This rather narrow project scope implies the following limitations:

- HF design of the display for the remote control system was excluded. This is likely to be a source of problems, as the implemented display comprised a 2-dimensional (2D) quad view of four corners of a container relayed by cameras, and would not support the task well as it required 3D perception.
- HF assessment of the task and operation scenarios (existing and new systems) was excluded. An account of the HF implications follows. First, the design of the control console would not account for transfer of learning effects, which might arise should the remote control system breakdown. Operators would then have to return to onsite physical control of the cranes. This situation would indeed have arisen, as the author found marked conflicts in the proposed layout and allocation of the controls to each of the operator's hands.

Second, consideration of the transfer of learning effects associated with the deployment in the new remote control system, of new and/or existing operators might similarly be overlooked with negative consequences for container handling operations.

Third, non-address of socio-technical design concerns might reduce the effectiveness of the new implemented semi-automated system. In particular, operators would no longer work alone inside a crane onsite and be the "king of their castle". Instead, they would be required to work in a remote control room with 22 other operators, all working directly under the "watchful eye" of a supervisor. The socio-technical change confronting the operators was therefore rather dramatic. To aggravate matters, "classroom" layouts of the control consoles were originally proposed (see Figure 3(a)), which would have heightened the already dramatic change. The author "softened" the

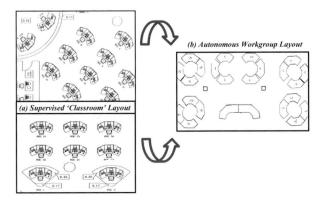

Figure 3(a) & (b). Alternative layout of consoles in a control room.

layout to the one shown in Figure 3(b). In addition, a proposal was made to consider a re-design of the work to optimize the new system, which involved moving away from an individual operator controlling a single crane to him/her controlling any number of cranes assigned by the computer. A possible design could be the re-organisation of individual work into autonomous group work. This solution would be consistent with the clustered layout of the control consoles proposed by the author.

Fourth, the exclusion of physical environment design from the project, might lead to the proposal of a less optimal design solution. In particular, the design of the control console proposed originally included audio feedback on completion of each container landing/pick up (job completed within a minute). This beeping repeated every minute multiplied by 23 consoles in the control room, would have made the noise intolerable.

Fifth, the exclusion of task design from the project, might lead to operator rejection and/or poor implementation of the new system. Indeed, the management of the company initially held the view that since the operator would now be performing only a part of the task and sit in a comfortable office environment (control room as opposed to a crane), their work cycle duration could be extended. In the author's view, the situation might actually be the opposite, as the intensity of the operator's task had actually increased with the new system, i.e. the work cycle duration might have to be shortened. The reason was because the new semi-automated system allocated to the computer the simpler task of trolleying/moving the spreader or container to a designated location. The skilled task component of container landing and pickup was left to the operator to perform. As such, the operator had to perform more skilled task actions per hour with the new system, since these tasks were no longer alternated with easier trolleying tasks. Thus, the operator had lost the "rest" period in between the intensive perceptual and psycho-motor tasks of container landing/pick up. With the new remote control system, the task might also be made more difficult by degraded visual perception (2D camera images only), variation in control–displacement ratio across different cranes assigned by the computer, and possibly less effective communication with prime mover drivers on the ground.

A brief review of the ISO standard for control room design (ISO 11064)

ISO 11064 will now be reviewed briefly in the context of the cases that the author has been involved with personally, namely: remote freight crane control (1 control room); water utility control (2 control rooms); electronic traffic management system (1 control room).

All of the projects above involve basic task analysis, physical workstation specification and control room layout. None of the projects involve display design although HF input is pertinent (especially in the case of remote crane control).

ISO 11064 is found to be very informative, well written and comprehensive in scope – it even covers the design of the building housing the control room (see EN ISO 11064-2:2000). Although ergonomic input would not involve such a wide scope in most cases, it is still very useful as a reference document to define the breadth of HF contribution to control centre design. Other comments on ISO 11064 are as follows:

- ISO 11064 provides a convenient formula to compute the distance from a large display screen/panel that a seated control console should be located (Figure 7, EN ISO 11064-3:1999). However, the basis for the recommendation seems to be confined to considerations on the field of view afforded by the console location. In particular, what an operator is expected to see in the display contents from that distance is not addressed. Although such a specification may be accounted for by the visual angle of display objects, this additional requirement might be overlooked by less experienced and/or meticulous practitioners, especially in a project that excludes display design. It is suggested that, at the very least, a cross reference should be made in EN ISO 11064-3:1999 to the requirements of visibility stipulated in EN ISO 11064-4:2004. The same should be done for the layout of control consoles relative to windows.
- ISO 11064's support for physical environment design (EN ISO 11064-6:2005) is good with the exception of lighting. In particular, explicit procedures are not given on how to determine an optimal location for lighting. The same inadequacy is found in HF texts on control room design (e.g. Ivergard 1989). The guidance provided by Grandjean (1992) is sufficient, provided the HF practitioner understands basic trigonometry. Such guidance should be incorporated into EN ISO 11064-6:2005.
- Although ISO 11064 suggests a wide variety of control room layout (Annex A, EN ISO 11064-3:1999), its guidance could be made more explicit concerning the implication of the proposed console layouts on team work and socio-technical factors.
- ISO 11064 should also include a section on how a design specification may be verified following implementation, e.g. measurement of a specified field of view.

References

EN ISO 11064. Ergonomic design of control centres. Parts 2, 3, 4 and 6.
Grandjean, E. 1992. Ergonomics in computerized offices. Taylor & Francis.
Ivergard, T. 1989. Handbook of control room design and ergonomics. Taylor & Francis.
Lim, K.Y. & Quek, S.M., 2002. Enhancing operator performance of remote container landing: an assessment of a 3d stereoscopic control and display system. In: Proceedings of the Australian HF'02 Conference, Melbourne, Australia, 2002.

MEDIA, SIGNS AND SCALES FOR THE DESIGN SPACE OF INSTRUMENT DISPLAYS

M. May[1] & J. Petersen[2]

[1]LearningLab DTU, Building 101
[2]Ørsted DTU, Automation, Building 326
[1,2]Technical University of Denmark DK-2800, Denmark

A systematic approach to the design space of instrument display components is described based on a semiotic extension of Cognitive Systems Engineering (CSE). A taxonomy of media types, representational forms, and scale types is proposed to account for the basic dimensions of the design space. The approach is applied to tailoring of soft instrument displays in supervisory control, showing how flexibility can be obtained by transformations of media, scale types and/or representational forms. These transformations correspond intuitively to an exploration of the design space.

Tailoring and support for interface flexibility

It is a generally accepted limitation in the design of automation systems and operator interfaces for safety-critical work domains (such as process control and transportation) that designers cannot fully predict all the situations and events that can occur within the larger system, for which the automation systems and interfaces have been designed. The operators (users) consequently have to adapt tasks and procedures and/or tailor the systems and interfaces in order to compensate for deficiencies in the design relative to the actual situations and contexts of use (Obradovich and Woods 1996; Vicente et al 2001). This tailoring includes ad-hoc modifications to instrument interfaces and display devices, but safe and efficient tailoring is usually not supported systematically through the design of flexible interface components.

The issue of flexibility does not only concern the *content of information* required by operators, but also the *media of presentation*, the *scale type* of data and presentations, the *representational form* and the *layout* (including spatial location and temporal extension of components). The need for flexibility can be observed indirectly through the workarounds and modifications introduced to compensate for the lack of flexibility. In Figure 1 three examples are given to illustrate the need for display component flexibility. To the left a wind direction indicator used on a car ferry have been turned upside down in order to support a coherent mental model given the visual orientation of the operators using it from a secondary aft bridge. This illustrates a change that only affects the layout. In the middle picture a pressure gauge in a power plant has been modified with a red pen to indicate an additional ordinal scale (normal–high pressure) on top of the ratio scale measurement. In the picture to the right an officer on an icebreaker, sailing in an archipelago with heavy traffic, have just received a VTS radio message about an upcoming passage of another ship: in order to communicate this to his fellow officers and to assist his own memory, he writes the information on a provisional whiteboard that has been placed over one of the preinstalled instrument panels.

Flexible generic soft instruments could provide better support for tailoring: rotating a wind indicator display would not have to rotate the characters as well, adding an extra

Figure 1. Ad-hoc modifications and tailoring of non-flexible interfaces.

scale to the pressure gauge (*scale superimposition*, Petersen and May 2005) could easily be obtained, and messages given in the transitory acoustic media could also be made available as text. Individual modifications will normally occur within a safe envelope of communication supporting the mutual construction of team situation awareness, but tacit "fights" over adequate instrument settings do in fact occur even in safety-critical domains.

There are two main strategies for interface-supported flexibility: *adaptation* and *tailoring*. Adaptive "intelligent" interfaces however suffer from a similar problem as the one characterizing the underlying automation, i.e. that designers cannot fully anticipate all the events and situations to which the interfaces should adapt. The alternative is to *design for tailoring through component-based generic interfaces* with "built-in" (but not automated) flexibility. Both strategies however will depend on the availability of a *systematic description of the design space* of these (automated or tailorable) interface components and this is the focus of the present paper.

On top of the *epistemic problem* of not knowing the full range of future situations and events, there is a *deontic dilemma* of flexibility implicated by the procedural violations involved in the workarounds and modifications through which expert operators effectively handle routine work and unforeseen events. The deontic or "moral" dilemma consists in the organizational and legal unwillingness to recognize the ability, opportunity and even necessity of operators to interpret implicit aspects of procedures, and to "fill in the blanks" in procedures considering the overall goals of the activity (De Carvalho 2005), and consequently the inability of designers to include support for this human flexibility in the design. Similarly (Polet et al 2003) have described the migration of expert operators from the safe field of use expected at design time to more efficient "border-line tolerated conditions of use", and they recommend that this migration should be taken into account during the design of systems and interfaces.

Limitations of Ecological Interface Design

Ecological Interface Design (EID), the main approach to *representation design* within *Cognitive Systems Engineering* (CSE) (Vicente and Rasmussen 1992), has been proposed as a design solution for complex work domains. EID will usually introduce design solutions based on a Work Domain Analysis (WDA) and the principle of *presenting constraints in the domain as affordances* for direct perception and manipulation. EID design solutions are backed up by empirical evidence indicating that they are more efficient that existing solutions. This procedure does however not exclude that there could be a whole range of other presentation design solutions that would support operators equally well or even better; the point being that *the design solutions of EID does not rest on an analytical understanding of the design space of interface components as a whole*.

Media, signs and scales for the design space of instrument displays 95

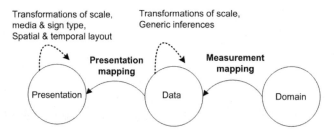

Figure 2. The measurement and the presentation mapping.

An example is given by the *configural displays* (mapping several variables to an integrated geometric form) that have been found to be more effective in supporting tasks than pure symbolic "digital displays" (Burns & Hajdukiewitcz 2004). Composite displays using *multiple forms of representation* was found to be even more effective (Bennett & Walters 2001), and the authors do not consider configural displays to be optimal designs. Coding conventions associated with each individual variable must be related unambiguously to represented system states. It is not enough that information can be obtained easily from the graphics, it "must also be semantically meaningful in the context of the domain task(s) to be performed" (Bennett & Walters 2001). Although a systematic approach to the design space of "ecological interfaces" have been outlined (Burns & Hajdukiewitcz 2004), we need to consider the explicit semiotic aspects of the syntax, semantics and pragmatics of interface components, cf. the "instrument semiotics" (May 2004) and feature-based multimedia semantics outlined in (May 2005, to appear).

Dimensions of the design space of interface components

Interface flexibility can only be supported systematically if designers have access to a coherent description of the basic dimensions and operations of the design space:

(I) the *scale types of data* according to the *measurement mapping* (Figure 2), the scale type of the corresponding physical *presentations* (cf. III), as well as the formal operations that can be applied to them (i.e. scale transformations and inferences from data constructing higher order data),

(II) the *forms of representation* (sign types) that can accommodate the types of information given by the *domain content* according to some domain *ontology*, as well as the *semantic properties* of these representational forms (i.e. the properties that makes the interpretation of images different from the interpretation of graphs or diagrams, for instance) and the operations they support (i.e. the things we are allowed to do with images as opposed to graphs or diagrams etc.), and

(III) the *media of presentation* (graphic, acoustic, haptic, or gestic), in which the representations are physically expressed, and their *properties and operations*.

In a formal sense measurements constitute data through a mapping assigning values to measured dimensions in the domain. Any data arising from this *measurement mapping* (Figure 2) will have to be on a certain *scale type*, i.e. *ratio, interval, ordinal,* or *nominal scale*. Zhang (1996) have stated a general principle for accurate and efficient presentation mappings: the scale type of the physical dimensions of the presentation (used to represent data) should match the scale type of represented dimensions in the domain (or rather as it is specified by the measurement mapping). Our *theory of scale transformations* (Petersen and May, to appear) modifies this view in a number of ways.

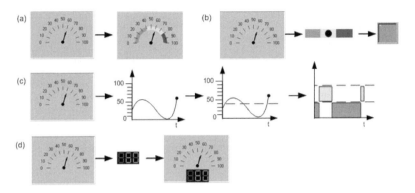

Figure 3. Transformations of scale type and representational form.

Generic inferences on data can be used to derive higher order data (such as integration of variables, derivation of higher order variables from constraints, variable state identification according to an ordinal scale etc.; Petersen 2004), and in some cases this implies *scale transformations*. Scale transformations can also be applied to the presentations as in the substitution or superimposition of graphical scales. Where Figure 3b exemplifies a series of *scale substitutions*, Figure 3a exemplifies *scale superimposition* by adding an ordinal scale to the ratio scale (cf. the tailored pressure gauge in Figure 1). Scale substitution and superimposition are both used in Figure 3c, where the implicit time of the dial instrument is made explicit by the history graph, i.e. by a change in the type of graph used.

Although data precision and operational support is lost in transforming data to a "weaker" scale, the *information value* of data can in fact be increased (Petersen and May, to appear). This is exemplified in the *instrument history* of tailoring shown in Figure 3b, where *ratio scale* data presented on a dial instrument is reduced first to data on an *ordinal scale*, and then further to a *nominal scale*, corresponding to a simple alarm. If the exact values are not relevant to present operations a weaker scale can be useful and informative.

The *presentation mapping* (Figure 2) is complex because it involves not only the assignment (cf. III above) of data to physical dimensions of a media (Zhang only considers the graphical media), each of which has its own scale type, but also the selection (cf. II above) of a *representational form* in which the data is semantically organized and expressed within the media. Furthermore presentations are often *combined* in the formation of "multimodal" forms, as in the example of the combination of the dial instrument display with its own transformation into a symbolic "digital" display in Figure 3d. This creates a macro-instrument from simple components based on superimposition and layout operations. A more complicated but effective form of combination is achieved when individual parts of a representation are substituted or superimposed by other representational forms, i.e. *embedding*. An example is the "multilayered" flow chart known as a "mimic diagram" in process control, where the simple nodes of a flow chart have been substituted by graphical symbols representing process plant components and furthermore superimposed with corresponding measurement data and relevant labels.

The whole description here only indicates *what is possible* (the basic design space). Further specifications are needed to determine *what should be actualised* under the constraints imposed by our knowledge of cognitive ergonomics and cooperative work in order to obtain a safe and efficient interface given a domain, and its tasks and goals.

Where the properties and operational possibilities of different physical *media of presentation* are based on the human sensory channels and their associated systems of communication (graphical, acoustic, haptic, gestic), the properties and operational possibilities of different *representational forms* are based on semantic differences (between iconic and symbolic) and on the difference between cognitive modules (visual perception, spatial reasoning, language). The main forms are *images and maps* (concrete iconic forms), *graphs and diagrams* (abstract iconic forms), and *language and symbols* (symbolic forms). These representational forms are constituted by their *invariant semantic properties across different media* although they are articulated differently in the different media and different media provide different operational support for their combination and manipulation (May 2005, to appear). These design space specifications of media, signs and scale are however still relevant for tailoring (i.e. even before we bring in the ergonomic specifications etc.) because they make a difference for distributed cognition and for the distributed work practices in terms of the *cognitive support* they can provide and in terms of the support they can provide for communication and control.

References

Bennett, K.B. and Walters, B. 2001, Configural Display Design Considered at Multiple Levels of Evaluation, *Human Factors,* **43(3)**, 415–434.
Burns, C. and Hajdukiewitcz, J.R. 2004, *Ecological Interface Design*, (CRC Press)
De Carvalho, P.V.R. 2005, Ergonomics field studies in a nuclear power plant control room, *Progress in Nuclear Energy*, in press.
May, M. 2004, Instrument semiotics: a semiotic approach to interface components, *Knowledge-Based Systems*, **14(8)**, 431–432.
May, M. 2005, Feature-based Multimedia Semantics: Representational Forms for Instructional Multimedia Design. In Ghinea, G. & Chen, S.Y. (eds): *Digital Multimedia Perception and Design* (Idea Group Publishing). To appear.
Obradovich, J.H. and Woods, D.D. 1996, Users as designers: how people cope with poor HCI design in computer-based medical devices, *Human Factors*, **38(4)**, 574–592.
Petersen, J. 2004, Model-Based Integration and Interpretation of Data. In Thissen, W., Wieringa, P., Pantic, M. and Ludema, M. (eds.), *2004 IEEE International Conference on Systems, Man & Cybernetics*, 815–820.
Petersen, J. and May, M. 2005, Scale Transformations and Information Presentation in Supervisory Control, *International Journal of Human-Computer Studies*, in press.
Polet, P. Vanderhaegen, F. and Amalberti, R. 2003, Modelling border-line tolerated conditions of use (BTCU) and associated risks, *Safety Science*, **41**, 111–136.
Vicente, K.J. and Rasmussen, J. 1992, Ecological Interface Design: Theoretical Foundations, *IEEE Transactions on systems, man, and cybernetic*, **22**, 589–606.
Vicente, K.J., Roth, E.M. and Mumaw, R.J. 2001, How do operators monitor a complex, dynamic work domain? The impact of control room technology, *International Journal of Human-Computer Studies*, **54**, 831–856.
Zhang, J. 1996, A representational analysis of relational information displays. *International Journal of Human-Computer Studies*, **45**, 59–74.

DEVELOPMENT OF A PROTOTYPE OVERVIEW DISPLAY FOR LENINGRAD NUCLEAR POWER STATION

Alexey Anokhin[1] & Ed Marshall[2]

[1]*Obninsk State Technical University of Nuclear Power Engineering (INPE), Studenchesky Gorodok, 1, Obninsk 249030, Kaluga Region, Russian Federation*
[2]*Synergy Consultants Ltd, Yewbarrow, Hampsfell Road, Grange-over-Sands, Cumbria, LA11 6BE*

An EU TACIS Project (R1.04/96A), entitled *Upgrading of Control Room Panels*, is in progress at Leningrad Nuclear Power Station in the Russian Federation. The main work of the project is to provide a prototyping and evaluation facility based on the existing full-scope training simulator. This facility will support the design and implementation of advanced control interfaces, based on computerised displays and taking into account modern ergonomics practice.

A key element of the ergonomics work has been to develop a prototype, generalized overview display which could replace the existing wall-mounted instrument arrays. This is expected to be in the form of a large screen display that can be viewed by all operational team members. The paper describes the prototype display design process which has provided a novel opportunity for co-operation between UK and Russian ergonomists.

Background

Computer screens are now the norm for the presentation of information in process plant control rooms whether the plants are for power generation, oil and gas production, or chemical processing. These plants can have a long life – power plants and oil refineries built in the sixties are still operational. The newest nuclear power plant in the UK first generated power in 1992 and the oldest began operation in 1965. This general picture is common throughout Europe and the rest of the world. Improving technology and obsolescence in instrumentation has meant that control rooms have been subjected to refurbishment and replacement throughout their working life.

The early control rooms comprised large arrays of panel-mounted instruments. Rows of dials, indicators and pen recorders displayed information while switches, knobs and levers were used to manipulate controls. Although conventions did evolve for the arrangement of these devices, there was little, if any, detailed ergonomics involved in the design and layout of these panels. As technology developed, controls were miniaturized and more sophisticated automation meant that more functions could be assigned to smaller displays. By the seventies, computer screens and printers were being exploited to display supplementary, detailed information, usually in the form of tables of alphanumeric text, which could not be shown on the fixed panels.

The increasing miniaturization also permitted more flexibility in the arrangement of instruments so that displays could be set out in functional groupings or in a topographic representation of the process, these are termed mimic displays. In the late seventies and eighties the computer screen began to displace the traditional arrays of analogue information and by the end of the eighties, the keyboard with mouse and trackerball had begun to replace the switches, levers and controls.

The overview

The Nuclear Industry was forced to focus sharply on the control room interface by events at the Three Mile Island (TMI) plant in 1979. At TMI, a near catastrophic accident was blamed largely on the fact that the experienced team of operators could not interpret plant condition from the presented array of information. One key ingredient was that operators concentrated on readings from a narrow range of instruments and did not take account of the total picture. The TMI control room was very much in the traditional style and the Nuclear Industry looked increasingly to computers to provide a solution. Research institutions started to address the benefits and potential problems as computerisation took hold in control rooms.

The use of computer screens, typically in a bank of several dedicated monitors accessed by a function keyboard, forced the operator to gather information in a serial manner, screen by screen, rather than being able to scan the wide array of information shown on the old panels. Operators, when asked, complained that they were losing the total view of the plant. This of course had already proved to be an issue at TMI. In the early eighties, the OECD Halden project first conducted experiments in which computer monitors, controlling a simulated nuclear plant, were augmented by a large (1.2 m by 1 m) projected image, generated by an original IBM PC. This display, called IPSO, was the first realisation of an overview display (Gaudio and Gertman, 1986).

Functions of the overview

Improvements in large screen technology has meant that overview displays are now the trend in control room design. In the process plant control room, the main objective of the overview is to provide an array of key information that the operator can scan in a single glance and then use to gain an appraisal of overall process status. This "feel" for plant condition is akin to the Aviation Industry's concept of "situational awareness". Typically the overview has the following features:

- It is permanently on view
- It is visible from all operator locations in the control room
- It has a large screen format
- Generates improvements in user acceptability
- It displays dynamic information in graphic, iconic and alphanumeric form.

The display usually aims to support the following operator requirements:

- The need for a permanent visualization of the process
- A common source of information for all staff
- The maintenance of the "broad picture" during disturbed conditions
- It allows visiting staff (e.g. managers) to gain an appraisal of plant condition without having to disrupt operational staff from their activities
- It reduces the need for continual searching through numbers of screen displays.

It should be noted at this point that in a recent survey of UK control room upgrades, looking at a range of industries, many had adopted the use of large screen overviews but in no cases had ergonomists been involved in their design.

Leningrad Nuclear Power Station

In Russia there has been a similar programme to prolong the life of plants and to upgrade, as necessary, the control room facilities. This paper describes part of a systematic test and evaluation programme to support the design and implementation of modern distributed control interfaces at Leningrad Nuclear Power Station. The main work of the project is to provide a prototyping and evaluation facility based on the existing full-scope training simulator. The project, supported by the EU TACIS (R1.04/96A) Programme, stresses the importance of taking account of modern ergonomics practice.

The station comprises two plants each with two RBMK 1000 Mw reactors, one first commissioned in 1974 and the second in 1981. In the wake of the 1986 Chernobyl accident, RBMK reactors were subjected to thorough design upgrades, much of the work supported by Europe and other Western countries. The control panels are typical of nuclear power plant rooms originally designed in the 1970s, the displays use a combination of traditional panel instrumentation and computer-based information. The majority of plant information is presented via large-scale arrays of analogue instrumentation that supports operation by a crew of four, who monitor the plant from a seated position, but must stand at control consoles when manipulating equipment. In any programme for replacing or upgrading this equipment, the provision of an overview, which would supplant the need for the back panel arrays, was considered to be essential.

Developing a prototype overview

It was a general assumption at the plant that such an overview would be a key feature in any major refurbishment of the existing unit control interfaces and further it was expected that any overview would be in the form of a large screen display that can be viewed by all operational team members.

Development of the prototype Global Overview Mimic Display (GOMD) has thus been a key issue in the Project. Designing the GOMD has presented a novel opportunity for co-operation between UK and Russian ergonomists working with experienced operational staff from the power plant. It was felt at the outset that any development of a prototype for the station should be based on a systematic consideration of experienced operator attitudes and opinions.

The work involved five main phases:

- A comparison of Western and Russian standards and guidelines
- Task analysis of selected routine and transient scenarios
- Structured interviews with operational staff regarding their information requirements
- Development of the prototype display
- Assessment of the proposed prototype display.

Western and Russian standards and guidelines

It is clear that the guidance and standards pertinent to computerised interface design are to be found in a large number of textbooks, guideline documents and official standards. This is equally the case for both Western and Russian sources. The problem for the interface designer and anyone assessing design proposals is finding an integrated single source for up to date human factors advice in this area. *DISC Plus* (Gregson et al, 2003) was an attempt to provide summarised guidance from a range of source documents used in the Nuclear Industry. Therefore, a comparison of the summarised guidance presented in *DISC Plus* was made with corresponding Russian advice. Few significant differences were identified. Generally, the Russian advice did not provide such specific guidance as was typical in the some of the Western source documents. There were some minor differences in the requirements for physical controls but the requirements for computerised information display and control were very similar. No particularly strong differences in stereotypes or expectations were found.

Task analysis

A task analysis was carried out of four routine operations and six fault scenarios. This was a hierarchical task analysis based on walk-through exercises undertaken on the existing full-scope simulator. The key decomposition categories were an examination of the interface actions, cognitive elements and team communication. It was a clear finding that bringing key parameters on to one co-ordinated display would aid team collaboration, both in routine activities and fault events. In addition, it became clear that the Supervisor, in particular, would be aided by such a display, as it would relieve him from continually looking from one operator station to another.

Structured interviews with operational staff

Structured interviews regarding the current control room facilities and ideas for future design improvements were conducted with ten experienced operational staff. The discussion was wide ranging and covered many control room issues. A general theme was a lack of integration between information sources and the need for instrumentation to be better grouped. There was considerable interest in the presentation of combined information to show for example overall steam and water flow balances which must currently be inferred from several instruments. There were a number of parameters which were required by all crew members but were only visible from one working position.

This initial survey was followed by a specific questionnaire aimed at eliciting systematic operator views on the contents of an overview display. This was based on drawing up a list of potential information items to be presented in the overview display. This list included consideration of the following sources of information:

- The most important RBMK operating parameters as listed by experts
- Information from the initial interviews
- Information from operating regulations
- An original display used by the instructors
- Results from the task analysis.

Eight plant experts then rated each of the listed parameters for inclusion in the overview display based on three criteria: frequency of use, information value and the parameter's value in determining the plant's operational safety status. Analysis of the ratings provided by the plant experts was used to determine which parameters were to be included in the GOMD.

Development of the proposed display

This information was then processed into a prototype format which could be projected as a 1.4 m by 1 m display using a projection system. The GOMD used standard symbols to represent equipment together with numeric values and text labels where appropriate. The proposals for the layout were discussed with a small team of senior instructors. The completed image had then to be coupled to the simulator computer model to become fully dynamic. The fully dynamic version of the display is still being completed.

Preliminary assessment of the prototype overview

Even without a fully dynamic version, it has been possible to carry out a number of evaluation exercises with the proposed display. In the first instance, the projected image has been subjected to initial ergonomics assessment.

Physical ergonomics issues

The screen has been checked for legibility at the maximum expected viewing distance. Although there are similar font-size requirements for Cyrillic characters there are key differences from the Roman alphabet. Various sets of characters are confusable like the number "5" and the letter "S" in Roman and there is little difference between the upper and lower case in Cyrillic fonts. It was soon discovered that the contrast properties of the projector were different from the normal LCD screens used for developing the display. This was demonstrated by numeric values shown in white which were difficult to see against the pale grey background, although they were clear on normal PC screens. These have been adjusted and necessary changes have been made to the style guide.

Prediction of symptoms

A useful exercise has been conducted with a group of four instructors. Each instructor was given a printout of the display in black and white. For each dynamic component in the display, they were asked to predict its value or appearance under a particular plant condition. These conditions were: full power operation, half-power operation, 30 seconds after a reactor trip and 30 minutes after onset of emergency shutdown cooling. Each instructor's results were discussed and a consensus display agreed. Versions of the overview were then prepared and projected. This immediately showed that some modification to the presentation was required if these conditions were to be readily discriminated on the basis of the overview alone.

Conclusions

Although the GOMD will benefit from dynamic evaluation trials planned for Spring 2006, its value and acceptability to operators has been enhanced by the ergonomics input in its early development. Initial exercises with the prototype have confirmed its potential usability.

References

Gertman D.I. and Gaudio P., 1986, *Integrated Process Status Overview* in Proceedings of the International Topical Meeting on Advances in Human Factors in Nuclear Power Systems, held in Knoxville, USA 21–24 April 1988.
Gregson D., Marshall E.C., Gait A. and Hickling E.M., 2003, *Providing Ergonomics Guidance to Engineers when Designing Human-machine Interfaces for Nuclear Plant Installations* Paper Presented at OECD Workshop on Modifications at Nuclear Power Plants – Role of Human Factors in Paris 2003.

OPERATIONS ROOM REDESIGN FOR THE AUSTRALIAN ANZAC CLASS FRIGATES

Stuart Sutherland, Suzanne Hanna & Susan Cockshell

Maritime Operations Division, Defence Science & Technology Organisation, Edinburgh, South Australia 5111

The Australian Defence Science & Technology Organisation (DSTO) has provided human factors expertise as part of an integrated project team (IPT) responsible for determining the feasibility, desirability and cost of redesigning the operations room of the Australian ANZAC class frigates. The process used was developed specifically for the project but drew upon ISO-11064, the DERA Human Factors Integration Guide and other handbooks. The needs capture involved a mixture of ship observations, interviews, questionnaires and an equipment audit of current and planned systems. The team developed room layout and console design concepts, which were driven by the physical ergonomics, communications, display and control needs of the personnel but were constrained by engineering limitations, crewing and cost. Computer visualisations allowed the IPT members and stakeholders to discover aspects of the design which were not otherwise apparent. A usability assessment process will include a ship team undertaking walkthroughs of critical activities.

Introduction

The Australian Defence Science and Technology Organisation (DSTO) was asked to undertake a human factors (HF) assessment of the Operations ("Ops") Room design of the Australian ANZAC class frigates. ANZAC ships have been in service since 1996 but there was anecdotal evidence from the ANZAC capability manager of dissatisfaction with the room's design (Griggs, 2004), notably from commanding officers (COs). The Ops room is the warfighting control room and the centre of activities at "Action Stations", when the ship faces imminent danger. At its heart is the combat system which displays information from sensors (radars, sonar etc) and allows the control of effectors (weapons and decoys) and other functions by operators. One of the important constraints placed on the study by the Navy was that, for cost and training reasons, crewing numbers and roles should not change.

This paper first describes the process used to incorporate human requirements into an improved Ops room design, then describes some of the HF techniques used by DSTO to develop concept designs, which are soon to undergo usability testing with a Navy team.

Review of previous Ops room design human factors

Hendy et al. (2000(a), 2000(b)) used an optimising layout tool (LOCATE) to position equipment based on visual, auditory, distance and tactile relationships between personnel. The LOCATE tool was applied to the layout of a US destroyer (DD21) Combat Information Centre (CIC) – the US equivalent of an Ops room. Hendy et al. (1989) previously used the LOCATE tool to evaluate ship bridge designs, based on an analysis of personnel movement. Edwards (2003) subsequently used LOCATE in redesigning the Canadian Halifax class frigates' Ops room. MacMillan et al. (2000) took a different approach to CIC design by matching the layout to an optimised team structure, based on ten principles, such as ensuring physical proximity among decision makers. TNO (2005) has undertaken such work, though no formal publications were found. Stanton et al. (2005) used observation and analysis techniques to understand RN Ops room activities in a simulation environment, though this analysis was not used to inform room design.

The design process

ISO11064 – Ergonomic design of control centres (ISO, 2000)

Part 1 of ISO11064 provides an excellent set of guiding principles, almost all of which were built into the process used in the Ops room project. These include an iterative user-centred design process in which ergonomic and technical requirements are jointly considered in an Integrated Project Team (IPT). The proposed ergonomic design process in ISO11064 was found to be useful but not entirely applicable. This was because crewing and roles were unable to be changed and because our project involved changing an existing capability, rather than designing a new one.

DERA Human Factors Integration process (DERA, 2000)

The match of the HFI process with our adopted process is fairly close, possibly due to the systems engineering approach taken by both. The HFI process appears to be better suited to new systems, for similar reasons to ISO11064. Consideration of the HFI process led the team to instigate registers for HF issues and risks. Items on the risk register were then transferred to the project risk register.

Other guidelines and standards

A search revealed 23 significant relevant standards. Of particular relevance were the UK DEF-STD 00-25 (MoD, 2000), US MIL-STD 1472F (DoD, 1999) and Australian DEF(AUST) 5000 (CoA, 2004).

Project nature and constraints

The "ANZAC Alliance" builds, maintains and upgrades the ANZAC ships. It uses a staged business model to reduce cost, schedule and technical risk by progressively iterating towards a final costed detailed design solution. The IPT, comprising staff from the industry partners (Saab Systems Australia and Tenix Defence), Navy and DSTO had to fit-in with this model by providing deliverables appropriate to the project stage even though these did not always neatly fit into the HF work sequence.

Process used by the HF team

The process used by the team is shown in Figure 1. The terms of reference and initial project team were established by the formal tasking statement from the ANZAC Alliance management board. The "initial" data collection and sharing has in fact been an ongoing iterative process as our understanding has improved with subsequent activities. Our team has employed a number of techniques, described below. The generation and assessment of candidate designs employed computer aided design (CAD) techniques, which were used to engage stakeholders whose views fed back into our understanding of the problem and the users' needs. The candidate designs have recently been mocked-up at DSTO and we

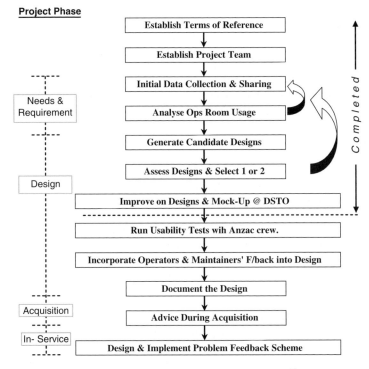

Figure 1. Process Used by Human Factors Team.

have designed a usability evaluation for use with a Navy team. The subsequent project stages are yet to be started.

The remainder of the paper describes the specific HF work undertaken.

User needs and system requirements analysis

Spanning the "Data Collection" and "Analyse Ops room Usage" steps of Figure 1, were a series of activities which defined the users, the teams, their tasks and their requirements.

Data collection

DSTO compiled a **questionnaire**, which was sent to all ANZAC COs. This used a mixture of operationally-based and user interface-based open questions to allow DSTO to understand some of the important issues and scope the work required to undertake a full assessment and room redesign. DSTO subsequently ran a **workshop** for more than 40 ANZAC ship personnel. This firstly identified the likes and dislikes of different ranks for the current room, then, using a forced trade-off constrained design process, determined priorities and ideas for the new room. From the questionnaires and the workshop, the HF team produced a substantial list of problems of the current room, which included: Poor situation awareness for the command team, inflexibility in personnel positioning and poor functional grouping of personnel.

DSTO **observed personal interactions and movements** of key personnel (CO, Principle Warfare Officer, PWO and Air Warfare Officer, AWO) at shore-based training facilities and at sea on two Navy ships. **Questionnaires** were employed after training exercises to gain data on interaction frequency between all Ops room personnel in different warfare modes and also on important personal interactions. **Structured interviews** were conducted with a wide range of personnel, with a particular focus on resolving ambiguous needs arising from the workshop and questionnaires. The team also used Navy **formal publications** which describe the roles and training of Ops room personnel.

An **audit** of current and planned Ops room equipment was used as the basis for a "bottom-up" analysis of user needs. It provided a useful checklist of items which might otherwise have been forgotten.

The users

The ANZAC Ops room typically contains from 5 to 25 personnel, depending on the threat regime in which the ship is operating. The personnel span up to five career categories and range in rank from Able Seaman to Captain.

Ops room functions

A high-level hierarchical task analysis (described in Kirwan & Ainsworth, 1992), derived from our data collection process indicates that there are four primary activities required during warfare: Planning, Administration, Whole of Ship Advice (primarily damage control and equipment failure management) and Warfighting/Tactical Operations. These form the four functional areas of the room. A fifth activity, Maintenance, usually occurs between warfighting. The warfighting/ tactical activities can be subdivided into four primary activities (Review/Change Posture, Air, Surface and Undersea Warfare).

Team structures and communications

From the above data we compiled: (a) Team organisation hierarchies and (b) visual representations of interactions between personnel during warfighting. A high-level task analysis of warfighting operations provided an indication of the importance of personnel, based upon their individual contribution to subtasks within each of the primary warfare activities. This resulted in high levels of importance for the room's primary decision-makers: The CO, PWO and AWO, who form the core of the 'command team'.

Microsoft Visio was used to visualise communications links between personnel. It allowed the people and equipment icons to be moved whilst maintaining the links.

Information and control needs

These we derived from the equipment audit, augmented with information from interviews and questionnaires. We would ideally have based this on a detailed task analysis but time limitations and our inability to modify crewing made this approach unattractive.

Needs validation

All 900 extracted user needs describing information, control, communications and personnel proximity were entered into a spreadsheet and divided into individual user's needs. These were sent to all ANZAC ships with a request to assess each user's needs in terms of frequency of use and importance. Six of the eight ANZAC ships completed the needs assessment, will be used by the project design team in making trade-off decisions.

Converting user needs into system requirements

User needs are usually converted into system functional and performance requirements. Since the room's functions are largely unchanged from the previous room, it is the design which must be changed to overcome the deficiencies of the current room and support the user needs which, for this project had a significant role in influencing design. The HF team also provided specific requirements in the form of reach envelopes, workstation physical ergonomics, criteria for information display and display luminance advice.

Conceptual designs

The IPT jointly provided input in the form of ideas and design constraints (cost, technical, physical, user needs and system capabilities). Some specific design principles emerged during this process:

- Make workstations more generic to provide greater flexibility in personnel location.
- Remove overhead (visually obstructive) equipment to allow the use of shared displays/stateboards and improve interpersonal visibility.
- Locate new and existing equipment to match individual user's needs.
- Address identified hazards.

Visualisations

The Visio communication link diagrams were used to assess new layouts and to explain design reasoning to stakeholders. Computer aided design (CAD) software was used to provide three dimensional visualisations and 'walk-throughs'. These not only engaged stakeholders more thoroughly but also allowed the design team to examine lines of sight and screen reflection problems. Several designs have been proposed.

Design mock-ups

The IPT will soon undertake a detailed costing analysis after which major changes to designs will not be possible. For this reason the HF team decided to physically mock-up the design options to allow a Navy team to participate in early usability testing.

Discussion

Using a variety of techniques, we have produced a validated list of user needs which, with existing HF guidelines and standards, have formed the basis of several operations room design concepts. The room's users have been engaged during data capture, needs validation and initial design. User feedback has then been incorporated by the team to produce well-received concept designs. A team usability evaluation has been designed and will be used to provide final feedback on the concept designs prior to the project's costs being set.

The processes and principles in ISO11064 and the DERA HFI process were used to design a custom process for the project. Customisation was necessary because the project was a redesign exercise, not a new requirement and because of business model constraints.

References

CoA. 2004: DEF(AUST) 5000: *Australian Defence Force Maritime Materiel Requirements Set.* Commonwealth of Australia. Canberra.
DERA, 2000: A Practical Guide to Human Factors Integration.
DoD. 1999: MIL-STD-1472F: Design Criteria Standard: Human Engineering.
Edwards, J. L. 2003: *LOCATE Analysis of Halifax Class Frigate Ops Room.* Document No.: DRDC Toronto CR 2003-124. Defence R&D Canada. Toronto.

Griggs, CAPT R, 2004: *Personal Communication. 11 May 2004.*

Hendy, K.C., Berger, J., and Wong, C., 1989: *Analysis of DDH280 bridge activity using a computer-aided workspace layout program (LOCATE).* DCIEM 89-RR-18.

Hendy, K.C., Edwards, J.L., Beevis, D. & Hamburger, T., 2000(a): Analysing advanced concepts for operations room layouts. *Proceedings of the HFES 44th Annual Meeting,* Santa Monica, CA, USA, Human Factors and Ergonomics Society.

Hendy, K.C., Edwards, J.L., Beevis, D., 2000(b): Modeling human-machine interactions for operations room layouts. P*roceedings of SPIE – "Integrated Command Environments" (Volume 4126),* Bellingham, WA, USA, SPIE, pp. 54–61.

ISO. 2000: ISO 11064: Ergonomic Design of Control Centres. International Standards Organisation. Geneva.

Kirwan, B., Ainsworth, L.K., 1992: *A Guide to Task Analysis.* Taylor & Francis. London.

MacMillan, J., Paley, M.J., Levchuk, Y.N., Serfaty, D, 2000: Designing the information space for a command center based on team performance models. *SPIE Conference 4126: Integrated Command Environments,* San Diego. SPIE, Bellingham, WA.

MoD. 2000: *Human Factors for Designers Of Equipment*: DEF-STD 00-25. UK Ministry of Defence. Glasgow.

Stanton, N. 2005: *Frontline: HFI DTC Newsletter no. 3.* at: http://www.hfidtc.com/public/pdf/frontlineedition03.pdf

TNO. 2005: *Operations room design for optimal team work.* At: http://www.tno.nl/

CCTV IN CONTROL ROOMS: MEETING THE ERGONOMIC CHALLENGES

John Wood

*CCD Design & Ergonomics Ltd,
95 Southwark Street, London, SE1 0HX*

The paper examines the ergonomic issues associated with the presentation of Closed Circuit Television (CCTV) images. Management expectations about users performance in using CCTV systems are compared to laboratory findings.

The development of a CCTV task taxonomy is discussed and the impact on operator performance of picture organisation is reviewed.

Introduction

Recent events in the UK have shown how much we now rely on the capture and playback of CCTV images. Scarcely a day passes without our news services screening images of robberies, acts of terrorism, dangerous driving or a town centre fight recorded on CCTV. Expectations have been raised that CCTV will provide a key to solving many of our society's problems. The media, however, give scant attention to those who have to work with these systems, monitoring their outputs and forensically analysing recordings. In this paper some of the ergonomic issues raised by the use of CCTV systems in control rooms are discussed.

Whilst all the ergonomic features highlighted in Figure 1 have a role to play in overall performance this paper will concentrate on some of the lessons we are learning

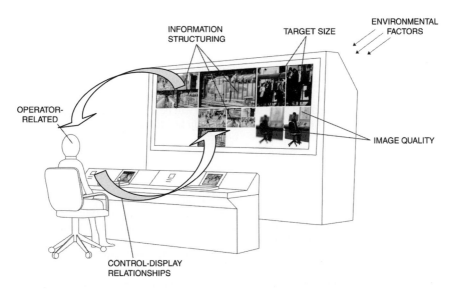

Figure 1. Ergonomics & the Control of CCTV Systems.

Table 1. Accuracy versus Monitor Numbers.

Monitor Numbers	1	4	6	9
Accuracy Scores %	85	74	58	53

about performance at CCTV picture observation, task taxonomies and the presentation of CCTV images.

Management aspirations & performance

Work by Tickner & Poulton, carried out in the 1970's, on the use of video images to monitor prisons highlighted operator limitations when looking at television scenes showing either a great deal of movement or little movement (Tickner et al, 1972 and Tickner & Poulton, 1973). Earlier work by the team had concluded, for short duration incidents on motorways, that the observation of more than one television picture would be likely to lead to failure to detect significant incidents (Tickner & Poulton, 1968) – note results reported in 1968!

The lessons from this early work seem to have been conveniently forgotten in the rush to install multiple camera systems whose output is typically presented on a wall of monitors. A bank of 30 monitors was recently counted during a visit to a town centre control room – to which must be added the 7 monitors on the workstation itself. The underlying potential for failure is compounded by management setting impossibly high standards where no faults are permitted – often by not actually specifying performance levels but then blaming operators when any failures occur!

Tickner & Poulton's work has been repeated by the Home Office, also under laboratory conditions, and examined the accuracy in target detection against the number of monitors being viewed. The target used was a man carrying crossing a busy road with an open umbrella at three possible locations in the picture – foreground, mid ground and in the distance. The open, large coloured golfing umbrella presented a clear and distinct target.

Subjects were seated 2 metres in front of a bank of 9 colour monitors. Subjects were asked to shout out when they observed a target, on any monitor and in any of the 3 locations in the picture.

Table 1 summarises the results from tests on 38 subjects who completed a 10 minute trial. Notably even with just a single monitor the best performance that can be achieved is 85% and this drops off rapidly to only 58% with 6 monitors (Wallace et al, 1997).

Management is deluded if it feels that operators will be able to reliably detect targets when viewing banks of 30 monitors for shifts of up to 12 hours!

CCTV task taxonomy

By grouping the observed use of CCTV from various applications CCD is developing a framework for a task taxonomy upon which measures of image "complexity" can be overlaid (Wood, J. & Cole, H., 2006). In ergonomic terms four primary uses have been identified for CCTV systems, Figure 2.

In a further breakdown of CCTV tasks activities such as "detection", "verification" and "recognition", are being linked to these primary tasks. The aim of the programme is to create a structure which will allow for the logical classification of CCTV tasks.

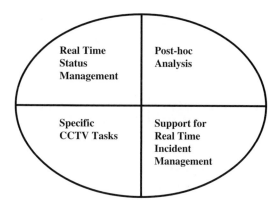

Figure 2. Primary Classification for CCTV Task Taxonomy.

Figure 3. Alternative CCTV Image Presentation Formats.

Picture organisation

Evidence is emerging that the way in which images are organised has an effect on operator performance.

In a programme to examine the most efficient way of looking at groups of CCTV images CCD tested different ways of presenting motorway pictures. The operator's task was to identify whether the "hard shoulder" was clear of obstructions before diverting traffic onto it. A 98% performance level had been set for this task. The ergonomist was tasked to determine the most efficient strategy for viewing images whilst also achieving the required levels of performance. With over 200 camera images to check a regime which required an excessively long time to complete ran the danger of hindering rather than aiding this congestion easing procedure.

The solution had to be based around the use of a single, control room operated monitor. Three alternative presentation methods were considered, see Figure 3.

Under laboratory conditions subjects were asked to view each image in turn and to record when a vehicle had been detected. The software was designed to record "response times" as well as "positive", "false positive" and "false negative" detections of vehicles.

The single image solution did provide very high levels of accuracy but the viewing sequences were excessively long. With 9 simultaneously presented images on a single screen the distraction caused by the surrounding images reduced detection performance. The "quadded" solution was that which was finally selected.

The final example of picture organisation is taken from a non-control room application. Train drivers are increasingly being asked to check that no one is trapped in the

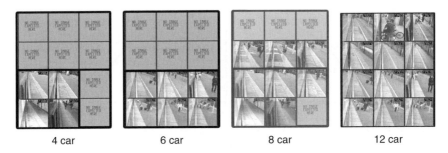

    4 car   6 car   8 car   12 car

Figure 4. Alternative arrangements of "Patches" on two Screens for 4, 6, 8 and 12 Carriage Configurations.

Table 2. Arrangement of CCTV Images and Detection.

Rolling Stock Configuration	Target detection reliability (%)
4 car	99
6 car	98
8 car	93
12 car	98
Overall	97

closing doors – a job that used to be carried out by the guard or station platform staff. This is made possible by bringing CCTV images of the platform – train interface into the cab itself. The train cab is an exceptionally constrained environment into which adding any extra equipment is a challenge. It was within this context that CCD was asked to examine the potential for introducing additional pictures ("patches") onto the two available flat screens.

The experiment examined the effects on driver performance of different "patch" arrangements on a set of cab-mounted CCTV screens. With two screens up to a 12 carriage formation could be catered-for providing the existing rail standards limiting the number on each screen to 4 could be relaxed. Without being able to prove the safety of this alternative arrangement only an 8 carriage train could be run within the existing constraints.

The images for train configurations of up to 12 carriages were presented in differing arrangements on the two monitors. "Patches" which were not being used were blanked off so that the driver could immediately identify a failed camera, Figure 4.

Under laboratory conditions drivers ran through simulations involving different length trains and using the CCTV images to detect a range of targets of varying difficulty. The study found that there would be no significant deterioration in target detection, as the number of scanned views increased to 12 though the time required to scan the images increased.

There was, however, an anomaly in the results which illustrates how the organisation of the patches could have a direct impact on accuracy in a detection task. It was found that with the 8 car arrangement of images performance was worse than for the 12 carriage presentation, see table 2. In retrospect it is clear that 4, 6 and 12 car presentation offered a tighter grouping and allowed for an easier series of fixations. It is proposed to examine this in more detail in future.

Mouchel Parkman and the Railway Safety and Standards Board (RSSB) are thanked for allowing work conducted on their research programmes to be cited.

References

Tickner, A.H., Poulton, E.C., Copeman A.K. and Simmonds D.C.V. 1972, *Monitoring 16 Television Screens Showing Little Movement*, Ergonomics, Vol. 15, No 3 279 – 291

Tickner, A.H. and Poulton, E.C. 1973, *Monitoring up to 16 Synthetic Television Pictures Showing a Great Deal of Movement*, Ergonomics, Vol. 16, No 4, 381 – 401

Tickner, A.H. and Poulton, E.C. 1968, *Remote Monitoring of Motorways Using Closed Circuit Television, Ergonomics*, Vol. 11, No 5, 455 – 466

Wallace, E., Diffley, C., Baines, E., and Aldridge, J., *Ergonomic Design Considerations for Public Area CCTV Safety & Security Applications*, Proceedings of 13th Triennial Congress of the IEA, Tampere, Finland, 1997

Wood, J., and Cole, H., *Image Complexity and CCTV operation*, Submitted to IEA Congress, 2006

DEFINING ERGONOMICS

ERGONOMICS ADVISORS – A HOMOGENEOUS GROUP?

Claire Williams & Roger Haslam

Health & Safety Ergonomics Unit, Department of Human Sciences, Loughborough University, LE11 3TU

A number of different professions apply "physical" ergonomics principles as part of their practice. In the drive to define and explain ergonomics as a field of practice, what professionals do in the name of ergonomics is important. This paper reports on a focus group study with professionals known to give advice on ergonomics issues as part of their work. Eight focus groups; two each of Ergonomists; Occupational Health Advisors; Health and Safety Advisors and Specialist Furniture/Equipment Suppliers were undertaken. Results show that there are differences between the professional groups in their aims, attitudes, perspectives and methods, when undertaking ergonomics activities. This paper focuses particularly on differences between the professions in their understanding of and their aims in using ergonomics. The implications of these differences are discussed.

Introduction

It is widely accepted that ergonomics is not solely the realm of Ergonomists but that a broad spectrum of different professions will apply ergonomics as part of their practice, and might be considered "Ergonomics Advisors". In the UK, it will certainly include Ergonomists, Health and Safety Advisors, Occupational Health Advisors and Physicians, Physiotherapists, Occupational Therapists, and Specialist Furniture Suppliers.

Some will carry out activities, such as work station assessments, or manual handling risk assessments as part of their ergonomics undertakings. This begs the question "will they all be delivering the same 'product', containing the same message, of the same quality and with the same outcomes?" Anecdotally, at least, the answer to this question is "no". In the ever present requirement to define the ergonomics field of practice (Wilson 2000; Ahasan & Imbeau, 2003) what other professions do in the name of ergonomics is an important aspect. In order to begin to build an answer to this question, the more fundamental issues underpinning the differences between these practitioners is also pivotal.

Given this situation, the aims of this study were twofold. First, to gain an appreciation of the different professionals' understanding of ergonomics and what ergonomics activities they carry out. Second, to understand the emphases and aims of the different professionals, by analysing their responses to a number of broad questions.

Methods

Eight focus groups were undertaken, involving a total of 54 participants. Sampling was stratified and purposeful, with each focus group containing participants from only one profession. The professions represented were: Occupational Health Advisors (OHAs) (n = 11); Health and Safety Advisors (HSAs) (n = 17); Specialist Furniture/Equipment Suppliers (SF/ESs)(n = 14); Ergonomists (n = 12).

The groups discussions were structured using a set of standard questions. The first focus group with each of the professions was recorded and fully transcribed. Template analysis (King, 2004) was the chosen method for analysing the groups' understanding of ergonomics. The template of themes was generated from the descriptions of ergonomics written by the IEA and the UK Ergonomics Society. The themes included the fact that ergonomics involves; understanding users, understanding jobs/tasks and the interactions between users and their jobs/tasks. It involves taking a user-centred, scientific, systems approach to design/assessment. It takes into account the cognitive, physical and organisational aspects with the aim of enhancing comfort, efficiency, productivity, safety and health, and can be applied in work and non-work environments. Any additional themes which emerged in response to the focus group questions were also noted.

The second focus group with each professional group was also recorded and analysed using the same template. This allowed for re-enforcement of themes which had already been covered by that profession, and for the addition of any new themes.

In addition, all 8 focus groups were asked about their aims when using ergonomics. Their responses were transcribed and the various aims given descriptive codes. These codes were then grouped under four headings. Each group was also asked to list the activities they undertook in which they used their ergonomics knowledge, in order to compare across the groups.

An external validator was present in each focus group to take notes of the key themes discussed and note emphasis. These notes were then used to check the trustworthiness of the findings represented here.

Findings

General

All of the focus groups responded to the question "What is ergonomics?" by giving a "text-book" definition of ergonomics, along the lines of *"fitting the task to the person"*. They all went on to make reference to most of the themes on the "ergonomics themes" template. Each of the professions cited a number of activities in common (Table 1) and each non-Ergonomist group had a strong sense of their limitations when it came to using ergonomics

> *"But again, when it's people with specific concerns, I have called Ergonomists in to come and have a look, because I've felt a little bit out of my depth...."* OHA

Health and safety advisors (HSAs)

For the HSAs, the workplace/employer organisation was the focus of much of the discussion and the target for the majority of their aims in using ergonomics.

> *"So in designing the plant right and making sure the kit's in the right place and you can operate it....then you can control it by having your control room set up in the appropriate way with the right amount of information... You're on-line, and if you're on-line in general, that's the safest condition you can be."*

The language tended to include "risk assessment" and technical terms and the goal often included finding "solutions".

> *"And we're doing a lot of work... trying to come up with solutions for designing out the risks. Or removing those risks... from the plant... trying to come up with*

Table 1. Sample of responses to "What activities do you do which you consider involve ergonomics?"

	Ergonomists	HSAs	OHAs	SF/ESs
DSE Compliance Assessments	•	•	•	•
Specialist Office Workstation assessment	•	•	•	•
Industrial Workstation Assessment	•	•	•	•
Site wide Ergonomics Assessment	•	•	•	•
Design	•	•	•	•
Product Sales	•	•	•	•
Tools/equipment advice	•	•	•	•
Management processes/Policy work	•		•	•

solutions on safe systems of work, basically, to prevent injury. "Cause this equipment is accessed on a regular basis, and… once you've accessed the equipment there's manual handling and ergonomic issues there. You're ducking under pipes and trying to turn valves that have all been put in at funny angles."

There was a strong emphasis on "systems thinking" in order fully to understand the nature of the jobs people were doing. There was less emphasis on understanding the person in the system, though the person was still important in their considerations.

"So what you're looking at is not just managing the routine, but managing the non-routine and looking at various scenarios, what sort of information would come to that person in what sort of time frame. What sort of time-frame they've got to react to that."

Occupational health advisors (OHAs)

The OHAs' spoke less about analysing the jobs people do, than the HSAs, though they did still cover this. However, their focus was very firmly placed on the person in the system. Much of the discussion and the majority of their aims were focussed around the individual.

"I want to get them… so that they can go home at the end of the day, not feeling any worse than when they came in."

The understanding of the user was broad ranging, and included taking into account the physical, psychological and the social attributes;

"I know I've been to places where… they're having some problems…from the way that …. they sit, but then when you actually start talking, there's a lot of office politics going on. They've been moved, they've done work and they actually end up in a smaller area to work in than they had before. So they're very unhappy."

Specialist furniture/equipment suppliers (SF/ESs)

Like the OHAs, the emphasis of the discussions of the SF/ESs was very much on the individual in the system, though the focus was almost completely on the physical. The aims were based on identifying solutions which, understandably, were always products in their specialist ranges.

"Primarily our work is reactive…to people who have musculoskeletal issues…part of what we do is go and assess the work environment as well as the

individual ... and then provide solutions by way of product to try and improve the way in which they work, but also their level of comfort."

Ergonomists

The Ergonomists had the most comprehensive discussions. They covered all of the template themes in both focus groups, with the exception of "productivity" which was not mentioned in either group. Only the Ergonomists covered the "scientific" nature of the field in both groups; *"it's the application of science....of knowledge about people"*; and talked explicitly about the interaction between the individual and their tasks, though this was implicit in the other groups' discussions. They were alone in discussing ergonomics as a philosophical approach, rather than just as a set of methods; *"it's a way of thinking about things ... it's an approach as well as a process"*. Overall their understanding had somewhat greater breadth, and much greater depth than the other professions, and their focus was not just at an individual level but higher up;

> *"....looking at the whole systems, not just the individual at a work station...you know looking at sort of the larger side of things. Looking at... "habitability" if it's a vehicle... It's not looking specifically at one individual, it's sort of all encompassing."*

In terms of their aims when using ergonomics, the Ergonomists' focus was relatively evenly spread over aims for themselves; *"it's about doing something I enjoy doing"* and aims for the user; *"making things better for people"*. However, the majority of their responses covered aims at a broader, "societal" level; *"I want to promulgate the idea that getting the ergonomics right is important"*.

Discussion

It is clear that the different professional groups represented in this study are neither homogeneous in their understanding, nor in their aims for using ergonomics. That said, it is noteworthy that all the professions demonstrated a relatively broad, if at times cursory, understanding of the key aspects of ergonomics.

The different emphases are likely to be the product of the role that each profession has "day-to-day". The SF/ESs have an overriding aim of selling products, and consequently their understanding and aims with respect to ergonomics are strongly biased around a physical, problem-solution model, in which their products are the solution for the individual's/employer's problems. However, for the most part, these professionals are brought into organisations when an HSA, OHA or Ergonomist has highlighted the need for an equipment based solution. Therefore, their physical model is likely to be perfectly adequate, and their superior product knowledge a great advantage.

The OHA's role in general, is one of facilitating the delivery of occupational health to individuals in an organisation, and will obviously be influenced by the fact that their background is in nursing, and is "patient-centred". This may explain why their ergonomics understanding focuses on diagnosing the problem in terms of the individual.

This contrasts with the HSAs, whose role often has more to do with ensuring the regulatory compliance and performance of an organisation or workplace as a whole. This, combined with the engineering/technical background of many of these professionals, may explain their emphasis on understanding the job and providing workplace engineering solutions, rather than on understanding the individual. There is the risk that this emphasis would prevent the true application of ergonomics with its user-centred imperative, however the HSAs participating in this study upheld this foundational tenet, whilst focussing on workplace changes.

The Ergonomists' greater depth and breadth of understanding, and more philosophical approach is likely to stem from the more intensive training they have received, and from the fact that ergonomics makes up all, not part of their role. However, rather than make the other professions seem inadequately equipped to carry out activities like DSE workstation assessments, the more extensive ergonomics knowledge of the Ergonomists seemed disproportionate to the requirements of these activities. Hignett (2000) discusses the fact that Ergonomists are skilled to deal with issues pertaining to working groups, organisations and general populations, rather than at the more individual level, and this was borne out in this study.

Some members of all of the professions stated that they might undertake broader ergonomics projects, such as site wide assessments. Given the understanding of ergonomics demonstrated during this study, these larger projects present the risk of under-representation of the field of ergonomics, thereby limiting the benefits accrued by its application, and showing ergonomics in a poor light. However, this concern should be mitigated by the fact that all groups had clear statements about when a problem required the input of an ergonomics specialist, and had mechanisms by which they sought this.

To be certain, the next step would be to look in the field at the assessments each of these groups made of similar situations, and see the outworking of their different understandings and approaches.

References

Ahasan, R., & Imbeau, D., 2003, Who belongs to ergonomics? an examination of the human factors community. *Work Study,* **52(3)**, 123–128.

Hignett, S., 2000, Occupational Therapy and Ergonomics. Two professions exploring their identities. *British Journal of Occupational Therapy.* **63(3),** 137–139.

King, N., 2004, Using templates in the thematic analysis of text, in C. Cassell and G. Symon (eds.) *Essential Guide to Qualitative Methods in Organizational Research*, (Sage Publications), 256–270.

Wilson, J. R., 2000, Fundamentals of ergonomics in theory and practice. *Applied Ergonomics,* **31(6),** 557–567.

ESTABLISHING ERGONOMICS EXPERTISE

Malcolm Jackson[2], David Hitchcock[1], Fari Haynes[2],
Dona Teasdale[2], Becky Hilton[2], Evelyn Russell[2],
Frances Flynn[2], Moira Coates[2] & Rachel Mortimer[2]

[1] *dh ergonomics & human factors, 96 Fairfield Road, Hugglescote, Leicestershire, LE67 2HG.*
[2] *Designing Workplace Solutions National Working Group, Psychology Division, DWP.*

The Work Psychology Division of the Department for Work and Pensions works with jobseekers/employees with a wide range of disabilities. To develop their skills, to increase awareness in the ergonomic discipline and improve confidence in its application, the Division embarked upon an internal programme of training and support for its work psychologists. This paper describes the approach taken, considers the benefits and difficulties encountered in the first 18 months and presents plans for the future. As such it offers a model of introducing and establishing ergonomics within an organisation.

Introduction and background

Work Psychologists (WPs) within Jobcentre Plus have a long-standing history, dating back to the early 80's in Employment Rehabilitation Centres to the present day, of delivering bespoke employment-related services to individuals with a disability/health problem who face complex barriers to finding work, or if in employment to help them retain their job. Currently WPs operate through a network of local Jobcentreplus offices and Access to Work Business Centres. More recently WPs have also become involved in supporting the delivery of the Government's Initiatives around improving the chances for sick or disabled adults to enter into the Labour Market. This builds on our long-standing experience of helping individuals on Incapacity Benefit find or retain suitable employment.

We provide support to individuals with a wide range of health problems/disabilities. For example an employee with memory difficulties resulting from a brain injury may need help in organising their workspace and in relearning current tasks.

We strive to apply our professional knowledge and skills of Occupational Psychology to help those who face considerable barriers in finding or retaining employment. As part of the consultancy and occupational assessment process, we may undertake an assessment of the individual's skills and abilities in relation to their job using a range of psychological approaches and methods, a workplace assessment utilising job and task analysis approaches and observational assessment, and a consideration of job re-design or redeployment options, depending on the particular needs of the individual and employer. We are then able to provide professional advice and guidance to the employer and employee on potential workplace support, solutions, strategies, and adaptations to help the individual retain employment. We are often able to identify creative, practical, and user-friendly solutions that meet both the employer and employees specific requirements.

One of the recommendations of the Government's Green Paper "Pathways To Work" is for advisors, including WPs, to "identify solutions in the workplace", to help those on Incapacity Benefit remain in employment. Given this new emphasis and our long-standing commitment to continually improving our practice, we identified a developmental need as practitioners to enhance our skills and knowledge in this field, and to disseminate this knowledge to other Jobcentre Plus specialist advisors.

As already mentioned, we have long recognised the lack of any formal internal training in this area for WPs within Jobcentre Plus. Most WPs are recruited with a Masters Degrees in Occupational Psychology that usually includes an Introduction to Ergonomics, but this is often not sufficient to adequately cover both the scope and depth of work psychology interventions in this field.

Strategic model

Adopting a consistent approach with other training and mentoring programmes, the Psychology Division of the Department for Work and Pensions approached an ergonomics consultant, David Hitchcock, to develop a three-tier model to meet our needs. This model had to include:

Tier 1 provision: "SURVIVAL GUIDE" to provide new starters with a quick reference pack to enable them to react positively, confidently and accurately to new ergonomics-related challenges which they may face in their new work.

Tier 2 provision: "TRAINING" on a national rollout basis to provide a rounded, holistic and comprehensive training course (and, as appropriate additional training modules) to substantiate and extend the "survival" provision. The purpose of the training was not to turn the attendees into fully-fledged ergonomists, but to add an ergonomics perspective and armoury to their existing skills base (which may already include some ergonomics, to a greater or lesser degree).

Tier 3 provision: "A DEVELOPING WORKPLACE SOLUTIONS" LEAD GROUP to further facilitate the development of skills and knowledge in this area. and promote ergonomics within the Work Psychology Service.

One of David's first tasks in order to satisfy both WPS and his own requirements, was to conduct a national training needs analysis of all Work Psychologists in Jobcentreplus. Twenty one WPs responded to his survey and it revealed that WPs in Jobcentreplus felt that the following topics or issues needed to be covered as core to any training package. They included: task analysis, assistive technology, disability, ageing and inclusivity, IT accessibility, ergonomic tools and techniques, holistic ergonomics, and posture and movement. (see Figure1 below)

The aim of this training package was to up-skill new WP's and provide experienced WP's with the opportunity to build on their knowledge and experience, providing further opportunities for continuing professional development.

Tier 1 – The survival guide

The guide was prepared on the proposition that effective ergonomics requires the person responsible for its implementation to have two things:

1. An understanding of the ethos of ergonomics – this provides a basis from which all solutions must be derived, without it, the application of details are typically unsuccessful. This is the part where the "correct questions need to be asked".
2. Access to details such as user expectation, size and equipment demands which can be considered and applied to any generated solutions. This is the part where "the correct answers need to be provided".

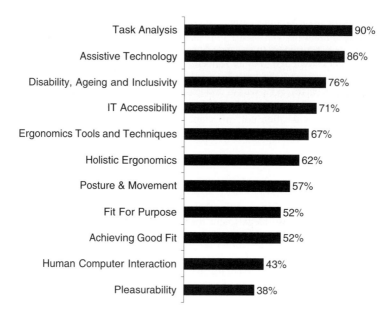

Figure 1. Training Course: Preferences of WPs.

Even at the "survival" stage these two elements were considered vital, albeit in a condensed, quick-easy-reference, and self-contained format. The guide was ultimately prepared as a "Flash" presentation and distributed via our internal intranet, accompanied by the introductory reference book "Ergonomics for Beginners"[1].

Tier 2 – The training course

The preferences revealed in David's survey formed the basis of what became a 5-day course introducing a number of issues/topics. The teaching sessions were complimented by a course project; the key elements of which were:

- It began and ended during the 5-day course, capitalising on the expert support of the tutor.
- Recipients were made aware that the project and practical work were not "fillers", but actually formed the major learning component of the course.
- It ran in parallel to the teaching, to enable immediate and in-context application of theory.

The response to the training sessions has been positive and enthusiastic from all delegates. The Work Psychologist (WP) participants have commented on the impact it has had on their professional development and practice.

Tier 3 – Provision

The "Designing Workplace Solutions National Working Group" was introduced some 12 months later, with the aim of further enhancing Work Psychologist effectiveness in

---

[1] Dul, J. and Weerdmeester, B. (2001). Ergonomics for Beginners – A Quick Reference Guide. Taylor and Francis, London. ISBN 0748408258.

managing employment interventions, contributing to national and regional understanding and thinking on "workplace solutions", and increasing operational performance. The terms of reference for the working group included the introduction of standard operating models in relation to policy, training, and new tools and solutions, and the evaluation of employer retention interventions. It was also charged with supporting the Ergonomics consultant, David Hitchcock, in delivering a master class within the next 12 months on some new ergonomic tools, techniques or models applicable to WP interventions, to a selected audience.

The future

Our viewpoint of our future would encompass three main strands or themes to the further development of Work Psychologist skills in these fields.

Firstly, there is a need to continue to develop the initial training course we have set up, in light of delegate and practitioner comments on perceived current and future needs. Future courses may need to include skills and processes elements such as customer engagement, relationship handling, and disengagement. There might also be a place for developing familiarity and confidence in tools and other techniques to further assess customer needs. The proposed programme of future master classes is likely to lead this developmental stream.

We have also identified a number of issues in regard to our handling of disabled employees and their employers. WPs will need to address issues around engagement, managing expectations, and disengagement when working with disabled employees. There is also a need to develop explicit guidelines and codes of ethical practice for WPs to follow including recognising our limits in regard to DDA and other employment law.

We feel that there is an important need to develop a "toolkit" of methods and techniques to help us usefully assess customers' abilities and their working environments, to identify miss-matches and suggest creative solutions to them. WPs recognise that in the new labour market, traditional job analysis techniques will be insufficient to meet this need (see Gosden and Birkin, 1999). There is a need on our part for new assessment tools/techniques, which combine both the assessment of people's abilities and their current working practices and environments. One possible solution may lie in further development of the Activity Matching Ability System (AMAS), a system originally developed for the steel industry, to support the placement of disabled employees in more effective and productive employment (Watson, Whalley and McClelland, 1990). This system has been later developed into an IT supported system, now known as AbilityMatch, that has been used with some success to enhance outcomes of employment assessments of disabled employees by WPs in our service (Birkin et al, 2004). Further developments required are likely to include the development of databases of assistive/adaptive technologies and the interaction of individual abilities and task/role design.

Lastly and very importantly there is a need for WPs like ourselves to look outside of our own immediate area of work, to communicate with others experts like ergonomists, rehabilitation specialists, etc., who have much to contribute to our further development, and to take up as many opportunities as we can, to network with as many as possible. That way we can ensure we make our full contribution to the engagement of disabled people in the future economic prosperity of our society.

References

Birkin, R., Haines, V., Hitchcock, D., Fox, D., Edwards, N., Duckworth, S., Gleeson, R., Navarro, T., Hondroudakis, A., Foy, T., and Meehan, M. (2004) Can the Activity

Matching Ability System Contribute to Employment Assessment? An Initial Discussion of Job Performance, and a Survey of WPs' Views. *Journal of Occupational Psychology, Employment and Disability*, Vol 6, 2, 51–66.

Gosden, D., and Birkin, R. (1999) *New Horizons for Job Analysis*. Occupational Psychology Division. Professional Note 207. Sheffield: Employment Service

Department for Work and Pensions (2002) *Pathways to Work – Helping People into Employment*. Cm 5690. London: TSO.

Department for Work and Pensions (2003) *Pathways to Work: Helping people into employment.* The Government's response and action plan. Cm 5830. London: TSO.

Watson, H., Whalley, S., and McClelland, I. (1990) Matching Work Demands to Functional Ability. In Bullock, M. (Ed.). *Ergonomics: The Physiotherapist in the Workplace*, 1990. (Churchill Livingstone, Edinburgh), 231–257.

PUTTING THE "DESIGN ENGINEERING AND SPECIFICATION" BACK INTO HUMAN FACTORS

Kee Yong Lim

Centre for Human Factors and Ergonomics, School of Mechanical & Aerospace Engineering, Nanyang Technological University, Nanyang Avenue, Singapore 639798

> This paper presents observations concerning human factors (HF) contribution to design that derive from the author's involvement in HF education and practice over fifteen years. The observations, taken together, support a number of assertions pertaining to the main underlying causes for the limited uptake of HF input. To address this poor state of affairs, the paper proposes 3 initiatives that should be advanced, namely the development of procedural HF knowledge; the definition of a broader design scope to support effective HF involvement in cross functional design development teams; and a review of existing HF texts to enhance their support for application, education and training in design specification.

Background

HF practitioners have continued to be unable to contribute effectively to product and system design (Stone 2005). Historically, HF practitioners have also gained an unflattering reputation of being concerned only with problem diagnosis (as opposed to the prescription of a design solution), and for being too vague in their design recommendations (see example in Figure 1). Specifically, its design recommendations and proposed modifications have been found to be difficult to accommodate due to two main reasons, namely:

- HF contributions often comprise recommendations pitched at the level of advice only rather than the specification of an explicit design solution (see Lim 1996).

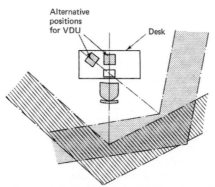

Ceiling lights must not be placed in the area marked with a diagonal stripe for VDUs directly in front of the operator, or in the stippled area for VDUs placed diagonally left of the operator.

Figure 1. "Phantom" values: how can one specify the exact location of the lights?

- due to its restricted involvement late in the design process (i.e. late evaluation as opposed to early design specification), problems relating to HF tend to be discovered too late and/or only after commitment to a specific solution. The outcome is the "too-little-too-late" problem of HF input (Lim & Long 1994).

Three initiatives need to be undertaken to address these problems and to ensure commercial relevance of HF in a world characterized by intense global competition. An account of these initiatives follows.

Advancing human factors contribution to design specification

To enable effective HF contribution to design, three initiatives need to be undertaken:

1. Advancement of procedural HF knowledge, in particular, to establish explicit and comprehensive HF design processes and methods to support design specification as opposed to summative/late design evaluation. This initiative has begun in the late eighties, culminating in the development of a number of comprehensive and explicitly structured HF methods that support design specification (see Damodaran, Ip & Beck 1988, Sutcliffe & Wang 1991, Lim & Long 1994). Further, efforts have since been made to integrate these HF design methods with similar methods from related disciplines. In this way, timely and sustained HF input may be facilitated throughout the process of design development.
2. Definition of a broader scope of HF design to support more effective involvement in cross functional design teams. In particular, requirements mandated by intense global competition have increased the emphasis on concurrent engineering and design, and thus the deployment of cross functional teams to ensure effective design development. A cross functional team typically comprises representatives from marketing, engineering and manufacturing (Clark & Wheelwright 1993). As HF concerns cut across all of these functions, support for its design input to these functions should be developed. Specifically, HF input to these functional areas would include the following:
    - *Marketing function:* product definition, user requirements of specific market segments, etc.
    - *Engineering function:* functional definition, usability and interaction design, design/selection of connectors etc. (e.g. press or screw fit) to facilitate product assembly and so reduce repetitive strain injury, design of the product architecture to ensure ease of maintenance, etc. HF design contributions would thus encompass concerns relating to both internal customers (manufacturing and maintenance staff) and external customers (end users of products).
    - *Manufacturing function:* operator workstations, process lines and work environment to increase productivity, quality of working life and occupational health & safety.

   In view of the broad concerns of product development above, it is no longer surprising that HF design contributions have been criticized for being too narrow in scope, and for failing to consider the product/system as a whole; e.g. its pre-occupation with design function and usability at the expense of aesthetics and commercial feasibility; the tendency of HF practitioners to specialize narrowly in their background disciplines instead of applying a wholistic and multi-disciplinary view on the design problem. To contribute effectively to design specification, HF practitioners need to address the concerns circumscribed by the Design For <X> or DFX, where <X> may represent design concerns relating to manufacturability (e.g. ease of assembly), safety, maintainability (e.g. ease of access) and sustainability. In other

Putting the "design engineering and specification" back into human factors 129

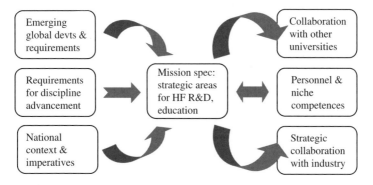

Figure 2. A scheme for developing strategic competences in human factors.

words, HF practitioners have to take an integral view and address both micro- and macro-level design concerns of product and system development. Failing this, the design envelope might be drawn too small, leading to an inappropriately narrow definition of the design problem. The end result might be the specification of sub-optimal design solutions, and thus ineffective product competition; or worse, a failed enterprise.

HF researchers and practitioners should also track emerging socio-economic trends and technological developments, so as to pre-empt or predict future customer needs and the socio-technical requirements and implications of nascent technologies. By tracking these developments, HF researchers and practitioners might be better positioned to capitalize on emergent opportunities, and thus ensure that their contributions remain state-of-the art, timely and relevant (see Lim 1999, 2002). Figure 2 shows a high level schematic of how such an initiative may be advanced. By looking ahead in this way, it might be possible for HF to move away from typical retroactive input and post-hoc rationalization of new designs (e.g. Cognitive Ergonomics has been used to explain the reason for graphical user interfaces being so effective (it exploits recognition instead of relying on recall), but played little or no role in its innovation), and make a case for pro-active and complete involvement in design from project inception.

3. Review and revamp (if necessary) existing HF texts to enhance their support for application, education and training in design specification. In particular, fundamental knowledge on how to "engineer" a HF design solution should be addressed and presented explicitly and not presumed. In this respect, HF texts and standards should be written meticulously to provide comprehensive support for design specification. In particular, they should go beyond design checklists and face value recommendations that might lead to "blind" or inappropriate application by designers with inadequate HF knowledge. A pervasive provision of face value recommendations would reduce HF to little more than a craft discipline. Unfortunately, established HF texts have been found to be inadequately meticulous and inconsistent in explicitness in their account of the design considerations and/or calculations involved in deriving a HF recommendation or value. For instance, Figures 3 and 5 show recommended values without due account of how the values have been calculated and differences across user populations. Not only must questions be raised concerning potential erroneous application of the values recommended, but also the validity of the cross comparison implied (see Figure 5). Without the necessary information, HF practitioners would not be able to make appropriate decisions, let

RECOMMENDATIONS FOR SEATED WORK-SURFACE HEIGHTS FOR VARIOUS TYPES OF TASKS

	Male		Female	
Type of task (seated)	in	cm	in	cm
Fine work (e.g., fine assembly)[1]	39.0–41.5	99–105	35.0–37.5	89–95
Precision work (e.g., mechanical assembly)[1]	35.0–37.0	89–94	32.5–34.5	82–87
Light assembly[1]	29.0–31.0	74–78	27.5–29.5	70–75
Coarse or medium work[1]	27.0–28.5	69–72	26.0–27.5	66–70
Reading and writing[2]	29.0–31.0	74–78	27.5–29.0	70–74
Range for typing desks[2]	23.5–27.5	60–70	23.5–27.5	60–70
Computer keyboard use[3]	23.0–28.0	58–71	23.0–28.0	58–71

Sources: [1] Ayoub, 1973; [2] Grandjean, 1988; [3] Human Factors Society, 1988.

Figure 3. Questionable collation without basis of data values from different sources.

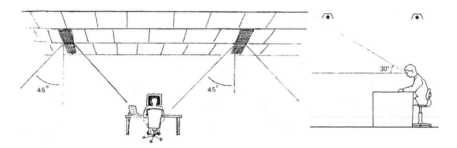

Figure 4. A design case where a working knowledge of trigonometry is necessary.

RECOMMENDATIONS FOR VDT WORKSTATION ADJUSTMENT RANGES FOR KEYBOARD AND TERMINAL

Dimension	Human Factors Society 1988	Lueder 1986a
Keyboard height (floor to home row)	23–28 in (58.5–71 cm)	24–32 in (61–81.5 cm)
Keyboard angle	0–25°	0–25°
Screen position		
Angle below horizon	0–60°	
Support surface height		24.4–35 in (62–89 cm)
Screen angle to vertical		
Ideal		+/− 20°
Minimum		+/− 7°

Figure 5. False precision and meaningless specification of values to decimals.

alone extend the design recommendations and/or modify the specified values as appropriate to suit alternative situations and contexts.

Basic HF texts should also include a working knowledge of mathematics, engineering, computing, industrial design and management of product development. Figure 4 shows a case of HF design specification that would require a working knowledge of mathematics (trigonometry). Without such knowledge, a HF practitioner would only be able to advise a client the value recommended, rather than specify explicitly the exact location of the lights in a given room. Further cases will be reviewed at the time of conference presentation.

A working knowledge of engineering and computing would be critical to a HF practitioner involved in accident reconstruction and simulation, and in user interface design and prototyping. Even more basic is a practitioner's knowledge of the accuracy that can be justified in a design specification. Qualifications (if any) of a specified value should always be indicated to ensure credibility. In particular, HF practitioners may not understand the meaning of significant figures in relation to the accuracy of a computed or measured value. A case in point is anthropometric design, in which data are tabulated with standard deviations for specific population percentiles corresponding to notional users (a person with such a set of anthropometric dimensions certainly does not exist). In applying these data, HF practitioners often fail to appreciate the limits of precision in relation to the significant figures that may be supported in a design specification. For example, it might not be meaningful to specify a design parameter to the lowest level of an integer unit (i.e. 0 to 9) let alone to decimal places (see Figure 5). The limit in accuracy justifiable in the specified value becomes clearer if one were to also consider the effect of notional design allowances (e.g. for shoes) and the variety of slump and seated postures adopted by users. To see a stark contrast in design viewpoint, the reader is referred to Grandjean (1992) and Pheasant (1996). In simple terms, anthropometric design by Grandjean would include behavioural considerations, while Pheasant's approach would be more deterministic. Nevertheless, Pheasant acknowledged that the accuracy of a specification is rarely tighter than an anthropometric "inch" of 25 mm.

It is the author's hope that by highlighting existing inadequacies, future HF texts may be written with keener consideration of the requirements for supporting HF design. This is the only intention of the paper.

References

Clark, K.B. & Wheelwright, S.C. (1993). *Managing New Product and Process Development.* Free Press.
Damodaran, L., Ip, K. & Beck, M. (1988). Integrating HF principles into structured design methodology: a case-study in the UK civil service. In: Bullinger H.J. et al (eds.), *Information Technology for Organizational Systems.* Elsevier Science.
Grandjean, E. (1992). *Ergonomics in Computerized Offices.* Taylor & Francis.
Lim, K.Y. (1996). Structured task analysis: an instantiation of the MUSE method for usability engineering. *Interacting with Computers,* 8(1), Elsevier Science.
Lim, K.Y. (1999). HF prospects in the new millennium: golden age or sunset? Keynote paper in: *Proc. Australian Ergonomics Society 1999 Conference,* Fremantle.
Lim, K.Y. (2002). Case file of HF practice: how well we do is a case of where we draw the system envelope for multi-disciplinary design. *Proc. International Ergonomics and Sports Physiology Conference,* Bali, Indonesia.
Lim, K.Y. & Long, J.B. (1994). *The MUSE Method of Usability Engineering.* Cambridge University Press, U.K.
Pheasant, S. (1996). *Bodyspace: Anthropometry, Ergonomics and Design.* Taylor Francis.
Stone, B. (2005). Ergonomist, heal thyself, or why is it we're still finding it hard to sell our human-centered wares? Donald Broadbent lecture, ES 2005 Conference, Hatfield, UK.
Sutcliffe, A. & Wang, I. (1991). Integrating human-computer interaction with Jackson System Development. *The Computer Journal,* 34(2), Cambridge University Press.

WHAT DID YOU DO IN THE WAR HYWEL? SOME FOUNDATIONS OF ERGONOMICS

R.B. Stammers

School of Psychology, University of Leicester, Leicester, LE1 7RH

The work of Hywel Murrell for the Army Operational Research Group during and just after the Second World War is overviewed in the light of his subsequent role in the founding of ergonomics. Over 20 of Murrell's reports dated between 1944 and 1946 have been located in the UK National Archives and their contents reviewed. Despite an emphasis on "time-and-motion" type activities, important features of equipment design, procedures and training were also being identified. At the same time, a recognition for the involvement of human science specialists early on in the design cycle was being promulgated and some successes were achieved.

Introduction

It is widely agreed, both in historical and autobiographical accounts, that the Second World War (WWII) was one of the major influences on the discipline of ergonomics. Individuals from a range of backgrounds became involved in work supporting the war effort. The subject of this paper, Hywel Murrell, began the War as an army officer. By the end of the period under study he was on his way to becoming a government civilian scientist, heading what was to become one of the first ergonomics groups.

Murrell's role in founding the Ergonomics Society and in coining the term *Ergonomics* has been widely reported (Waterson & Sell, in press). His developing interest in the subject came about through his war time work as a serving officer with the Army Operational Research Group (AORG). Early accounts of ergonomics have made reference to this work, but its nature has never been reported in detail. Most accounts of ergonomics in the UK draw attention to the year 1949 and the formation of the Ergonomics Society and its first meetings. However, it is interesting to explore what went on before this date, to see what provided the grounding for the discipline that was subsequently to emerge. The present author has a particular interest in the development of ergonomics before 1949. When the period immediately preceding 1949 is explored, the role of Murrell emerges as a key one.

Murrell has left us with a number of autobiographical accounts of his role in the development of ergonomics (Murrell, 1953; 1980; 1985). He graduated in chemistry from Balliol College, Oxford in 1930. He then worked in the field of printing for a few years in London, moving in 1935 to a carton making company in Bristol. He learned about time-and-motion study with this company and also about photographic techniques for industrial printing. The outbreak of WWII led to work at the company ceasing and subsequently he volunteered for the Army. He was commissioned into the Royal Engineers in January 1940 and became involved in map printing activities. A subsequent move to a training role was less satisfactory for him and he responded to a call that went out for officers with experience of time-and-motion study. Thus, in 1944 he became a member of the Motion Study Wing of the Army Operational Research Group. The Wing was headed by a psychologist, with at least one other psychologist from the National Institute of Industrial Psychology as a member. The group were tasked with improving the efficiency of human users of military equipment *etc*. Murrell worked with this group through to just after the end of WWII. At that time, work carried out for the Navy, whilst still a member of the AORG, was the first phase of his involvement in the subsequently emerging Naval Motion Study Unit.

This paper will focus on what it has been possible to discover of his work between 1944 and 1946 at the AORG. Murrell's own accounts are relatively brief and it appears that his work for the Admiralty was seen by him as more mainstream in its contribution to ergonomics. Although the title of this paper indicates consideration of work carried out during WWII, some of the activities took place after the end of the War. The paper therefore focuses on work carried for AORG and its successor. Documentary sources have been found in the form of over 20 reports. These reports, authored or co-authored by Murrell between 1944 and 1946, have been located in the National Archives. Review of their content gives an indication of the direction that his work was taking at this time.

The reports

These reports have to be taken for what they were. They were written with a specialist focus and for a restricted audience. They were also written under, in the main, wartime pressures and as such did not closely resemble the technical reports of more recent times. Their importance lies, not so much in the particular tasks and equipment that they focus on, but as records of the range of work situations encountered, the way that the work was approached and reported, and the way that state-of-the-art film technology was used to collect field data. These reports were probably only released into the public domain in the 1970s, as many of them carried a high security classification. However, they may have been available to the professional community before then. Some of the reports (the memorandum type) would have had only limited circulation. It is possible that there are other reports by Murrell during this period, but they have not been located in the National Archives. However, it is felt that those that have been found provide a reasonable representation of the type of work that was conducted.

The studies, carried out under a "motion study" guise are, on the surface, just aimed at improving procedures and times for activities. However, from quite early on it is clear that issues to do with equipment design and location *etc.*, were also being routinely commented on. The earliest reports are based on straight forward observational work of military personnel in various simple tasks (Murrell, 1944a; 1944b; 1945a). A move was quickly made to a more detailed series of studies of personnel carrying out ammunition loading tasks in armoured fighting vehicles (Murrell, 1945b; Murrell and de la Riviere, 1945a; 1945b). In these studies, advanced techniques of film recording were used. Whilst the focus was still on procedures, such topics as anthropometric fit of users is routinely commented on. Additionally, opportunities are taken to compare different sets of equipment.

A standard approach is adopted utilising film recording of three or more participants, individually carrying out the loading tasks. The film records are used to generate data that enable comparison between different equipment fits and different vehicles. In addition, drills are examined for ways in which they could be made more effective. Attention is also placed on the importance of training in those drills to enhance performance. Opportunities arose for the evaluation of equipment at prototype stages (Murrell and de la Riviere, 1945c; 1945d). Under these circumstances, quite detailed recommendations on redesigning equipment and improving drills are provided. The hope was, presumably, to influence the final design.

Other issues that emerged in the reports include the recognition of the importance of seating height in vehicles (Anonymous, 1945); trade-offs in design between freer movement, but less armour protection (de la Riviere and Murrell, 1945); and the location of controls for ease of operation (Murrell and de la Riviere, 1945e). A report over viewing these studies (Murrell, 1945c), is an early example of drawing out general design principles on the basis of a number of related studies.

The team may have felt that some progress was being made, as some later reports are based upon the evaluations of prototype or mock-up equipment (Murrell and de la Riviere, 1945f; Murrell and Edwards, 1945a; Murrell and Edwards, 1946a). As a result of these studies, suggestions were made for re-design. In one study, physical injury to participants resulted from one set of prototype equipment, leading to a very strongly worded recommendation! The reports increasingly used photographic illustrations of awkwardness or the effects of bad equipment fit on the users.

The importance of effective task sequences or drills is stressed in a number of reports. Demonstrations of time improvements for improved drills are provided, as well as an emphasis being placed on the importance of training (Cooper and Murrell, 1945; Murrell and Edwards, 1945b; Murrell, 1946).

One report on the interior design of a vehicle draws attention to the importance of early involvement in the design cycle, pointing out that a number of features could not be modified given the stage of development that had been reached (Murrell et al, 1946). This report is also noteworthy for collaboration with Medical Research Council seating specialists. The point about the early involvement in the design cycle was picked up in a subsequent report (Murrell and Edwards, 1946b), where a mock-up and a prototype vehicle were available for study. A number of features emerged and the authors indicate: "it must be emphasised that the vehicle was still in the hands of the designers when we inspected it and that some of the defects mentioned ... may be remedied before the vehicle is officially handed over" (p. 1). A final report in this series is notable in that it involved filming of participants whilst a vehicle was in motion. In the study, some very sophisticated photographic equipment was used to collect data for subsequent analysis (Murrell et al, 1946).

A key point in Murrell's professional life was his move from working for the Army to working for the Admiralty. This began with an invitation to carry out a study of naval gun drills. Using a complex filming method, detailed recordings were taken on board a warship. A number of suggestions for improving drill and equipment were made as a result of this study (Murrell and Carr-Hill, 1946). This evolved into the subsequent secondment of Murrell to the naval gunnery school. Whilst still officially working for the Army, a final report was produced on a naval command-and-control centre (Murrell and James, 1946). This report is very different to any other that had been produced, focusing on psychological demands of tasks, in addition to their physical demands. This aspect of his work marks a transition point between the Army and Navy work and is a useful point to conclude.

Conclusions

The work Murrell undertook at the AORG provided a strong basis for his subsequent development of ergonomics. Hands-on experience of people carrying out a of range complex and demanding tasks was gained. Technical experience in the collection and analysis of data was also important. A recognition of the central importance of the user can be summed up by the following quotation. It concerns the aim of a study which was to determine whether, "stowage was satisfactory from a *user viewpoint*, and whether the controls and seating were correctly positioned" (Murrell and Edwards, 1946b, p. 1, italics added). In addition, the importance of the involvement of human sciences specialists early in the design cycle, nowadays a cliché for ergonomists, comes through strongly in the reports. The ground was thus set for his transfer to the Scientific Civil

Service, as head of the Naval Motion Study Unit. It was in this context in the late 1940s and early 1950s that more specifically ergonomics studies were conducted. But that is another story.

References

(Notes: to save space, full publication details for reports have not been given. The abbreviations AORG and MORU stand for, Army Operational Research Group and Military Operational Research Unit respectively. Place of publication is not indicated on the original reports, but can be assumed to be London. The numbers in square brackets at the end of the report listings refer to the document numbers in the UK National Archives, Kew, London)

Anonymous, 1945, *Report on ammunition stowage and loading of 105mm gun in the Sherman 1B*, AORG Report No. 278. [NA WO 291/258]

Cooper, J.S.P.C. & Murrell, H. 1945, *Report on ammunition stowage and loading of 75mm gun in Churchill VII,* AORG Report No. 295. [NA WO 291/273]

de la Riviere, D. & Murrell, H. 1945, *Report on ammunition stowage and loading of the 90 mm gun in the General Pershing (T26) tank,* AORG Report No. 290. [NA WO 291/268]

Murrell, H. 1944a, *Studies of working methods employed in ammunition depots: Study No 3,* AORG Memorandum 428. [NA WO 291/752]

Murrell, H. 1944b, *Studies of working methods employed in ammunition depots: Study No. 6,* AORG Memorandum 433. [NA WO 291/755]

Murrell, H. 1945a, *Studies of working methods employed in ammunition depots: Study No. 10,* AORG Memorandum 459. [NA WO 291/779]

Murrell, H. 1945b, *Report on ammunition stowage and loading of the 77mm gun in Comet tank,* AORG Report No. 267. [NA WO 291/248]

Murrell, H. 1945c, *Report on stowage of ammunition in tanks, with special reference to 17 pr.,* AORG Report No. 281. [NA WO 291/261]

Murrell, H. 1946, *Memorandum on the method of loading the 77mm gun in the Comet*, AORG Memorandum No. 502. [NA WO 291/817]

Murrell, K.F.H. 1953, Fitting the job to the sailor, *Occupational Psychology*, 27, 30–37.

Murrell, H. 1980, Occupational psychology through autobiography, *Occupational Psychology*, 53, 281–290.

Murrell, H. 1985, How ergonomics became part of design. In N. Hamilton (ed.) *From Spitfire to Microchip: Studies in the History of Design From 1945,* (The Design Council, London), 72–76.

Murrell, H., Campbell-Jones, J.B. & Mound, S.H.C. 1946, *Motion study of proposed ammunition stowage in A45 with particular reference to loading on the move in a vehicle stabilised in azimuth*, MORU Report No. 24. [NA WO 291/966]

Murrell, H. & Carr-Hill, E.A. 1946, *Time and motion study of six inch naval gun drill*, MORU Report No. 11. [NA WO 291/954]

Murrell, H. & de la Riviere, D. 1945a, *Report on ammunition stowage and loading of the 77 mm gun in modified Comet tank,* AORG Report No. 270. [NA WO 291/251]

Murrell, H. & de la Riviere, D. 1945b, *Report on ammunition stowage and loading of the 17 pr. gun in Sherman Vc,* AORG Report No. 273. [NA WO 291/253]

Murrell, H. & de la Riviere, D. 1945c, *Report on ammunition stowage and loading of the 17 pr. gun in Black Prince,* AORG Report No. 276. [NA WO 291/256]

Murrell, H. & de la Riviere, D. 1945d, *Report on ammunition stowage and loading of the 17 pr. gun in A41,* AORG Report No. 277. [NA WO 291/257]

Murrell, H. & de la Riviere, D. 1945e, *Report on ammunition stowage and loading of the 17 pdr. gun in the SPI,* AORG Report No. 291. [NA WO 291/269]

Murrell, H. & de la Riviere, D. 1945f, *Report on ammunition stowage in mock-up of SP4,* AORG Report No. 293. [NA WO 291/271]

Murrell, H. & Edwards, N. 1945a, *Ammunition stowage and loading of 17 pdr. gun in SP2,* AORG Report No. 301. [NA WO 291/279]

Murrell, H. & Edwards, N. 1945b, *Report on ammunition stowage and loading of the QF 25pdr. Mk II gun on ram chassis (SP 25pdr. Sexton),* MORU Report No. 1. [NA WO 291/945]

Murrell, H. & Edwards, N. 1946a, *Report on experiments to determine the optimum conditions for the stowage and loading of a gun in a proposed medium SP,* AORG Report No. 302. [NA WO 291/280]

Murrell, H. & Edwards, N. 1946b, *Motion study of stowage and controls in A39 (the Tortoise),* AORG Report No. 315. [NA WO 291/291]

Murrell, H. & James, J.B.C. 1946, *Motion study of an action information centre (Cruiser/Battleship),* report prepared for HMS Excellent. [NA ADM 219/581]

Murrell, H., Tunnicliffe, G. & Foss, B. 1946, *Report on stowage and fittings of carrier CT20,* AORG Report No. 314. [NA WO 291/290]

Waterson, P. & Sell, R. in press, Recurrent themes and developments in the history of the Ergonomics Society, *Ergonomics.*

DESIGN

MAXIMISING VMC: THE EFFECTS OF SYSTEM QUALITY AND SET-UP ON COMMUNICATIVE SUCCESS

Orsolina I. Martino, Chris Fullwood, Sarah J. Davis,
Nicola M. Derrer & Neil Morris

University of Wolverhampton, School of Applied Sciences, Psychology, Millennium Building, Wolverhampton WV1 1SB, UK

Video-mediated communication (VMC) is becoming more widely used in workplace and conference settings, with a variety of different systems available. Whilst there are methods of improving VMC regardless of the system used, the efficacy of any system is largely mediated by its quality and set-up. This review considers the different forms of VMC, discussing the impact that factors such as audio quality, delay, and image size and quality have on communication. Recommendations are made for good practice when working with video-mediated technologies, using theoretical and empirical evidence to suggest ways of maximising available resources for optimal communication.

Introduction

The increasing popularity of video-mediated communication (VMC) has led to a growing body of literature examining the efficacy of such technology. Two distinct forms of VMC are becoming popular: desktop videoconferencing (DVC) and videoconferencing supported by specialist conference centres. In DVC a camera is placed near the user's computer monitor and a telephone line or Local Area Network (LAN) is used to connect with other users; this method also enables other functions to be performed simultaneously, e.g. sending files or images. Specialist conferencing equipment comes in various forms, ranging from purpose-built rooms equipped with cameras and monitors to the more standard "Rollabout" systems, which include an integrated camera and monitor mounted onto a cart so that they can be moved to wherever the system is required. The common aim of these different VMC systems is, in most cases, to replicate the benefits of face-to-face communication as fully as possible. However, there are limitations to any form of collaboration taking place over a distance (Fullwood & Doherty-Sneddon, in press). These limitations can impact upon both physical and psychological factors to influence the efficacy of the communication system. This paper reviews the effects of system quality and set-up, discussing how VMC can be used to its full potential to maximise communicative success.

Audio quality

Although good audio quality is essential for successful videoconferencing (Angiolillo et al, 1997), even the most sophisticated equipment can be let down by poor room acoustics and bad microphone placement. Unwanted echoes and unpleasant "boomy" sounds make it difficult for both speakers to talk at once, disrupting free-flowing, two-way conversation. This can be ameliorated by the use of sound absorbing wall coverings and other types of acoustical treatment (Angiolillo et al, 1997). The quality of the signal itself can also have important psychological and communicative effects. Watson and Sasse (1996) found that raising the quality of an audio signal led to the video signal from a high-definition monitor being perceived as better by participants. This effect was also noted by Hollier and Voelcker (1997), with quality ratings even more distinct when the image was of a person speaking. As well as the quality of the actual audio signals and acoustics, the type of audio channel used can influence communication between speakers. A full duplex ("open") channel enables both speakers to talk simultaneously, whereas a half duplex channel transmits only one voice at a time using either voice-activated switching or a "click to speak" mechanism. Sanford et al (2004) compared both channel types with face-to-face interaction. While patterns of speech were more "natural" in the open channel condition, with more interruptions and overlapping speech, performance on a collaborative task (the Map Task; Brown et al 1984) was worse than in the half duplex condition, in which participants performed as well as those in the face-to-face condition. The authors suggest that the open channel users made insufficient allowances for the restrictions of the system, assuming that they could communicate as they would face-to-face. The half duplex users tended to say more in their dialogues, using longer turns and taking greater care in managing turn-taking procedures (Sanford et al, 2004). These findings may also indicate a context-dependency in the appropriateness of audio channel choice; very task-orientated dialogues, such as those used in the Map Task, may be more efficient when speakers are made aware of the constraints of VMC and are able to adapt.

Delay

Delay is a major technical limitation of VMC and varies between systems (see Angiolillo et al, 1997). Audio and visual signals may be delayed from one location to another, or may even be processed at different speeds so that audio signals and video images are not synchronised. This may be overcome by introducing a similar delay in the audio channel to maintain lip synchronisation; however, moment-to-moment changes in signal make it difficult to predict the length of delay. This is especially an issue in DVC; although the introduction of high-speed networks and compression techniques has made Internet conferencing more widely used (Sasse, 1996) problems with bandwidth, which can be defined as "a measure of the speed of the connection" (Videotalk, 1998), may ensue as a result of the unpredictability of Internet transmission performance. Whilst a complete disregard for lip synchronisation has a detrimental impact on communication (Angiolillo et al, 1997), the fact that users are more frustrated by audio delay than by the absence of lip synchronisation suggests that attempts to synchronise sound and image should be abandoned (Tang and Isaacs, 1992). People who experience delay are usually unaware of it, and tend to assume that the delay is due to the other speaker (Kitawaki et al, 1991). Co-ordinating conversation relies on refined timing and without this, a breakdown in communication may occur (Angiolillo et al, 1997). Delay in video particularly has adverse effects on communication and collaborative task performance (O'Malley et al, 1996). Nonetheless, in the event of a breakdown in communication people adapt by adopting a more formal conversational style, with longer turns and fewer interruptions. This fits the view of Sanford et al (2004) that speakers can adapt more effectively when they are aware of system constraints.

Image size and quality

The implementation of VMC systems is based largely on the assumption that visual cues improve human interaction (Fullwood and Doherty-Sneddon, in press). With this in mind, the quality of the image being transmitted is an important consideration. Again this becomes a problem in DVC; using compression techniques to reduce the number of frames per second means faster and cheaper communication, but results in low picture definition and jerky pictures – which in turn may impair communication. Barber and Laws (1994) found that presenting a video at 12.5 rather than 25 frames per second reduced ability to perform tasks such as lip-reading and repeating back messages. Another factor is the size of the screen. Smaller screens may improve the appearance of low-resolution images (Angiolillo et al, 1997), but users generally prefer larger displays up to life size (Inoue et al, 1984). The reason for this is thought to be a need for social co-presence, as a larger screen gives the instinctive impression that the remote speaker is actually present in the room (Angiolillo et al, 1997). Indeed, research has indicated that subjective ratings of social co-presence are significantly enhanced with the use of a large, life-like screen (e.g. De Greef & Ijsselsteijn, 2001). Using the full screen to focus on each speaker close-up also makes it easier for nonverbal cues to be picked up (see Davis et al, this volume, for a more detailed review). Despite the obvious advantages of larger screens, most VMC systems are limited by cost and bandwidth. Specialist conferencing centres are more likely to employ larger screens so will be more able to benefit from the communication of nonverbal behaviour (Tiffin & Rajasingham, 1995); with DVC, however, a trade-off may be necessary for practicality. Whilst it is essential to maintain the convenience of the system, it is also important to preserve the advantages of using video in the first place (Angiolillo et al, 1997).

System set-up

How the VMC system is set up has a bearing on not just the quality of the information being transmitted, but also on what information is made available to participants. The room chosen for videoconferencing should ideally have professional direct and indirect lighting available to avoid the dark and blurred images of speakers often associated with a usual office environment (Angiolillo et al, 1997). The physical layout of the room will determine how monitors can be placed, so care should be taken to avoid viewing being restricted. Even with high video quality, the way in which remote participants can be observed may influence perceived satisfaction with the exchange. Sellen (1995) asked participants for their views on communicative success after using one of three different set-ups: picture-in-a-picture (PIP), Hydra and LiveWire. PIP displays simultaneous pictures of all participants on one monitor, with a single speaker broadcasting all their voices. Hydra has each person facing three screens, with a camera below each monitor and a speaker below each camera. LiveWire is voice activated to show the current speaker to all but the current speaker, who is shown the image of the last speaker. Two thirds preferred Hydra and the remaining third preferred PIP; reasons included the Hydra set-up supporting selective gaze and simultaneous conversations, and PIP enabling them to see themselves on the monitor and converse without having to turn their heads. Overall, the results indicated that people prefer systems that are set up to provide a similar spatial layout to face-to-face meetings.

In some situations, though, focus on the faces of speakers is not so high a priority. Whether users prefer video links (i.e. images of people) or video data (i.e. images of relevant objects) depends very much upon the context of the conversation or task. Anderson et al (2000) found that participants tested in two scenario simulations – the "Travel Service Simulation", where participants plan a holiday itinerary and the "Financial Service Simulation", where they choose a property and arrange a mortgage – ranked video data more highly in terms of what was most useful, and what was the most important feature to preserve and improve. In other words, images such as maps and video clips of holiday destinations or houses were seen as more important than images of the sales representatives. Satisfaction with the outcome of the tasks and with many aspects of the communication process was equal for both the video link and video data. These results could be seen as contradictory to findings that a view of the face is important for bargaining (e.g. Whittaker, 1995); alternatively, as Anderson et al (2000) point out, it appears that the nature of the task to be accomplished will determine which types of multimedia data are most appropriate. One additional component to VMC is referred to as the "shared whiteboard". This allows remote participants to draw and type in the same workspace simultaneously, in conjunction with audio communication or audio and video communication. The efficacy of each set-up was examined by Gale (1990) in a series of group work tasks designed to represent various types of office work. Using the audio and video with whiteboard system improved task performance, ratings of co-presence and perceptions of productivity. However, the results also suggest that the need for video is task-specific; difficult assignments requiring higher levels of communication appear to be more video dependent than assignments that are easier to complete.

Conclusion and recommendations

System quality and set-up factors in VMC have important practical and psychological implications, not only individually but also interactively. Ideally, a VMC system will have good audio and image quality, with minimal delay and clear views of speakers to create the best possible sense of face-to-face communication. Nevertheless, it is often necessary to impose limitations due to cost, convenience and available resources. With this in mind, the following recommendations are made with regards to minimising difficulties with VMC and maximising communicative success:

1) The restrictions of such technology should be made clear to users so that they have sufficient opportunity to adapt;
2) Compromises may be necessary to maintain the benefits of VMC alongside an economical and efficient design;
3) The context of the interaction should be considered in deciding which factors should take precedence.

References

Anderson, A.H., Smallwood, L., MacDonald, R., Mullin, J., Fleming, A., & O'Malley, C. (2000). Video Data and Video Links in Mediated Communication: What Do Users Value? *International Journal of Human Computer Studies*, **52**, 165–187.

Angiolillo, J.S., Blanchard, H.E., Israelski, E.W., & Mane, A. (1997). Technology Constraints of Video-Mediated Communication. In K.E. Finn, A.J. Sellen & S.B. Wilbur (Eds.), *Video Mediated Communication*. New Jersey: Lawrence Erlbaum Associates.

Barber, P., & Laws, J. (1994). Image Quality and Video Communication. In R. Damper, W. Hall, & J. Richards (Eds.), *Proceedings of the IEEE International Symposium on Multimedia Technologies and their Future Applications* (pp 163–178). London: Pentech Press.

Brown, G., Anderson, A.H., Yule, G. & Shillcock, R. (1984). *Teaching Talk*. Cambridge: Cambridge University Press.

Davis, S.J., Fullwood, C., Martino, O.I., Derrer, N. M. & Morris, N. (This volume). Here's looking at you: A review of the nonverbal limitations of video mediated communication.

De Greef, P., & Ijsselsteijn, W.A. (2001). Social Presence in a Home Tele-application. *Cyber-Psychology and Behaviour*, **4**, 307–315.

Fullwood, C. & Doherty-Sneddon, G. (In press). Effect of gazing at the camera during a video link on recall. *Applied Ergonomics.*

Gale, S. (1990). Human Aspects of Interactive Multimedia Communication. *Interacting with Computers*, **2**, 175–189.

Hollier, M.P., & Voelcker, R. (1997). Towards a Multimodal Perceptual Model. *BT Technological Journal*, **15**, 162–316.

Inoue, M., Yoroizawa, I., & Okubo, S. (1984). In *Proceedings International Teleconference Symposium* (pp 66–73). International Telecommunications Satellite Organisation. Reprinted in D. Bodson & R. Schaphorst (1989). *Teleconferencing* (pp 211–218). New York: IEEE.

Kitawaki, N., Kurita, T., & Itoh, K. (1991). Effects of Delay on Speech Quality. *NTT Review*, **3**, 88–94.

O'Malley, C., Langton, S., Anderson, A., Doherty-Sneddon, G., & Bruce, V. (1996). Comparison of Face-to-Face and Video-Mediated Interaction. *Interacting with Computers*, **8**, 177–192.

Sanford, A., Anderson, A.H. & Mullin, J. (2004). Audio channel constraints in video-mediated communication. *Interacting with Computers*, **16**, 1069–1094.

Sasse, M.A. (1996). Noddy's Guide to Internet Videoconferencing. *Interfaces*, **31**, 6–9.

Sellen, A. (1995). Remote Conversations: The Effects of Mediating Talk with Technology. *Human-Computer Interaction*, **10**, 401–444.

Tang, J.C., & Isaacs, E.A. (1992). *Why Do Users Like Video? Studies of Multimedia-Supported Collaboration* (SMLI Tech. Rep. No. 92-5). Mountain View, CA: SUN Microsystem Laboratories.

Tiffin, J.W. & Rajasingham, L. (1995). Telelearning in cyberspace. In J.W. Tiffin and L. Rajasingham (Eds.), *In Search of the Virtual Class: Education in an Information Society.* London: Routledge.

Videotalk: Retrieved on February 17, 1999 from the World Wide Web: http://www.videotalk.co.uk/home.html. © 1998 Wincroft Inc.

Watson, A., & Sasse, M.A. (1996). Evaluating Audio and Video Quality in Low Cost Multimedia Conferencing Systems. *Interacting with Computers*, **8**, 255–275.

Whittaker, S. (1995). Video as a Technology for Interpersonal Communication: A New Perspective. *IS&T SPIE Symposium on electronic imaging science and technology*, **2417**, 294–304.

A COGNITIVE STUDY OF KNOWLEDGE PROCESSING IN COLLABORATIVE PRODUCT DESIGN

Yuan Fu Qiu, Yoon Ping Chui & Martin G. Helander

School of Mechanical and Aerospace Engineering,
Nanyang Technological University, Singapore

This study explores the understanding of product designers' knowledge processing in collaborative design. Designers' knowledge is classified. Experiments on experienced designers are conducted for capturing designers' knowledge processing behaviors. Using protocol analysis, designer's cognitive needs, communication and design behaviors are analyzed and described. A cognitive study of knowledge processing will result in a natural integration of knowledge management tools into the design process.

Introduction

Design is a knowledge driven process. Designers handle large amounts of knowledge during the design process. However, existing tools for knowledge management integrate poorly with the design process. It has been argued that there is often a fundamental mismatch between the way human process knowledge and the way it is processed by technology support system (Kushniruk, 2001). Pioneering studies in cognitive science examined the conceptual design process from a variety of perspectives, such as problem solving (Goel, 1991). Most studies are lab-studies with short-term tasks assigned to small sample of subjects. Protocol analysis is the main technique for data analysis (e.g. Goel, 1991). However, little research has addressed the cognitive and social psychological factors within knowledge activities, such as human cognitive needs, abilities and limitation when engaged in knowledge acquisition, sharing and utilization in product design. This paper studies the knowledge processing in collaborative design. It studies three consecutive design phases, namely conceptual design, system-level design and detail design (e.g. Ulrich & Eppinger, 2004). A knowledge coding framework for cognitive study is presented. Four experienced designers participated in the experiments and results are illustrated.

Methodology

Knowledge in design

Nonaka & Takeuchi (1995) classified knowledge as tacit and explicit knowledge. Tacit knowledge is that which is contained within a person's head, and is difficult or impossible to express, write down and codify. Explicit knowledge, on the other hand, is that which can be readily articulated, written down, codified and shared. The knowledge in design is illustrated in Figure 1.

Shown as Figure 1, knowledge in design space can also be defined as "explicit knowledge" or "tacit knowledge", which is divided by the dashed line in Figure 1. The part above the dashed line refers to "explicit knowledge" in design space, whereas the part under the dashed line refers to "tacit knowledge" in design space. The designer's

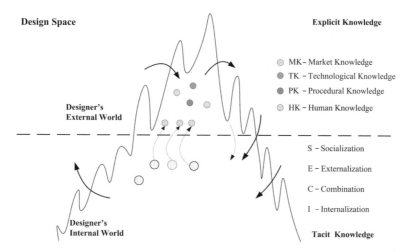

Figure 1. Knowledge in design space.

internal world refers to human's cognitive world, whereas the external world refers to the external reality. The explicit knowledge in designer's internal world is easy to capture by analyzing the verbal communication. However, the tacit knowledge in designer's internal world is difficult to capture. Knowledge management in design should motivate the designer to transfer the tacit knowledge into explicit knowledge.

Figuratively, let the tacit knowledge be the sea-based, and the explicit knowledge be the part above the sea surface level. The designer internal world is like a piece of ice on the sea (shown in the Figure 1). The ice that is on top of the sea surface is the designer's explicit knowledge. The ice that is under the sea surface is the designer's tacit knowledge. The knowledge processing in design space is a spiralling process of interactions between explicit and tacit knowledge

In this paper, in order to study the knowledge cognitive processing of designer's internal world, the knowledge in designer's internal world is classified to market knowledge (MK), technology knowledge (TK), procedural knowledge (PK) and human knowledge (HK). MK, TK, PK and HK is mainly explicit. MK represents the knowledge interacting with the external interface like customers, partners, suppliers and other stakeholders. TK refers to the knowledge about technologies. PK refers to the knowledge about how to accomplish an end. HK refers to the knowledge that is externalized from the tacit knowledge in the designer's internal world.

According to Nonaka & Takeuchi (1995), knowledge creation is a spiralling process of interactions between explicit and tacit knowledge. The interactions between the explicit and tacit knowledge lead to the creation of new knowledge. The combination of the two categories makes it possible to conceptualize four conversion patterns, namely socialization (S), externalization (E), combination (C) and internalization (I).

- Socialization: It refers to the exchange of tacit knowledge between the designer and the external world in order to convey personal knowledge and experience.
- Externalization: It describes the conversion process of tacit into explicit knowledge within designer's internal world.
- Combination: It refers to the exchange of explicit knowledge between the designer and the external world in order to convey more complex and more systematized explicit knowledge.
- Internalization: It is a process in which explicit knowledge becomes part of tacit knowledge.

Figure 2. Data analysis procedure.

Table 1. Knowledge fragment coding framework.

Time Stamp	Knowledge Activity	Knowledge Class	Description		Level of Detail
Start time	Generation Sharing	Market knowledge	Geometry Material	Assembly Manufacturing	Unlabeled Labeled
End time	Evaluation Utilization	Human knowledge	Position Shape/Size	Ergonomics Esthetics	Associative Qualitative
		Technological knowledge	Attribute Linkage	Maintenance Costs	Quantitative
		Procedural knowledge	Function Kinetics	Safety Market	
			Software	Miscellaneous	

Experimental design

The experiments were designed for understanding how designers handle different kinds of knowledge in design. The experiments were conducted in a test room, where observers can watch without disturbing the events in the test room. Onscreen "actions" of the designers were relayed with picture-in-picture video to a monitor and video recorder. A team with four experienced designers was tasked to design a portable entertainment device in three consecutive 2-hour sessions, namely conceptual design, system-level design and detail design. Task description provided background information of the task, and no correct answers exist for the task. Every participant was provided with a booklet of task description, a brief presentation about tools to be used. Self-assessment and short interviews were conducted after each experiment. The dialog, designers' emotion, the sketch actions and so forth were recorded for analysis and reference.

Data analysis technique

This study uses verbal protocol analysis method for data analysis. The data analysis procedure is illustrated in Figure 2.

The knowledge fragments will be coded by a knowledge fragment coding framework (as illustrated in Table 1).

Shown as Table 1, knowledge fragment is defined based on its time stamp, knowledge activity, knowledge class, description and level of detail.

- Knowledge activity: it is referred to the activity that designers perform with knowledge during design.
- Knowledge class: it is the class to which the knowledge belongs.
- Description: it identifies those properties of the design object which the designers are dealing with (Gunther, Frankenberger & Auer, 1997; Pahl & Beitz, 1993).
- Level of detail: it is a knowledge measure which is being qualitatively defined (Baya, 1996). These levels are defined from the perspective of evolution of a design concept.

A cognitive study of knowledge processing in collaborative product design 147

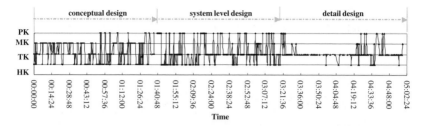

Figure 3. Processing of different knowledge classes with time in three design phases.

Table 2. Frequency distribution of different descriptions in design.

Knowledge class	Knowledge activity	Level of detail					
		Unlabeled	Labeled	Associative	Qualitative	Quantitative	Total
Human		15	61	27	60	35	198
	Generation	7	32	2	7	15	63
	Sharing	1	5	0	2	1	9
	Evaluation	4	7	16	47	7	81
	Utilization	3	17	9	4	12	45
Market		9	28	11	40	39	127
	Generation	1	0	0	0	1	2
	Sharing	6	18	4	10	15	53
	Evaluation	1	7	6	26	21	61
	Utilization	1	3	1	4	2	11
Procedural		37	32	4	14	11	98
	Generation	1	4	0	2	0	7
	Sharing	22	17	0	4	8	51
	Evaluation	4	3	2	3	1	13
	Utilization	10	8	2	5	2	27
Technology		22	149	27	52	85	335
	Generation	3	32	1	8	14	58
	Sharing	9	62	6	20	20	117
	Evaluation	8	41	17	23	37	126
	Tilization	2	14	3	1	14	34
Total		83	270	69	166	170	758

Results

The factors in knowledge fragment can be analyzed respectively according to the time stamp. Much understanding can be obtained by combining or comparing several factors. Some of results are shown in this section. The knowledge processing process is illustrated in Figure 3. The results show that the period of time using human knowledge decreases from 33% in conceptual design, to 31% in system level design, and to 3% in detail design. The period of time using market knowledge decreases from 35% in conceptual design, to 3% in system level design, and to 1% in detail design. The period of time using technology knowledge increases from 25% in conceptual design, to 49% in system level design, and to 94% in detail design. The period of time using procedural knowledge increases from 7% in conceptual design to 17% in system level design, but decreases to 2% in detail design. It shows that designers' knowledge is used differently in different design phases. Table 2 summarizes the frequency distribution of various descriptions in design.

Table 2 shows that the usage of different class of knowledge varies with different knowledge activities. Knowledge is expressed by designers in different levels of detail. Different description in design is related with different class of knowledge. Knowledge of such a priority will be helpful in developing representations for different description, especially when the description is correlated with different knowledge class, knowledge activity and level of detail. Technology support according with designers' knowledge processing process will result in better performance.

Discussion

This research is only an initial investigation into the cognitive aspect of knowledge processing in design. Understanding of other issues needs to be addressed further, such as how knowledge is shared and utilized for decision making. It is essential to use this understanding to develop a method of support for knowledge management.

References

Baya, V. 1996, *Information handling behaviour of designers during conceptual design: Three experiments,* Unpublished Doctoral Dissertation, Stanford University

Goel, V. 1991, *Sketches of thought: A study of the role of sketching in design problem solving and its implication for the computational theory of mind,* Unpublished Doctoral Dissertation, University of California, Berkeley

Gunther, J., Frankenberger, E., & Auer, P. 1997, Investigation of individual and team design processes. In N. Cross, H. Christiaans & K. Dorst (Eds.), *Analysing Design Activity* (John Wiley & Sons, New York), 117–132

Kushniruk, A. W. 2001, Analysis of complex decision-making processes in health care: Cognitive approaches to health informatics, *Journal of Biomedical Informatics,* **34**(5), 365–376

Nonaka, I., & Takeuchi, H. J. 1995, *The Knowledge-Creating Company: How Japanese Companies Create the Dynamics of Innovation* (Oxford University Press, New York)

Pahl, G., & Beitz, W. 1993, *Konstruktionslehre* (3rd Edition ed.), (Springer-Verlag, Berlin)

Ulrich, K. T., & Eppinger, S. D. 2004, *Product Design and Development* (3rd Edition ed.), (McGraw-Hill, Boston)

COMPARISONS BETWEEN USER EXPECTATIONS FOR PRODUCTS IN PHYSICAL AND VIRTUAL DOMAINS

Bahar Sener, Pelin Gültekin & Çigdem Erbug

Department of Industrial Design – BILTIR/UTEST Product Usability Unit
Middle East Technical University, Faculty of Architecture,
06531 Ankara, Turkey

This paper focuses on the expectations that users form when they are confronted with audio products. It presents the results of a comparative study conducted with twenty users to reveal similarities and differences in their expectations for products in physical (i.e. hardware) and virtual (i.e. software) domains. The results are suggested to be valuable on two accounts. First, for the design of products that can mediate between increased technological possibilities and increased user demands. And second, in relation to methodological issues, for the systematic assessment of user expectations for new products.

Introduction

In the past few years, developments in digital technologies have emerged to create an almost immediate impact on user lifestyles, leading to new ways of interacting with products at home, on the road, and at work. The prevailing effects of the developments in digital technologies are not limited to the transformations they bring into everyday life, but also to changes in the way users interact with products, and their expectations from them. Consumers are increasingly embracing digitally integrated everyday products, such as audio devices, video recorders, and microwave ovens. From a design perspective, the quality of the interaction between such products and their users is highly dependent on designers having a firm understanding of the triangular relationship that exists between users, products and the environments in which they are utilised.

This paper focuses on the expectations that users form when they are confronted with audio products. It presents the results of a comparative study conducted to reveal the similarities and differences of user expectations from portable audio devices versus audio player computer applications. The study is targeted at contributing to the advancement of knowledge about expectations and preferences of users during the process of buying and using a new product.

Elicitation of user expectations

Data collection method. Initially developed in the clinical psychology field by Kelly (1955), the Repertory Grid Technique (RGT) represents a widely used set of tools for studying personal and interpersonal systems of meaning. Because of their flexibility, repertory grids have been used in studies of a broad variety of topics. However, the application of RGT within human-computer-interaction (HCI) studies has been scarce,

---

Tel: +90 312 210 4219, Fax: +90 312 210 1251, bsener@metu.edu.tr – pelin.gultekin@gmail.com – erbug@metu.edu.tr

although not completely inexistent (Fallman and Waterworth, 2005). For this study, the RGT was adopted as an applicable and potentially very effective method for systematically generating and analysing product evaluation data.

Participants, products, procedure. A total of twenty undergraduate Industrial Design students from Middle East Technical University participated in the study. The mean participant age was 21, ranging from 18 to 27 (11 male and 9 female). The study was conducted by dividing the participants into two groups of ten. A different product type was evaluated by each group: one evaluated five portable digital audio devices; the other evaluated five audio player software applications. In order to elicit user expectations on a wide range of designs and configurations, as much as possible both product types consisted of samples that were visually and functionally diverse. The studied products were: Apple iPod, Apple iPod Shuffle, Sony D-VJ85 VCD player, Sports Radio, Premier MP3 player, Foobar 2000 V.083, Apple iTunes V.4, Lycos Sonique Player V1.96, MusicMatch Jukebox V.10, and WinAmp V.5.091.

Participants were introduced to the research project and given an explanation of what was expected at the end of it. They were then given written explanations of the order of activities within their evaluation sessions. Each session with each participant was carried out separately. Sessions comprised four parts, with the same procedure followed for all sessions.

(i) *Questionnaire.* Prior to the session, participants completed a two-part questionnaire. In the first part, demographics, user characteristics (e.g. amount of time spent with a computer), computer expertise level, and interest in technological products were assessed. The second part collected information on music related activities, including the usage patterns of audio devices (e.g. preferred playback device, frequency of use, duration of use), and satisfactory and unsatisfactory experiences during its use.

(ii) *Product familiarisation.* On completion of the questionnaire, participants were presented with their allocated set of five products. The portable audio devices were laid on a table, whereas the audio player applications were cascaded on a 17″ computer screen. The participants were left alone for up to 10 minutes in order to familiarise themselves with the products.

(iii) *Personal construct elicitation.* A randomly generated triad of audio players was presented to participants. Similarly, a random triad of software was displayed on the computer screen. Participants were then asked to think of ways or "dimensions" (i.e. personal constructs) in which two of the products (i.e. elements) were similar, but distinct from the third (i.e. a bi-polar construct). Participants then verbally labelled the bi-polar construct using a word or phrase and assigned a negative or positive meaning to the poles. For example, two products may be identified as "big" whereas the third is "small", for a product where "small" receives a positive association and "big" a negative. In all cases, the researchers took notes of the polar constructs on a pre-formatted chart. After collecting data for the triad, the elicitation process was repeated with subsequent random triads. Participants were eventually unable to identify a dimension that they had not mentioned before.

(iv) *Product rating.* The participants were asked to grade each product according to the personal bi-polar constructs they had created in the previous stage. A seven-point scale from "1" (for negative connotations) to "7" (for positive connotations) was used as the grading criteria.

Results and analysis of personal construct elicitation

The analysis was undertaken separately for products in the physical and virtual domains, with the findings for both domains then cross-compared in order to elicit information on users' expectations. Each participant's construct list was used as the basis for the analysis. Across domains, a total of 63 personal constructs were elicited. Of these, 24 were elicited for products only in the physical domain; 11 for products only in the virtual domain; and 27 common to both domains. The analysis was made both at multi-participant and cross-comparison levels. The multi-participant analysis aimed to reveal patterns or other kinds of relationships between different participants' constructs within a single domain, whereas the cross-comparison analysis aimed to compare data across the product domains. It is important to note that the results should be regarded as relational rather than absolute values. The analysis stages were as follows.

1. *Rephrasing of constructs:* Constructs were reviewed by two researchers. Different constructs judged to have the same meaning were rephrased as a single new construct.
2. *Translation of constructs:* As the study was carried out in the participants' native language (Turkish), all constructs were translated into English.
3. *Tabulation of constructs according to originated domain:* Constructs within each domain were tabulated separately. This data included the participant ratings of each product against their personal constructs.
4. *Grouping of constructs:* Two levels of classification were adopted. The first level of classification (construct group-1) was undertaken to study the nature of the constructs within two broader themes: those that were predominantly pragmatic, and those that were predominantly hedonic (Voss et al, 2003; Hassenzahl, 2000). The classification could then be used to determine if products in either domain were especially linked to achieving practical goals (pragmatic) or experiencing pleasure in use and a personal affinity to design (hedonic). The second level of classification (construct group-2) placed all the constructs under subject headings using content analysis procedures, to aid subsequent analysis and discussion. Three main subject headings were identified: visual content, functional content, and usability.
Visual content. This group consisted of constructs related to visual criteria and aesthetic considerations. Examples included: "bulky-elegant", "unprofessional look-professional look".
Functional content. This group consisted of constructs related to functional properties and attributes of products. Examples included: "low capacity-high capacity", "insufficient features-sufficient features".
Usability. This group consisted of constructs related to interface elements and quality of interaction. Examples included: "unclear interface-clear interface", "difficult to use on the move-easy to use on the move".

   A particularly large number of constructs fell under the usability heading, demanding that a re-classification into five sub-groups be made: efficiency, clarity, consistency, feedback, and satisfaction. Constructs within the satisfaction sub-group included degrees of: interest in use; motivation; enjoyment; and sense of achievement.
5. *Product ranking:* The scores for product ratings were processed, with the highest total scores indicating the most favoured products. Accordingly, amongst the products in the physical domain, P1 was rated the highest with a score of 747, followed by: P5 (667), P2 (666), P3 (520), and P4 (480). Amongst the products in the virtual domain, S4 was rated the highest with a score of 627, followed by: S2 (618), S1 (514), S3 (392), S5 (369).
6. *Correspondence analysis:* "Product-construct", "product-construct group-1" and "product-construct group-2" relationships were analysed using SPSS Software V.11, a commonly-used application for statistical analysis (Hair et al, 1998). Separate analyses were made for the data arising from physical and the virtual domains, with the aim of revealing similarities and differences between user expectations across the two domains. The data set for product-construct analysis comprised the product ranking scores for each product. The data set for product-construct group-1 and

group-2 analyses comprised the total ratings for the classified groups in relation to all products. The outcome of this stage was graphical representations of the participants' perceptions of the evaluated products, displayed in the form of two-dimensional "perceptual maps", showing how products in each domain relate to: i) individual constructs; ii) construct groups 1 and 2.

Discussion and conclusions

This section builds on the results and analysis section by discussing similarities and differences of users' expectations for products in physical and virtual domains. Table 1 provides a breakdown of the hierarchical product rankings in both domains.

To aid the interpretation of the perceptual maps, and to identify the subject heading (i.e. product construct group-2) associated with a given construct, each construct (represented by an x-y coordinate on the perceptual map) was assigned a "subject heading" colour. The colour coding helped to visualise how each product (and more generally each product domain) was affected by the interaction of constructs. Then, the perceptual maps representing product-construct groups 1 and 2 were studied to reveal trends within and across the two product domains. Two example perceptual maps representing product-construct group-2 relationships, for both domains, can be studied in Figure 1.

General observations. The findings of the study show "usability" to be the most dominant criteria when considering user expectations of products in both domains, followed by visual aspects and then functional content. When considering only the product domain, construct groups-2 results were found to range in dominance/importance (from the most to the least): usability, visual content, and functional content. Whereas for the software domain only, the order was: usability, functional content, and visual content. The dominance of usability across both domains may be attributed to the fact that products in general increasingly have integrated software and digital components, which in turn results in interfaces similar to those of "pure" software. Therefore, as the interaction of products becomes increasingly similar to software interaction, so the evaluation criteria for products might also become similar. As a passing observation, a greater

Table 1. Ranking order for products in both domains.

Ranking order for Physical (P) Products					Ranking order for Virtual (S) Products				
1	2	3	4	5	1	2	3	4	5
P1	P5	P2	P3	P4	S4	S2	S1	S3	S5

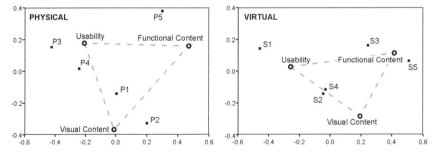

Figure 1. Perceptual maps for (a) physical domain; (b) virtual domain.

number of constructs were generated for physical products than virtual. This suggests that users are more critical and demanding, when assessing physical products.

Physical domain. P1 is the highest ranked product and appeared to have a hybrid interaction with the construct groups-2. In other words, as indicated by the perceptual map in Figure 1, it is located at the heart of the three construct groups. P1 is also surrounded with *hedonic* constructs, which suggest that hedonic qualities had a persuasive role in evaluating and differentiating the product. In fact this idea is supported by the observation that P5 (ranked in second order) is also very close to hedonic constructs. This suggests that users' expectations of physical domain products are multi dimensional: apart from satisfactory functional and visual content, and usability, they expect the products to provide hedonic satisfaction. P4 is ranked last in order, and had a weak interaction with the both construct groups.

Virtual domain. The two highest ranked software (S4, S2) were most strongly linked to the visual content and usability constructs within construct group-2. Although the visual content group was the smallest set of constructs in this domain, S4 and S2 were distinguished by their visual qualities. It may suggest that users find the level of functionality satisfactory enough and do not question it further. But when software is embedded in a physical product, users raise more concerns about the functionality, because software has higher influence on the user-product interaction. Similar to the physical domain, hybrid interaction of products with the construct group-2 increased their ranking.

Methodology. Elicitation of the personal constructs by the RGT technique and the subsequent analysis procedure described, required substantial exertion of time and effort by the research team. However, it is important to note that the RGT had not previously been used for comparing products in physical and virtual domains, and its use in product design and evaluation generally was underdeveloped. The application of the RGT in this paper, combined with the described analysis methods, show great potential for wider application in new product development and product evaluation programmes. The specific results presented in this study are suggested to be valuable for the design of a broad range of digitally enabled products, whether in physical or virtual domains.

Acknowledgements

The authors would like to thank Tolga Levent and Dr Owain Pedgley for providing useful suggestions.

References

Fallman, D. and Waterworth, J.A. *2005*, Dealing with user experience and affective evaluation in HCI design, *Proceedings of ACM CHI 2005*, 2–7 April, Portland, 1–5.
Hair, J.F. et al *1998*, *Multivariate Data Analysis*. New Jersey: Prentice-Hall, Inc.
Hassenzahl, M. *2000*, Hedonic and Ergonomic Quality Aspects Determine Software's Appeal. *Proceedings of ACM CHI 2000*, 1–6 April, the Netherlands, 201–208.
Kelly, G.A. *1955*, *The Psychology of Personal Constructs*. New York: Norton.
Voss, K.E., Spangenberg, E.R. and Grohmann, B. *2003*, Measuring the Hedonic and Utilitarian Dimensions of Consumer Attitute. *Marketing Research*, Vol. XL, 310–320.

DESIGN – ENGAGE PROJECT

INTRODUCTION TO PROJECT ENGAGE: DESIGNING FOR EMOTION

Miguel Tito[1] & Thomas H.C. Childs[2]

[1]*Instituto de Biomecánica de Valencia, Edificio 9C, Universidad Politécnica de Valencia, Camino de Vera, s/n, 46022 Valencia, Spain*
[2]*The School of Mechanical Engineering, University of Leeds, Leeds, LS2 9JT, UK*

> This short paper is to introduce what is the ENGAGE project that has brought together the researchers and practitioners, with a diverse range of backgrounds, who have contributed to this section of the Proceedings. It describes ENGAGE's rationale and objectives and the expected results from its activity.

Introduction

ENGAGE is a Coordination Action project funded by the European Commission under the 6th Framework Programme. The project is steered by a consortium made up of 21 partners from 9 European countries, including many key organisations in the field of design and emotions.

Project ENGAGE is focused on the field of design and emotions, following a practical approach. The main goal is to gather and exchange knowledge that is useful for practitioners and companies. Thus the ENGAGE motto: designing for emotion (Figure 1).

Rationale and objectives

Quality, human factors and other traditional product development disciplines have expanded their reach with the aim of including and trying to fulfil users' "soft", subjective, needs. This situation has led to designers and manufacturers demanding methods and tools to include those new requirements into the design process and transform them into product specifications.

On the other hand, researchers are developing studies in a great variety of cross-related fields, ranging from statistics or semantics to cognitive psychology or fuzzy logic. In many cases, little exchange of information takes place between those fields.

Figure 1. The combined ENGAGE and Information Society logos.

It is the ambition of the ENGAGE project to produce a major impact in the area of design and emotion, improving the described situation by:

- Creating a knowledge community in the field of design and emotion, in which researchers and practitioners will be able to meet, exchange information, organise joint events and, in general, benefit from a rich global synergy.
- Collecting, documenting and assessing a wide variety of methods and tools for designing for emotion. This will provide designers and companies with a host of ways to put emotional design to practice, together with examples of good practice, case studies and the tools' corresponding theoretical backgrounds.
- Identifying gaps in current research lines and pointing new directions for further investigation.

Expected results

By the end of the project it is expected to have produced the following outcomes:

- An analysis of the state of the art in the field.
- Template-based, collectable information about over 200 methods and tools for designing for emotion.
- An agreed glossary, containing the basis of a common language.
- 9 events of national and international scale, featuring lectures and workshops by key persons in the field of emotions and design.
- An active, free-of-charge knowledge community, in which members can access and publish information, discuss different issues with experts and take part in Special Interest Groups.
- Examples of good practice when designing for emotion.
- A document with guidelines for future research.

Some of those goals have already been reached and are publicly available. More information and free ENGAGE membership may be obtained by visiting http://www.engage-design.org. The following papers are examples of the interests of leaders in the consortium.

Acknowledgement

We wish to thank the officers and staff of the Ergonomics Society for their support and enthusiasm for developing this ENGAGE symposium within their annual conference programme and the EC for funding the ENGAGE action through its Thematic Priority IST-2002-2.3.3.1 "Products and service engineering 2010", Project no. 520998.

IMPLEMENTATION OF AFFECTIVE DESIGN TOOLS IN THE CLOTHING INDUSTRY

María-José Such, Clara Solves, José Olaso, Begoña Mateo, José Montero, Carlos Chirivella, Pedro Vera & Ricardo Dejoz

Instituto de Biomecánica de Valencia (IBV), Universidad Politécnica de Valencia, Edificio 9C. Camino Vera s/n, 46022 Valencia, Spain

> This paper presents the results obtained for three case studies in the clothing product domain (baby, children's and women clothing), where the influence between the emotional dimension of the product design and the *purchase decision* is studied. The methodology used is firstly based on the extraction of the semantic space in each sub-domain and secondly in the application of a statistical regression model able to quantify not only the global influence of the emotional dimension of the product design but also the contribution of each individual semantic axis, in the purchase decision. These results are finally discussed on the direction of appropriateness of affective design tools in this specific product domain highly guided by the temporary nature of fashion.

Introduction

A new field mainly known as *Affective Design* has recently emerged aimed at the provision of objective methods and tools (García et al, 2005) to support the conceptual generation, specific definition and validation of **pleasurable products** as a key contributor to the **competitive advantage of a firm** (Spillers, 2004).

The potentiality of the existing methods and tools in each industrial sector depends not only on the nature of the user-product interaction during the product life but also on the nature of the consumer-market interaction in the moment of purchase decision. In this sense, peculiarities of the application of Affective Design in the clothing industry is affected by a user-product interaction where social recognition and personal values are prominent factors and a consumer-market interaction where user is overwhelmed with a wide and variable offer. Therefore, the identification of the garment that best fits with the emotional expectances of the consumer requires mostly a time consuming process that is alleviated when the consumer can rely on a commercial brand as giver of the emotional properties desired by a person.

In both senses, the Institute of Biomechanics of Valencia (IBV) has been applying affective tools in several industrial sectors (Alcántara et al, 2005).

This paper presents the results obtained for three case studies in the product domain (Schütte, 2005) of clothing. These results aim to bring up the possibilities of affective design tools in this specific product domain highly guided by the temporary nature of fashion.

Background

Three sub-domains were considered, baby clothing, children's clothing and women's clothing, all of them in the context of the Spanish population.

In that context, the first sub-domain stands out because of its high decorative component. Baby clothing is generally used as present after a baby birth. Baby footwear was included in this sub-domain since it is considered, for this target user as another piece of attire without its functional load of product for walking.

In the second sub-domain (children's clothing), the consumer and the user are again different agents, but now the functional component of the good turn to be more relevant. Finally in the women's clothing sub-domain, the garment has a higher component of aesthetics in comparison with children's, but this time for the user and consumer as coincident.

This background is here presented as introduction for the interpretation of the presented results below.

Methodology

The study took place into two phases for each sub-domain. The objective of the first stage was to identify the semantic space (Osgood et al, 1957) as instrument to explore in a second stage, the influence between the emotional dimension of the product design and the purchase decision.

Product representation used in the case studies were real products arranged over an horizontal surface at 78 cm. from the floor in order to provide to the participants as much visual information as possible without tactile interaction. Any identifier of the manufacturer or distributor was hidden. Participants in the product evaluations were representative of the consumer profile, as detailed in table 1. In the baby clothing sub-domain the participant women fulfilled the added characteristic of being in close relation with babies at present or in a past not further than 5 years ago.

Semantic space

For each sub-domain, the Semantic Space was obtained following the procedure documented in several other sources (Nagamachi, 1997, Ishihara, 2001).

A Factorial Analysis with Principal Components extraction method was applied to the resulted scores for each set of words per product domain. As a result, a reduced space of orthogonal factors was identified with a representativeness of the total variance of 62–66%. Table 2 presents the semantic identifier given to each product domain axis. They are listed following an order of percentage of total variance explained and translated from the Spanish semantic identifier. The symbolic and affective meaning could be distorted by translation.

Influence of the emotional dimension of the product design in the purchase decision

An experimental testing was conducted where a product sample was punctuated over a questionnaire with five point scale for each orthogonal factor obtained in the first stage. An added question was included in the questionnaire referring to the purchase intention (positive versus negative). Table 3 contains the size of the experimental testing. Product samples were selected with the criteria of covering the product design variability in the target market.

A logistic regression model was applied with the *purchase intention* as dependent variable and the *semantic factors* as quantitative independent ones, with the only consideration of their principal effect. The Nagelkerke R square resulted for each model was considered as a measure of the portion of the *purchase intention* that is caused by the emotional dimension of the product design gathered by the whole set of semantic axes.

Table 1. User target and consumer profile for each clothing sub-domain.

	Target user:	Consumer profile:
Baby clothing:	0–24 months old (all genders)	Women (20–50 years old)
Children's clothing:	3–8 years old (all genders)	Mothers (25–45 years old)
Women's clothing:	Women 40–60 years old	Women 40–60 years old

Table 2. Semantic space for each product domain.

Baby clothing	Children's clothing	Women's clothing
Beautiful	Elegant – charming	Elegant – of quality
Innovative – original	Basic – everyday	Youthful
Elegant	Innovative – original	Basic – everyday
Secure – good fastening	Adult style	Inconspicuous
Handmade	Elastic – flexible	Exclusive
Unisex	Suitable for all kind of situations	Retro
Light	Comfortable	No temporary
	Easy to combine	
	Sporty	

Table 3. Design of experiment for each product domain.

	Baby clothing	Children's clothing	Women's clothing
Sample size:	51 women	26 women	26 women
Product sample:	56 pieces of attire	31 outfits	23 outfits

A second data treatment was done with the aim of quantifying the individual influence of each semantic axis in the purchase decision. The coefficient odds (exp β) obtained for the previous logistic regression model was used as indicator. This indicator, (Peña, 2002), cannot be interpreted linearly but the greater the absolute value of the odds for a semantic axis, the greater is its influence in the purchase decision. A positive odds means a preference by the consumer of the affirmative stimulation of the semantic axis.

Results

The weight of the emotional dimension of the product design in the *purchase intention* was significantly different for each clothing domain, being relevant for the baby and children's domain but poor for the women's (table 4).

For the product domains where the global emotional dimension takes part in the purchase decision, the relevance of the individual semantic axes was also significantly different. It is remarkable the prominence of the individual semantic axis *beautiful* (aesthetic semantic axis) in the product domain "baby clothing". This aesthetic component highly promote the purchase intention (table 5), and therefore justify the interest on future research for the identification of the design items that improve this specific emotional component.

Table 4. Portion of the purchase intention that is caused by the emotional dimension of the product design (Nagelkerke R square for the applied logistic model).

Baby clothing	Children's clothing	Women's clothing
60.2%	55.2%	0.032%

Table 5. Individual influence of each semantic axis in the purchase decision of *baby clothing* (odds for the significative semantic axes in a logistic regression model).

Semantic axes	Influence in *purchase decision* (odds)
Beautiful	2,329
Secure – good fastening	0,790
Innovative – original	0,472
Handmade	0,334
Light	0,279

Table 6. Individual influence of each semantic axis in the purchase decision of *children's clothing* (odds for the significative semantic axes in a logistic regression model).

Semantic axes	Influence in *purchase decision* (odds)
Elegant – charming	0,826
Comfortable	0,679
Easy to combine	0,658
Innovative – original	0,587
Basic-everyday	0,379
Suitable for all kind of situations	0,203

A more complex contribution of the set of semantic axes for the children's clothes domain is obtained, with no such a strong dominance of an individual axis over the others (table 6).

Discussion

The results obtained for the three case studies presented, highlighted the differences between the effect of the emotional dimension of product design in purchase decision for different clothing domains.

The low effect obtai ned for the case study on women's domain, do recommend future research on the same effect when taken into consideration the person factor. It seems reasonable to think that each individual has a personal rule about which emotions motivate his purchase decision, and it seems also reasonable to think that these rules are not the same for all the Spanish women. Personal style or values should be taken into consideration for future applications of affective design tools in this subdomain. In this women's clothing domain, affective design tools could be used in the direction of fitting the emotional content promoted by the product design of a company to a specific personal style or set of values.

On the contrary, a common relation between the product emotional dimension and the purchase decision is obtained for the baby and children's clothing case studies. This result justifies the application of affective design tools such as Kansei Engineering to infer emotion by means of design items, and therefore to conclude design criteria that could improve the market acceptance.

Finally, some considerations should be taken into account regarding the experimental settings. Product sample was built with real garments. Nevertheless tactile interaction with textiles was considered out of scope. This decision was taken following the interest on exploring the use of affective design tools where the user could only have virtual interaction with the product. More research efforts should be addressed in this direction.

Acknowledgements

The authors wish to thank IMPIVA (Ref. IMPYPF/2004/2) for their support.

References

Osgood, C.E., Suci, G.J., Tannebaum, P.H., 1957, *The Measurement of Meaning*. University of Illinois Press.

Ishihara, S., 2001, Kansei Engineering procedure and Statistical Analysis, Workshop at the International Conference on Affective Human Factors Design, Singapore.

Nagamachi, M., 1997, Kansei Engineereing: The framework and Methods, in: Nagamachi, M (ed.), *Kansei Engineering 1*, Kaibundo Publishing co. LTD, Kure pp. 1–9.

Spillers, F., 2004, *Emotion as a Cognitive Artifact and the Design. Implications for Products That are Perceived As Pleasurable*. In Proceedings of the Forth Conference on Design and Emotion, Middle East Technical University, Ankara.

Schütte, S., 2005, *Engineering Emotional Values in Product Design, Kansei Engineering in Development*: doctoral dissertation. Department of Mechanical Engineering, Linköping University, Sweden.

Alcántara, E., Artacho, M.A., González, J.C., García, A.C., 2005, Application of product semantics to footwear design. *International Journal of Industrial Ergonomics*, 35-8, 727–735.

García, A.C., Such, M.J., Tito, M., Wouters, C.H., Mateo, B., Olaso, J., Durá, J.V., Vera, P., 2005, *Emotional Design. ENGAGE Project*. 2nd INTUITION International Workshop.

Peña, D., 2002, *Análisis de datos multivariantes*. Mc Grau-Hill, Madrid.

VALIDATION STUDY OF KANSEI ENGINEERING METHODOLOGY IN FOOTWEAR DESIGN

Clara Solves[1], María-José Such[1], Juan Carlos González[1],
Kim Pearce[2], Carole Bouchard[3], José Mª Gutierrez[1],
Jaime Prat[1] & Ana Cruz García[1]

[1]*Instituto de Biomecánica de Valencia, Universidad Politécnica de Valencia
Edificio 9C. Camino Vera s/n, 46022 Valencia, España*
[2]*Industrial Statistics Research Unit, University of Newcastle upon Tyne,
Kings Walk, Newcastle NE1 7RU*
[3]*Laboratory Product Design and Innovation, 151 boulevard de l'Hôpital,
75013 Paris, France*

> This paper describes a validation study developed in the footwear sector for the practical application of products semantics and kansei engineering at the industry. This work is part of the Kensys Project, which aim is the integration of the user's preferences in terms of emotional requirements in the process of product development through products semantics and kansei engineering. Results obtained will be implemented into a new computer system that will support companies in the process of development of user-oriented products by means these methodologies. Three research and development centres, a design consultancy and three footwear manufacturers from France, Spain and United Kingdom constitute the consortium of the Kensys Project.

Introduction

Considering the current situation of European traditional industrial sectors like footwear, in a more and more competitive market in which the commercial fight based only in prices seems to be lost in favor of countries with lower manufacturing costs, the manufacturer must react providing their products with added values for users. These added values involve looking further the functional and technical requirements of a product to its associated symbolic meaning, which means to integrate the users' preferences in terms of emotional requirements in the product design.

This paper presents part of the results obtained in the project, a comparative study of the application of Product Semantics and Kansei Engineering in the footwear sector in two European countries: Spain and UK. The study has contributed to the validation of the procedures and tools resultant of the Kensys Project for the development of products fulfilling users' emotional requirements.

Product Semantics, developed by Osgood et al (1957), is used for analyzing the symbolic value of objects. This tool analyses the universe of product qualifiers and extracts a set of independent concepts (semantic space), a more reduced semantic structure able to characterize the users' perception of products. Although product semantics was proposed more than thirty years ago, it is still one of the most powerful quantitative techniques for analyzing the emotional meaning of products.

The users' perception is related to the design parameters of products by means of Kansei Engineering (KE), which was crated in Hiroshima University about 30 years ago by Mitsuo Nagamachi (1995). KE allows the design and evaluation of products according to the users subjective perceptions and it has been applied in many different fields (Jindo et al, 1995; Ishihara et al, 1997; Jindo and Hirasago, 1997).

Methods

The first stage was the application of Product Semantics for obtaining the semantic space (SS) of man footwear. Two parallel processes were followed independently in the two countries, since there were considered the different symbolic and affective meanings depending on the cultural differences (Maekawa and Nagamachi, 1997; Matsubara et al, 1998).

The process started compiling from several sources (final users, experts and promotional literature) the words and expressions used to describe the perception regarding all-day man footwear (Kansei words). Only Spanish sources were used in Spain and English sources in UK. Using all this information, words were reduced to a reasonable number to avoid too long questionnaires for respondents, which may have resulted in low reliability due to tiredness. The most repeated terms and those related to marketing strategies where selected; synonyms and antonyms were grouped together and very specialized terms were discarded. Finally, 48 terms in UK (converted into 36 semantic differential scales of adjective/antonym or term/ negative) and 50 in Spain (converted into 50 semantic differential scales of term/ negative) constituted the Reduced Semantic Universe (RSU) of man footwear in the two countries.

Two samples of subjects took part in the evaluation of samples of shoes (43 models in Spain and 27 in UK) by means the semantic differential scales using randomized questionnaires based on five point semantic differential scales. Between 55 and 60 sets of responses were collected for each product in Spain and between 23 and 30 in UK. The semantic spaces for all-day man footwear in Spanish and English were obtained applying factor analysis to the respective evaluation results. SPSS 12.0 was used for the statistical analysis.

Once the semantic structure of the all-day man footwear was elicited, the Kansei Engineering was applied to obtain the design rules of the product by exploring the user's emotional responses and relating them with the design features. A common process was followed for the definition of the design elements and categories and for the selection of the product sample.

To define the design elements for man footwear, a group of designers defined and weighted the elements and categories and a group of users selected the final list of 10 design features depending on their influence in the preference regarding the product.

A representative sample of products was selected to extract the effects of the design elements on the consumers' response. To obtain a balanced combination of design categories, a sub-sample was selected using a designed experiment array with the 5 most important DEs in order to obtain independent evaluation of at least some of the design elements and a complementary sub-sample was added trying to represent as well as possible the variability of the other 5 design elements. A set of 99 images of products was finally selected.

During the Kansei evaluations in Spain and UK, samples of users evaluated the experimental sample of products in the respective set of five-point semantic differential scales corresponding to the respective semantic spaces. Being ordinal the response variables, ordinal logistic regression (Smets and Overbeeke, 1995; Hsiao and Huang, 1995; Overbeeke et al, 2000) was used as statistical modeling technique. One ordinal logistic regression model was generated for each English and Spanish semantic axis for the prediction of each characteristic. All of the design elements were considered as potential predictors and each model was finally constituted by those elements with their estimated coefficients for the model different from zero.

Results

After applying factor analysis to the results obtained from the application of Product Semantics, the semantic spaces corresponding to "man footwear" in Spain and UK were obtained. They are shown in Table 1. Each axis is formed by the set of terms of the RSU which are the most correlated to the concept.

After obtaining the semantic concepts to characterize the product perception, design rules were defined applying Kansei Engineering. These rules relate the design elements of the product with the semantic axes previously obtained.

The design elements and their corresponding categories summarize the most important characteristics that provide the product with its identity and appearance. Table 2 shows the design elements finally obtained for man footwear. The design features found to be significantly influencing the consumers' perception for each of the semantic scales are shown in Table 3 (UK Semantic Axes) and 4 (Spanish Semantic Axes), indicating the values of the coefficients in the logistic regression models.

Discussion and conclusion

The man footwear semantic space constitutes the base to analyze the users' perception associated to the product. The cultural differences affect on one hand to the composition of the semantic spaces and on the other hand to the product perception. However, although terms of the respective RSUs are grouped in different way, results are qualitatively comparable. In both cases there are axes defining the temporality of the product: Modern/Innovative and Up-To-Date in Spanish and Contemporary/Original and Fashionable in English. There are also axes concerning the use and performance of the

Table 1. Semantic Space for all-day man footwear.

Spanish Semantic Space	Modern/Innovative, Elegant/Well-Dressed, Comfortable/Practical, Sober/Classic, Good Quality/Well Finished, Up-To-Date, Casual/Sporting	60.32% variance explained
English Semantic Space	Contemporary/Original, Casual, Attractive/Tasteful, Fashionable, Comfortable, Hard Wearing/Practical, Easy/Convenient, Expensive/Good Quality	61.6% variance explained

Table 2. Design elements for man footwear.

Design Element	Categories
Last Wideness	Loose/Tight
Top Profile	Square/Rounded
Material	Leather/Nubuck, Suede, Split
Colour	Dark/Light
Seams distribution	Plain Toe/Moccasin/Longitudinal/Straight-winged/Sporting
Collars	Without Collar/Reinforced Collar/Bagged and foam filled
Fastenings	With Fastening/Without Fastening
Sole type	Soft-flat/Angular
Sole Material	Leather/Synthetic
Method of Application	Cemented-moulded/Welted (sole stitched on)

Table 3. Design features significantly influencing the consumers' perception.

	Top Profile	Material	Colour	Fastening	Sole Type	Sole Material	Seams Distribution	Last Width	Method of Applications	Collars
English SS										
Casual	–	3	2	2	4	–	2	3	1	–
Contemporary–Original	1	–	–	–	1	–	4	–	1	–
Attractive–Tasteful	–	–	1	–	–	2	3	1	–	2
Fashionable	–	1	–	–	1	–	4	–	1	1
Practical–Hardwearing	1	3	–	1	–	3	1	–	–	2
Comfortable	–	1	1	1	1	4	–	1	–	1
Easy–Convenient	–	–	1	4	–	2	–	1	–	–
Expensive–Good Quality	–	1	1	1	1	–	3	2	1	–
Spanish SS										
Elegant–Well Dressed	–	3	1	1	4	–	3	3	–	2
Good Quality–Well finished	–	–	1	–	1	–	2	–	–	1
Up-To-Date	1	2	1	–	2	1	2	1	–	1
Casual–Sporting	–	4	2	2	4	2	2	3	1	1
Modern–Innovative	1	1	2	–	4	2	3	1	–	1
Sober–Classic	–	2	2	1	4	3	3	–	–	2
Comfortable–Practical	–	3	2	1	2	3	2	1	1	–

Coefficient value (c.v.):
1: c.v. < 0,4; 2: 0,4 < c.v. < 0,7; 3: 0,7 < c.v. < 1; 4: 1 < c.v.

product: Comfortable/Practical in Spanish and Comfortable, Hard Wearing/Practical and Easy/Convenient in English. As regards aesthetic: Sober/Classic, Elegant/Well-Dressed and Casual/Sporting in Spanish and Casual and Attractive/Tasteful in English. And finally, referred to the quality of the product: Good Quality/Well Finished in Spanish and Expensive/Good Quality in English.

The ordinal logistic regression models show that strong predictive ability is only reached by particular axis (Casual in both countries), this has been related to a general low consensus shown by respondents and to a limited definition of the design elements of the product. A more accurate user profile definition, considering additional factors apart from the most generalist considered in this study (age and gender), should improve the predictive ability of the models. On the other hand, footwear is a complex product with design elements difficult to define and this lack of perfect definition of the products also affects the model quality.

In this paper we present the results obtained from the application of products semantics and kansei engineering to man footwear in two European countries, which have led to characterize the way users perceive and assess products and about how this perception is related to the design components of the products. Results obtained are directly applicable by European footwear manufacturers for integrating the users' emotional requirements into the process of product development.

The study has proven that cultural and demographic characteristics will have to be considered when planning future studies, which will depend on the companies' target markets. A more accurate definition of the users' profile will provide more robust models with better predictive ability.

Acknowledgements

Kensys Project is funded by the European Community under the "Innovation & SMEs programme (IPS-2000)", VFP. This work was supported by the Laboratoire de Conception de Produits et Innovation (Société d'Etudes et de Recherches l'Ecole Nationale Supérieure des Arts et Métiers), the Industrial Statistics Research Unit (University of Newcastle UPON TYNE), Pikolinos, Diedre Designs, Bjarni's Boots and County Footwear.

References

Osgood, C.E., Suci, G.J. and Tannenbaum, P.H. 1957, *The measurement of meaning*, University Illinois Press

Nagamachi, M. 1995, Kansei Engineering: a new ergonomic consumer-oriented technology for product development, *International Journal of Industrial Ergonomics*, **15(1)**, 3–11

Ishihara, S., Ishihara, K., Nagamachi, M. and Matsubara, Y. 1997, An analysis of Kansei structure on shoes using self-organizing neural networks, *International Journal of Industrial Ergonomics*, **19**, 93–104

Jindo, T., Hirasago, K. and Nagamachi, M. 1995, Development of a design support system for office chairs using 3-D graphics, *International Journal of Industrial Ergonomics*, **15(1)**, 49–62

Jindo, T. and Hirasago, K. 1997, Application studies to car interior of kansei engineering, *International Journal of Industrial Ergonomics*, **19**, 105–114

Matsubara, Y., Wilson, J.R. and Nagamachi, M. 1998, Comparative study of Kansei engineering analysis between Japan and BRITAIN, *Proceedings of the 6th International Conference on Human Aspects of Advanced Manufacturing: Agility and Hybrid Automation*

Maekawa, Y. and Nagamachi, M. 1997, Presentation system of forming into desirable shape and feeling of women's breast, *Kansei engineering-I: Proc first Japan-Korea Symposium on Kansei Engineering -Consumer- Oriented product development technology* (Kaibundo), 37–43

Smets, G.J.F and Overbeeke, C.J. 1995, Expressing tastes in packages, *Design Studies*, **16 n.3**

Hsiao, H. and Huang, H.C. 2002, A neural network based approach for product form design, *Design Studies*, **23**, 67–84

Overbeeke, C.J., Locher, P. and Stappers, P.J. 2000, Try to see it from the other side: A system for investigating 3-D form perception, *Proceedings of the 16th congress of the International Association of Empirical Aesthetics*, 17

SAFETY SEMANTICS: A STUDY ON THE EFFECT OF PRODUCT EXPRESSION ON USER SAFETY BEHAVIOUR

I.C. MariAnne Karlsson & Li Wikström

Chalmers University of Technology, Department of Product and Production Development, Division of Design, SE-412 96 Gothenburg, Sweden

This paper presents a study in which relation between the expression of a product (a bathing chair for small children) and users' behaviour was investigated from a safety point of view. The results show that the evaluated products were perceived as expressing "safety" to different extents and that this expression could be attributed different details in the product's design (material, shape etc). The results show furthermore that a strong expression of "safe" may result in an unsafe behaviour, contradictory to any warnings provided.

Introduction

Each year a large number of accidents occur where people are injured. In Sweden, approximately 1/4 of these accidents happen at work or in traffic while 3/4 occur in people's homes and during leisure activities, games or sports (www.vardguiden.se). A similar distribution has been reported from, e.g., Finland (www.redcross.fi) while statistics from the Netherlands indicate that 50% of all accidents occur in or around the home (Weegels 1996). In many of the accidents is at least one consumer product involved.

Different explanations exist to the emergence of accidents. Accidents are considered to have one cause (e.g. the risk homeostasis theory), to have multiple causes (e.g. the domino theory), or to be the result of the interaction between agent – host – environment (e.g. Heinrick 1959, Haddon et al 1964, Wagenaar & Reason 1990 in Weegels 1996). Yet another approach is to consider accidents from a strictly technical viewpoint (e.g. technical failures) while behavioural approaches entail models of information processing, decision-making and cognitive control (e.g., Rasmussen 1983, 1986).

With the purpose of understanding and preventing accidents that involve consumer products, the different approaches briefly mentioned above have met critique. For instance Weegels (1996) argues that the approaches address other situations than the use of consumer products, that the diversity of the user population in such situations have not been considered, nor the amount of freedom as to how, where and when to use a consumer product. Furthermore, the interplay between user and product has not been taken into consideration other than from a more traditional "human error" point of view.

The appearance of a product may, however, be just as relevant in the occurrence of accidents in that it may provoke different behaviours and ways of use (cf. Weegels 1996). The product can look safe when it is not, or it can invoke a safe way of use (Singer 1993).

This paper summarizes an explorative study investigating the relation between product appearance and users' actions from a safety point of view, in particular the feasibility of product semantics analysis as a tool to further understand this relationship.

Product semantics

The theoretical basis for the study is product semantics. Product semantics has been defined as "*.... the study of the symbolic qualities of man-made forms in the cognitive and social context of their use and application of knowledge gained to objects of industrial design.*" (Krippendorff & Butter 1984). Product semantics concerns, thus, the relationship between, on the one hand, the user and the product and, on the other, the importance that artefacts assume in an operational and social context.

Monö (1997) has, based on product semantics theories, chosen to describe the product as a trinity. The first dimension, the *ergonomic whole* includes everything that concerns the adjustment of the design to human physique and behaviour when using the product. The *technical whole* stands for the technical function of the product, its construction and production. The third aspect, the *communicative whole*, designates the product's ability to communicate with users and its adjustment to human perception and intellect (Monö 1997). Through the product gestalt, i.e. the totality of colour, material, surface structure, taste, sound, etc. appearing and functioning as a whole, the product communicates a message which is received and interpreted by the customer/user. This message is, according to Monö (1997), "created" by four semantic functions. The semantic functions are:

- *To identify*. The product gestalt identifies, e.g., its origin and product area. A bowl can be identified as part of a specific china set; a company can be identified by its trademark or by a specific design philosophy apparent in its products.
- *To describe*. The product gestalt can describe the product's purpose and its function. It can also describe the way the product should be used and handled. For instance, a doorknob can describe the way it should be gripped and turned.
- *To express*. The product gestalt expresses the product's properties, for instance "stability", "lightness" or "softness".
- *To exhort*. The product gestalt triggers a user to react in a specific way without contemplating or interpreting the product's message. For instance the user is triggered to be careful and to be precise in his/her operation of the product.

From a product safety point of view, at least the last three of the above functions may play an important role in the way users interact with and use a product. The study described in this paper has, however, focused in the function "to express".

The empirical study

Introduction

Small children are often involved in accidents. From a product semantic perspective, one reason may be that the children's parents have assumed that the products used in caring for the children, e.g. nursing tables, bathing chairs etc., are safe products. The products may express "safety" in a way that make the parents less aware of possible risks associated with the products' use, the products may communicate a false message of safety.

One product, a particular type of bathing chair, was noted to have been involved in a number of drowning accidents. Investigations showed that the parents had left the child in the chair, placed in the bathtub, for a short while to, e.g., open the front door. They had done so even though a warning label read: "Warning! To prevent drowning never leave the child unattended!" During the parents' absence, the chair had tilted and the child had fallen forwards, face down in the water.

Table 1. Number of subjects who indicated what were the desired properties, undesired properties, and properties of no consequence (n = 24).

Property	Desired	Undesired	No of consequence
Active	16	–	8
Functional	23	–	1
Durable	22	–	2
Clinical	–	19	5
Amusing	15	–	9
Soft	24	–	–
Unpleasant	–	23	1
Frightening	–	24	–
Safe	23	–	1
Restful	12	1	10

Method, procedures and results

The fundamental method used for the evaluation and comparison was Product Semantic Analysis (Karlsson & Wikstrom 1999, Wikstrom 2002).

The first step in the analysis is the identification of key concepts and the construction of a semantic scale and evaluation instrument. The assumption is that desired and undesired product expressions can be verbalized by consumers. The words used for describing the qualities of a design are, however, for each type of product. Therefore the words describing the particular product were generated in two focus group interviews, the first consisting of eight parents with children aged 0–1 year old, the second consisting of individuals who did not have any small children. According to the focus groups, bathing a small child was associated with "play" and "activity". At the same time one could never feel altogether "certain" or "safe", one had to be "careful". Almost identical lists of words were generated in the two groups, including (in translation from Swedish) "safe", "useful", "comfortable" and "soft". In order to acquire the individual user's appraisal of the product expression, two semantic instruments were constructed on basis of the results from the focus group interviews. One was used for the assessment of desirable and undesirable qualities of the product while the second was used for the assessment of the specific product's expression (see Table 1). The semantic scale used was a visual analogue scale ranging from 0–100. The one pole indicated the maximum value of the property, expressed in terms of a word describing this certain property; the midpoint designated a "neutral" value for the property, while the opposite pole indicated the opposite of the maximum value (the meaning left to the individuals themselves).

The second step concerned the evaluation of different product designs. Five different bathing chairs (products A-E) were evaluated (see Figure 1). Altogether 24 subjects participated in the study of which 16 were parents with children aged 0–1 year old and 8 had no children of their own or had grown up children. Eight were men and 16 were women, their average age was 33 years. The five chairs were evaluated, one at a time in a randomized order, in two different contexts; in a neutral environment (white background, on white cardboard) and in the intended use environment (placed in the subject's own bath tub, in 15 cm deep water). The subjects were allowed to see and touch the product but not try it out by placing a small child in it.

The subjects' indications of desired and undesired product expressions show that "safe" was a desired expression, as was, e.g., "soft" (see Table 1). No differences could be noticed between subjects with or without children.

Figure 1. Products A and B. Product A (left) is the product that was documented to have been involved in a number of drowning accidents.

Table 2. Ratings of product expressions (median values, n = 24). The scale ranged from 0 = not at all to 100 = maximum.

Property	Product				
	A	B	C	D	E
Active	**85**	61	27	50	52
Functional	67	67	74	59	**81**
Durable	81	76	**88**	50	79
Clinical	18	40	**86**	56	44
Amusing	**80**	56	21	20	50
Soft	57	59	76	75	**81**
Unpleasant	50	50	54	50	22
Frightening	21	24	50	12	22
Safe	*81*	56	64	50	60
Restful	38	41	**82**	78	**82**

The subjects were asked to rate the expression of each product (see Table 2). The different designs were perceived differently in terms of what they expressed and how "strong" the expression was. The ratings did not appear to be influenced by context.

The subjects were also asked to describe what in the product gestalt that caused a specific expression. For instance, products A and B were considered to express the properties "functional" and "durable". The expression "functional" in product A was achieved by the fact that the child was considered to be able could sit on its own in the chair without any support from the parent. Product C was considered to have a "clinical" expression, i.e. a not desired expression. Associations were clinical experiments and medical examinations, the product was therefore considered "frightening". Product D was, on the other hand, considered to express desired properties, such as "soft" and 'soothing. These expressions originated mainly from the choice of material (terry cloth). The results show further that product A was perceived to have a strong expression of safety. Compared to the other four products the rating of "safe" was the highest and with the least deviations. The product was explained to be safe mainly because of its form; a large flat surface on which the child sits, a T-bar to keep the child in position, and large suction cups in each corner of the chair to keep it firmly attached to the bathtub. Only one out of the 24 subjects commented on the possible risk of the suction cups becoming unattached.

In an additional study participated another 12 parents with small children aged 0–1 years. Four were men and 8 women, their average age was 30 years. In this study, the subjects was first asked to rate the expression of product A. Then two different scenarios were presented and the subjects were asked to describe the way they would act given the described situation. The expression of "safe" was given a median value of 87. Ten out of the 12 subjects would consider doing "other things" in the bathroom while the child was having a bath placed in the bathing chair. Six would feel safe enough to leave the child alone for a few minutes. The reasons given corresponded to those in the previous study.

Conclusions

The study shows that a product's appearance may influence a user's safety behaviour. Different product designs may result in different expressions, including more or less "safe". An expression of "safety" may, through the product's gestalt, carry stories of use in which no risks or reasons for caution are present. These stories may be what determine people's behaviour and use of products. Product semantics may be an important complementary tool when designing products and when analysing the causes for accidents.

References

Heinrick H.W. 1959, Industrial accident prevention. A scientific approach. McGraw-Hill, New York.
Haddon W., Suchman E.A., & Klein D. 1964, Accident research. Methods and approaches. Harper & Row, New York.
Karlsson M. & Wikstrom L. 1999, Beyond aesthetics! Competitor advantage by a holistic approach to product design. In proceedings from the 6th international product development conference, EIASM, Cambridge, July 5–6.
Krippendorff K. & Butter R. 1984, Product Semantics. Exploring the symbolic qualities of form. The Journal of the Industrial Designers' Society of America. Spring.
Monö R. 1997, Design for product understanding. Liber, Stockholm.
Singer L.D. 1993, Product safety and form. In: Interface '93., 84–88.
Wagenaar W.A. & Reason J.T. 1990, Types and tokens in road accident causation. Ergonomics, 33 (10/11), 1365–1375.
Weegels M.F. 1996, Accidents involving consumer products. Delft Technical University, Delft.
Wikstrom L. 2002, The product's message. Methods for assessing the product's semantic functions from a user perspective. Department of Product and Production Development, Chalmers University of Technology. Gothenburg. (In Swedish)

A SEMANTIC DIFFERENTIAL STUDY OF COMBINED VISUAL AND TACTILE STIMULI FOR PACKAGE DESIGN

Brian Henson, Donald Choo, Cathy Barnes & Tom Childs

*The School of Mechanical Engineering, University of Leeds,
Leeds, LS2 9JT, UK*

People's subjective responses to products depend on the messages they receive from different product features. A semantic differential study is reported here, to establish the subjective effects of combining tactile and visual stimuli. Respondents were asked to complete semantic differential questionnaires after touching surface textures; after touching surface textures printed over images; and after looking at but not touching the images. A principal components analysis identified two components; the first a reaction to the images and the second a reaction to the surface textures. With one exception, the component scores from the combined visual and tactile tests were weighted averages of the scores of tactile and visual tests alone.

Introduction

Consumers' emotional engagement with products depends on the combination of messages they receive from different product stimuli. The stimuli include, amongst others, visual cues such as branding and colour, and tactile cues such as surface texture. Messages from each of the stimuli should be complementary to ensure a coherent product message.

This paper reports a preliminary study to establish the subjective effects of combining tactile and visual stimuli. Although there has been much previous research into how humans integrate information from different senses, this has concentrated mainly on cognitive aspects, such as whether visual cues can improve tactile discrimination, rather than on the affective influence of the senses. For example, Zampini et al (2003) researched the combination of auditory and tactile; and Zellner and Whitten (1999) investigated visual and olfactory stimuli.

In the visual and tactile domains, Guest and Spence (2003) have shown that combining visual and tactile stimuli does not enhance perception of surface texture. They cite previous work (such as Jones and O'Neil (1985)) that suggests that visual and tactile inputs lead to weighted averaging of the information from the different senses. Guest and Spence conclude that visual and tactile inputs act as independent sources of information which can both contribute to the decision process. Others have obtained similar results (Lederman et al, 1986). The nature of the averaging that occurs in these studies is not fully characterized.

In the area of affective integration of the senses, Schifferstein (2004) used self-report questionnaires to investigate which senses dominate consumers' interaction with products. The relative importance of the senses was found to depend heavily on the particular product, for example on whether the product was a vase or a television. Schifferstein's approach depends on the ability of subjects accurately to report their experiences of using products.

If a person's ability to perceive a stimulus is related to how they feel about the stimulus, then the literature suggests that combining visual and tactile stimuli will result in an affective response which is a weighted average of the responses to the two stimuli separately and that the average should be weighted in favour of the visual stimulus.

This paper reports an experiment to test this hypothesis. The tests reported here were carried out without specifying a product context, but the results may have application to the specification, for example, of consumer goods' packaging.

Methodology

Test item preparation and word selection

The tactile test items for the semantic study were 4 by 4 arrays of 20 mm by 20 mm patches of transparent surface textures printed on to 180 mm by 155 mm transparent acetate sheet. The surface textures were made from polymers screen-printed onto the sheet. The surfaces were printed with two different polymers resulting in two types of surfaces designated "Polymer A" and "Polymer B". A was elastic to touch while B was slightly sticky. Patterns of dots were printed, characterized by the % of surface area covered by the polymer and the number of lines per inch (LPI). The nominal details of the surfaces are shown in Table 1. Stylus surface profilometry showed that for coverages of 75% and 100% individual dots merged and could not be resolved and that patches 5 and 6 were patterns of holes rather than bumps.

Four of the printed acetate sheets were placed against backgrounds made from a selection of smileys chosen by the researchers (Figure 1) and were surrounded by a neutral-coloured card border. Smileys are images that are inserted into email messages to convey the tone intended by the sender. Another four were place against white backgrounds and surrounded by identical neutral-coloured card. These duplicate four pairs of test items were used in the psychophysical experiments. Figure 1 reproduces the smileys at the same resolution as was used in the experiments, but their original colour was yellow, with some variations of shade.

A focus group of ten students was held to identify the adjectives used for the semantic questionnaire. The group was shown the smileys and was asked to touch the selection of surface textures. 31 descriptive words used by the group were noted. Fifteen of them were selected by the investigators, and converted to bi-polar pairs (Table 2), to cover the dimensions of the evaluation, potency and activity classification proposed by Osgood (1957).

Psychophysical testing

Sixty three students were recruited to complete the questionnaires. The questionnaires asked them to indicate their subjective response to stimuli on a five point semantic scale between two polar opposite adjectives (Figure 2). Three experiments were run. The participants were asked to 1) touch the square patches without the smileys. 2) touch the patches with smileys as a background and 3) respond to the questionnaire by looking at, but not touching, the smileys.

Analysis

All three of the experiments, touch alone, touch and visual, and visual alone were grouped as a single data set. For every test item and word the 63 questionnaire responses were converted from ticks to numbers between -2 and $+2$ (the example in Figure 2 is -1) and averaged. A principal component analysis was carried out on the n (items) $\times$ m (words) matrix of average numbers using SPSS version 11.0. First of all the semantic

Table 1. Nominal surface texture characteristics.

Patch	Polymer	Coverage %	LPI	Patch	Polymer	Coverage %	LPI
1	B	45	60	9	A	15	20
2	A	75	80	10	B	15	40
3	B	45	20	11	A	100	60
4	A	45	20	12	B	15	20
5	A	75	20	13	B	100	20
6	B	75	20	14	A	100	20
7	A	15	40	15	B	100	60
8	B	75	80	16	A	45	60

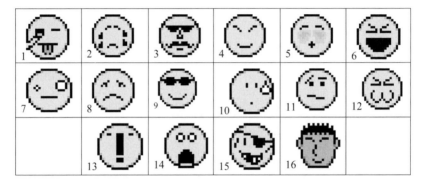

Figure 1. Smileys 1 to 16.

Table 2. List of bi-polar adjectives for semantic questionnaire.

Funny-Stern	Unique-Ordinary	Smart-Stupid	Simple-Complex	Relaxed-Tensed
Like-Hate	Smooth-Coarse	Vulgar-Subtle	Artificial-Natural	Domestic-Industrial
Witty-Serious	Cool-Snobbish	Fashionable-Retro	Practical-Impractical	Happy-Sad

	Strongly agree	Agree	Neutral	Agree	Strongly agree	
Unique		✓				Ordinary

Figure 2. Structure of self-report questionnaire form, with an example tick response.

word space was derived, with the loadings of the words in that space. Then the items were scored into the space. Details of the procedure, using Varimax rotation and Kaiser normalization, but for a different study, have been published elsewhere (Barnes et al, 2004).

Results

During analysis, an error was found in the way that item 13 had been assembled, and this may have affected the responses. Item 13 was therefore removed from the analyses.

Table 3. Rotated component matrix.

Variable	Component 1	Component 2	Component 3	Variable	Component 1	Component 2	Component 3
Funny	**0.95**	−0.04	0.01	Like	0.66	0.53	0.44
Unique	0.60	−0.61	−0.27	Domestic	0.27	0.78	0.11
Happy	0.90	0.13	0.33	Vulgar	0.12	**−0.85**	−0.25
Smart	0.39	0.20	**0.75**	Artificial	−0.21	−0.53	0.59
Simple	0.21	**0.93**	0.06	Fashionable	0.68	0.14	0.52
Smooth	0.10	**0.92**	0.09	Cool	0.75	0.26	0.48
Practical	0.19	0.34	**0.81**	Witty	**0.94**	0.07	−0.03
				Relaxed	0.75	0.51	0.31

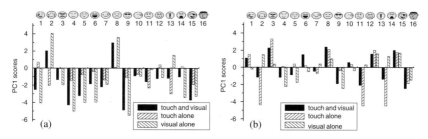

Figure 3. Histogram of items' (a) PC1 and (b) PC2 scores.

The components extraction resulted in three components with eigenvalues greater than 1. The first component accounted for 49% of the variance, the second component a further 23% and the third 11%, 83% in all. The rotated component matrix containing the word loadings on the principal components is shown in Table 3. The adjectives pairs which loaded most highly on principal component 1 (PC1) are funny/stern and witty/serious and these loadings have been picked out by bold italics. The adjective pair happy/sad loaded highly on PC 1 but also loaded on PC3 with a correlation value of 0.33. The adjectives which loaded highly on PC2 alone were simple/complex, smooth/coarse and vulgar/subtle. There were no adjective pairs which loaded highly on PC3 alone. The pair smart/stupid loaded highly on PC3 but also loaded on PC1 (with a correlation of 0.39) and practical/impractical loaded highly on PC3 but also loaded on PC2 (with a correlation of 0.34). No adjectives can be clearly identified that load into PC3 alone and consequently PC3 is not considered further.

The projections of the three sets (touch, touch and visual, and visual) of 15 items' scores on to the PC1 and PC2 directions are plotted as histograms in Figures 3a and b respectively.

Discussion

Inspection of Figure 3 suggests that it is highly likely that PC1 is a reaction to the smileys and PC2 is a reaction to the surface textures. The visual items have the highest component scores in PC1 and the tactile items have the highest scores in PC2. The adjectives that loaded most highly on PC1 are funny/stern and witty/serious. Those smileys depicting happy faces receive large negative scores (indicating agreement with the first word of each word pair) whereas those smileys that appear to be sad or crying receive large positive scores. The adjectives that loaded most highly on PC2 are simple/complex, smooth/coarse and vulgar/subtle. Those surface textures which are smoothest, items 2, 11 and 14, received large negative scores (again, indicating agreement with the first word in each pair). Those which felt roughest, items 3, 8, 12 and 15, received high scores. The visual items appear to dominate the responses to the combined stimuli in PC1 and the tactile items dominate more of the responses to the combined stimuli in PC2.

In most cases the component scores of the combined visual and tactile test items are weighted averages of the scores of tactile and visual test items alone. For some of the test items though, the scores for the combined stimuli were larger than either of the tactile and visual stimuli alone. For all of these (items 7, 12, 16 for PC1 and 8, 10, 15, 16 for PC2) the difference between combined stimuli score and the visual score (for PC1) or the tactile score (for PC2) is, given the confidence interval of the measurement, not significant. Only item 6 for PC2 shows a combined stimuli score significantly departing from a weighted mean value.

Conclusions

This experiment confirms that for the visual and tactile senses the effects of combining two stimuli results in a subjective response that is usually a weighted average of the two stimuli combined. In some cases, the combined stimuli score is dominated by one or other of the visual and tactile stimuli. But in others, if one stimulus is incongruent to the other, a clear dilution of score occurs. For surfaces that are intended to be both seen and touched, both their visual and tactile effects need to be considered.

References

Barnes, C.J., Childs, T.H.C., Henson, B. and Southee, C.H., 2004, Surface finish and touch – a case study in a new human factors tribology, *Wear*, **257**, 740–750.

Guest, S. and Spence, C., 2003, What role does multisensory integration play in the visuotactile perception of texture?, *International Journal of Psychophysiology*, **50**, 63–80.

Jones, B. and O'Neil, S., 1985, Combining vision and touch in texture perception, *Perception & Psychophysics*, **37**, 66–72.

Osgood, C., Suci, G. and Tannenbaum, P., 1957. *The measurement of meaning*. (University of Illinois Press, Urbana).

Lederman, S.J., Thorne, G. and Jones, B., 1986, Perception of Texture by Vision and Touch: Multidimensionality and Intersensory Integration, *Journal of Experimental Psychology*, **12**, 169–180.

Schifferstein, H.N.J., 2004, Sensing the Senses: Multimodal Research with Applications in Product Design, *Fourth International Conference on Design and Emotion*, 12–14 July 2004, Ankara, Turkey.

Zampini, M., Guest, S. and Spence, C., 2003, The role of auditory cues in modulating the perception of electric toothbrushes, *Journal of Dental Research*, **82**, 929–932.

Zellner, D.A. and Whitten, L.A., 1999, The effect of color intensity and appropriateness of color-induced odor enhancement, *American Journal of Psychology*, **11**, 585–604.

DRIVERS' EXPERIENCES OF MATERIAL QUALITIES IN PRESENT & FUTURE INTERIORS OF CARS AND TRUCKS

Lena Sperling & Per Eriksson

Division of Industrial Design, Lund University, P.O. Box 118, SE-221 00 Lund, Sweden

The aim was to study how professional drivers experience hard surface materials of their present vehicle interiors, "green" materials of and other future materials. Truck and car drivers participated in a pilot study. Material quality experiences of their present vehicles were studied with a customer satisfaction interview. Drivers were in general satisfied with material qualities of present vehicles but gave lower ratings and many comments about cleaning aspects. The User Compass Chart (UCC) was used for characterisation of 37 different material samples. Vectors were "more professional" – "less professional" and "more natural" – "more synthetic". Drivers were asked to position material pieces on the UCC and to adjust positions if needed. Some material samples were more often positioned in the project's desirable North-East sector. The UCC proved to be a useful mediating tool for identification of user's experiences.

Introduction

The main objective of the BIOAUTO project was to explore the possibility of industrially viable applications of renewable resource based materials for the manufacture of automotive interior components. In order to minimise impact on the environment, hard surface materials of ecological properties should be used in an as natural shape as possible. Vehicle interiors are important for pride of professional drivers and for their daily well-being, but it was not known which qualities of surface materials they appreciate in today's vehicles, which materials will be valued by them in the future and to which degree ecological materials may be visible in their real qualities. Experiences of car interiors were previously investigated with the quantitative semantic environment description, SMB (Küller 1991), and it was concluded that with qualitative techniques this method would provide more interesting results (Karlsson et al 2003). The present subproject aims at identifying professional drivers' positively perceived qualities of existing and future hard surface materials of vehicle interiors. Values beyond material aspects were to be investigated.

Methods

Company members of the consortium suggested some specific vehicles to be included in the study. Four male truck drivers and four male taxi drivers participated together with a female private car driver, all of them 46–63-years-old. Truck and taxi drivers had between 19 and 40 years experience of professional driving and their yearly service mileage was between 60, 000 and 250, 000 kilometres. One truck driver and two taxi drivers were owners of their vehicles.

Interviews about hard surface materials in existing vehicles

The following qualities with explaining associations were to be evaluated by the drivers in their present vehicles: (1) Durability; (2) Hygiene; (3) Touch; (4) Look; (5) Sound; (6) Smell; (7) Hard surfaces of the interior as a whole and (8) Interior as a whole. Drivers were asked to judge qualities by means of the VOICE customer satisfaction scale (Volvo Car Corporation 1986), ranking from "Very dissatisfied" (1–2) to "Very satisfied" (9–10), which is easy to use as a mediating tool for generating comments. Subjects were for each question urged to comment on their ratings. If a subject was dissatisfied with a quality he/she was also asked to rate the degree of severity of the problem.

Interviews about material samples of present and future vehicles

A selection of 37 material samples, including ecologic materials, was compiled, among them specimens from vehicle companies of the consortium as well as new materials of the project. Other samples were selected in order to complete an interesting mix of various material qualities. Samples were presented in pieces of about 50 × 50 mm and without frames or other fittings. Five different material categories were represented: (1) Plastic-like; (2) Rubber-like; (3) Wood-like; (4) Stone-like and (5) Future-like. A User Compass Chart (UCC) 500 times 500 mm was designed for the experiment (Sperling and Eriksson 2005). The idea was based on the compass or sector chart used in industry to demonstrate the target and direction of strategic development. Labels of the four chart points were decided after discussions within the project group. Labels were: (North) *More professional* (with the associations adequate, efficient); (South) *More unprofessional* (associations inadequate, amateurish), (West) *More synthetic* (associations in-natural, artificial) and (East) *More natural* (associations environment-friendly, ecologic). In the central part of the UCC was a neutral zone (neither nor). The – from the project's point of view – desirable combination was located in the North-East sector (Figure 1).

Procedure

The aim of the project and the experimental procedure were presented before each interview. It was pointed out to the subject that he/she was regarded the expert of the experiences about the vehicle's materials and that all comments consequently were

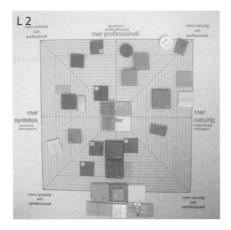

Figure 1. A truck driver's positioning of material samples and markers representing materials of his present vehicle (black) and dream vehicle (white).

most important. The first part of each interview took place in the actual vehicle's cockpit. The experimenter ticked the driver's ratings on the scales and noted even rather peripheral comments in the questionnaire. The subject had full overview of the ticking and the notes in the protocol. For practical reasons, the interview with the material samples took place indoors, with the UCC game board on a table. The use of the UCC was presented by means of a polystyrene sample, which the experimenter positioned in the South-West corner of the game board. The subject was then asked to position each sample in the sector which he/she regarded as most adequate. It was pointed out that the characterisation was about the subject's emotions and not a test in knowledge about materials. Material categories were presented in the same order to all subjects but the order within each category optional. When all pieces were positioned, the subject got the opportunity to adjust the positions of the samples. Finally, the subject was asked to position a black round piece representing the hard surface materials of his/her present vehicle and a white piece representing materials of a "dream vehicle". Subjects' reflections were at the first experiments documented with a digital recorder but later on with video in order to document the connections between the subjects' verbal reflections and positioning of each sample. Each UCC experiment resulted in an individual image board of experienced material qualities which was documented with a digital camera. The subjects were awarded with cinema-cheques.

Results and discussion

Experiences of materials of existing vehicles

Drivers were rather or very satisfied with qualities of the hard surface materials of their present vehicle interiors. Ratings by truck drivers were mostly lower than those by car drivers, probably due to the higher demands of the truck drivers' working environment. Neutral and lower ratings were only given by three truck drivers and one taxi driver regarding hygiene (difficulties to clean fittings and narrow dust-collecting compartments), touch and hard surfaces of the interior as a whole. Ratings by the female car driver were within the range of those of male taxi drivers. Focusing on drivers' positive comments on material qualities of their present vehicle interiors, and converting experienced problems into potentials, a series of user requirements could be formulated for hard surface materials in future vehicle interiors. With the UCC, all professional drivers except one characterised hard surface materials of their present cars as professional, while one truck driver was neutral regarding both professional and natural. One truck driver regarded materials of his present interior as somewhat more natural (Scania day cab 2004), while two characterised the hard materials of their present interiors as *neutral,* and one as somewhat more synthetic (Volvo FH12 1999). Differences between judgements were small. Two taxi drivers judged the hard surfaces of their present car interiors as most professional (Mercedes Benz and SAAB 95). One taxi driver positioned the marker of his present car corresponding to *rather professional* as well as *natural* (Volvo V70). The Mercedes Benz driver was neutral regarding the degree of natural. Two taxi drivers regarded their cars as somewhat more synthetic (Chevrolet Transport and SAAB 95). The marker of the dream car was by all except one of the truck drivers positioned within the North East sector. It is reasonable to believe that the deviating driver positioned his marker on the text *more natural* without further reflections about the degree of professional.

Positioning of material pieces on the User Compass Chart

Differences in individual positioning of material samples were large. An example of a truck driver is presented (Figure 1). Of specific interest was to study the material samples positioned in the, from the project's point of view, favourable North East sector (more professional as well as more natural). In summary, *wood-like* materials demonstrated the highest percentage presence in the N.E: sector, followed by *stone-like* and *plastic-like* samples. Compared to truck drivers, the participant taxi drivers positioned some more *plastic-, stone-* and *future-like* samples in the North East sector. Among the wood-like samples, *Oak* was in total the most frequent one, followed by *Ash* and *Oak imitation*. It is perhaps noticeable that two drivers positioned the sample *Flax composite hard* from the BIOAUTO project in the N.E sector. Among the stone-like, *Grey granit* composite was the most frequent sample (n = 5).The plastic-like sample *Petrified wood* was regarded as professional by all subjects who decided to position it on the UCC (n = 6). The future-like *Carbon sandwich, shiny* sample was regarded synthetic by all positioning subjects (n = 7), five of whom however also as professional. The sample *Wood fibre in PLA matrix* of the BIOAUTO project was characterised as unprofessional by all subjects. All material samples positioned in the North-East sectors were of neutral colour, except for a red stone-like and a blue future-like sample. In the South-West sector, future- and plastic-like materials were most frequent. Truck-drivers were more negative regarding future-like materials than car-drivers.

In the present study, all drivers, except for one subject, were professional drivers had long-term experience of driving. Consequently their expressed experiences of interior materials are respectable and most important. They were in general very satisfied with the material qualities, but problems related to cleaning need to be solved as soon as possible. New materials should be introduced together with improved cleaning possibilities to enhance their impact. Owners of vehicles seemed to be more positive to the qualities of their vehicles than others. Only one less severe problem was reported by the truck-driver who owned his vehicle. Interviews about present vehicle interiors made it evident that the VOICE scale was an efficient mediating tool for eliciting of user requirements. At the UCC sessions, all professional drivers positioned their dream vehicles in the favourable North East sector. As subjects were informed about the "green" aims of the project at the introduction, it may be possible that this have biased their viewpoint. Natural wood (oak and ash) proved to be experienced as natural as well as professional by a majority of the nine drivers, followed by imitated oak, although the unframed sample made it evident that it was synthetic. This is an important indication that not only dark and more traditional wood materials and their imitations may be used in future car interiors. However, also plastic-like materials were experienced as natural as well as professional. If ecological materials are functionally or emotionally unacceptable as surface materials, they could be provided with more appealing surface materials of a professional image. The possibility to introduce stone-like surfaces should be further considered and investigated. Especially professional drivers were reluctant regarding future-like materials, and it would be interesting to repeat the experiment with them in some years in order to investigate if they would be more positive in the future.

The UCC proved to be a stimulating mediating tool for communication with users. The method gives associations to a game and was easy to understand for all subjects, although positioning of a few samples seemed to be somewhat occasional. Material samples were presented as cuttings of larger pieces and the visual impression of the lateral sides of the samples may have biased the subjects. However, the piece of imitated oak had visible plastic structures but was nevertheless one of the material pieces regarded as natural. In future UCC experiments samples should be fitted into some kind of frames in order to make them more uniform.

Conclusions

Hard surface materials contribute to several important qualities of vehicle interiors. Interviews about material surfaces demonstrated that cleanability of materials and compartments must be improved. All drivers regarded the hard surface materials of their present vehicles as professional and, in total, as somewhat more natural than synthetic. In future cars, drivers seem to like surface materials experienced as considerably more natural than today's materials. Wood and wood-like materials of lighter shades should be further investigated as surface materials in vehicles. The User Compass Chart proved to be a useful mediating tool for identification of user's experiences of qualities of materials but should in the future include video-recording of positioning of samples.

Acknowledgements

The project was supported by VINNOVA and the Swedish automotive industry, through the BIOAUTO consortium coordinated by IFP Research AB, Sweden.

References

Karlsson B., Aronsson N. and Svensson K. 2003, Using semantic environment description as a tool to evaluate car interiors. *Ergonomics*, **46**, 1408–1422.

Küller R. 1991. Environmental assessment from a neuropsychological perspective. In T. Gärling and G.W. Evans (Eds.) *Environment, Cognition, and Action. An Integrated Approach,* (Oxford University Press, New York), 111–147.

Sperling L., Eriksson P. 2005, Drivers' experiences of interior materials in vehicles. Report to the BIOAUTO project consortium. Division of Industrial Design, LTH, Lund University, Sweden.

Volvo Car Corporation, 1986. *Some questions to you as a Volvo driver.* Quality Staff, Volvo Car Corporation, Göteborg, Sweden.

OF THE INTERACTION OF CULTURAL AND EMOTIONAL FACTORS ON INDUSTRIAL DESIGN

Pierre-Henri Dejean, Marilia de Souza & Claudia Mourthe

Université de Technologie de Compiégne, Centre Pierre Guillaumat, Compiégne 60206, France

This survey aims at crossing the results of previous researches on the involvement of cultural factors (CF) in industrial design (ID) with emotional engineering (EE) a new topic developed by the European ENGAGE consortium. We will apply to EE the method adopted for CF. The purpose of this survey is to cross the results of previous researches focused on the involvement of cultural factors in industrial design with emotional engineering that is a new topic studied by the ENGAGE consortium. We will apply to EE the method adopted for CF. Exploration of a bibliography, proposition of a model, elaboration of an experimental protocol to test it, the aim being to obtain a strong and reliable system of references in order to develop specific tools to deal with the questions aroused by this crossed study of the interaction of cultural and emotional factors on industrial design.

Introduction

This working paper proposes to compare the impact of emotional and cultural factors. Actually we set as a working hypothesis that CF influences emotions. What type of influences are they? And how can consumers' cultural and emotional behaviours be predicted? And what kinds of data, tools, references could be provided to designers so that cultural emotions are involved in the design process? The answer is to get a strong and reliable system of references to develop specific tools adapted to the different questions aroused by the crossing of emotional and cultural factors.

The product design process fluctuates between two attitudes. The artistic one gives total freedom to the designer, considered as a creative master. The industrial one, held back by constraints, is more directive. Actually, a midway attitude prevails, much influenced by existing knowledge and tools in different fields. Technical fields provide knowledge and tools that give to the product objective data respected, accepted and often understood by many. While for human factors fields, dealing with emotion or usability before the situation is quite different. Confronted by these fields designers have to adopt subjective approaches and they have trouble to explain and justify objectively their results. Our objective is to conceive the knowledge and provide the tools to turn objective or more accurately implicit design into objective and measurable results. Nine years ago cultural factors were our challenge (de Souza, 2001). To day we have the same challenge concerning emotion.

We organise our paper in three parts, first is a survey about relevant cultural factors in design, second is some results about a survey in progress about emotion and design, third exploration of influences and links between this two fields. We conclude with an anticipation of some tools that could be interesting to improve the emotional quality of products.

Emotion

We have considered that we were starting from scratch, which obliged us to adopt an exploring approach, as wide as possible. One good reason was that currently the meaning of the concept and the word emotion varies according to the different contexts and situations. We have adopted different ways to define the different meanings of emotion. First, the academic definitions how "emotion" appear in fundamental works; secondly the application, how "emotion" is used in different professional fields where this factor is recognised as important, such as in arts, marketing, human factors; third discussions with our research focus group and design expert, the fourth is this kind of conference where we can hear papers that have relationships with emotion, or propose a paper and hear the reactions.

The results are very instructive and place emotion in a complex way. Emotion is opposed to rationality from the academic point of view, it is an immediate feeling that influences the mood, the cognitive behaviours from the professional people's point of view. Emotion can modify the result of the normal reasoning process. Airlines, the nuclear industry, the army would like to avoid this problem. They would like to suppress "emotional situations" and ask for this the help of human factors to maintain emotion at a reasonable level, preventing loss of control. Emotion in this context is negative. On the contrary in the arts, fine arts, music, movies, games the goal of the artist is to communicate emotion. Performance without emotion is nothing and emotion in this context is positive and really a goal in the design process. The product design field closer to marketing, business, communication and advertisement pays a great attention to emotion in order to communicate a "positive emotion" to the consumer.

From this outlook we retain the research that gives a generic framework that could help the observer or the designer to identify and classify the emotions. Russell's circumplex model (1980) introduces a dynamic system with two dimensions: valence (pleasure, displeasure) as horizontal axe and arousal (arousal-sleep) as vertical axis. Ekman (1992) proposes a list of universal basic emotions, Lang, (1980) proposes an affective rating system in order to get self-assessments of the dimensions of valence and arousal, the Self-Assessment Manikin (SAM). The interesting thing is that these references were used in different applications like we could see in HCI 05 (Oertel, 2005).

We must improve this part of our work to progress and to clarify some questions that were raised in discussions with the focus group. Emotion can be considered in some studies like a brief feeling, an impulse, such as flashing on a product in a shop or over reacting to a problem in an emergency. But can it be considered in other like a lasting state of consciousness? An important question regarding the product field is: can we consider that the immediate emotion that the consumer felt for the product will became an emotion that last as long as the product itself. This question could permit us to apply the Moles (1972) very interesting psychological theory about the client cognitive justification after an affective purchase.

Culture

The starting point of our team's involvement in "cultural factors in design" was the perceived lack of designers' knowledge in mastering the design process in relation to this major question. We only present an outline of the full research programme. The aim of the research was to develop a theoretical basis and tools. It was necessary to borrow knowledge developed by disciplines outside that of Design. The approach adopted combined a bibliography review in which the perspectives developed by each discipline were confronted and compared in order to try to apply the knowledge of each to product design, and present the result to professionals and experts.

Design, culture, intercultural studies, anthropology were all explored in relation to this subject. Each discipline has developed its own point of view leading to different fields of application. The most important points retained are the following. 1) Cultural studies (Berry, 1997, Cuche, 1996) which allowed us to see how culture, civilisation, nationalism and regionalism are related, and their relationships and influences with innovation and conservatism. This type of knowledge is very important in order to define the field of applied of culture to product design. 2) Ethnology and anthropology with in particular, the reciprocal influences of object and thought (Leroi Gourhan, 1964) the relation to the body (Warnier, 1999) and its current prolonged effect on ergonomics (Rabardel, 1996). This kind of knowledge is very important in order to understand the dynamic links existing between products and culture. 3) Intercultural studies (Demorgon, 1996; Hofstede, 1994; Hall, 1971,1984; Trompenars, 1993) which play a central role in isolating cultural factors. This knowledge enables us to propose a model and discuss this.

Here we present just some main ideas. The concept of culture remains the idea of work leading to a transformation. Culture is not innate but acquired, and also developed in a system and shared by a same human group (Cuche, 1996). Culture does not substitute itself for personality but contributes to its formation. Culture moves with time, it is alive. Lastly a main result of the research was to identified 7 cultural factors as bearings for use in analysis or design of products: order, time, communication, space, belonging, regulation, and gender. This list is certainly not exhaustive, as other factors may be identified and the above factors may be redefined. We give an example based on these factors.

A questionnaire introduced the seven factors and was launched in France, Singapore and USA. The first step showed us that the major factor which was relevant for persons living in Singapore was risk management. Singapore attributes a high degree of attention to risk prevention and in particular to the control of uncertainty. What is shown in the questionnaire results is confirmed by examination of different products, and everyday behaviour of Singapore citizens. Regulation was studied by Hofstede (1994), and Usunier (1992). The central idea concerns how one culture considers and manages the control of uncertainty and risk management.

The same study was conducted in the USA at Orlando, University Central Florida. In this way it was possible by means of comparison to confirm the importance of risk management as a factor in user's behaviour towards products in Singapore. Some Singapore people were surprised, astonished or shocked by the proverb "somebody who never takes risks, never gets anything". often achieved a score of 1 ("it is stupid statement") or 2 ("the statement implies danger"). In Orlando, the response was often 5 ("it is an obvious statement"). Furthermore a similar proverb figures on a sport picture in an office, at the ergonomics department of University of Central Florida "Win without risk is to conquer without glory", which in effect is the translation of "A vaincre sans perils, on triomphe sans gloire" (Pierre Corneille, Le Cid, XVII century). It is easy to find a strong relationship between emotion and this cultural factor. It is interesting too to see that human factors in this case pay attention to the emotion and to propose training exercises toward avoiding more and more the fear and the loss of control.

Emotion and culture

This part of the paper is more speculative and needs experiment in order to confirm the ideas that are explained. We are just proposing some hypothesis of applications of each cultural factor about emotion. We build on our 7 cultural factors.

Firstly communication. Two tendencies exist in communication (Hall, 1979; Usunier, 1992): 1) A strong reference to the context, that is the implicit style. Your interlocutor assumes that you know perfectly what he is talking about. Reminding you of it seems unnecessary and may even seem to be offensive. 2) A feeble reference to the context is an explicit communication style. Your interlocutor then supposes that you are "starting from scratch". He must bring you all the elements that will allow you to understand. This could be a main factor: emotion will be visible or hidden depending on implicit or explicit style.

Then belonging. Trompenars (1993) refers to the relationship between individuals and the community of society. It is important to know what are the different objects and values corresponding to a personal usage or a usage more pertaining to a group. Customisation appears as an emotional factor.

Then gender. Hall (1979) and Hofstede (1994) express the variations in the delegation of roles within a culture from an another. We can define a society that tends to differentiate the roles and activities according to gender to be masculine and if the roles are interchangeable, the society is defined as being more feminine. If products are very different for males and females, emotionally it will not be comfortable for a man to use a device designed for a female.

Then time. The important question about this factor is time management in the organisation of the present moment: monochronism refers to only accepting one task or action at a time, while polychronism is to accept actions and tasks at the same time. The loss of control could be evident when a monochromic user has to manage a polychronic device.

Then order. This is the way adopted to ordering and subordinating things, values, needs, and relations between individuals (Hofstede, 1994; Usunier, 1992). This factor applies to organisation, order, and importance.

Then space. Hall (1959), describes the proximity and distance accepted between individuals and objects as well as the opening or enclosure of a space. Every traveller knows the emotion provoked by the adaptation to a new space style.

Finally hierarchical distance and the logic of organisation. This refers to the way in which cultures organise and accept relations of power.

Emotion could affect the outlook on the products, organisation of tasks, roles and freedom of individuals, subordination of the tasks in the organisation of use, handling the product and first use.

Conclusion

The importance of some phenomena like culture, emotion, in product design has been established for a long time. However designers lacked theoretical bases and tools to approach their influences. In these situations managers are often limited to employ "experts" who can sometimes take an aura of "gurus". Nowadays Design cannot avoid subjects like cultural factors, emotion. Product design can become an important topic in research policy of many disciplines that are involved in these subjects, psychology, anthropology, philosophy.

Culture and emotion are extremely complex notions. It would be risky to give a more superficial definition of the words. It is therefore important to be able to refer to the background and to underline some elements of complexity here. It saves confusing the effects and their manifestations with the causes and the superficial elements, with those which are more profound, a means of recognising a living culture. Following on from this, what we call Swedish blue is a colour that makes one think immediately of Sweden and in the same way the Scottish kilt instantly evokes Scotland, both are culturally recognised. However here it is more a question of signs of recognition of cultural identity than the living expression of a culture. Behaviour in relation to an everyday product, in relation to space (body language, movements, accepted distances, arrangement and disposition of things amongst themselves) and its conception can bring us to the heart of culture and to what the product maintains or doesn't maintain in regard to formal symbolic values. In other words it is not the habit which denotes the monk nor the kilt which denotes the Scotsman. Therefore, the approach to understanding a product, the organisation of activities to use it, whether they are functional, social or informal and the gestures and thoughts which are connected to the object in question and the dynamics of all of these combined need to be considered.

References

Berry J W., Cegal M H. & Kagitciba C.1997. *Handbook of cross-cultural psychology: vol3 social behaviour and application*, second edition (Editors Cambridge Press).
Cuche D.1996. *La notion de culture dans les sciences sociales* (Paris, La dècouverte).
Dejean P-H. & De Souza M. 2001. *Integrer les facteurs culturels dans la conception de produits.*(Les techniques de l'ingénieur Paris) http//www.techniques-ingenieur..fr
Demorgon J. 1996. *Complexite des cultures et de l'interculturalite*. (Paris. Anthropos.)
Ekman P. 1992. *An argument for basic emotions*. Cognition & Emotion, 6, 169–200.
Hall E T. 1959. *Le langage silencieux*. (Paris. Seuil.)
Hall E T. 1979. *Audela de la culture*. (Paris. Seuil.)
Hofstede Geert 1994. *Vivre dans un monde multiculturel. Comprendre nos programmations mentales*. (Paris. Les editions d'organisation)
Lang PJ., Bradley MM. & Cuthbert BN. 1997. *International Affective Picture System (IAPS): Technical Manual and Affective Ratings*. Retrieved February 20, 2005 from: http://www.unifesp.br/dpsicobio/adap/instructions.pd
Leroi-Gourhan A. 1964. *Le geste et la parole*. (Paris. Albin Michel.)
Moles A. *Théorie des objets* (Editions Universitaires Päris 1972).
Oertel K., Schultz R., Blec M., Herbort O., Voskamp J. & Urban B. 2005. *EmoTetris for Recognition of Affective States* (HCI 2005 proceeding Las Vegas US).
Rabardel P. 1996. *Activites avec instruments*. (Paris. PUF)
Russell J. 1980. *A circumplex model of affect*. Journal of Personality and Social Psychology, 39(6), 1161–1178.
de Souza FM. 2001. *Culture et Design: application de l'interculturalite à l'évaluation et à la conception de produits*. Memoire de These. Universite de Technologie de Compiegne.
Trompenars 1993. *Riding the waves of culture- Understanding cultural diversity in business*. (Economic Books)
Usunier J-C. 1992. *Commerce entre cultures, une approche culturelle du marketing international*. (Paris. PUF)
Warnier J-P. 1999. *Construire la culture materielle: l'homme qui pensait avec ses doigts*. (Paris. PUF)

UNDERSTANDING USER EXPERIENCE FOR SCENARIO BUILDING: A CASE IN PUBLIC TRANSPORT DESIGN

Gülay Hasdoğan, Naz Evyapan & Fatma Korkut

Middle East Technical University (METU), Department of Industrial Design, 06531 Ankara, Turkey

Scenario building is widely used as a resourceful method to explore and foresee the experiential aspects of user-product interaction in the product design process. It is commonly agreed that an initial research stage that explores the existing situations is essential to be able to envisage imaginary situations. The paper discusses the nature of scenario building process in general and how scenarios contributed to the design process in the case of an educational project on inner-city public transportation design. It focuses on how user experience was explored through research in context and how the findings shaped the scenarios and design concepts in the process.

Introduction

Scenario building is a widely used method to explore the dynamic nature of user-product interaction within a variety of contexts and conditions in the product design process. It is an act of conceiving the product in a story line that involves users, activities, events and the environment. Designers commonly use scenarios in a *spontaneous* manner as part of their cognitive and external communication activity. As the complexity of consumer products increased with the introduction of information technology, the role of design has shifted from creating static forms to facilitating dynamic interactions. This change in the nature of design activity nourished new efforts to use scenario building in a *systematic* manner.

Traditional ergonomics uses scenarios as part of evaluating and modelling the human performance in work conditions. Accident scenarios and task analysis are used both to analyse the efficiency of workplaces and to foresee possible failure conditions in order to increase human performance. As complex interfaces became part of consumer products, the role of human operator was replaced with ordinary untrained user. User satisfaction, rather than human performance, has become the main focus. Thus, in addition to performance modelling, experience modelling came into the scene of ergonomics. Pleasurability was introduced as a new design criterion, besides conventional ones such as functionality and usability. New methods such as participatory design and ethnographic methods have emerged to explore the subjective measures of user satisfaction and emotional responses. Scenario building was among those that proved to be a resourceful method to explore and foresee the experiential aspects of user-product interaction (Fulton Suri and Marsh, 2000; Zoels and Gabrielli, 2002).

Scenario building is mainly used as a method in the concept generation phase of the design process. It is commonly agreed that an initial research stage that explores the existing situations is essential to be able to envisage imaginary situations. Scenarios are also favoured as a communication tool between the members of the design team (Fulton Suri and Marsh, 2000) and as a powerful presentation tool to convey dynamic interaction ideas to the clients (Zoels and Gabrielli, 2003). Scenario building supports the widely accepted phases of the design process, namely need identification, concept generation, design, and communication through (1) exploring existing situations and conditions; (2) conceiving imaginary situations and future conditions; and (3) presenting and contextualising the design idea. With reference to these three phases, the paper will discuss the nature of scenario building process in general and how scenarios contributed to the phases of design process in the case of an educational project in particular. It will focus on how user experience was explored through research in usage context and how its findings shaped the scenarios and design concepts.

Scenario building process for an inner-city public transportation vehicle design

A senior year project carried out in 2004 involved collaboration between METU Department of Industrial Design and Otokar, a mini and midi bus manufacturing firm in Turkey. Students were required to develop a concept design for minibuses used primarily as an inner-city public transportation vehicle called *dolmuş*. The word *dolmuş* means "full" in Turkish. It is a public transportation vehicle that only starts its journey when it is full with passengers. Therefore it is not bound to a specific time schedule. The drivers of *dolmuş* are small entrepreneurs who aim to obtain maximum profit from maximum capacity. Customer research in Otokar revealed that almost all *dolmuş* drivers take the vehicle to maintainers just after the purchase to modify and decorate its exterior and interior. Considering the widespread use of this vehicle in daily life, the students were required to explore the needs and expectations of both the driver and the passengers, and to respond to them with holistic design solutions suitable for the production line of the firm. The project was carried out in four groups and involved 15 students.

Research: Exploring existing situations and conditions

Initial research phase which supports scenario building involves exploring existing situations and conditions by focusing on real contexts in which people use the product or service under consideration. Particular techniques that explore user experience include interviews that let users think aloud in the usage context; creating mood boards; telling stories and narratives to elicit user's expectations, aspirations, needs, experiences and emotional responses. Video-ethnography and shadowing are among the observational techniques that aim to gain an understanding of people's attitudes, behaviours and values within their social context. When it is not possible to talk to users, other stakeholders are used as user surrogates (McQuaid et al, 2003); "walking in the shoes of users" or role-playing in a real context with real people are among the other techniques to empathise with users (Buchenau and Fulton Suri, 2000).

In the *dolmuş* project, the process involved a user research phase after which the groups determined their design goals and the specific issues they would address for achieving them. The students made field observations and interviews, and presented the immediately observed problems and critical situations related to the usage of the vehicle from the perspectives of drivers and passengers. Being a passenger in *dolmuş* was a familiar experience for them; however, they needed to explore the variety of needs of other types of passengers. One of the groups made field observations by taking journeys in the vehicle and playing the role of passengers. They took photographs during the activities of the driver and passengers in order to identify problems in getting on and off, sitting and moving in the vehicle, paying the fare, changing the route label (Figure 1) etc. They also documented additional accessories such as fare box that do not come with the vehicle but added later by the drivers. Another group interviewed the drivers to elicit their opinions about the efficiency, usability and maintainability of

Figure 1. User research: Field observation for identifying usability problems (left) and drivers' modifications (right).

the current vehicle. They also inquired drivers' aspirations by asking their brand choice of a private car. The same group took photographs of the interior and exterior of the vehicles in use to document the accessories added and the modifications made by drivers for driving comfort, passenger comfort, appearance, personalising the vehicle and achieving aspirations. Figure 1 shows the accessories and logos added on the exterior of the vehicles to make them look more powerful.

Concept generation: Conceiving imaginary situations and future conditions

Scenarios enable designers to hypothetically test and contextualise a design idea during the concept generation and development phase of the design process. Initial research phase prepares the ground for characters, setting and events of the scenario. Being informed of the characteristics, needs, aspirations and expectations of potential users, it becomes possible to characterise typical and untypical users. These fictional characters created as part of scenarios are variously referred to as personas, user profiles and user archetypes in the related literature (Pruitt and Grudin, 2003). Possible activities, situations, events and consequences are also identified together with personas in order to allow multiple narratives in various combinations. Multiple narratives give the designer the opportunity to envisage unexpected situations and conditions in relation to the design concept. The elements of a scenario are sometimes identified in a systematic way that is called concept mapping (McQuaid et al, 2003). When the user experience is concerned in particular, the characters are matched with various critical usage situations, activities and events in relation to their needs, expectations and/or aspirations, and then their consequential emotional responses are envisaged. Scenarios cultivate and make design concepts flourish; as new concepts emerge, scenarios provide them with new grounds to test.

In the *dolmuş* project, based on user research and the project goals, the groups prepared scenarios that represented the issues they wanted to address. The scenarios were developed to describe various usage situations related to the vehicle, and the users in context. This stage helped the groups to determine their final design concepts. One of the groups continued the field observation by identifying four different travel lines, and created four driver profiles. They described travelling conditions arising from the travel lines such as the length of journey, intensity of the traffic and the needs of the various passenger profiles such as students, elderly and workers. This helped them to combine driver profiles, passenger expectations and conditions of travel lines to explore and identify a variety of critical usage situations. Another group also continued fieldwork by revisiting the drivers in context in order to inquire more private or less visible usage situations. They asked a driver how a typical working day begins, how he drives

Figure 2. Scenario building: The real user acting out the usage situations.

to and queues in the main stop, and asked him to describe and at the same time act as if he is wiping off the snow, taking a nap and having breakfast in the vehicle while queuing (Figure 2). This helped them to identify the problems and situations the driver experiences when not driving passengers. The same group developed scenarios by role-playing in context that helped them to envisage situations such as night drives, morning traffic and late working hours from the perspectives of drivers and passengers. They also developed future scenarios involving better-educated drivers or increased security problems, and envisaged how travel comfort and interaction between drivers and passengers would be affected. A third group developed scenarios envisaging the difficult situations the passengers encounter such as the problem of getting on and off the vehicle when it is full, especially for disabled and elderly.

The initial ideas developed by the groups showed that it was necessary to extend the ideas to the entire vehicle, not only focusing on a particular area. An interaction matrix exercise was conducted to let the groups discuss system components and project objectives in terms of the stakeholders (driver, passengers, other vehicles and pedestrians). After observing the entire problem area and envisaging a wide range of possible usage situations, the groups were better able to have a holistic perspective of the design problem.

Communication: Presenting and contextualising the design idea

Scenarios help to contextualise the design ideas in the form of narratives and make them easy to understand for all parties both during the design process and as a final output. Storyboards, snapshots, videos and animations are among the commonly used techniques to communicate.

The *dolmuş* design process continued with a preliminary evaluation jury for which the groups elaborated scenarios describing usage and users in context, and drawings and models of their final concept. The final jury requirements included scenarios describing the design concept, drawings and a scaled final model. Scenarios were used to communicate the project outcomes to the course tutors and the collaborating firm. This time scenarios described the final concepts from various perspectives and how it related to the overall system it was envisaged to fit.

The overall design strategies followed by the students can be grouped under six main categories: (1) *Exterior appearance:* maintaining a strong and powerful look that preserved the traditional cubical form of *dolmuş* in accordance with the identified aspirations of the drivers. (2) *Passenger's feeling of comfort and privacy:* spacious interiors with larger window areas, comfortable and individualised seats, providing private space for passengers, homogeneous ventilation. (3) *Driver's feeling of comfort, privacy and security:* personal storage facility within driver's reach, reclining seat for sleeping posture, wider driver space allowing other activities such as reading and eating. (4) *Ease of use for passengers:* electronic route labels on the exterior for increased visibility, providing information on current location of the vehicle in the interior, ergonomic and flexible handrails for both sitting and standing passengers, ease of circulation between seats, illuminated stairs for safety during night travel, double sliding doors for ease of getting on and off. (5) *Ease of use for the driver:* ease of changing route information by electronic labels, ease of access to the engine, a fare box design for driver's arranging cash and coins easily. (6) *Adaptability to the needs of different travel lines:* modular construction for accommodating different user needs in different lines such as leaving luggage space for the central train station line.

Conclusion

The case revealed that the overall design strategies developed by the students were strongly influenced by the design process based on the exploration of user experience and scenario building. Scenario building in the design process requires extensive knowledge about the problem area. The case showed that some of the student groups preferred to go back to the field to continue user research in the scenario building phase of the project, because they needed to further develop their understanding of the context and users. Therefore the initial research phase is crucial to understand the user experience and context especially when the designer is not familiar with the problem area. The case also indicated that user modifications and interventions on the products could be a valuable source of information to understand their expectations and aspirations. Letting *real* users to act out usage situations can also be a rich source of information to understand user experience especially for unfamiliar or less exposed usage situations. Creating multiple narratives by combining variety of users, usage situations and conditions is crucial in order to envisage new user experiences.

References

Buchenau, M. and Fulton Suri, J. 2000, Experience prototyping, In *the Proceeding of the DIS2000, Designing Interactive Systems*, (ACM Press, New York), 424–433.

Fulton Suri, J. F. and Marsh, M. 2000, Scenario building as an ergonomics method in consumer product design, *Applied Ergonomics,* **31**, 151–157.

McQuaid, H.M., Goel, A. and McManus, M. 2003, When you can't talk to customers: Using storyboards and narratives to elicit empathy for users. In *Proceedings of the Designing Pleasurable Products and Interfaces Conference*, Pittsburgh, PA, (ACM Press, New York), 120–125.

Pruitt, J. and Grudin, J. 2003, Personas: Practice and theory. In *Proceedings of the 2003 Conference on Designing for User Experiences*, (San Francisco, ACM Press).

Zoels, J. C. and Gabrielli, S. 2003, Creating imaginable futures: Using design strategy as a foresight tool. In *Proceedings of the 2003 Conference on Designing for User Experiences*, (San Francisco, ACM Press) 1–14.

SUBJECTIVE ASSESSMENT OF LAMINATE FLOORING

Simon Schütte, Anna Lindberg & Jörgen Eklund

Quality and Human-Systems Engineering, Department of Mechanical Engineering, Linköping University, SE-58183 Linköping, Sweden

The Pergo group is a major manufacturer of laminate flooring. One major goal is to design attractive products for the contractor market. A general model of Kansei Engineering was used for evaluating subjective impressions of laminate flooring. The results showed that such flooring can be described with four basic semantic descriptions and seven main properties. It could also be seen that craftsmen and architects had different impressions of laminate flooring. It was concluded that laminate flooring sometimes is perceived as inferior natural flooring material despite the fact that many of its properties objectively are superior. Hence, selective marketing may improve its image.

Introduction

The Pergo Group has its headquarters in Sweden. Pergo's business idea is to develop, produce and market laminate flooring. Its main outlet markets are Scandinavia, Western Europe and the USA. Pergo is well known in the consumer market, where the products are mainly sold through flooring retailers and home improvement chains. However when it comes to the contract market, i.e. selling through flooring entrepreneurs and architects, the same products don't work as well (Pergo, 2004).

One reason for this is that the competition is not the same. In the consumer market, laminate flooring is well established and therefore Pergo is mostly competing with other brands of laminate flooring. On the contract markets, Pergo's products are mostly challenged by hardwood or ceramic flooring, which often are considered as "the real thing". Hence, laminate flooring with wood-like decor is compared to real hardwood flooring and laminate with ceramic look to real ceramic flooring instead of comparing laminate to other laminates.

In many cases laminates are generally considered to be inferior and a low budget alternative, despite the fact that laminate in many aspects is better then the original material regarding mechanical properties, application and design. This way of thinking also affects Pergo's attempts selling their product to contractors and architects.

Apparently, offering good quality products is not enough. Products must also correspond to the customer's subjective expectations in order to be successful on the market and to be considered as an original. Therefore Pergo was interested to understand the customer's needs and integrate them into the product development work.

Laminate flooring

Laminate flooring consists of three main layers (see Figure 1). All three layers serve specific purposes, in order to give the flooring the desired characteristics. On top there is a laminate layer, made of paper material. The middle layer constitutes the core and is usually made of HDF (high-density fibreboard) or MDF (medium-density fibreboard). On the bottom, an impregnated backing layer protects against moisture, balances the

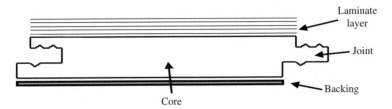

Figure 1. Laminate flooring consists of three layers.

flooring and keeps it from cupping. In modern laminate floorings this layer is often combined with a sound-reducing layer.

The different layers can be bonded through two different methods, HPL (High-Pressure Laminate) and DL (Direct Laminate). HPL is an older method where a surface laminate is first created from several sheets using pressure and heat. In a second step the laminate and the backing are glued onto the core. Using DL, all parts of the three layers are bonded simultaneously using heat and pressure. During the last couple of years most flooring companies have changed technology from glued joints to click joints, which are fit together without any glue.

Aim/Limitations

The study described in this paper was part of a bigger project carried out at Pergo with the purpose to map customer needs, to integrate Kansei Engineering methodology into Pergo's R&D processes and to give Pergo a basis from which future studies can proceed.

However, this study was only a part of it and its aim was to determine how different groups of professionals experience laminate flooring and which properties they considered to be the important. Moreover, the connection between subjective impression and physical properties should be described. Outgoing from this, improvement suggestions should be given.

Method

Kansei Engineering (Nagamachi, 1989) was chosen as the prime method due to its unique ability to grasp subjective impressions translate them into concrete product solutions and describe the linkages. Regarding the structure the researchers followed Schütte's general model on Kansei Engineering which had been used for several previous studies in European Kansei Engineering research. This structure is displayed in Figure 2.

The idea behind a product can be described from two different perspectives: The semantic description and the description of product properties. These two descriptions each span a kind of vector space. Subsequently these spaces are analysed in relation to each other in the synthesis phase indicating which of the product properties evokes which semantic impact and vice versa. After these steps have been carried out, is it possible to conduct a validity test, including several types of post-hoc analyses. The two vector spaces are updated and the synthesis step is run again if necessary. When the results from this iteration process appear satisfactory, a model (qualitative or quantitative) is built describing how the Semantic Space and the Space of Properties are associated.

Research structure

The product domain was determined with special regard to the customer group of professional users, such as contractors and architects.

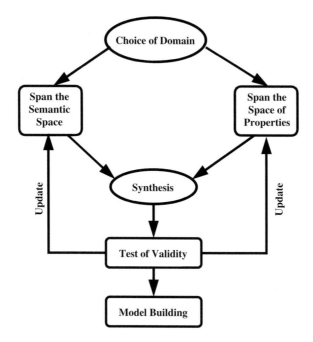

Figure 2.   Kansei Engineering structure (Schütte, 2005).

Semantic descriptions were collected from various sources and Pergo staff. Also a brainstorming was done involving only experts at Pergo. Using affinity diagram as an alternative to factor analysis the collected descriptions were grouped and ranked. The Kansei words were chosen for further evaluation in customer questionnaires (Osgood et al, 1957).

Samples of laminate and wooden flooring available on the Swedish market were collected. The product characteristics were analysed and used to create a Pareto chart in order to rank the importance of the identified characteristics to the intended user groups. Focus and expert groups were used to get product attribute information. Nine different samples were extracted, including prototypes, which were presented in parts of 400 mm × 170 mm. This implies that the interaction was limited to viewing and touching a product sample but not walking on it.

In the synthesis phase the product samples were presented to architects and contractors from different companies and varying backgrounds. The participants were supposed to rate the product samples on semantic differential scales against the selected Kansei words. The data distribution and quality was checked prior using e.g. K-S test. Also, the questionnaire data was treated with factor analysis. Regression analysis with Quantification Theory Type 1 (QT1) (Komazawa, and Hayashi, 1976) was conducted, linking the Semantic Space and the Space of Product Properties.

Result

The collection of semantic expressions describing laminate flooring gave 116 words; the list was completed by another 66 words from the brainstorming at Pergo. Affinity analysis grouped these words into 14 groups, of which 16 words were selected as Kansei words for further evaluation.

The product properties were selected from originally over 70 items of which 3 are unique for Pergo's image. However, it is expected that these attributes will make Pergo's flooring distinctive and act as a unique selling point. Table 1 presents the final selection.

The questionnaire data provided the base for factor analysis. Four basic factors were extracted, explaining the variances of all 16 Kansei words as displayed in Table 2. Factor 1 determines the impression if the floor "can do its job". Factor 2 focuses on appearance of the floor. The third factor describes the durability of the appearance. Finally, factor 4 represents the impression of the floor in its environment.

Joining together the Semantic Space represented by the Kansei words and the Space of Properties represented by product attributes in Table 2 revealed the relative importance of the properties. Table 3 presents the properties according to their affinities to a certain factor and in order of their relative importance. It is seen that the two respondent groups disagreed regarding factors 1 and 2, but had similar opinions for factors 3 and 4.

Table 1. Product properties chosen.

Properties from existing products	Properties from new concepts	Properties reflecting the company image
Format (plank, tile) Thickness Type of décor (wood-like, other) Surface (matt oiled, wood texture, smooth or ceramic) Type of wood décor	Texture (synchronic; other) Tone	Carrier (regular, moisture resistant, soundblock®)

Table 2. Kansei words and clusters.

Factor 1: "Reliable/practical"	Factor 2: "Modern design"	Factor 3: "Classic style"	Factor 4: "Nice and solid"
Reliable Practical	Designed Modern Expressive	Pure style Timeless Fresh-looking Classic Exclusive	Snug Solid Welcoming Natural Safe Quality

Table 3. Result of QT 1.

	Factor 1: "Reliable/ practical"	Factor 2: "Modern design"	Factor 3: "Classic style"	Factor 4: "Nice and solid"
Craftsmen	Core/Carrier	Surface Thickness	Type of décor Surface Format	Surface Format
Architects	Surface Type of decor	Type of décor Tone Format		

Discussion

Since the collection of the Kansei words and product properties was mostly done by Pergo staff, this might have resulted in biased starting data. However, the involvement of external experts and constant verification should have avoided this.

Before the participants filled out the questionnaires they were asked to test the product samples provided. Since these samples were presented as a tile, the participants had to image how the flooring would look like in a room. This might have influenced the results. However, previous studies show that the order and subjective impression often is equivalent to the real product.

Conclusions

It was concluded that architects apparently prefer natural flooring, but appreciate the availability of varying properties such as tone, structure and pattern type according to their individual preferences. Craftsmen, on the other hand, set most value on properties, which related more strongly to mechanical attributes rather than visual appearance. Selective marketing may help improving acceptance of laminate flooring. Moreover, it can be assumed that craftsmen are very aware of the specialties of a laminate floor since they are concerned with those properties, which relates to the Pergo company image. Architects, on the other hand, are not aware of these differences and probably cannot easily recognize a laminate floor. The tone, which was introduced as a property from a new concept, was only recognized by architects as an important feature. This is probably due to their more extensive background regarding trends and competing products.

References

Pergo, 2004, Pergo Homepage, www.pergo.com.
Nagamachi, M., 1989, *Kansei Engineering* (Kaibundo, Tokio).
Schütte, S., 2005, *Engineering Emotional Values in Product Design – Kansei Engineering in Development*, Doctoral Thesis (Linköping University).
Osgood, C.E., Succi, G.J. and Tannenbaum, P.H., 1957, *The Measurement of Meaning*, (University of Illinois Press, Illinois).
Komazawa, T. and Hayashi, C., 1976, A statistical method for quantification of categorical data and its application to medical science. In de Dombal, F.T. and Gremy, F. (ed.) *Can Information Science Help* (North Holland Publishing Co, Amsterdam).

REALPEOPLE

Shayal Chhibber, Sam Porter, J. Mark Porter & Lynda Healey

Department of Design and Technology, Loughborough University, Loughborough, LE11 3TU

Traditionally, ergonomics input into design has focused on the physical and cognitive effort placed on the user of a product or system. This input often occurs late in the design process, in the form of prototype evaluations, task analyses, and so on. However, modern user centred design strategies have broadened to consider a more holistic view of the user, attempting to understand their emotional and "pleasure" needs to a greater extent. To date, few of the methods that have been developed to investigate user "pleasure" are available to the designer from the very start of the design process, and crucially, represented in a suitable format for the manner in which designers work. The RealPeople project is an attempt to fill this void; giving designers' data about the "pleasure" needs of different consumer segments from the start of the design process, in a format that they find useful and usable.

RealPeople is an AHRC (Arts and Humanities Research Council) funded research programme that has led to the development of an interactive DVD based database. It is designed to inspire and inform designers in the early stages of the design process; highlighting the key "pleasure" needs of different market demographics and promoting greater empathy with the user. The resource is not intended to replace first hand research, but designed to inspire initial concept generation, and to guide and focus preliminary research activity.

The resource has been developed through investigating designers' needs, and their desires from tools that aid the design process. The interface allows designers to specify a user group through selecting certain variables e.g. gender, and view statistically validated data on that groups' generic attitudes towards products. They can also browse individual people's data, where there is more in-depth information about products that they find pleasurable, providing a more intimate portrait of different market segments. These data take the form of short video clips, high quality still images, and detailed lifestyle information.

The data in the resource has been collected through questionnaires and interviews with a quota sampling strategy used to take an inclusive view of the population, seeking participants from both sexes, and a broad range of ages. The database has been evaluated by designers, receiving positive feedback about the manner in which it operates, the quality of the information in it, and the format in which it is presented.

For further information please contact:

c.s.porter@lboro.ac.uk
j.m.porter@lboro.ac.uk

HADRIAN

Russell Marshall[1], J. Mark Porter[1], Ruth Sims[1], Diane E. Gyi[2] & Keith Case[3]

[1]Department of Design and Technology, [2]Department of Human Sciences, [3]Mechanical and Manufacturing Engineering, Loughborough University, Loughborough, LE11 3TU

"Design for All" or inclusive design is an approach to product, environment or service design that aims to maximise usability, not to tailor designs to the user in a bespoke fashion, but to provide a single solution that accommodates the needs of all users including those who are older or disabled. Key to this is establishing empathy between designers and the people who would primarily benefit who are often older, in poor health and unable to achieve all the tasks they would like to with ease and confidence (if at all).

The use of ergonomics evaluations in design rely heavily on data concerned with body size, joint mobility, strength and other anthropometric or biomechanical characteristics. However, there are concerns with existing databases: the commonly accepted view of designing for 5th to 95th percentile users, the lack of empathy in designing for a disembodied set of numbers, the poor applicability of metrics taken in highly standardised postures, for all right-handed people, to maximal (beyond comfort) limits, and avoid of any task related context, or only related to a task devoid of any realism. Furthermore most data focuses on the physical and does not convey how people can be excluded from using a product or service because of factors such as the cognitive and emotional. For example, a person may be able to reach a control but not able to manipulate it as required; able to use a lift but not able to locate it in a noisy and cluttered train station; able to walk unaided but not confident enough to cross a busy road, and so on.

HADRIAN is our inclusive design tool created and developed as part of ongoing research into Design for All. HADRIAN consists of a unique database of 100 individuals whose data are kept intact rather than broken down into percentile tables. Data is included on a broad range of size, shape, and abilities and is presented in a highly visual manner including images and video of the individuals performing a variety of task such as placing items on kitchen shelves, removing a baking tray from an oven with the use of oven gloves, getting into and out of in a variety of seats, or entering/exiting from a range of public transport.

The database can also be used in a task analysis environment that allows assessments of design exclusion. Using HADRIAN's task analysis tool a design can be assessed in a virtual user trial by the individuals in the database. The analysis presents a visualisation of any problems encountered by the individuals and results in the identification of the number of people who fail an element of the task and are consequently "designed out". Designers are then encouraged in exploring design solutions that are more socially inclusive.

For further information please contact:

j.m.porter@lboro.ac.uk
r.marshall@lboro.ac.uk

HCI SYMPOSIUM – KNOWING THE USER

VALIDATING DIAGNOSTIC DESIGN KNOWLEDGE FOR AIR TRAFFIC MANAGEMENT: A SUCCESSFUL CASE-STUDY

Becky Hill & John Long

Ergonomics & HCI Unit,
Now at: UCL Interaction Centre, University College
London, 31-32 Alfred Place, London WC1E 7DP

This paper reports research, validating design knowledge for air traffic management (ATM). The knowledge is applied to an ATM simulation to diagnose design problems, associated with controller planning horizons. The case-study is judged a success. The design knowledge is correctly operationalised, tested and generalised to a simulation, more complex than that used to develop the knowledge.

Introduction

Cognitive Ergonomics researchers have been criticised for not building on each other's work (Newman, 1994). Elsewhere, Long (1996) claimed that poor discipline progress resides partly in the failure of research to validate its design knowledge. This paper reports a successful case-study, validating diagnostic design knowledge, applied to air traffic management (ATM).

Reconstructed Air Traffic Management: ATM is the planning and control of air traffic. Operational ATM manages air traffic, e.g., Manchester Ringway Control Centre (UK). The Centre manages a terminal manoeuvring area configured by: 9 beacons; more than 2 airways; 1 stack; and 2 exits. The management involves track and vertical separation rules. Planning is supported by paper flight progress strips and controlling by radar. Dowell (1998) developed a simulation of the Centre – termed "reconstructed air traffic management" (rATM). It comprised: 5 beacons; 2 airways; and no stack. Traffic was limited to 8 aircraft and sector entry was staggered. There was a single controller. Dowell (1993) also developed a domain model, comprising airspace and aircraft objects, consisting of attributes having values. Transformation of these attribute values results in aircraft "safety" and "expedition", which express performance as "task quality".

Diagnostic Design Knowledge: Timmer (1999) has developed a Theory of Operator Planning Horizons (TOPH). It consists of: a set of frameworks (domain; interactive worksystem (operator and devices); and performance (Dowell, 1998)) and a method for diagnosing design problems, associated with operator planning horizons. Timmer applied TOPH to rATM to produce a set of models, which he used to diagnose design problems. TOPH is to be validated here.

Design Knowledge Validation: following Long (1996), design knowledge validation comprises: conceptualisation; operationalisation; test; and generalisation. Here, Timmer's conceptualised TOPH is operationalised, tested, and generalised over a more complex ATM simulation.

Features of a Correct Operationalisation: following Stork, Middlemass and Long (1995), the features of a correct operationalisation of TOPH are: 1. Diagnosis completeness; 2. Diagnosis consistency; 3. Application of domain, worksystem and performance models; 4. Rationale for model application; 5. Features of diagnostic method, embodied in diagnosis.

Case-study Success: following Middlemass, Stork and Long (1999), case-studies of design knowledge can be successful or unsuccessful. These two types of outcome together establish the scope of the design knowledge. Design scenarios are considered to vary in their: definition; complexity; and observability.

Case-study Scenario: TOPH was applied to an ATM simulation, "reconstructed validation air traffic management" (rvATM) (Debernard and Crevits, 2000). rvATM simulates an en-route sector in the region of Bordeaux (France). It is configured by: 21 beacons; multiple airways and multiple exits. The traffic is heavy, up to 40 aircraft on sector at any one time and flight patterns are very varied. Track and vertical separation rules are close to operational. There are two controllers – planning, responsible for electronic flight strips and radar, responsible for control. The validator (first author here), applying TOPH, was trained in HCI. She had no previous experience in TOPH application. The validation study was "managed" by the second author. All difficulties in applying TOPH were documented. rATM and rvATM are similar as concerns definition and observability. However, rvATM is more complex, having a more complex sector configuration and a more extensive and varied traffic profile. The flight strips are electronic. There are two controllers. These differences are the basis for the generalisation process.

Design knowledge application and evaluation

An observational study was conducted, using four video cameras. Videos recorded both planning and radar displays and their controllers, including verbal communications. The controllers were practised in rvATM. The validator later produced a protocol of the synthesised data (PSD) and constructed a table of controller interventions (TCI). Ambiguities were resolved with the controllers.

Examples of interventions follow: "Worried about conflict between KLM051 and N7225U. Plan change KLM051 after AFR543. KLM051 turn right. IBE712 change heading direct TERNI". The TCI includes: the aircraft; beacons; controller plans; and the validator's comments.

An integrated model for rvATM (rvIM) is now constructed, using the TOPH frameworks. Table 1 shows extracts from the rvIM for aircraft IBE550. It integrates work system-related models (Columns 1–5) with domain related-models (Columns 6 and 7). Column 1 models the goals of the worksystem ((planning and radar) and devices (flight progress strips and radar)). Column 2 models the controllers' behaviours. The model uses the TOPH operator architecture – physical ("head" and "hands") and mental ("working and long-term memory, and goal store"). It also includes "process structures" ("search for" and "form goal") and "representation structures" (categories of aircraft – "active", "expeditious" and goals – "establish", "amend", and "intervene"). Physical behaviours can be observed on the video recording (a controller head movement towards the radar, indicating a "search for" (aircraft) behaviour). Mental behaviours are inferred. Column 3 shows a model of the controllers' representation of the domain. The model uses TOPH mental categories for managed aircraft ("incoming/active"; "safe/unsafe"; "expeditious/ unexpeditious"). Categories in turn derive from domain attribute values, such as aircraft; radar position; altitude; speed; heading etc. Column 4 shows the controllers' representation of the devices, i.e. flight strips and radar. Column 5 shows a model of device behaviours, with which the controller's behaviours (Column 2) interact. A comparison between Column 2 and Column 5 indicates appropriateness of the interactions for achieving the goals (Column 1). Column 6 shows a model of the product goal achievement, expressing the effect of an intervention on an aircraft. The achievement relates to the worksystem's goals. Column 7 shows a domain model of the state of each aircraft. The two highest states are "safe" (not in conflict with other aircraft) and "expeditious" (moving through the sector in a timely manner). The rvIM is now complete.

Table 1. Extracts from rvATM Integrated Model for aircraft IBE550.

Worksystem goals	Controller behaviour	Controller rep (domain)	Controller rep devices	Device behaviour	Product goal	Aircraft transformation
(A) Intervention IBE550 Heading 39 at ENSAC	HIGHLIGHT: IBE550 FPS PULLDOWN: change heading SELECT: 39 CATEGORISE: IBE550	IBE550, heading 45 changing IBE550, (from) active safe expeditious to active safe unexpeditious (heading) aircraft	IBE550 FPS selected IBE550 FPS, heading 39	IBE550 FPS highlighted IBE550 FPS heading 39 Radar BTZ, IBE550, heading 39	–	IBE550 Progress worse Fuel use worse Safety same Exit worse
Planning/ Executions	POP GOAL: (B)					

Table 2. Planning horizon for aircraft IBE550.

Encode	Intervention	Category	Plan/Execution
FPS	–	Incoming aircraft	–
Radar trace	–	Active aircraft	–
Heading 45 Altitude 310	–	Active safe expeditious aircraft	Change heading 39 at ENSAC
Position Ensac	IBE550 Heading change 39	Active safe aircraft unexpeditious (heading)	Leave IBE550
Position SAU Alt 310 Heading 39	–	Active safe unexpeditious (heading)	Give Heading Terni after KLM358 passed
–	–	Lapse	Lapse
Position Velin Alt 310 Heading 39	IBE550 Heading change POI	Active safe expeditious (heading)	Change heading to POI
–	–	Active aircraft exit	–

Before diagnosing design problems, the controller's planning horizons need to be constructed. Following TOPH, controller tasks comprise: administration; monitoring; and planning/execution. Planning horizons can be constructed only for planning/execution tasks. A plan is a mental representation structure, associated with mental process structures ("form"; "discard"; "decay" etc), giving rise to planning behaviours. Plans can have three different outcomes: "plan and decay"; "plan and discard"; and "plan and execute". Planning horizons are constructed on information, associated with: the controller; the devices; the plan; its extension (over time) and its adequacy (to achieve worksystem's goals). The data are extracted from the PSD and the rvIM. The planning horizon for IBE550 is shown is Table 2.

Column 1 shows the controller's encoding of IBE550. Column 2 shows the controller's interventions. Column 3 shows the aircraft category. Column 4 shows the plan/execution.

Given the PSD, the rvIM and the planning horizon, the diagnosis method can be applied in its four stages: 1. Identify problem; 2. Analyse planning horizon; 3. Extract data from the integrated model; and 4. Generate causal theory. A design problem exists, when actual and desired performance differ. The planning horizon provides an overview of the design problem. It supports causal theory generation. The theory suggests possible design solutions.

In the case of IBE550: an intervention has produced poor quality of work (unexpeditious aircraft with respect to heading). The problem arose, due to an intervention to make the aircraft safe. The controller later made a plan to rectify the unexpeditious problem by changing heading. This plan decays (or is discarded). To solve this design problem the controller would have to take more account of aircraft progress and fuel use, when planning. In rvATM, safety is of primary importance and little attention is paid to fuel use etc. Information on fuel use could be displayed and a prompt issued, indicating such reductions in performance. The prompt might be to direct the controller to return an aircraft to its original airway more quickly, thus enabling performance parameters to become as desired and so solving this particular design problem.

The rvATM diagnosis is evaluated here analytically. First, the diagnosis is considered to be of rvATM, as supported by the controller interventions, observed by video and documented in the PSD and rvIM. Second, the diagnosis is of design problems, as identified by Column 7 of Table 1. Last, the diagnosis relates to planning, as supported by the planning horizon and the causal theory. The rvATM diagnosis meets the requirements of being a design problem, associated with controller planning.

TOPH is judged to be correctly operationalised. First, the diagnosis is complete, as it corresponds to the application of the diagnosis method. Second, the diagnosis is consistent with the planning horizon, which is consistent with the rvIM, which is in turn consistent with the controller's interventions and the PSD. Third, the domain, worksystem and performance models of the rvIM are applied to the planning horizon construction and so to diagnosis formulation. Fourth, the rationale for the application of the models has been (selectively) exposed. Last, features of the diagnostic method are embodied in the diagnosis (plan extension and adequacy).

The TOPH application is considered to meet the validation requirements. First, the design knowledge was operationalised, that is, the already conceptualised TOPH was applied in the case-study to a more complex simulation. Second, the knowledge was tested, in that it resulted in the identification of design problems, associated with operator planning. The test, however, also identified difficulties in the application of the knowledge, experienced by the validator, which must count to some extent, against the validation. For example, the syntax for representing interventions in the rvIM was found difficult to apply (Table 3). The validation can, then, be considered only partial.

The case-study is considered a success. That is, rvATM, more complex; but equally well-defined and observable as rATM, is judged to fall within the scope of TOPH. Although the case-study is successful, the validation of TOPH is only partial, because of the validator's difficulties in its application.

Discussion and conclusions

The aim of this paper is to report a successful case-study, validating TOPH diagnostic design knowledge, as applied to ATM, in the form of rvATM. The case-study is considered a success, as rvATM, more complex than rATM, is judged to fall within the scope of TOPH. The latter is partially validated, inasmuch as it was operationalised, tested and generalised. The validation, however, was only partial, because the validator experienced difficulties in the application of TOPH. These difficulties constitute design problems for TOPH and their solution is a requirement for future research.

Table 3. Difficulties, experienced by the validator in the application of TOPH.

From Page	From section/ paragraph	Diagnosis of problem	Solution to problem	Comments	Speculations
85	6.3.1.1.2 3rd paragraph	In rATM Operator physical behaviour hand movements correspond to radar (highlight; pulldown; select) whereas in rvATM, these behaviours correspond to both radar and FPSs.	Analyse hand movements corresponding to Radar (highlight; pulldown; select) and to FPS (highlight; pulldown; select)	Implemented as the flight strips are electronic and thus the corresponding hand movements in rATM for FPS (move, delete; write) do not apply here	Warn users of the method that the physical architecture will change with changes in the simulation being analysed

Last, this paper began with a critique of Cognitive Ergonomics researchers for not building on each other's work (Newman, 1994; Long, 1996). It is hoped that the research, reported here, of a successful case-study, which partially validated design knowledge for ATM, suggests how this criticism may be met.

Acknowledgments

This research was carried out in collaboration with the Laboratoire d'Automatique et de Mécanique Industrielles et Humaines (LAMIH) (Valencienne, France), under its director of research Jean-Michel Hoc and his team. CNRS (France) and the Royal Society (UK) funded the collaboration.

References

Debenard, S., and Crevits, I. 2000, Projet Amanda – note intermediaire 1.2 CENA/N12v1/.
Dowell, J. 1993, Cognitive Engineering and the rationalisation of the flight strip. Unpublished PhD Thesis, University of London.
Dowell, J. (1998). Formulating the design problem of air traffic management. *International Journal of Human-Computer Studies*, **49**, 743–766.
Long, J. 1996, Specifying relations between research and the design of human-computer interactions. *Int. Jnl. of Human-Computer Studies*, **44 (6)**, 875–920.
Newman, W. 1994, A preliminary analysis of the products of HCI research, using pro forma abstracts. In Proc. CHI '94, Boston, Mass., 278–284.
Middlemass, J., Stork, A., and Long, J. 1999, Successful case study and partial validation of MUSE, a structured method for usability engineering. In Proc. INTERACT 99, Edinburgh UK, 399–407.
Stork, A., Middlemass, J. and Long, J. 1995, Applying a structured method for usability engineering to domestic energy management user requirements: a successful case study. In Proc. HCI '95, Huddersfield, UK, 367–385.
Timmer P. 1999, Expression of operator planning horizons: a Cognitive Approach. Unpublished PhD Thesis, University of London.

HUMAN FACTORS EVALUATION OF SYSTEMS THAT RECOGNISE AND RESPOND TO USER EMOTION

Kate Hone & Lesley Axelrod

School of Information Systems, Computing and Mathematics, Brunel University, Uxbridge, UB8 3PH

There is now growing interest in the development of computer systems which respond to users' emotion and affect. Our research has been exploring the use of emotion detection at the interface from the user's perspective. It has empirically tested some of the assumptions about human behaviour which underlie the use of the technology. Through controlled experiments, using the Wizard of Oz methodology to simulate emotion recognition, we have assessed the impact of the application of emotion recognition on measures of usability and user satisfaction. This paper will summarise our main findings and explain their relevance to the development of future systems that can meet the needs and expectations of real users.

Introduction

There is now growing interest in the development of computer systems which respond to users' emotion and affect. Researchers around the world are working on systems to detect expressed user emotion at the human-computer interface, many with the intention of using this input to improve the quality of human-computer interaction. However, arguments for how this might actually be achieved are sketchy and there is little hard evidence of a benefit to users. Furthermore it is not known how people will react to the technology. These vital Human Factors questions have been neglected in favour of technological development. Our research has been exploring the use of emotion detection at the interface from the user's perspective.

Most researchers working on automatic emotion detection hope that their work will eventually lead to improved human-computer interaction. Picard (1997) for instance, maintains that giving computers the ability to recognise, react to and express emotions will make them more effective at communicating with their human users. Often the argument presented appears to rest on two premises. First, that emotional recognition and expression are important for human-human communication. Second, that humans behave towards computers as they do towards other humans. This second argument is based primarily on the controversial findings of Reeves and Nass (1996) who, in a series of empirical studies, concluded that humans respond socially and naturally to media (including computer interfaces) and that these responses are unconscious. While the work of Reeves and Nass (1996) is frequently cited by researchers in the field of affective computing, none of the experiments which they conducted pertain directly to the question of whether systems which recognise human emotion can improve human-computer interaction. At best their work suggests that users may unconsciously alter their emotional expressions (facial and voice) when communicating with a computer, since this is natural, social behaviour. It is worth noting that not all human emotional signals are unconscious, in fact it is very common for humans to deliberately send emotional expressions which do not necessarily reflect their inner state (Argyle, 1988).

It is therefore also possible that users will learn to use emotional expressions in their interactions with computers if they see a benefit in doing so.

Assuming that users will provide appropriate cues to their emotional state that systems can recognise, the next question is how that information might be used in shaping the human-computer interaction. A number of applications have been proposed which might benefit from emotion recognition components, including intelligent tutors, ubiquitous computing applications such as an intelligent CD player, entertainment applications and help systems. However, few authors have specified in any detail how emotional input would be used in these applications. Furthermore, despite widespread interest in the development and use of emotion recognition technology, minimal research effort has gone towards actually testing the potential benefits in any systematic way. Even the basic assumption that users will display recognisable emotions towards computer systems, in the same way as in human-human communication, remains largely untested. A key reason for the lack of such research has been the relative lack of robustness in the technological solutions currently under development. Most do not yet approach the level where they can be applied in real situations. Our work overcomes this limitation by using the Wizard of Oz paradigm (Fraser and Gilbert, 1991). This is a technique for simulating future interactive technology. The user interacts with a computer system, but (unbeknownst to them) the "computer's" recognition ability is provided by an experimenter. This approach has proved useful in the design of speech recognition technology and similarly has great potential for the study of future affective technology. Just as experimenters can recognise human speech in place of a computer, so a trained experimenter can recognise the visible or audible manifestations of user emotion when interacting with a system. This recognition can then be used as input to adjust the computer's responses. We took this approach to investigate some basic Human Factors questions about the benefit and use of emotion recognition technology.

The Wizard of Oz study

We were interested in determining user responses to affective computing interventions. Specifically we were concerned with whether such interventions could (a) improve performance and (b) improve user satisfaction with the system. We were also interested in the degree to which users would emphasize their emotional expressions in situations where they believed that the computer could respond to this. In order to investigate these issues an experiment was designed in which an affective computing application was simulated through the use of a Wizard of Oz (WOZ) set up. Affective interventions were provided within the context of a game involving problem solving. This used a word ladder task where the challenge is to change one word into a target word by changing one letter at a time (making a new word each time). The design of the experiment was such that in some conditions the system appeared to vary its response on the basis of recognized emotional expressions at the interface, for instance providing clues when the user displayed negative affect. We hypothesized that such interventions would lead to improved task performance and improved satisfaction. We also varied the conditions according to whether the user was explicitly told in advance that the system might react to their emotional expressions. This design allowed us to separately examine the effects due to the system response itself and those due to users' expectations. We predicted that those anticipating an affective application would show increased emotional expression during their interactions. An overview of the method is provided here. More detail can be found in Axelrod and Hone (2005a, 2005b).

Experimental design

Sixty participants took part in the study (42 male and 18 female). The experiment had a 2 × 2 between-subjects factorial design. The factors were:

1. ***Acted affective*** (with two levels; "standard" vs. "system appears affective"). This refers to whether the system appeared to act affectively. In the "standard" condition clues and messages appeared only in response to the user clicking the "help" button. In the "system appears affective" condition, if the participant was observed via the one way mirror to use emotional expressions, the game was controlled so that appropriate clues or messages appeared to them.
2. ***Told affective*** (with two levels; "expect standard system" vs. "expect affective system"). This refers to whether the participant expected the system to act affectively. In the "expect standard system" condition participants were told they would be testing the usability of a word game. In the "expect affective system" condition they were told that they would be testing the usability of a word game on a prototype system, that might recognize or respond to some of their emotions.

There were therefore four experimental conditions in total, representing the factorial combination of the two factors.

Measures

The main measures in the study were (1) task performance (2) affective behaviour and (3) subjective satisfaction. Task performance was measured by counting the number of completed "rungs" on the ladder during a 10 minute period. Behavioral responses were coded from the recorded observational data. Common behaviors such as smiling, frowning, resting chin on hand, shifting posture, grooming (e.g. adjusting hair) and blinking were identified, counted and rated for valence and intensity. More in-depth analysis of the behavioural data drew upon Sequential Interaction Analysis methods. A novel coding system was developed to identify units of emotionally expressive behaviour, "affectemes" (see Axelrod and Hone, in press, for more detail). Subjective satisfaction was assessed by asking participants to rate their affective state on a nine point scale using the Self Assessment Manikin (Bradley and Lang, 1995) which uses stylized figures to illustrate emotional state. Participants were also asked to complete a questionnaire which rated various aspects of their interaction, including whether they believed that they had shown emotion during the experiment.

Results

Task performance

ANOVA showed a main effect of system affective response on task performance ($F(1,56) = 7.82$, $p < 0.01$). Participants were able, on average, to complete significantly more rungs of the puzzle with the affective intervention.

Affective behaviour

When using an apparently affective system, users' affectemes were rated as having significantly more positive valence (with a main effect of system affective response on valance, $F(1,56) = 12/63$, $p < 0.01$). There was also an interaction effect such that the most positively rated emotional expressions were from participants who had been told the system might respond to their emotional expression and where the system did act affectively.

There was a main effect of being told the system was affective on blink rate ($F(1,56) = 4.57$, $p < 0.05$). Blink rates were significantly higher when participants were told the system may respond to emotional expressions. Those told that the system they would be using would act affectively were also rated as having significantly more intense arousal ($F(1,56) = 4.74$, $p < 0.05$).

Subjective satisfaction

There was a main effect of affective response on user subjective ratings of their affective state after the interaction ($F(1,56) = 10.25$, $p < 0.005$). Participants reported themselves as significantly happier, on average, after interaction with the system that responded to their emotional expression.

Discussion

The results described here demonstrate that significant improvements in both task performance and users' subjective feelings can be achieved as a result of adapting an application on the basis of affect recognition. Participants performed significantly better and felt more positive when using the adaptive affective recognition version of the application. This confirms suggestions that affective computing has the potential to improve human-computer interaction. Previously empirical support for this claim was limited due to the relative immaturity of the technology to enable emotion recognition. In this experiment we were able to simulate (using a Wizard of Oz set up) the proposed capability of future systems to recognize affect from observed user behavior. The results are notable as they support the importance of ongoing work to enable emotion recognition at the user interface.

The behavioural results suggest that people do adapt their emotional expression (showing more positive valence) when using an affective system. However, this result was only found for participants who (1) were told the system was affective and (2) experienced a system that did adapt to their emotional expression. This result is potentially important, since it suggests that user adaptation to affective technology depends to some extent on the user being aware that the technology is there, i.e. that the valance of expressions is being consciously controlled by users. However, it is not enough that they are told about the system, the affective response also has to take place, presumably acting to reinforce the user behaviour.

There was also evidence that users were more emotionally aroused when using a system that they had been told was affective. Both ratings of arousal and blink rate (which is often linked to arousal and stress) were higher in the "told affective" conditions. This result may be indicative that participants who are told the computer is going to observe them, experience increased anxiety as a result. Rickenberg and Reeves (2000) previously found that the presence of an onscreen animated agent watching user performance could affect user anxiety and performance in a similar way to the presence of a human observer. The results here suggest that similar effects may apply when the "observer" is even less obviously human in character (apparently involving only a video recorder and recognition software).

Overall these results suggest that affective systems have promise in terms of improving human-computer interaction. Our results have confirmed some tangible benefits of such systems. They also indicate that people do adapt their level of emotional expression as a result of interacting with such systems. The types of emotional expressions that we have observed could help developers of future affective systems by highlighting the kinds of emotional expressions that are used at the interface, at least during problem solving type tasks. The research also highlights that both the knowledge that a system is affective and whether it actually responds as such both play a role in influencing user behaviour.

Acknowledgements

This work is supported by EPSRC grant ref: GR/R81374/01.

References

Argyle, M. 1988, Bodily Communication, 2nd Edition. New York: Methuen.

Axelrod, L. and Hone, K. in press, Affectemes and allaffects: a novel approach to coding user emotional expression during interactive experiences. For publication in *Behaviour and Information Technology*.

Axelrod, L. and Hone, K. 2005a, Emotional advantage: performance and satisfaction gains with affective computing. *CHI 2005 Extended Abstracts on Human Factors in Computing Systems*, Portland, OR, 1192–1195.

Axelrod, L. and Hone, K. 2005b, Uncharted passions: user display of positive affect with an adaptive affective system. In J. Tao, T. Tan, and R.W. Picard (Eds.): *ACII 2005, Lecture Notes in Computer Science*, **3784**, 890–897.

Bradley, M.M. and Lang, P.J. 1994, Measuring emotion: the Self-Assessment Manikin and the Semantic Differential. *J Behav Ther Exp Psychiatry* **25**, 49–59.

Fraser, N.M. and Gilbert, G.N. (1991) Simulating speech systems. *Computer Speech and Language*, **5**, 81–99.

Norman, D. 2004, *Emotional Design: Why We Love (Or Hate) Everyday Things*. Basic Books.

Picard, R.W. 1997, *Affective Computing*. Cambridge, MA: The MIT Press.

Reeves, B. and Nass, D. 1996, The Media Equation: How people treat computers, television and new media like real people and places. Stanford: CSLI Publications.

Rickenberg, R. and Reeves, B. 2000, The effects of animated characters on anxiety, task performance and evaluations of user interfaces. *Chi Letters*, **2**, 49–56.

HCI SYMPOSIUM – USABILITY AND BEYOND

EYE-CENTRIC ICT CONTROL

Fangmin Shi, Alastair Gale & Kevin Purdy

Applied Vision Research Centre, Loughborough University, Loughborough LE11 3UZ

There are many ways of interfacing with ICT devices but where the end user is mobility restricted then the interface designer has to become much more innovative. One approach is to employ the user's eye movements to initiate control operations but this has well known problems of the measured eye gaze location not always corresponding to the user's actual visual attention. We have developed a methodology which overcomes these problems. The user's environment is imaged continuously and interrogated, using SIFT image analysis algorithms, for the presence of known ICT devices. The locations of these ICT devices are then related mathematically to the measured eye gaze location of a user. The technical development of the approach and its current status are described.

Introduction

The availability of low cost systems which measure eye gaze behaviour has led to an increasing number of viable opportunities for utilising eye movement recording as a tool for interacting with the environment (see for instance, Istance and Howarth, 1994).

Similarly, there are an increasing number of ICT and electrically operated devices in our environment that have the potential to be operated remotely, as well as add-on products that are available for automating typical manually operated items. For the purposes of this paper all such items are simply referred to as "objects" or "controllable devices".

Using eye gaze information, people can achieve efficient interaction with their surroundings (e.g. Shell et al, 2003). For instance, various commercial eye-typing systems are available, to help disabled people interact with a computer and users can be trained to achieve fast typing speed by selecting soft keys displayed on the PC screen (c.f. Ward and MacKay, 2002).

However, using eye gaze as a selection device can be problematic as sometimes what people look at is not what they are actually attending to and the issue of attention is ever intriguing (for instance, see Wood et al, 2005). Such an involuntary selection can result in a direct operation of the target object, leading to unpredictable false reactions. We are currently investigating a selective attention control system, using eye gaze behaviour, which overcomes such an issue. It does this initially by supplying a Graphical User Interface, which allows people to confirm their intentions consciously, so that any control executions are actually based on their needs. This system is known as Attention Responsive Technology – ART (Gale, 2004).

Eye-centric control system

System integration

A laboratory-based prototype system has been set up to carry out the ICT device control using eye point of gaze. It consists of four main components:

- An eye tracker – to record users' eye movements and monitor eye fixations on objects in the environment

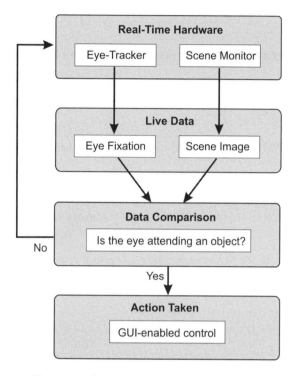

Figure 1. System work process illustration.

- An object monitor – to observe the user's environment with a view to identifying any object within the user's field of attention
- A user-configurable panel – to provide a GUI for the user to confirm his/her attention and initiate control of an object
- A controller – to enable the actual control of an ICT device upon any PC-based command.

The integrated system runs under our specially developed software, which receives raw data from the first two units and issues commands to the last two units after performing extensive computational process. The work flow is illustrated in **Error! Not a valid bookmark self-reference.**

Mini cameras for eye tracking and object monitoring

The system development starts with a commercially available head-mounted eye tracking unit (ASL 501 system). This contains two mini compact video cameras. One is the eye camera, which records the eye pupil and corneal reflection and outputs the eye's line of gaze with respect to the head mounted system. The other is the scene camera which has a wide field of view lens. This is mounted facing the environment in front of the user and monitors the frontal scene.

These two cameras are linked together after being calibrated. The user's point of gaze can then be directly mapped to its corresponding position in the scene image.

System calibration

The eye monitoring system needs to be calibrated for each user. This process only takes a short time. Calibration entails the user sequentially looking at a matrix of nine points

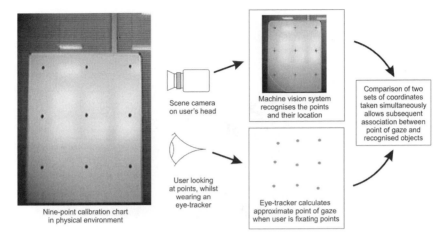

Figure 2. System calibration process.

as shown in the left image of Figure 2. The eye fixation data for each point are recorded with reference to the eye camera system as illustrated at the bottom middle frame of Figure 2. At the same time, the image coordinates of the nine points with reference to the scene camera system are extracted by our image processing algorithm, which are known as target points and highlighted as crosses at the upper middle image of Figure 2. Through comparing these two sets of coordinates, a point of gaze recorded by the eye camera can be directed to the falling point on the scene image simultaneously.

Fixations can then be traced. Whether they fall on any target object is dependent on whether the scene image contains any recognisable object of interest on that point.

Object identification

The above approach makes the complexity of 3D object recognition and location reduced to 2D object recognition only. Algorithms can then be developed and applied to the scene camera output to try to recognise objects in the scene. An efficient and reliable object identification method is under development in the research project. This performs image feature matching between a scene image and an image database that collects images of target objects. The image feature detection algorithm is based on the SIFT- Scale Invariant Feature Transformation, approach proposed by Lowe (2004). SIFT features are adopted because they have advantages over other existing feature detection methods in that their local features provide robustness to change of scale and rotation and partial change of distortion and illumination. An example showing the SIFT matching result for identifying an electric fan in a scene is given in Figure 3. At the right of Figure 3 is the image, which contains an electric fan, taken by the scene camera. The image of the electric fan to the left of the figure is one of the reference images in the database. The lines between the two images indicate the points of matching between the reference and the real object images and shows how an environmental object is recognised.

GUI-enabled ICT control

Having recognised an object in the scene, then the system needs to allow the user to choose whether to operate it. Assume a user keeps looking at a target object for a certain period of time, e.g. 0.5 seconds, and then a GUI will be enabled. The concept is that the interface will "ask" the user to confirm whether or not it is his/her intention to operate

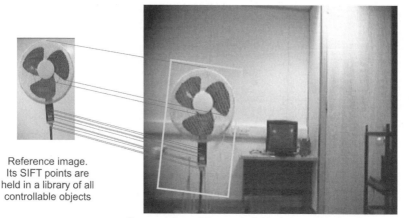

Reference image. Its SIFT points are held in a library of all controllable objects

Scene picture of the user's field of view within their environment

Figure 3. An example image showing the SIFT matching result.

the object. To conduct the control of the devices, for example, to switch on/off a TV, our system employs the X10 control protocol popularly used in the home automation community. A control command is issued from the PC via an X10 adaptor. Any object to be controlled is connected to the normal mains electrical supply by plugging into an X10 module first. There are no wires required for connecting between the objects and the PC – interaction is through wireless communication or so-called X10 signals.

Conclusions and future work

A system is described which enables a user to select and control ICT objects by using their eye gaze behaviour. The system entails using a head mounted eye movement recording device. Currently, the overall framework of the ART system has been achieved. ICT objects can be identified in the user's environment by the algorithms designed and built into the ART system. Additionally we have interfaced the eye tracking system to the object monitoring and identification system. At present the ART system components work separately and current research effort is focussed on integrating the separate modules into a fully cohesive ART system which will work in real time.

References

Gale A.G., 2005, Attention responsive technology and ergonomics. In Bust P.D. & McCabe P.T. (Eds.) Contemporary Ergonomics, London, Taylor and Francis, 273–276.

Istance H. and Howarth P., 1994, Keeping an eye on your interface: the potential for eye-gaze control of graphical user interfaces, Proceedings of HCI'94, 195–209.

Lowe D.G., 2004, Distinctive image features from scale-invariant keypoints. *International Journal of Computer Vision*. **2**, 91–110.

Shell J.S., Vertegaal R. and Skaburskis A.W. EyePliances: Attention-Seeking Devices that Respond to Visual Attention, CHI 2003: New Horizons, 771–772.

Ward D.J. and MacKay D.J.C., 2002, Fast hands free writing by gaze direction. *Nature*, **418**.

Wood S., Cox R. and Cheng PCH., 2006 Attention Design: Eight issues to consider. Computers in Human Behavior. Special issue on Attention-aware systems.

APPLYING THE KEYSTROKE LEVEL MODEL IN A DRIVING CONTEXT

Michael Pettitt, Gary Burnett & Darius Karbassioun

School of Computer Science and I.T., University of Nottingham, Nottingham, NG8 1BB

The Keystroke Level Model (KLM) was developed to predict the performance time of expert users on desktop computing tasks. In recent years, computing technology has been implemented in more diverse environments, including the safety critical driving situation. To evaluate the potential use of the KLM as a means of predicting task times for in-car interfaces, 12 fully trained participants carried out a series of tasks on two in-car entertainment (ICE) systems. Results showed high positive correlations between KLM predictions and observed task times (R = 0.97). Fitts' law was used to make more accurate predictions for the homing operator, in place of the values given in the original KLM method. Further work is necessary to investigate the KLM as a means of assessing the visual demand of in-car systems.

Introduction

Developed in the 1980s, the Goals, Operators, Methods and Selection Rules (GOMS) family of modelling techniques allow designers of computing systems to produce qualitative and quantitative predictions of how people will use a proposed system (John, 1995). The most striking benefit of GOMS techniques is their ability to provide designers with quantitative information about proposed systems without the need to run costly user trials, which may come too late in a design process to be of substantial benefit (Green, 2003). The simplest variant of GOMS, the keystroke level model (KLM) is used, specifically, to predict the time taken by expert users to complete tasks on a system (Raskin, 2000). KLM techniques involve decomposition of a task into its primitive operators, for example, pressing a key, pointing with a mouse, and so on. A time value is assigned to each operator, with the predicted task time being the sum of these values (Card et al, 1983). A mental operator (M) is included to account for the time users take to mentally prepare for an action.

In today's world, computing technology is becoming increasingly ubiquitous, and there may be concerns that ergonomic techniques such as the KLM, which was developed for desktop computing situations, are no longer valid. In this respect, there have been attempts to assess the KLM in more novel computing environments. Myung (2004) assessed the use of the technique for Korean text message entry on mobile phones. It was found that, by substituting new values for M, the KLM was a useful and reliable technique for text entry user interface evaluation. Stanton and Young (2003) assessed the methods could be taught to non-ergonomics specialists. The KLM was found to be easily taught and the most reliable method. The study, involving evaluation of in-car entertainment (ICE) systems, reported good correlations between KLM predicted task times and actual task times. However, the results are not reported in depth. A much more detailed example of the KLM being applied in a driving context is found in Manes et al (1998). The KLM was assessed against an empirical study of drivers performing destination entry tasks on a navigation system. The authors found strong correlations between observed and predicted times. However, they also proposed new values for a number of operators, and suggested new levels of complexity to the original method. For example, different values for second presses of a button than for initial presses.

A significant strand of research in the area of in-vehicle technology has suggested that the time associated with a task performed on an in-vehicle device is directly related to safety (Green, 2003). The applicability of the KLM to such devices is therefore of interest. The focus on task times has led to the development of the 15-second rule for driver-interface usability and safety (Green, 2003). Originally developed for assessment of destination entry tasks on navigation systems, but thought to be more generalisable, the rule broadly states that tasks that can be completed in 15 seconds or less when in a stationary vehicle can be safely performed whilst driving. There has been some criticism of the rule, with claims that it neglects to address the visual demands placed on the user when performing a task whilst driving (noted in Green, 2003). However, Green (2003) argues that static task time is well correlated with task time whilst driving, which is in turn well correlated with total eyes-off-the-road time, a well regarded measure of distraction.

Although the KLM has previously been assessed in a driving context, it is felt that there is scope for further validation. Of particular interest is the application of Fitts' law to accurately predict the value of the homing operator (H), which is the time taken to move the hand between controls, (e.g., in the driving context, between the steering wheel, buttons, rotaries). Fitts' law allows the time to travel to a target to be calculated as a function of the target's distance and size (Raskin, 2000). It has previously been used to more accurately predicting pointing operators (P), the time taken to point with a mouse, and has also been used for predicting H operators with mobile phones (Raskin, 2000; Myung, 2004).

Methodology

An empirical study was conducted to assess the applicability of the KLM in a driving context. User trials were conducted with twelve participants (mean age = 24; range: 22 to 28; 50:50 gender split) on two in-car entertainment (ICE) systems in a counterbalanced order. For each system, eleven tasks were performed in order (see Table 1). All users were fully trained in all tasks before the trials. Training consisted of verbal instructions, followed by the opportunity to practice various tasks. As KLM predictions are associated with expert performance, it was required that participants were suitably skilled in the use of the ICE systems. Consequently, an assessment of each participant's competence was made prior to recording actual data. Trials took place in stationary vehicles, and participants were filmed performing the tasks to ensure that accurate task times were recorded.

Table 1. Tasks.

Task no.	Task	Task no.	Task
1	Switch on radio	7	Save frequency to pre-set 2
2	Switch on tuner	8	Decrease volume
3	Tune into Radio1 (FM 97.9)	9	Insert cassette
4	Save frequency to pre-set 1	10	Increase base to full
5	Increase volume	11	Eject cassette
6	Change station to BBC Nottingham (FM 103.8)		

KLM models of each task were created using the original rules set out by Card et al (1983) and task times were predicted. Fitts' law (Raskin, 2000) was used to predict the value for homing between the steering wheel and the ICE system and for movement between buttons.

Results

The original value for the H operator was 0.4 seconds (Card et al, 1983). Using Fitts' law, three values for H were determined for this study: homing from the steering wheel to the ICE system; homing from the ICE system to the steering wheel; and, homing between controls on the ICE system. Table 2 provides a summary of the different H values

Correlations between observed task times and KLM predictions were high (see Figures 1 and 2). The user trials did, however, show some small differences in task times that were not predicted by KLM. For system 2, in particular, there was greater variability with user trials in the shorter tasks, although the lengthier tasks were well correlated with the KLM predictions. For all tasks, the system with the shorter task time was accurately predicted by the KLM (for a selection, see Table 3). The difference in value between the systems was less well predicted, with the KLM generally more conservative. This may be due to the relatively small sample group in the user trials, and is possibly further influenced by the use of operator values from the original KLM, which were based on desktop computing actions (Card et al, 1983).

Discussion and conclusions

The strong correlations between the predicted task times and observed task times are encouraging. They corroborate other research, which has suggested that KLM techniques

Table 2. Fitts' law predictions for Homing operators (secs).

Movement	System 1	System 2
Steering wheel to ICE	0.969	0.998
ICE to steering wheel	0.621	0.65
Between controls	0.2	0.2

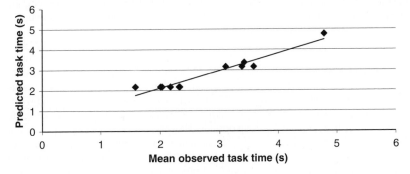

Figure 1. Comparison of mean observed task time and predicted task time for system 1 (r = 0.97).

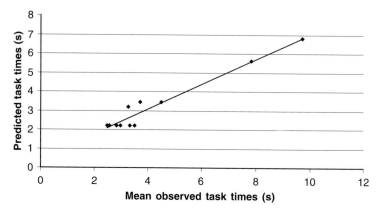

Figure 2. Comparison of mean observed task time and predicted task time for system 2 (r = 0.97).

Table 3. Preferred system for selected tasks according to KLM and user trials.

Task	Preferred system – KLM (% diff. between systems)	Preferred system – trials (% diff. between systems)
Switch on radio	System 1 (47%)	System 1 (49%)
Save frequency to preset 1	System 1 (66%)	System 1 (128%)
Decrease volume	System 1 (1%)	System 1 (42%)
Tune to Radio Nottingham	System 1 (42%)	System 1 (103%)
Insert cassette	System 1 (1%)	System 1 (87%)

can be applied to computing environments that do not fit the desktop paradigm for which they were designed (Manes et al, 1998; Myung, 2004). In developing the keystroke level models of the tasks in this study, Fitts' law was used to determine the values of the homing operator (H) for hand movements. The original keystroke model used an average value for H that was based on the movement between the keyboard and the mouse. Fitts' law was felt to be particularly useful for predicting movements between the steering wheel and the ICE system, however there was some doubt over its applicability to movements between controls on the ICE system; it is possible that participants employed different strategies, for example, using multiple fingers to manipulate adjacent buttons, instead of actually moving their hands.

There has been considerable debate within the research community regarding the usefulness of measuring static task times of in-vehicle devices. The chief concern being that in-vehicle tasks are often performed whilst driving, adding a new dimension to safety requirements. Of particular concern is how visually demanding a system may be, a characteristic not assessed by the original KLM. Critics of the 15-second rule claim that it fails to discriminate between systems that have short task times but may require the user to glance towards the system for long periods. However, the developer of the 15-second rule, Paul Green, has argued that the primary cause for safety concern is task length (Green, 2003). Green refutes the suggestion that the 15-second rule ignores systems that promote long glances; "it is difficult to think of driver tasks ... that have short total task times but long glance durations" (Green, 2003: 856). It has been proposed that an occlusion component could be included in the protocol for the 15-second rule (Green,

2003). The occlusion technique assesses in-car systems based on the ability of a user to complete tasks in a series of manageable chunks.

The greatest drawback to methods that rely on user trials to make judgements on safety is the requirement for highly developed prototypes; at this late stage in the design process, it can be difficult or impossible to make significant adjustments to a design (Green, 2003). It is therefore worth pursuing methods, such as the KLM, that can be applied in early design stages (Green, 2003). What is now necessary is to investigate potential extensions of KLM methods that seek to address the perceived shortcomings in measurements of static task time alone, particularly the relationship between tasks and their potential visual demand. Such efforts may focus on modelling the measures reported by the occlusion technique.

References

Card, S.K., Moran, T.P. and Newell, A. 1983, *The Psychology of Human-Computer Interaction*, (Lawrence Erlbaum Associates, London).
Green, P., 2003, Motor vehicle driver interfaces. In J.A. Jacko and A. Sears (eds) *The Human-Computer Interaction Handbook*, (Lawrence-Erlbaum Associates, UK), 844–860.
John, B.E., 1995, Why GOMS? *Interactions*, October 1995: 80–89.
Manes, D., Green, P. and Hunter, D. 1998, Prediction of destination entry and retrieval times using keystroke level models. Report no. UMTRI-96-37, UMTRI, University of Michigan, Michigan.
Myung, R., 2004, Keystroke-level analysis of Korean text entry methods on mobile phones. *International Journal of Human-Computer Studies* **60**: 545–563.
Raskin, J., 2000, *The Humane Interface*, (Addison-Wesley, London).
Stanton, N.A. and Young, M.S., 2003, Giving ergonomics away? The application of ergonomics methods by novices. *Applied Ergonomics* **34**: 479–490.

THE PHYSICAL WORLD AS AN ABSTRACT INTERFACE

Darren Edge, Alan Blackwell & Lorisa Dubuc

*University of Cambridge Computer Laboratory, William Gates Building,
15 JJ Thomson Avenue, Cambridge CB3 0FD*

> We describe an approach to understanding the way in which people can use physical objects as if they were components of an abstract language. This arose from our study of common devices that cannot support direct manipulation, because users are controlling future operations such as VCR recording. We apply an adaptation of the Cognitive Dimensions of Notations framework, a vocabulary for the analysis and design of information devices, to the digitally augmented physical objects of tangible user interfaces. Our specific focus is on assisting collaboration, awareness and communication between individuals. We describe three case studies: rhetorical structure for children, shared information spaces within organizations, and assistive technology for individuals with dementia.

Introduction

Whenever we arrange physical objects or manipulate mechanisms, we rely on immediate perceptual feedback in order to adjust our actions and confirm their effects. For over 20 years, this experience of "direct manipulation" has inspired the design of graphical user interfaces (GUIs), in which visual elements of the computer screen are made to behave as if they were arrangements of physical objects. However physical systems themselves do not always support direct manipulation. As soon as the earliest electromechanical controls enabled physical action at a distance, it became necessary to design physical indicators for the human operator. Even more challenging is the situation in which physical systems include internal state, such that their future behaviour may vary. An operator then needs to anticipate that future behaviour, taking additional precautions to specify, test and review the effects of current actions. This has been described as the irony of automation (Bainbridge, 1987). We describe it in terms of an "attention investment" equation in which the mental effort of specification test and review must be offset against the expected convenience of automatic operation (Blackwell, 2002).

The relationship between physical action, perceptual feedback, system state and future behaviour has become more critical in computer science as we develop technical infrastructures for pervasive or ubiquitous computing. In these fields, we anticipate design scenarios in which many physical objects may be augmented with computation and communication facilities. In some scenarios, system users may interact with pervasive network facilities via a conventional screen interface (perhaps made more portable via mobile phones, PDAs, wrist displays or compact head-mounted displays). However we are particularly interested in those circumstances in which the augmented physical objects themselves can be manipulated in order to control the system. This is described as a *tangible user interface* or TUI (Ishii and Ullmer, 1997).

A critical question for the designers of TUIs is whether they are able to apply the design principles of direct manipulation. At first sight, it seems that this might be the primary advantage of a TUI. Even the best GUI is only a simulation of object behaviour in the physical world, whereas a TUI is composed of real objects. However this fails to account for the centrality of internal state to augmentation. An augmented object, almost by definition, is one that is associated with additional state information. The benefit of this additional state is that manipulating one object may have effects on others, or that a manipulation may have some effect in the future. In systems terms, these can be considered varieties of automation. In computational terms, they can be considered varieties of programming. In order to specify multiple actions or future actions, it is necessary to describe those actions using some abstract representation. This is the precise opposite of direct manipulation, in which there is no abstract intermediary between action and result.

In our analysis of TUIs, we therefore distinguish direct actions on the world (perhaps mediated by a physical control device) from abstract actions that manipulate representations of internal state. The first can be analysed using conventional usability approaches, whereas the latter should be analysed using approaches from the psychology of programming. Our design approach emphasises the manipulation of abstract notational systems (of which programming languages are one example). In the case of TUIs, this leads us to think of the ways in which the physical world can express an abstract information structure, and an arrangement of physical objects can become a notation, perhaps expressed as a *manipulable solid diagram* (Blackwell et al, 2005).

The Cognitive Dimensions Framework

The Cognitive Dimensions (CDs) framework (Blackwell and Green, 2003) is a tool for the design and evaluation of notations, where a notation is the means of representation and control of an underlying information structure. Notations are evaluated against a set of interrelated criteria – cognitive dimensions – with the result being compared against the desired profile of the activities to be performed with the notation. This procedure can also be followed in reverse to aid the design of notations, by looking for solutions whose profiles fit closely with that of the intended activities. The CDs framework was designed to be applicable for any notation – virtual (as in GUIs) or otherwise – and as such the names of the dimensions reflect high level concepts that can often be interpreted in a number of ways. By restricting the domain of analysis to TUIs, some of these dimensions have interpretations that are particularly salient and worthy of more role expressive names. We call such interpretations the *tangible correlates* of the cognitive dimensions.

We apply the cognitive dimensions and their tangible correlates at the earliest possible stage of TUI design, as a precursor to creating prototypes with modelling materials. Although prototyping allows "hands-on" creativity, an inherent problem is the vast number of alternative prototypes which can be produced when the mind is allowed to wander freely. The benefit of potentially finding novel and surprising solutions is balanced against the cost of creating numerous unsuitable designs. The definition of a "suitable design" is one whose profile provides a close match with the profiles of the activities to be performed with the design. A preliminary CDs analysis can therefore be beneficial in placing constraints on the design space, such that only suitable designs are considered in the prototyping phase. The rigour of CDs analysis ensures that more of the possible design space is considered, and the constraints it imposes ensure that more of the suitable design space is explored creatively during prototyping.

To illustrate the kind of reasoning typical of a CDs analysis we will present examples from our past and current work, using the typographic convention of cognitive dimension$_{<CD>}$ to refer to the traditional CDs, and tangible correlate$_{<TC>}$ [cognitive dimension] to refer to their tangible correlates. For a full description of the Cognitive Dimensions framework, see http://www.cl.cam.ac.uk/~afb21/CognitiveDimensions.

Case Studies

Design of TUIs to support collaboration between collocated users

In the WEBKIT project, we designed a physical representation of the abstract structure of arguments (Stringer et al, 2004). Our goal was to help schoolchildren make a critical assessment of material they had found during research on the Web, as they selected and presented evidence in support of some proposition. We used tangible tokens (with embedded RFID tags) to maintain the relationship between specific claims formulated by the students, and the evidence from which those claims were derived. The tokens could be ordered into physical structures that both illustrated and enforced the abstract argumentation structures of classical rhetoric. This process was carried out in teams, where all children could see the emerging abstract structure, and collaborate in constructing and refining it. When the argument was eventually presented to the whole class, the ordered tokens provided private cues for a speaker, while also offering a simple control interface to the projection of multimedia illustration material for the talk.

Each stage of an argument is represented by a token holder that provides two parallel racks for the linear arrangement of tokens. Tokens can slide freely within these linear constraints and be moved freely from one rack to another. This is an example of low viscosity$_{<CD>}$, or resistance to change, in terms of both the manipulation of tokens within a rack, and the movement of tokens between racks. We say that the interface has both low rigidity$_{<TC>}$[viscosity] and low rootedness$_{<TC>}$[viscosity] respectively – desirable characteristics for a collaboration interface. These correlates commonly trade-off against the shakiness$_{<TC>}$[error proneness] of the physical notation, or its proneness to accidental and unrecoverable damage. This is also the case with the WEBKIT interface – tokens can be easily knocked out of the shallow racks. However, the consequences of this are partially offset by the use of LEDS on each token to indicate detection status – this feedback, which eliminates hidden augmentations$_{<TC>}$[hidden dependencies], makes it easier for users to detect and correct such errors.

Design of TUIs for to support awareness within distributed teams

One of our current projects is a tangible interface to support shared awareness within distributed teams. In this situation, rather than having a single focal interface to support active collaboration, each user needs to have their own personal interface for passively communicating information to other users, and monitoring the changing state of other users' information structures in return. Furthermore, as users' focus is their existing work, such an interface needs to operate on the periphery of users' attention and complement the traditional monitor, mouse and keyboard setup.

The structural approach we have devised involves a division of labour between the two sides of the workspace – tokens representing items of interest, or *information entities*, are arranged and manipulated on an interactive surface positioned by the side of the keyboard not occupied by the mouse or other pointing device. The opposite side – where the pointing device is located – is used to group together all of the tools of the interface that are used to navigate and set attributes of information entities. The positions and identities of tokens are detected using computer vision techniques, and attributes of the associated information entity are dynamically displayed around the token on the interactive surface.

The types of information entity envisaged include people, documents, tasks, reminders, timers, and so on. Users will be able to use their interface to "physically" pass information entities between interactive surfaces, and associate tokens with others' entities to monitor their evolution, e.g. progress on a task. A shared web space will be used to set up token-entity associations and publish in real time the changing information structures of all team members.

In this kind of TUI, the role expressiveness$_{<CD>}$ of physical tokens must be traded-off for increased adaptability$_{<TC>}$[abstraction] in terms of the different information entities they can be used to represent. In accordance with the "peripheral" requirement, the interactive surface in our TUI has sufficiently low bulkiness$_{<TC>}$[diffuseness] that it can be relocated by the user to their home office, client site, etc. – the interface as a whole has low rootedness$_{<TC>}$[viscosity]. The physical tools also have a degree of permanence$_{<TC>}$[visibility] in their location, which means that the user can learn to operate them by touch alone. Combined with the parallel, bimanual method of interaction, this decreases the overall rigidity$_{<TC>}$[viscosity] of the notation.

Design of TUIs to support communication for individuals with dementia

Another current project we are working on is a TUI to assist older individuals with dementia, in maintaining a conversation through the virtual management of conversational topics. TUIs were a good candidate for use by this user group, as older people are often more comfortable working with concrete physical representations than abstract virtual ones. In the interface we are developing, physical objects are used to denote individual conversational topics and placed on a shared tabletop. The spatial arrangement of objects may be used to represent planned conversational topics and their anticipated flow, but this arrangement can be altered dynamically at any point during the conversation. A camera on a mobile phone is used by a person with dementia to select a physical object (and thereby a conversational topic); as topics are selected using the phone, the software tracks which topics have been discussed and relays this information back to the individual when repetition occurs. Externalising information in this way reduces hard mental operations$_{<CD>}$ by reminding users of what has been discussed previously, and providing visual cues for possible future topics.

As individuals with dementia may be easily distracted by their environment and have a limited attention capacity, the interface also requires a high level of structural correspondence$_{<TC>}$[closeness of mapping] for ease of understanding, and needs to keep hidden augmentations$_{<TC>}$[hidden dependencies] to a minimum so as not to surprise users. This is achieved in our interface by using special tags that have a one-to-one mapping with conversational topics; the user with dementia having control over which tag designs to use and how they should be attached to familiar, role expressive$_{<CD>}$ physical objects. This control is important for maintaining individuals' sense of identity, the loss of which can become a barrier to communication (Dubuc and Blackwell, 2005).

These tags are detected and identified by computer vision software on the mobile phone. The phone is packaged in a casing designed to reduce unwieldy operations$_{<TC>}$[hard mental operations] for users with impaired motor control, and has good purposeful affordances$_{<TC>}$[role expressiveness] for pointing due the real-time camera display on the screen of the phone. The tag recognition software only functions when the camera is held pointed directly at a tagged object, making it easy to intentionally capture a conversational topic, and difficult to accidentally capture one. Hence both the shakiness$_{<TC>}$[error proneness] and rigidity$_{<TC>}$[viscosity] of the capture mechanism are minimised.

Conclusion

Conventional GUI design guidelines, based on direct manipulation, are satisfied trivially by TUIs. As a result, they give little assistance for TUI design. We have therefore outlined a new approach to help designers analyse the needs of diverse user populations, in order to construct innovative digital augmentations of the physical world.

Acknowledgments

This work is sponsored by Boeing Corporation. The WEBKIT project was funded by European Union grant IST-2001-34171.

References

Bainbridge, L. (1987). Ironies of automation. In J. Rasmussen, K. Duncan and J. Leplat (Eds) *New Technology and Human Error*. Chichester: Wiley, 271–284

Blackwell, A.F. (2002). First Steps in Programming: A Rationale for Attention Investment Models. In *Proc. IEEE Symposia on Human-Centric Computing Languages and Environments*, 2–10

Blackwell, A.F. and Green, T.R.G. (2003). Notational systems - the Cognitive Dimensions of Notations framework. In J.M. Carroll (Ed.) *HCI Models, Theories and Frameworks: Toward a multidisciplinary science*. San Francisco: Morgan Kaufmann, 103–134

Blackwell, A.F., Edge, D., Dubuc, L., Rode, J.A., Stringer, M. and Toye, E.F. (2005). Using Solid Diagrams for Tangible Interface Prototyping. *IEEE Pervasive Computing*, **4**(4): 74–77

Dubuc, L. and Blackwell, A. (2005). Opportunities for augmenting conversation through technology for persons with dementia. In *Proc. Accessible Design in the Digital World (ADDW)*, Dundee, Scotland

Ishii, H. and Ullmer, B. (1997). Tangible bits: towards seamless interfaces between people, bits and atoms. In *Proc.CHI'97 Conference on Human Factors in Computing Systems*. New York, NY: ACM Press, 234–241

Stringer, M., Toye, E.F., Rode, J.A. and Blackwell, A.F. (2004). Teaching rhetorical skills with a tangible user interface. In *Proc. ACM Interaction Design and Children*

HCI SYMPOSIUM – ACCESS AND INCLUSIVITY

ACCESSIBILITY VS. USABILITY – WHERE IS THE DIVIDING LINE?

Sambhavi Chandrashekar[1] & Rachel Benedyk[2]

UCL Interaction Centre, University College London, London (U.K.)

Web accessibility for people with disabilities may be looked upon as the overcoming of barriers that make it difficult for them to use Web resources. Design criteria for Web usability have evolved over the past decade but guidelines for Web accessibility are still emerging. Issues such as whether accessibility guidelines also encompass the usability needs of users with disabilities are being debated. This study analyses the relationship between accessibility and usability and concludes that accessibility includes both access and use of resources by people with disabilities, which could be supported by a combination of a set of accessibility guidelines and a set of usability guidelines for them. It proposes a criterion of distinction between these two types of guidelines and tests the idea on an experimental website with a sample of users with visual disabilities.

Introduction

The Web today is a large repository of resources. Web accessibility for people with disabilities may be looked upon as the overcoming of barriers that prevent them from using these resources. Web accessibility can be improved through the use of assistive technologies in some cases by enabling access through alternative or enhanced modes. For example, people with visual disabilities can use a screen reader software program that reads out the computer screen contents using its internal speech synthesizer and the computer's sound card. However, the content on the Web has to be in a form that can be interpreted by such technologies. A study in the USA of 50 websites found that more than 50% of them were still only partly accessible or inaccessible in 1999 (Sullivan and Matson, 2000).

Web Accessibility Initiative

The World Wide Web Consortium (W3C) launched its Web Accessibility Initiative (WAI) in 1997 to develop strategies, guidelines and resources to make the Web accessible to people with disabilities. This would mean that people with visual, auditory, physical, speech, cognitive, and neurological disabilities can perceive, understand, navigate, and interact with the websites, and that they can contribute to the Web (WAI, 2005). Web Content Accessibility Guidelines (WCAG) 1.0 (Chisholm et al, 1999), are considered the current de facto standard for content accessibility on the Web. The next version, WCAG 2.0 (Caldwell et al, 2004), is published as a working draft which is being revised periodically. This study used the draft version dated June 30, 2005.

Accessibility and Usability

According to Coyne & Nielsen (2001), with current Web design practices, users without disabilities experience three times higher usability than users with visual disabilities.

---

[1] sambhavic@gmail.com
[2] r.benedyk@ucl.ac.uk

A study in the United Kingdom of 1000 websites by Petrie et al in 2004 (Petrie, 2004) reported that 81% of the websites did not satisfy the most basic accessibility guidelines and even where they did, there were problems in the usability of the resources by people with disabilities. Leporini & Paterno (2002) and Theofanos and Redish (2003) stated, after studies of accessibility and usability of websites with blind screen reader users, that meeting with accessibility standards does not necessarily make a website usable by people with disabilities. These studies have published usability guidelines for people with disabilities. However, none of them has analysed the distinction between accessibility and usability. This study analyses the relationship between accessibility and usability in order to characterise the guidelines required to support Web accessibility. It attempts to find a distinction between accessibility guidelines and usability guidelines for people with disabilities, to better clarify their role in Web design.

Method and results

The method adopted for the study comprised of: (1) theoretically analysing the relationship between accessibility and usability, (2) proposing a criterion for determining the relative scope of accessibility guidelines and usability guidelines in Web design for people with disabilities, (3) user testing the idea with a sample of visually disabled users and (4) relating the outcome to the support of accessibility.

Theoretical analysis

Several possible relationships between accessibility and usability are discussed below. In the diagrams, A denotes "Accessibility" and U denotes "Usability". To begin with, accessibility and usability could be considered as mutually exclusive. There is a perceived dichotomy between the two among HCI practitioners – some claiming to be accessibility professionals and some to be usability professionals. However, there is no available basis nor evidence for such a division. Hence the argument that accessibility and usability are two unrelated issues (as in Fig. 1) may be considered untenable.

There is the diametrically opposite school, which would prefer to merge accessibility with usability to form a single entity of Universal Usability (as in Fig. 2). As envisioned by Shneiderman (1998), "Universal Usability will be met when affordable, useful, and usable technology accommodates the vast majority of the global population". While this is a desirable dream, it may be difficult to achieve in practice.

The definition of accessibility by the International Standards Organisation as "the usability of a product, service, environment or facility by people with the widest range of capabilities" [ISO TS 16071: 2003, definition 3.2] portrays accessibility as having a larger scope than usability. But literature does not support the possible conclusion derivable from this, that accessibility is a super-set of usability (as in Fig. 3). In fact, there is evidence to the contrary as can be seen below.

Henry et al (2002) consider accessibility as a subset of usability. Web usability has been popular for longer than Web accessibility and theory, techniques, and guidelines relating to Web accessibility are more recent than those of Web usability. Accessibility could thus be considered as having a narrower scope than usability (as in Fig. 4).

Having said this, it is still for consideration whether accessibility is entirely contained within usability or it merely overlaps with it. The concept of accessible Web design promoted by WAI introduced aspects into Web design that were not a regular part of usability practice and as such could be considered outside the sphere of current usability. In particular, accessibility requirements as per the WAI cover technical as well as non-technical design aspects. Usability generally focuses on non-technical user interface design issues. So accessibility does not fit neatly within usability and a relationship of overlap between the two (as in Fig. 5) appears meaningful. Boundaries

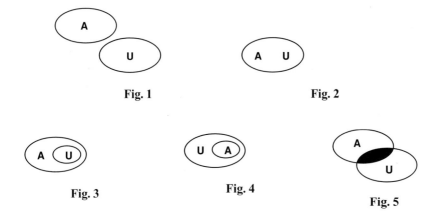

Fig. 1 Fig. 2 Fig. 3 Fig. 4 Fig. 5

defining this overlap may be dynamic and could depend on a variety of factors like nature of disability, nature of resource, content creation method, assistive technology used, user's individual capabilities, etc.

Guidelines for accessibility

The conclusion from the above analysis is that the broad concepts of accessibility and usability overlap and that the notion of accessibility includes both a technical component and a specific usability component for people with disabilities. The fact that research-based usability guidelines for such users are being published supports this idea. This study proposes that access and use of Web resources by people with disabilities could be supported by the combination of a set of accessibility guidelines and a set of usability guidelines for them, instead of a single set of accessibility guidelines to cover the entire gamut of accessibility. It becomes necessary, however, to define the scope of accessibility guidelines unambiguously so as to distinguish them from usability guidelines that may be required specifically for people with disabilities. Accessibility measures could be seen as those that help them cross the barrier to use. Usability measures for them could be seen as those that help them use the resource more easily beyond the barrier. Based on this, the following definitions could be considered:

A guideline that recommends a design feature which enables access of underlying Web content by people with disabilities, *in the absence of which design feature it is not technically possible for them to access that content* – would be an accessibility guideline.

A guideline that recommends a design feature *in whose absence the underlying content would still be technically accessible to* people with disabilities while its presence would make it easier for them to use that content – would be a usability guideline for people with disabilities.

The two sets of guidelines would together work towards ensuring Web accessibility for people with disabilities.

User testing

The above idea was subjected to user testing. Six visually disabled users performed a set of six tasks on an experimental website that was designed with features in conformance/non-conformance with selected accessibility guidelines from WCAG 2.0 that applied to users with visual disabilities. Observation, think-aloud technique and pre/post-test interviews were used to gather data. Accessibility was measured in terms of success or failure of users in performing assigned tasks designed to test their ability to access content. Usability was recorded qualitatively in terms of their expressed opinions about the ease of use of the content accessed. The results of these exploratory user tests showed that non-conformance with the selected accessibility guidelines resulted in inability to access underlying resources while conformance resulted in access. It also showed that the usability of tables and forms varied based on the extent of application of usability guidelines (although this was not provable empirically).

Discussion

Barriers to access are relative and subjective; they are non-uniform across people and environments. It is difficult to determine which accessibility measures will remove the barrier to Web access for a user with a particular disability and which may merely make access and use easier. The "user" here is actually a combination of the person and the assistive technology used. With the same disability, say visual impairment, the measures required for overcoming the accessibility barrier for a screen reader user may not apply equally for a Braille output user who is not listening to the audio output.

However, some barriers, like those in respect of perception of content and operation of interface controls, can be viewed in relation to modes of computer operation. It is possible to map perception of content with the output modes of the computer, viz., visual (content seen through the monitor); audio (content heard through the speakers); and tactile (content felt through the Braille output). Again, it is possible to map operation of interface controls with the input modes of the computer, viz., visual (using mouse); audio (using microphone); and tactile (using keyboard). Barriers associated with visual impairment are the inability to perceive content through visual output mode (monitor) and the inability to operate controls using the visual input mode (mouse). Barrier associated with hearing impairment is the inability to perceive content through the audio output mode (speakers).

This also illustrates the compensating role played by assistive technologies in providing alternative or enhanced modes. A screen reader program enables a person with visual impairment to perceive visual content through audio output (speakers) and operate interface controls through the tactile input (keyboard). A screen magnifier program enables a person with low vision to perceive visual content through enhanced visual output.

However, the kind of discreteness associated with perception and operation does not exist with understanding of content, which represents a continuum and so better matches with the idea of usability. This prompts the suggestion that the existing content accessibility guidelines could be enhanced/regrouped to include a set of usability guidelines for people with disabilities. The above discussion does not suggest that current guidelines under the accessibility principles mentioned above are classifiable as accessibility or usability guidelines. Rather, it is to say that a regrouping/expansion of the existing guidelines which differentiates between accessibility and usability guidelines may be more meaningful. As an example, provision of ALT Text would be an accessibility guideline; that the ALT Text should be contextually meaningful would be a usability guideline.

Conclusion

Accessibility of Web resources by people with disabilities, which involves both access and use, is best supported by a combination of a set of accessibility guidelines and a set of usability guidelines. Enhancement of the existing content accessibility guidelines into accessibility and usability guidelines for people with disabilities will make them more meaningful. This would enable better implementation of accessibility by Web developers and hopefully better use of Web resources by all.

References

Caldwell, B., Chisholm, W., Vanderheiden, G. and White, J. (2004) "Web Content Accessibility Guidelines 2.0", http://www.w3.org/TR/WCAG20/ (accessed 18th July 2005)

Chisholm, W., Vanderheiden, G., and Jacobs, I. (eds.) (1999) Web Content Accessibility Guidelines 1.0 http://www.w3.org/TR/WCAG10/ (accessed 10th March 2005)

Coyne, K. P. and Nielsen, J., (2001) "Beyond ALT Text: Making the web easy to use for users with disabilities," Nielsen, Norman Group http://ada.ucsc.edu/beyond_alt_text2002.pdf (accessed 25th May 2005)

Henry, S. L., et al (2002) *Constructing Accessible Websites*, Birmingham: Glasshaus

Leporini, B., Paternò, F. (2002) Criteria for Usability of Accessible Web Sites, 7th ERCIM Workshop "User Interfaces for All", Springer-Verlag, (2003), pp. 43–55

Petrie, H., Hamilton, F. and King, N. (2004) Tension, what tension? Website accessibility and visual design, Proceedings of the international cross-disciplinary workshop on Web accessibility, May 18, 2004, New York City, New York, USA

Sullivan, T. and Matson, R. (2000) "Barriers to use: Usability and content accessibility on the web's most popular sites," In *Proc. of CUU'00*, pp. 139–144, ACM Press

Shneiderman, B. (1998) *Designing the User Interface: Strategies for Effective Human-Computer Interaction (3rd ed.)*. Addison-Wesley Longman

Theofanos, M. F. and Redish, J. C. (2003) *Interactions, [X. 6]*, November–December 2003, 38–51 http://www.redish.net/content/papers/InteractionsPaperAuthorsVer.pdf (accessed on 20th August 2005)

WAI (2005) http://www.w3.org/WAI/intro/accessibility.php (accessed 10th April 2005)

A TECHNIQUE FOR THE CLIENT-CENTRED EVALUATION OF ELECTRONIC ASSISTIVE TECHNOLOGY

Gordon Baxter & Andrew Monk

Centre for Usable Home Technology, Department of Psychology, University of York, Heslington, York YO10 5DD

Electronic Assistive Technology (EAT) provides assistance and assurance for an increasing number of elderly and disabled people who wish to live independently. The technique described here aims to optimise the use of EAT by ensuring that it impedes as few aspects of everyday life as possible. The Post Installation Technique (PIT) is designed to be used by people with little technical or human factors knowledge to provide a client-centred evaluation of a recently installed EAT application. It systematically probes for aspects of their daily life that have been negatively affected by the technology. These problems are prioritised and passed to the EAT service provider so that it can be better tailored to client's needs.

This paper describes the development of the PIT through application in two small field studies and an expert evaluation.

Introduction

The potential benefit of assistive technologies to increase the quality of care of older people and reduce the associated costs is widely accepted within the UK (The Audit Commission, 2004). Electronic Assistive Technology (EAT) is increasingly used to enable older and disabled people – clients – to live independently in their own homes. Clients' are assessed on their ability to perform routine daily activities, such as eating, bathing, and moving around, to determine if EAT could help them. The provision of EAT is often technology-led, however, rather than needs-led (Sixsmith & Sixsmith, 2000). Although the prescribed EAT may make it easier or give clients more confidence to do some activities, after installation it may not properly support the task at hand – e.g., window opener switches located nowhere near the relevant window – and it can adversely affect other tasks in ways that are irritating, or problematic, and are not immediately obvious. When a client washes their kitchen floor, for example, they need to move the flood sensors beforehand – to avoid a false alarm – and remember to replace them afterwards. Such small details can affect the successful operation of the overall system.

Solving the problems described above requires a detailed knowledge of the minutiae of the client's everyday life. The obvious person to provide this knowledge is the client, although they cannot be expected to have a deep understanding of the technological constraints. In a post installation evaluation, however, this is not necessary. The client can provide details of any aspects of their daily life that have been negatively affected by the EAT which can then be passed to the EAT provider to make appropriate adjustments.

Instantiating the risk management framework

The Post Installation Technique (PIT) is an instantiation of a framework for investigating the risks of introducing and using technology in the home (Monk et al, Accepted for publication). The framework gives social and psychological harms the same level of importance as physical harms (injury, cost and so on.). The risk is assessed by considering the likelihood of occurrence (*high, medium, low*) of a generic harm (*injury, untreated medical condition, physical deterioration, dependency, loneliness, fear or costs*) arising from everyday activities, and the generic consequences of that harm (*distress, loss of confidence, a need for medical treatment, death*). The importance of the harm (e.g., injury) is conditioned by its consequences (e.g., distress, medical treatment).

The PIT allows clients (possibly with the help of a carer) to systematically consider how EAT affects their everyday living. It embodies a client-centred approach, identifying problems from the client's point of view in a four step process.

Step 1 lists the EAT installed in the client's home. This list of equipment determines which checklist type questions are asked in Step 2, to elicit the benefits and problems of the EAT when clients perform everyday activities. The client simply ticks the appropriate box for each question.

Once all the relevant questions have been answered, Step 3 summarises the benefits, and details the problems for further analysis. For each problem the client: selects the potential generic harm that could result from the problem; assesses the chances of that harm occurring; and decides what the consequences of that harm might be. Clients are also asked where the problem arises and how important it is to fix it.

In Step 4 clients suggest how they would like to see each of the problems solved. The completed forms are then handed on, nominally to the EAT service provider, for action.

Evaluating the post installation technique

The PIT is being iteratively developed and has been evaluated three times so far. For the first and third evaluations the PIT was used with clients in the field; for the second a semi-structured interview technique was used during an expert review of the PIT.

Field study 1: West Lothian

The PIT was initially trialled in West Lothian with three participants, all elderly females. The purpose of the visit and the PIT questionnaire were first explained to the client, before asking the questions in the order that they appear in the questionnaire.

The first participant lived in her own flat, and used a walking stick to get around her home. She had had EAT installed for about a month, and mainly used the carephone to set the security alarm when she went out. She had had one false alarm which she had cancelled without any problem.

The second participant lived in her own sheltered accommodation flat. She could walk a little using a Zimmer frame, but generally used a wheelchair. The EAT had been installed for just over a year when the client moved into the flat.

The third participant lived in her own home, and had limited mobility. She had had EAT installed since September 2003, but had had similar equipment installed in her previous house. She had had a couple of false alarms (once when her granddaughter leant on the pendant button, and once when she burnt the toast!).

Results and discussion

No problems with the EAT were uncovered, but the interviews revealed some problems with the PIT. The main problem concerned the original intention of working through the PIT walking round the home with the client. The mobility problems of the clients in West Lothian suggests that this may often not be a viable option.

The PIT systematically poses simple (yes/no) questions to identify problems. Without a full appreciation of the bigger picture, however, clients were concerned about the correctness and usefulness of their answers. They often answered in general terms, rather than focusing on specific activities and particular technology in a particular room.

Just having the EAT made all the clients feel more safe and secure. Even though no problems were identified for these clients, discussions with the support worker highlighted the existence of problems with EAT, such as extreme temperature sensors placed too close to the cooker, and the potential problem of clients falling if they rush to get to the carephone to clear any false alarms within the 15 second time limit.

The PIT was changed so that clients do not have to walk around the house. The structure and purposes were made clearer to the clients by revising the preamble for the PIT. The questions were also amended to focus first on client activities, and then on the room(s) where they are performed.

In addition, the last three questions in each section, which related to the aesthetics and general defectiveness of the system, were placed into a single general section at the start of the PIT. Getting these general issues out of the way first should help clients to focus their attention better on the specific issues raised in subsequent questions.

Expert review: Belfast

Copies of the PIT were passed to a service manager from Belfast. This was followed up with a semi-structured interview focusing on the PIT and the assessment process that they used.

After installation a follow-up visit is conducted about two weeks later, mainly to check that clients understand how to use the EAT. The other purpose is to identify any initial problems such as sensors that need to be relocated, or to add extra devices to the system. It is rare for any devices to be removed at this stage: assessors try to persuade clients to at least attempt to work with the EAT. Further reassessments are carried out at six month intervals. As part of the process, a record for each installed system is kept, covering: the client's call history; which sensor initiated any call; what type of response was generated; and details of any system maintenance or upgrades.

The service manager suggested that clients would respond better if questions were asked face to face, rather than having them complete the forms themselves. The multi-part questions were regarded as a little too long, and could possibly be reduced to two or three parts. It was also pointed out that questions about washing and toileting will have been asked during the original assessment; it may be a bit too personal to ask clients about them again.

It was felt that the PIT could be useful during the six monthly reassessments. The listed categories of harm, chances of harm and possible consequences could also be used with their existing assessment process. More generally, it was suggested that some of the material from the PIT could be used in training staff.

The preamble to the PIT was revised to make it clear that only the relevant parts of questions should be answered. The answer boxes for the questions were also flagged as relating to either Benefits or Problems, to make it clearer which form they should be copied to for step 3 of the PIT.

Field study 2: Durham

The first client had a basic EAT package installed in her home, and had recently been given a fall detector, and a bed sensor with a lamp attached. She had stopped wearing the fall detector – a common problem, especially among women – and had instead become more reliant on her pendant, wearing it all the time. She had had a couple of problems with the carephone. One was attributed to a fault on the line, although none were found subsequently. The other was its loudness when the phone dials through to the call centre.

Any other problems appeared to be isolated incidents, often light-heartedly dismissed by the client, apart from the bed sensor which she said could be removed as far as she was concerned. The second part of the interview therefore focused solely on this issue.

The first problem was that the lamp never came on when she got out of bed in the night; she had resorted to using the bedroom light instead. The second problem was that the client did not think that the sensor was working. She reported that the device had been programmed for her being in bed by midnight. One night, however, she said she was sat in her living room at 12:15 and no alarm was ever raised.

The second client had a trial lifestyle monitoring system installed (in August 2004). Such systems collect data from strategically placed sensors and upload them for analysis so that inferences can be made about the client's state of health. She identified four problems with the system. The first two are really installation problems (making sure that sensors are securely fitted so that they cannot be dislodged accidentally or otherwise by the client or their pet(s), and making sure that door sensors are fitted to doors that are likely to be regularly opened by clients). Such problems can easily be avoided by talking more to the clients prior to installation.

The third problem was that the electrical plugs for the equipment are large and heavy. The fourth problem was that the flood detector was not properly sensing when a flood occurred, because it was not positioned on a level floor. These two problems are more indicative of possible design flaws that would have to be addressed by the equipment manufacturer, although the client viewed them as irritants, rather than major problems.

The client had also had problems with the bed sensor failing to detect that she did not get up during a period of illness. This was attributed to the device's timing parameters having been incorrectly set.

Results and discussion

This was the first time the PIT had been used in analysing the identified problems, and it uncovered some shortcomings in how the client does this. The main concern is how the PIT deals with harms and the likelihood of harm. In industrial risk assessment methods, such as HAZOP (Kletz, 1999), the harms and the associated likelihood of those harms occurring are assessed by a panel of qualified experts. The client-centred nature of the PIT means that judgements about possible harms and the likelihood of the occurrence of those harms are made by the client (and carer).

The first client was very explicit about the difficulty of determining the likelihood of the harm occurring, saying, "Your guess is as good as mine, dear." The second client mostly regarded the problems as largely unquantifiable irritants, although she had deeper concerns about the problem with the electrical plugs. A better method is therefore needed to enable the client (and carer) to appropriately assess the likelihood of harms occurring.

In thinking about the problems, the clients seemed to focus on a specific incident. One-off incidents may be perceived as temporary glitches, whereas persistent incidents are more likely to be considered *real* problems. Clients may still not be able to express the likelihood of the identified harm occurring, however. The PIT was therefore revised to ask clients to focus on their personal experiences to identify any real harms that have been "caused" by the equipment. Clients are now also asked to consider whether anything worse could happen given the same problem, e.g., "What do you think is the worst thing that could happen to you if this problem happened again?"

The problem of identifying the likelihood of harm was rephrased to reflect the client's personal experiences. So clients are now asked how often the problem has arisen or does arise (Does the problem arise daily/weekly/monthly? and so on.). This can subsequently be translated into a qualitative equivalent (high, medium and low).

Summary and future work

Our experiences of developing the PIT to date suggest that it is a useful and worthwhile exercise. The PIT was designed to be used either as a standalone instrument, or as part of the client reassessment process. Most EAT service providers routinely reassess clients approximately six months after installation. One service provider has already expressed interest in using the PIT as part of this reassessment.

Evaluating the PIT has been a lengthy process. The main reason for this is the need for access to suitable clients with appropriate EAT. This often requires delicate negotiations with care providers or social services. Although clients should ideally be randomly selected, opinionated and loquacious clients tend to provide more extensive feedback on the EAT and the PIT, which helps to improve both.

Whilst the PIT has proved to be useful for identifying and analysing benefits and problems of EAT, further evaluation is required to test out the latest revisions which should improve the analysis of problems in particular. Once the PIT reaches a steady state, the intention is to release it for use by clients. The latest revision of the PIT can be downloaded from http://www-users.york.ac.uk/~am1/ftpable.html.

References

Kletz, T. A., 1999, *HAZOP and HAZAN*, 4th edition, (Institute of Chemical Engineers, Rugby, UK)

Monk, A., Hone, K., Lines, L., Dowdall, A., Baxter, G., Blythe, M. and Wright, P. Accepted for publication, Towards a practical framework for managing the risks of selecting technology to support independent living, *Applied Ergonomics*

Sixsmith, A. and Sixsmith, J. 2000, Smart care technologies: meeting whose needs?, *Journal of Telemedicine and Telecare*, **6**, (Supplement 1), 190–192

The Audit Commission 2004, *Assistive technology: independence and well-being 4*. (The Audit Commission: London)

SYNERGY OF ACCESSIBILITY, USABILITY AND ACCEPTANCE: TOWARDS MORE EFFECTIVE AND EFFICIENT EVALUATION

Chandra M. Harrison & Helen L. Petrie

*Department of Computer Science, University of York,
Heslington, York YO10 5DD*

Guidelines and heuristics are available to assist in the design and evaluation of usable and accessible websites. Emotional or experiential design, or acceptance, is also receiving research attention to encourage satisfaction in design. Despite the availability of usability, accessibility and acceptance guidelines designers and developers do not user-test or adhere to the guidelines often enough to ensure usable, accessible and pleasurable products for all users.

By identifying the synergy and/or autonomy of accessibility, usability and acceptance issues and the most relevant aspects for quality design, more effective and efficient evaluation methods are possible, encouraging greater user guideline compliance. To unpack the various issues and determine an optimum methodology, a preliminary study comparing eCommerce and eGovernment websites, reveals that there is indeed an overlap for some problems. In addition, inexperienced and/or disabled users are better at identifying certain types of problems, than mainstream users.

Introduction

Usability and accessibility are becoming an expectation rather than an exception, partly because of greater consumer awareness and legislation such as the Disability Discrimination Act (DDA) (1995) in the UK and similar legislation in other countries. The DDA states that providers of goods and services, including the internet, must make reasonable adjustments to make them accessible to everyone. However, it is clear from recent research than many websites still have accessibility problems (DRC, 2004) and usability is still low. Guidelines have been developed to encourage usability and accessibility, for example the Web Accessibility Initiative Guidelines (Chisholm et al, 1999) and government organisations provide useful usability guidelines based on established procedures (Usability.gov, 2005). Many guidelines focus specifically on either accessibility or usability rather than evaluating both simultaneously, making them less efficient. In addition, the focus in evaluation is often placed on a particular group of users, potentially excluding other groups.

As well as a focus on usability and accessibility, pleasurability, emotional and experiential design, or acceptance, are also becoming more relevant and need to be incorporated into user testing to encourage user satisfaction. Satisfaction is an often cited aspect of good usability (Nielsen, 1993), while others suggest that pleasurability is an important element of design in its own right (e.g., McCarthy & Wright, 2005; Jordan, 1997).

While a pleasurable experience is likely to ensure acceptance, a negative or frustrating experience is likely to have the opposite effect. Many users do experience difficulties

---

chandra.harrison@cs.york.ac.uk and helen.petrie@cs.york.ac.uk

with using websites. These difficulties are caused by factors such as the design of the website, the aesthetics, the utility of the technology, the skills of the user and any disabilities that may be apparent. Assistive technology for people with sensory, physical and cognitive impairments, such as screen readers, complicate user acceptance further, adding another layer to interfaces and creating negative emotion through frustration. In addition, while a website might be accessible, the user's disability or assistive technology may highlight usability problems. Satisfaction is one dimension of usability and there is debate regarding whether accessible technology can also be usable and aesthetically pleasing (Petrie et al, 2004), highlighting another overlap between usability, accessibility and acceptance. This all suggests a potential synergy between usability, accessibility and acceptance is apparent.

Despite the availability of guidelines for both usability and accessibility and the increase in information available on improving the user experience there are still unusable, inaccessible and unfriendly websites. The lack of compliance with guidelines and testing may be due to a lack of knowledge on the part of the designers and developers or may be due to limited resources or time. It is important then to ensure that evaluation tools or guidelines are as effective and efficient as possible to encourage their use and compliance with legislation. The purpose of this study is to unpack the concepts of usability, accessibility and acceptance, in an effort to determine the level of synergy, rather than to identify and enumerate all possible problems with a site. Larger issues that are easily identifiable by non-accessibility or usability experts are targeted rather than employing fine grained analysis of violations of existing guidelines. By identifying if there is synergy and/or autonomy of accessibility, usability and acceptance and which aspects are the most relevant to users and which are most severe, it may be possible to determine more effective and efficient evaluation methods to encourage greater user testing by reducing the variety of testing that has to be conducted.

Method

Participants

Six participants (2 visually impaired, 2 dyslexic and 2 non-disabled matched age control), including three males and three females with a mean of 32 years of age participated in the study. Self report of previous diagnosis was used to classify participants in the visually impaired and dyslexic groups and these participants were included to help unpack the synergy between accessibility and usability. All but one participant were human-computer interaction (HCI) novices with little or no specific knowledge of usability or accessibility or web design.

Materials

eGovernment and eCommerce websites are regularly used by the general public. Previous research supports testing this type of site as eGovernment web sites perform reasonably well on accessibility while eCommerce sites are less compliant (DRC, 2004). Specific details of the sites used in this study have been avoided to maintain anonymity. However, three eCommerce websites were chosen on the basis of mainstream popularity and familiarity to the researcher, including a bank, supermarket and a retail catalogue store. Three eGovernment websites were selected on the basis of perceived general need, including a local council, a government employment centre and a licensing authority. The websites were classified as either highly usable on the basis of Nielsen's usability heuristics (Nielsen, 1993), as accessible using WCAG priority 1 criteria and/or as pleasurable on the basis of the researcher's subjective opinion of aesthetics.

Due to the nature of the findings the classification of the sites is not detailed here. Consistency of websites was checked between tests. Order of evaluation was counter balanced and each site was assessed by two participants who had little or no previous exposure to that particular site. Each participant tested two sites only.

Two tasks, one simple and one more complex, were established for each website to assess and unpack the concept of usability. All tasks required participants to leave the homepage and find information on other pages. Simple tasks included finding the nearest branch/store of the website. More complex tasks involved seeking less readily available information such as specific information relating to a particular service, i.e. finding a home mortgage rate. Participants were required to complete the task without using any generic search function, with the shortest pathway to the goal noted to check absolute success. Participants were encouraged to complete the task alone, with guidance offered only if there was visible frustration or excessive time had been spent.

Procedure

Testing was conducted in the HCI laboratory at the University of York. Following a pre-test questionnaire to gather demographic information, participants spent a few minutes looking at the homepage of first website without clicking on any links to unpack acceptance issues. Their initial reaction was gathered before the participant undertook the two tasks. The researcher recorded the path taken, comparing this to the optimal pathway to assess level of success of the task. Participants were encouraged to "think aloud" as they completed the task and if problems were encountered an assessment was conducted. Problems were issues encountered in pursuit of their goal rather than identifying as many violations of guidelines and good practice as possible. An open question about the cause of the problem was followed by a specific question whether the problems encountered were usability, accessibility or aesthetic/enjoyment issues. This was to avoid assumptions that any issues encountered by assistive technology users are accessibility issues and to determine whether the concepts of usability, accessibility and acceptance were understood. Participants were asked to rate the severity of the problem as cosmetic, minor, major or catastrophic. The researcher also rated the problems as to cause and severity classifying catastrophic as unable to continue, major as severe frustration, minor as mild annoyance and cosmetic as a problem noticed but not relevant. After each task, an assessment was completed before the participant moved onto the next. Once both tasks were complete an overall reaction to the site was gathered and an engagement questionnaire (Petrie, 2005) assessing factors such as the degree to which participants believed they had fun was completed, before repeating the procedure for the second site.

Results and discussion

Across the six websites tested, 41 individual problems were identified, with 16 of these having between two and seven multiple occurrences (total 71). The most common problems were poor or no headings, unclear categories/links and large lists of links or text. On average, 12 problems were identified per website. Based on participant categorisation, accessibility problems accounted for 18.3%, usability/design problems for 62.0% and acceptance/pleasure issues 19.7%. Using the criteria detailed below, the majority of problems (77.5%) could be classified in more than one category. The overall success rate for the tasks was 62.5%, slightly less than in the DRC report. In addition, the pre-determined optimal pathway was followed in only two cases, indicating that the shortest route was not always easy to find. Most participants rated the severity of problems equally or less harshly than the researcher. For example, very few of the individual problems (five) were rated as catastrophic by participants. The researcher classified nine problems as catastrophic, the number of tasks not completed.

Accessibility

For the purposes of this study, accessibility issues were those associated directly with the assistive technology being used by the visually impaired participants (VIP) and problems specific to the dyslexic/VIP participants. There were eight problems classified as accessibility issues (including missing skip navigation tags and use of poor colour contrast), two of which could also be classified as reducing usability and/or acceptance for mainstream users. For example, large amounts of white space or text reduce access to information for dyslexics and can be aesthetically displeasing. One participant stated that the long lists of links "…overwhelmed me". In addition, parts of one site were completely inaccessible to a user of the screen reading software Jaws, highlighting that some issues are more severe than others.

Several of the problems classified as accessibility issues potentially also caused usability issues. One site included a "skip navigation" option, mainly used to skip to main content, which caused major difficulties for the user as they skipped over the left hand navigation and the main content had nothing to explain what options were available. The use of the word "quote" on the banking website caused problems for screen reader users. Because Jaws says "quote" when it encounters speech marks the user was ignoring the word quote and followed a false lead.

Acceptance

Participants expressed many positive and negative reactions to the websites both during their initial viewing and whilst completing tasks. Positive statements such as "I felt like I was going somewhere" suggests that even on utilitarian pages positive feelings are important. Negative reactions included obvious frustration and annoyance, indicating a low level of acceptance and engagement with the websites. Comments included statements such as "that really irritates me" and "does wanting to throw the monitor out the window count as a problem?" While all problems cause some degree of negative reaction and could be categorised as acceptance issues, only a few were purely acceptance issues mainly to do with aesthetics, trust and welcoming, with comments such as the site is "a visual muddle," "unwelcoming," "naff," "bland and boring," all indicating that not only can participants identify acceptance issues easily but that these types of problems are apparent and pertinent to users.

Measuring the level of engagement for utilitarian websites may be optimistic, but not impossible. However, feeling welcomed to the site, feeling supported in the task and belief that the user is moving towards their goal are all aspects of pleasure that lead to engagement, even when buying a toaster or doing online banking. The absence of negative emotion such as frustration and annoyance would also lead to a more pleasurable and engaging encounter with technology. The engagement scale used in this study provided valuable feedback but may need refining to clarify reactions. For example, the extent to which participants experienced surprise could be high but either positively or negatively, which was not measured.

Usability/design

When participants were asked open questions about the cause of the problem, the majority stated that the cause was "poor design". When asked specifically whether they believed the problems they encountered were usability, accessibility or aesthetic/enjoyment issues, most stated that design issues were usability related. These usability issues included things as simple as not using headings appropriately (which could also be an accessibility issue) and more complex issues involved with categorisation of products and services listed on the site.

Experienced users and those without visual impairment did highlight usability issues, for example one of the VIP stated "its [the site] accessible but not usable." Those with the greatest difficulties highlighted the greatest number of issues, whether they were accessibility or usability. The least experienced internet user (also dyslexic) highlighted the most problems. The fact that participants with disabilities find more problems reinforces the concept that they make the best test participants. The less experienced, more disabled, user is not as used to work arounds as able experienced users and can therefore highlight more problems.

Conclusions

This study is a preliminary venture in a series designed to empirically investigate the synergy of usability, accessibility and acceptance in an effort to determine more efficient and effective guidelines to encourage better design and more user testing. While this study establishes that there is some cross over between accessibility, usability and pleasure in using websites, the degree of synergy and severity weighting have yet to be empirically determined. The study provides evidence that some problems are category specific but most can be cross classified. In addition, it is clear that seeking information about these three factors can best be conducted using less experienced internet users and disabled people as they tend to identify more issues under each category, including those that would be identified by experienced mainstream users. Measuring engagement, recording negative emotional expression and asking open questions about the look is also useful in drawing out acceptance issues. It is also clear that while broad categories of usability, accessibility and acceptance are useful, these categories can be further broken down for greater clarity. For example accessibility could involve issues such as programming errors, the specific condition or assistive technology, while acceptance could contain issues of negative emotion, trust and engagement.

References

Chisholm, W., Vanderheiden, G., and Jacobs, I. 1999. *Web Content Accessibility Guidelines 1.0*, http://www.w3.org/TR/WCAG10/ website.

Disability Discrimination Act (DDA), 1995, *Part III: Discrimination in other areas* (London: The Stationery Office).

Disability Rights Commission (DRC), 2004, *The web: Access and inclusion for disabled people.* (London: The Stationery Office).

Jordan, P. W. 1997, Human factors for pleasure in product use. *Applied Ergonomics*, **29**(1), 25–33.

McCarthy, J. and Wright, P. 2004, *Technology as experience.* (Cambridge, MA: MIT Press).

Nielsen, J. 1993, *Usability engineering.* (Boston, MA: Academic Press).

Petrie, H., Hamilton, F. & King, N. 2004, *Tension, What Tension? Website Accessibility and Visual Design*, W4A at WWW2004, pp. 13–18.

Petrie, H. 2005, Report *on evaluation of the Primary Film Online first prototype with disabled children.* (York: Department of Computer Science, University of York).

Usability.gov (2005). *Your resource for designing usable, useful and accessible websites and user interfaces.* http://www.usability.gov/ website.

A FRAMEWORK FOR EVALUATING BARRIERS TO ACCESSIBILITY, USABILITY AND FIT FOR PURPOSE

Suzette Keith[1], Gill Whitney[2] & William Wong[1]

[1]Interaction Design Centre, [2]Collaborative International Research Centre in Universal Access, School of Computing Science, Middlesex University, The Burroughs, Hendon, London NW4 4BT

> In just over a decade, the availability of information through the world wide web has revolutionised the way many of us find information. There is, however concern that those who are older, disabled or socially disadvantaged may be excluded from this resource. Many of these users have experienced difficulties navigating through a website, using the search functionality provided, or understanding the retrieved content. Accessibility by disabled people is one of many potential barriers affecting successful information retrieval. This paper proposes a fit for purpose framework which brings together accessibility, usability and the user context.

Introduction

Websites have become important information resources providing users with information to solve various information seeking needs. However not all users find the experience successful and concern is being expressed about the "digital divide" whereby some users are excluded from this opportunity. While poor usability of the interface can affect all users, accessibility specifically addressed the needs of disabled people to access web content. A survey of 1000 UK websites by the Disability Rights Commission (DRC 2004) reported that only 19% of web pages passed the first priority level checkpoints described by the World Wide Web Consortium (W3C) Web content accessibility guidelines (Chisholm et al, 1999). Sullivan and Matson (2000) reported finding many barriers to overcome when comparing website accessibility using the W3C guidelines. They further identified a challenge to information design to maximise information access and all its contents to an increasingly diverse user group.

In Wong et al (2005), we reported the application of a fit for purpose study with a public information terminal intended for use within a socially disadvantaged community. This study went beyond the interface design issues and examined the user context in order to determine "fit for purpose" and "whether the purpose for which the system was commissioned translates into the physical design and implementation of the delivered system in meeting the needs of its target users". This study applied complementary techniques "to provide a multi-perspective approach to problem diagnosis". The study concluded that the notions of usability and accessibility "do not usually address the deeper, semantic compatibility requirements of the target users."

The aim of the second case study (Whitney et al, 2005) was to compare ten transport websites taking account of the needs of older and disabled people when planning a journey using passenger ferries. The evaluation applied a three-part methodology to take account of accessibility guidance, usability issues, and the travel planning task. The study concluded that in addressing the task of travel planning "the web pages studied appeared to fail disabled travellers in all three ways … they did not contain all the information needed by disabled travellers, the information that they did contain was not easy to find, and the web pages were not accessible". As in the previous study, the barriers to information access went beyond the presentational issues of interface design and extended into the structure and delivery of the content.

Information seeking is not an easy activity and is affected by the complexity of the information needed and the user skills. In developing a framework to support an evaluation of fit for purpose we need to take account of the user context in relation to the task, design of the content and the delivery mechanism.

Defining the user context

Our two studies found no suggestion that the less skilled user is likely to have simpler information needs. Instead our results indicate that the less skilled user might be faced with more complex problems as summarised in Table 1 (adapted from Wong et al, 2005).

Information seeking can be described as a complex process of refinement of the information goal as the user becomes more familiar with the topic and the resources available (Kuhlthau, 1988). Initially the user may find it difficult to articulate an information need and may become anxious. We created a representation of two levels of skill and three levels of problem complexity, introducing a very early stage which we called "don't know what they need to know" which can affect both high and low search skills.

Table 2 summarises three evaluation techniques which can be combined to explore different aspects of fit for purpose, and specifically whether a user can complete specified tasks.

Table 1. Information need and skills in relation to information content.

	Don't know what they need to know	Simple needs	Complex needs
High search skills	e-brochure & kiosk ware (awareness, articulation)	Independent web search e-brochure & kiosk ware (bite-size take-away)	Independent web search Advisor required
Low search skills	e-brochure & kiosk ware (awareness, articulation)	e-brochure & kiosk ware (bite-size take-away)	Advisor required

Table 2. Identifying barriers to fit for purpose.

Techniques	Issues examined
Accessibility	*Can the user access the web content?* Barriers to interaction may prevent the user getting to the home page or beyond it to significant content.
Usability	*Can the user use the interface?* Barriers to interaction may leave the user confused and lost within the navigational structure of the system.
Co-discovery User trial	*Can the user complete the tasks?* Barriers to completing information seeking tasks may result from overly complex or insufficient information or poor structure of the content making it difficult to find.

We selected these different evaluative techniques in order to identify the barriers between the user goal and the information content: (i) an examination of web content accessibility using a checklist approach, (ii) an examination of usability issues by applying a heuristic evaluation, and (iii) co-discovery user trial applying representative tasks.

Can the user access the web content?

The W3C Web Content Accessibility Guidelines focus on whether the content is accessible to a person with a disability. These guidelines offer 14 principles of accessibility of web-content design, describing both the problems and possible design solutions. They address system level elements such as the use of mark up and style sheets as well as presentational issues of text and graphics. A checklist tool included with the Guidelines is arranged according to three levels of priority. Automated checking tools are available through the internet and simplify the accessibility assessment; however some manual checks are also essential.

The accessibility assessment reported in our second case study considered conformance with Level 1 checkpoints. These include, for example, the provision of text equivalents for non-text elements, that the content is understandable without the use of colour, and that the organisation of the documents can be understood without style sheets. None of the ten sites assessed met all the Level 1 checks. This would suggest that people with visual impairments, especially those using screen reading devices, will be unlikely to be able to interact with the sites. For example, in one of our evaluations, it was not possible for a visually impaired person using a screen reader to find out what time the ferry departed because the timetable was available only as a scanned image.

Can the user use the system?

Usability is a complex concept and is concerned with the effectiveness, efficiency and satisfaction of the user and system interaction [ISO 9241-11, 1998]. Expert usability evaluation methods have been developed from human computer interaction principles in order to identify and explain user problems. One of the most commonly available expert usability methods is the heuristic evaluation developed to support a quick, low cost assessment of usability issues. Nielsen (1994), for example, offers ten usability heuristics which can be used to examine the design of any interface including website design.

The usability heuristics identify common classes of user interaction problems that are likely to lead to confusion, errors or task failure. Unlike the accessibility guidelines the heuristics do not offer solutions or an agreed prioritisation. Usability issues have the potential to affect all users and are of particular importance to new users and those with limited skills or insufficient experience to overcome errors.

Can the user complete the task?

From the user perspective the important issue is whether they can complete a task or whether they meet an insurmountable barrier of whatever cause. Nielsen proposes the use scenarios and simple think-aloud protocol analysis as a useful additional tool in discount usability engineering. The think-aloud test subject needs to maintain a commentary about what they are doing and why, providing a researcher the opportunity to understand the choices made. In considering the needs of the less able user, this approach, however, can overload their capacity to perform the task.

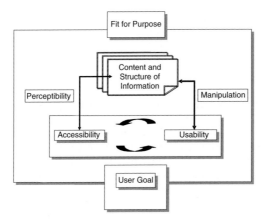

Figure 1. The accessibility, usability and fit for purpose framework.

Instead we have included a variation of the think-aloud technique called the co-discovery technique which allows a pair of users to work together and discuss their actions in a naturalistic and supportive manner. The users can also be interviewed following completion of the task. This offers a more holistic insight into the difficulties faced by the users in completing typical task scenarios.

Applying the fit for purpose framework

Any one of the three techniques may reveal barriers between the user goal and successful task completion. Figure 1 adopts a user perspective of the fit for purpose framework. The user goal to find information and the context defines the user requirements. Accessibility and usability factors act as barriers to the primary goal of interaction with the information content, especially problems of perceptibility of the display and manipulation of control elements for which there may be some interaction effects. The design of the information content and structure may also interact with the user goal – information may be undiscoverable or unintelligible and thus still present a barrier to success.

Conclusion

This fit for purpose evaluation framework provides an opportunity to include the contextual factors that reflect the requirements of older, disabled or otherwise less skilled users. The combination of evaluation strategies using familiar and commonly available techniques goes some way towards discriminating between barriers derived from the system design and the content design by focussing on the user goals and context. It is unlikely these barriers will respond to quick fixes to the interface design and problems attributable to design decisions affecting the structure of the system or content may need extensive rethinking. The fit for purpose framework offers the potential to support the design of web content by predicting possible barriers to a proposed design concept and to the mechanisms of the information delivery. The framework offers an opportunity in which to secure improvements in e-accessibility to web content especially for non-expert users as well as offering better informed comparative assessments of website quality. The framework is currently being applied to a further study of travel planning by older users.

References

Chisholm, W., Vanderheiden, G., Jacobs, I., (eds) (1999) Web content accessibility guidelines 1.0. Retrieved December 2005 from www.w3.org/TR/WAI-WEBCONTENT

Disability Rights Commission. (2004) Formal Investigation report: web accessibility. TSO

ISO 9241-11:(1998) Ergonomic requirements for office work with visual display terminals (VDTs) – Part 11: Guidance on usability

Kuhlthau, C. (1988) Longitudinal case studies of the information search process of users in libraries. *Library and information science research* 10 (3) pp. 257–304

Nielsen, J., (1994) Heuristic evaluation. In Nielsen, J., Mack, R., *Usability Inspection Methods*. John Wiley and Sons Inc (also Retrieved December 2005 from www.useit.com/papers/heuristic)

Sullivan, T., & Matson, R. (2000) Barriers to use: usability and content accessibility on the Web's most popular sites. *Proceedings 2000 conference on Universal Usability* pp. 139–144 ACM Press New York, NY, USA

W3C Web Content Accessibility Guidelines 1.0. May 1999. Retrieved March 10, 2005 from W3C site: http://www.w3.org/TR/WAI-WEBCONTENT/

Whitney, G., Keith, S., & Kolar, I. (2005) Electronic Travel Booking, the Evaluation of Usability, Accessibility and Fitness for Purpose by Older People and People with Disabilities. *Human Computer Interaction International Conference*, Las Vegas 2005

Wong, B. L. W., Keith, S., & Springett, M. (2005). Fit for Purpose Evaluation: The case of a public information kiosk for the socially disadvantaged. In *People and Computers XVIV, Proceedings of HCI 2005*. (Vol. 1, pp. 149–165): Springer Verlag

HCI – APPLICATIONS

THE DECAY OF MALLEABLE ATTENTIONAL RESOURCES THEORY

Mark S. Young & Neville A. Stanton

School of Engineering and Design, Brunel University, Uxbridge, Middlesex UB8 3PH, UK

Malleable Attentional Resources Theory (MART) purports that attentional resources degrade in underload conditions, such that an operator is less able to cope when demands increase (e.g., in an automation failure scenario). Thus far, there has been considerable empirical and theoretical support for MART, and it has the potential to define an underload threshold for a given scenario. However, in order to realise this potential, it is first necessary to determine the decay and recovery curves for attentional resources. This paper represents the next step for MART, as data are reported that plot the time-decay curve for the attentional resources of participants in a vehicle automation experiment. We reveal that underloaded participants experience significant resource decay with the first minute of the task. The findings are discussed with respect to the elusive "redline" for mental underload.

Introduction

In 1999, we introduced the concept of *malleable attentional resources* as a parsimonious explanation for the effects of mental underload on performance (Young and Stanton, 1999). Whilst most authors agree that underload is equally detrimental as overload (e.g. Hancock and Parasuraman, 1992; Neerincx and Griffioen, 1996), there remains some debate as to the actual mechanism for this effect. The underload problem has been variously attributed to complacency (de Waard et al, 1999), trust (Rudin-Brown and Parker, 2004), situation awareness (Kaber and Endsley, 1997), or effort regulation (Matthews and Desmond, 2002).

Malleable attentional resources theory (MART; Young and Stanton, 2001; 2002a; 2002b) was offered as a way of reconciling these explanations by appealing to existing resource theories of attention (e.g. Kahneman, 1973; Wickens, 2002). It is suggested that, just as attentional capacity can fluctuate in the long term with arousal, mood, or age (cf. Kahneman, 1973), task demands can directly influence the size of resources in the relatively short term. As resources degrade, the operator becomes less able to cope if demands should suddenly increase.

Since its inception, MART has gathered momentum in the literature on human interaction with automated systems. In our own laboratory, we gathered empirical data on mental workload and allocation of visual attention to demonstrate that capacity is directly influenced by workload in automation conditions (Young and Stanton, 2002b). Elsewhere, Bailey and Scerbo (2005) gave credence to MART in a study of human supervision of reliable automation, while Straussberger et al (2005) invoked MART as a possible consequence of monotonous air traffic control tasks.

Nevertheless, there remain a number of questions and areas of further development for MART before it can become a truly workable theory. One of these concerns the precise nature of resource degradation during underload tasks. Young and Stanton (2002b) reported that during a vehicle automation experiment, underload led to a significant reduction in resources in just 10 minutes. However, they did not give any further information about the level of decay over that 10-minute trial, so it is difficult to predict exactly when performance may begin to suffer. If we can plot a continuous curve of attentional resources against time, we can analyse just how quickly capacity shrinks.

Last year, Young and Clynick (2005) attempted to do just that in a flight simulator experiment. Moreover, they continued to run the experiment after demands had increased, to determine how quickly capacity recovers after underload. Unfortunately, the study failed to produce the required underload effect, thus it was impossible to plot an informative time-decay curve for attentional resources. The authors offered a number of explanations for this, the most likely of which was that MART may be an automation-specific theory, as the flight study did not involve an automation condition.

In the present paper, then, we return to the vehicle automation experiment of Young and Stanton (2002b) to see if the data can be analysed in enough detail to plot the time-decay curve. As we know that an underload effect was observed in that study, we should be able to derive the relevant plot for the 10 minutes of the experimental trials.

Method

Design and procedure

Since this is essentially a re-analysis of existing data, what follows is a summary of the key points from the Young and Stanton (2002b) experimental design.

A medium-fidelity driving simulator was used to provide four levels of vehicle automation as a within-subject factor: Manual, Adaptive Cruise Control (ACC), Active Steering (AS), and ACC + AS. ACC is an extension of conventional cruise control, in that it uses a radar to detect slower vehicles in the path of the subject vehicle, and adjusts speed to maintain distance. AS is a lateral control device, designed to keep the subject vehicle in its lane without any steering input from the driver. The experimental task used a "follow-that-car" paradigm to standardise non-manipulated demand across the independent variables, and each trial lasted for 10 minutes.

Dependent variables included a visuo-spatial secondary task as an indirect measure of mental workload, and allocation of visual attention measured via a low-light miniature camera installed on the dashboard. Number of correct responses on the secondary task, and total time spent looking at the secondary task, were the variables recorded. Thirty participants, who had held a UK driving licence for at least one year, were recruited to take part. However, due to difficulties in data recording, only a subset of 23 participants was used in the video analysis of eye-movements.

Data reduction

The dependent variables on the secondary task and the allocation of visual attention were combined to arrive at an *attention ratio* score – that is, an inferred measure of attentional capacity. The assumption is that we can assess capacity by dividing the number of responses on the secondary task by the amount of time spent attending to it. A null hypothesis of fixed attentional resources should result in a reasonably consistent attention ratio score – one should spend no more time than is necessary to respond to the secondary task. If resources fluctuate with mental workload, then the attention ratio score should decrease in line with workload reductions. In order to plot the attention ratio curve over time, the data were divided into 10 one-minute blocks.

Results

To recap Young and Stanton's (2002b) mental workload results, generally speaking ACC did not reduce mental workload on its own, but AS significantly reduced workload, and there was a further significant reduction in the ACC + AS condition.

Combining the secondary task data with the visual attention results, the pattern of attention ratio scores across the automation conditions mirrored that for mental workload. In other words, the inferred size of attentional resources declined in exactly the same way as reductions in mental workload when using automation.

Now, to determine the temporal decay of resources, we first need to establish the pattern of mental workload over time. For this paper, we have restricted the timeline analysis to the ACC + AS level of automation, as this condition represented the lowest level of workload, and hence the greatest amount of resource decay. Figure 1a shows the pattern of secondary task responses across the duration of the trial (bear in mind that the secondary task score is a measure of spare capacity, and thus a higher score implies a lower level of workload on the primary task). A visual inspection of the curve suggests that mental workload gradually decreases until around the fourth minute of the trial, when it stabilises for the rest of the drive. Using a repeated measures analysis of variance (ANOVA), with the first minute as the reference condition (to act as a baseline), actually reveals that the workload reduction in the second minute is significant ($F_{1,22} = 5.63; p < 0.05$). Thus the actual reduction in workload is immediate.

Looking at the timeline plot of the attention ratio data now (Figure 1b), we again see that resource capacity mirrors mental workload in this condition. Indeed, the effects on capacity are even more dramatic than those for workload, with the graph displaying an immediate drop in capacity before stabilising for the duration of the trial. The attention ratio scores when underloaded appear to be around one-quarter of that in the baseline section. In other words, participants are spending four times as long per response as they would do when under normal conditions.

Again, a repeated measures ANOVA confirmed that the initial drop in capacity was significant, when comparing the second minute against the first baseline minute ($F_{1,21} = 4.37; p < 0.05$). (Note that data for one participant were discarded in this analysis, due to their not making any secondary task responses during the first minute of the trial. This gives them an attention ratio score of zero, artificially deflating the baseline measure).

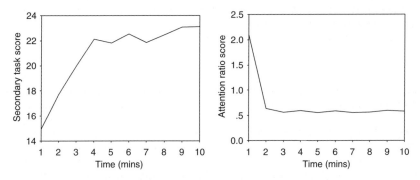

Figures 1a and 1b. Timeline plots of secondary task responses and attention ratio scores for the underload condition.

Discussion

The point of this analysis was to begin to delineate the temporal nature of resource decay as predicted by MART. By analysing the attention ratio score within a proven underload condition, we have demonstrated that presumed resource capacity appears to shrink directly in line with reductions in mental workload. Moreover, to all intents and purposes, this shrinkage appears to be virtually instantaneous.

That resource decay occurs so quickly was surprising, to say the least, as it was anticipated that there may be some lag as attentional capacity adapts to the task demands. Nonetheless, the results provide further clarification of the mechanism by which MART explains the relationship between underload and performance.

Notwithstanding these conclusions, we must accept that there may yet be issues with the dataset which weaken this interpretation of the results. This particular association between attention ratio and secondary task responses may actually be an artefact of the way these data are measured. Again, more sensitive measurement techniques (e.g. by using eyetracking equipment), and comparing the data to a more appropriate baseline would help to rule out this criticism.

Whilst these suggestions for the experimental design and statistical analysis would undoubtedly strengthen the case for MART in this context, we still believe that the existing design was as solid as it could be. If it were indeed an artefact of the measurement procedure, one would expect it to be relatively simple to reproduce the underload/decay relationship. However, as Young and Clynick (2005) demonstrated last year, the association is by no means clear cut yet.

Thus there is a good deal of research still to be conducted on MART, both on the existing datasets as well as more empirical studies. The most promising avenues of research are in further exploration of these decay and recovery curves. The decay curves can be elucidated by treating the other automation conditions (i.e., Manual, ACC, AS) in the same way as has been achieved here, in order to establish a more robust baseline for attentional capacity. More experiments need to be conducted in the vein of Young and Clynick's (2005) study, in extending the trial into a recovery phase of normal workload.

Eventually, we would hope to identify a threshold of resource decay beyond which true underload (i.e. in terms of degradation of performance) can be predicted. If we accept the absolute figures on the attention ratio curve in the present paper, we can venture to state that a shrinkage factor of 75% was necessary to produce underload. Now, intermediate levels of workload need to be analysed to establish other relative shrinkage factors, but if we can plot similar curves under a number of other conditions, it may well be possible to isolate a cut-off point for underload. Levels of workload and attentional resources could then be used in an *a priori* manner to predict performance given a level of task demands – and thus putting us in a strong position to establish the elusive "redline" of mental workload research.

References

Bailey, N., and Scerbo, M. 2005, The impact of operator trust on monitoring a highly reliable automated system. *Proceedings of HCI International 2005, Las Vegas, July 22–27 2005*.

de Waard, D., van der Hulst, M., Hoedemaeker, M., and Brookhuis, K. A. 1999, Driver behavior in an emergency situation in the Automated Highway System. *Transportation Human Factors*, **1**, 67–82.

Hancock, P. A., and Parasuraman, R. 1992, Human factors and safety in the design of Intelligent Vehicle-Highway Systems (IVHS), *Journal of Safety Research*, **23**(4), 181–198.

Kaber, D. B., and Endsley, M. R. 1997, Out-of-the-loop performance problems and the use of intermediate levels of automation for improved control system functioning and safety. *Process Safety Progress*, **16**(3), 126–131.

Kahneman, D. 1973, *Attention and effort*, (Prentice-Hall, Englewood Cliffs, NJ).

Matthews, G., and Desmond, P. A. 2002, Task-induced fatigue states and simulated driving performance. *Quarterly Journal of Experimental Psychology*, **55A**(2), 659–686.

Neerincx, M. A., and Griffioen, E. 1996, Cognitive task analysis: harmonizing tasks to human capacities. *Ergonomics*, **39**, 543–561.

Rudin-Brown, C. M., and Parker, H. A. 2004, Behavioural adaptation to adaptive cruise control (ACC): implications for preventive strategies. *Transportation Research Part F*, **7**, 59–76.

Straussberger, S., Kallus, K. W., and Schaefer, D. 2005, Monotony and related concepts in ATC: A framework and supporting experimental evidence. *Proceedings of HCI International 2005, Las Vegas, July 22–27 2005*.

Wickens, C. D. 2002, Multiple resources and performance prediction, *Theoretical Issues in Ergonomics Science*, **3**(2), 159–177.

Young, M. S., and Clynick, G. F. 2005, A test flight for Malleable Attentional Resources Theory. In P. Bust and P. McCabe (eds.), *Contemporary Ergonomics 2005*, (Taylor and Francis, London), 548–552.

Young, M. S., and Stanton, N. A. 1999, Miles away: A new explanation for the effects of automation on performance. In M. A. Hanson, E. J. Lovesey, and S. A. Robertson (eds.), *Contemporary Ergonomics 1999*, (Taylor and Francis, London), 73–77.

Young, M. S., and Stanton, N. A. 2002a, Attention and automation: New perspectives on mental underload and performance, *Theoretical Issues in Ergonomics Science*, **3**(2), 178–194.

Young, M. S., and Stanton, N. A. 2002b, Malleable Attentional Resources Theory: A new explanation for the effects of mental underload on performance, *Human Factors*, **44**(3), 365–375.

USER RESPONSES TO THE LEARNING DEMANDS OF CONSUMER PRODUCTS

T. Lewis, P.M. Langdon & P.J. Clarkson

Engineering Design Centre, Engineering Department, University of Cambridge, Trumpington Street, Cambridge CB2 1PZ

Designers are typically male, under 35 years old and unimpaired. Users can be of any age and currently over 15% will have some form of impairment. As a result a vast array of consumer products suit youthful males and in many cases exclude other demographics (e.g. Keates and Clarkson, 2004).

In studying the way a range of users learn how to use new products, key cognitive difficulties are revealed and linked back to the areas of the product causing the problems. The trials were structured so each user had to complete a specific set of tasks and were consistent across the user spectrum. The tasks set aimed to represent both everyday usage and less familiar functions. Whilst the knowledge gained could provide designers with valuable guidelines for the specific products examined, a more general abstraction provides knowledge of the pitfalls to avoid in the design of other product families.

Introduction

Inclusive Design aims to challenge designers to consider every user with an impairment in the design of products for the mass market. Designers typically consider how they will use a product and as the majority of designers are male, young and highly able, large sectors of the overall population are often excluded (e.g. Keates and Clarkson, 2004). Whilst much has been achieved in tackling how designers can consider those with sensory and motor impairments through simulation kits and literature, there is relatively little research on how designers can empathise with those who have cognitive disabilities (e.g. Cardoso et al, 2004).

Symbols and labelling used in consumer products are vital to provide the user with suitable cues to remember the sequence that controls need to be activated in order to achieve their overriding goals (e.g. Norman, 1998). When a button can have more than one function, highlighting the current mode of the device is clearly important but it is also vital to link each mode with the symbol or label that corresponds to the function of the button in each mode (e.g. Neilsen, 1993, Ryu and Monk, 2004).

A previous study had examined the recognition of symbols, in a group of users over 60 years in age, from Video Cassette Recorders (VCR's). Older models used none, if any, computer interfaces but more modern VCR's and the next generation electronic devices and DVD players, use modality and new, confusing symbologies extensively. Despite this, fewer than 20% of users were able to recognise the symbols for the most commonly used functions of "play" and "stop" (e.g. Lewis and Clarkson, 2005).

Background

The cognitive walkthrough method analyses the stages through which a user forms goals and executes actions in using a product or interface. The use of goal-action connections is similar to analysis from other methods, e.g. GOMS (e.g. Card et al, 1983). As a product analysis method, its main advantage is avoiding the need for recruiting users and running lengthy trials whilst it has also been used in conjunction with trials to predict likely errors (e.g. Polson et al, 1992).

Ryu and Monk (2004) use a low-level analysis that simplifies the method to 3 phases and combines ideas with their notion of Cyclic Interaction Theory. This examines the loop of interaction between the user, the product and the effects they have on the world. Goal-Action analysis is extended to include the connections with Effect giving three categories of potential error areas; this extends to the development of an error-classification scheme (e.g. Ryu and Monk, 2004). Norman's scheme of catagorising human slips also has relevance in any development of new scheme of errors (e.g. Norman, 1988).

The simple cognition model (Fig. 1), a simplification of standard models in use in Human Factors research (e.g. Wickens and Hollands, 2000) provides a simplified understanding of how the mind responds to changes in the environment that can be brought to bear on product demand. Sensory organs perceive stimuli from the real world, many are ignored, flowing straight through the central executive with little, if any, attention resulting in no action. If the central executive decides to react; some reference to memory will be made.

Long-term memory stores episodic and autobiographical experiences from the past. The central executive can first check whether a stimulus has been experienced before, what the resulting actions were and whether the effects these had on the real world were intended. The similarity of the situation to what has been experienced before will affect the speed and efficiency of the retrieval of responses from the long-term memory.

Working memory is for storing items whilst the central executive deals with them. Both processing in the central executive and the use of the working memory have limits to both the amount of information they can handle as well as the speed they work. Attempts to exceed either of these will normally result in cognitive problems such as

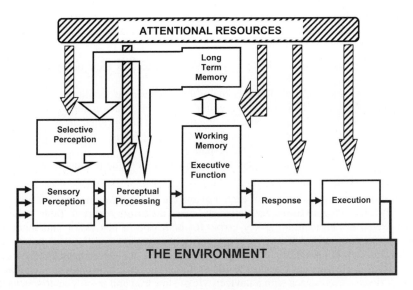

Figure 1. Simple cognition model.

incorrect conclusions, inappropriate actions and ultimately making errors (e.g. Reason, 1990; Langdon et al, 2004).

Method

A number of products were tested as pilot studies to develop a procedure and examine potential problems before embarking on a bigger trial, including both a mobile phone and an mp3 player. Extended trials were performed with the car as this was easier to film and subsequently more accurate in recording button presses. Initially 15 users, 8 male and 7 female varying from 20 to 60 in age were tested on the car using functions and tasks that involved no motion of the vehicle for safety reasons. As such, this restricted the task list to include exercises pertaining to lights, wipers, etc.

Users were provided with a task list beforehand to reveal any comprehension problems but were told that questions relating to how a particular task would be achieved would not be answered. From the pilot studies, various methods of user commentary had been explored. No commentary provided insufficient data in the analysis stage as to whether the user knew they were making errors. Whilst live commentary avoided the users forgetting what mistakes they had made, it was considered this style would interfere in how they were using the device. A protocol analysis was adopted that allowed the user to immediately watch the video of the trial and then provide a retrospective commentary.

Analysis and discussion

The users' performance and ability to learn the product was measured by recording the time taken to complete individual tasks. This provided an understanding of which users had most problems but, to develop this further, the errors experienced were classified. An ad-hoc categorisation scheme was first developed from the pilot trials based on the errors that had been observed. However, this lacked completeness and had no formal theory to support it. Using the cognitive model shown in Fig. 1, 7 as an overview, broad categories were selected to identify which cognitive areas were active in segments throughout the trial. Identification of when the errors occurred and knowledge of the cognitive area active could then be linked to the task and identify the product feature failing the user.

The 7 categories were as follows:

- Perception; as the user initially perceives and processes their perception of the product's interface
- Executive Function; further processing of the perceived information
- Working Memory; if the user was required to store information during attention
- Long Term Memory; past knowledge or experienced being called upon
- Response Selection; deciding upon the most likely action to achieve the goal
- Execution; performing the selected response
- Feedback; anticipating and receiving the feedback from the product or world as a result of the action.

Analysis of the video footage of each trial and the protocol commentary was used to identify which cognitive area was active and when the errors occurred. Table 1 shows abbreviated results identifying key errors for both (top) a non-driver and (bottom) driver for a range of tasks.

Low-level walkthrough analysis would predict that the non-drivers would incur more errors than the drivers and that they would occur during perception and executive function periods in accordance with a novice style of slow trial and error. The drivers were more likely to err during Long Term Memory reference when proactive interference from

Table 1. Example results from a non-driver (top) and a driver (bottom) for tasks.

Time (s)	Task	User action	Perception	Exec. Func.	Working Memory	Long Term Mem.	Response Select.	Execution	Feedback
30	Rear screen wash	Turns on front wipers	•					•	
34	Rear screen wash	Activates front screen wash	•					•	
61	Beam	Turns sidelights on	•		•				
73	Lights off	Passes on beam			•	•			
95	Rear heated window	Looks at rear window			•	•			•
10	Front wipers	Examines lights stalk	•				•		
23	Headlights	Examines wipers stalk	•				•		
62	Open/close front passenger window	Opens front driver window			•			•	

their own models was occurring. Table 1 confirms both these tendencies for these users and further highlights the tasks during which they were occurring.

The experienced drivers were slowed by having to identify which stalk was assigned to the wiper control and which was for headlights. Whilst the convention of using multi-function stalks for these purposes is common, the arrangement of left and right stalks is varied throughout the car market. Their experience also made them less attentive to the tasks and occasional errors such as opening the wrong window without noticing were observed. The non-drivers suffered from requiring more feedback to confirm they had completed the task. Experience of the heated rear window function and knowledge of the differences between sidelights, headlights and hi-beam was lacking and time was spent trying to link the term to the relevant symbol. Often a haphazard, "trial and error" approach was adopted.

Other error schemes highlight additional problems that were not discovered through the simplistic cognitive approach described above. The Ryu and Monk (2004) combination of cognitive walkthrough and cyclic interaction theory describes four further different "goal-action" errors. For example, the heated rear window control is very similar in shape to the fog lamps control, which also provides very limited feedback; this is classed as "weak affordance to a correct action" problem.

Conclusions

The method developed clearly has its merits and revealed many of the errors the users experienced during the car trials. As a development of the previous method it is successful but is not complete and misses some problems that other schemes and methods would have classed. Future work needs to consider an amalgamation of the method used, with those discussed and others to develop a categorisation system that encompasses most errors, not just from the car trials, but also over other products.

With an improved method and scheme, further products will be tested so trends can be drawn across different products families and guidelines can be provided to designers of many different product types as to how to better design products for ease of learning.

References

Card, S.K., Moran, T.P. and Newell, A. (1983), *The Psychology of Human-Computer Interaction*, (Hillsdale, NJ: Erlbaum).

Cardoso, C., Keates, S. and Clarkson, P.J. 2004, Comparing Product Assessment Methods for Inclusive Design. In: *Designing a More Inclusive World*, Springer-London.

Keates, S. and Clarkson, P.J. 2004, *Countering Design Exclusion – An Introduction to Inclusive Design*. Springer-Verlag (London, UK).

Langdon, P. and Adams, R. 2005, Inclusive Cognition: Cognitive Design Considerations, Proceedings of UAHCI, International Conference of Human, Computer interaction, Las Vegas.

Lewis, T. 2003, Understanding the User – Masters thesis. University of Cambridge.

Lewis, T. and Clarkson, P.J. 2005, A User Study into Customising for Inclusive Design. In: *Proceedings of Include 2005* (London, UK).

Nielsen, J. 1993, *Usability Engineering* (Academic Press, London, UK).

Norman, D. 1988, *The Design of Everyday Things* (Doubleday, New York, USA).

Polson, P.G., Lewis, C., Rieman, J. and Wharton, C. 1992, Cognitive Walkthroughs – a method for theory-based evaluation of user interfaces. *International Journal of Man-Machine Studies*, 36(5), 741–773.

Reason, J. 1990, *Human Error* (Cambridge University Press, Cambridge, UK).

Ryu, H. and Monk, A. (2004), Analysing Interaction Problems with Cyclic Interaction. Theory: Low-level Interaction Walkthrough. *PsychNology Journal* 2(3): 304–330.

Wickens, C.D. and Hollands, J.G. (2000), *Engineering Psychology and Human Performance*, 3rd Ed. (Prentice Hall, Inc. London).

E-LEARNING SUPPORT FOR POSTGRADUATE STUDENTS

Andree Woodcock, Anne-Marie McTavish, Mousumi De, Leith Slater & Ingrid Beucheler

Coventry School of Art and Design, Coventry University, Priory Street, Coventry CV1 5FB, UK

An action learning investigation was undertaken to determine the requirements of a virtual learning environment (VLE) that would provide a meeting place and document repository for taught postgraduate (MA: Design and Digital Media) and PhD students within Coventry School of Art and Design. A commercial, collaborative support system was piloted to determine the needs of the students, stakeholder and functionality issues that might influence wider uptake of such a system. The experiences are explained in terms of the Technology Acceptance Model.

Introduction

Postgraduate students differ from their undergraduate counterparts in terms of their support requirements, course structure, ways of working and informational needs. The postgraduate (PhD) community at Coventry School of Art and Design (CSAD) comprises around 20 full and part time students working on a diverse range of projects, many of which involve collaboration. Students engaged in the development of multimedia installations or visual artefacts need to be able to show large files, of different formats to their peers and tutors for review. Such functionality is not routinely offered in virtual learning environments (VLEs) designed to support and assess undergraduate students. Additionally most postgraduate students spend time working from home or on field studies, so it is difficult for them to maintain contact with, and manage supervisory teams, have a continued awareness and sense of belonging to the immediate and larger community and to share information.

Although the university has an educational, e-learning support system (WebCT), this has been designed primarily to serve the undergraduate population. Informal monitoring of its usage by CSAD PhD students in previous years, showed zero uptake. A commercial system (eStudio, www.same-page.com) was selected as being potentially more useful. same-page.com was founded in 1999 to provide SMEs, government agencies and non-profit organisations with Internet based collaborative technology. eStudio offered the desired functionality namely: shared calendar, noticeboard, bookmarks; chat and discussion boards; emailing; project management functionality – a task tracker; public and private areas into which over 50 file types could be imported, accessed and critiqued; an archival vault; account management facilities for monitoring usage and assigning privileges. Potentially, such functions could be useful for augmenting the postgraduate community. For example, through notification of events (performances, calls for papers and research bids etc.), providing a research training material repository and on-line resources that students could access when appropriate, a task tracker to aid in project management, and private areas for asynchronous tutorials which could be accessed by the supervisory team.

However, providing an additional VLE is not something that can be undertaken lightly. Therefore, a pilot study was conducted in which two eStudios were developed; one to support the PhD and MA groups (over 50 students) and the other to support beneficiaries (from SMEs) enrolled on a short course to enhance soft management and IT skills. As part of the latter group, Beucheler evaluated eStudio as a system for users with cognitive processing difficulties. This paper draws on our experiences of setting up, using and maintaining these sites with especial attention given to the evaluation by all stakeholders. The results will be explained in relation to current theories on the uptake of technology.

E-learning and VLE's... ...

E-learning may be described as learning that takes place wherever and whenever the student chooses to interact with electronically delivered teaching material. This is achieved through the delivery of (structured) material over communication networks or devices. In such cases, the relationship and communication between the teacher and learner becomes in-direct. This has led to the need for additional mechanisms to enhance individual and group learning experiences (*e.g.* discussion groups, interactive learning material). E-learning encompasses a number of standard tools such as:

1. teaching material – reading lists, module notes, handouts and multimedia content
2. communication tools – e-mail, newsgroups, mailing lists and bulletin/discussion boards
3. assessment tools – electronic submission of assignments, self-tests, assessed tests.

... ...and postgraduate student requirements

Many VLEs have been developed primarily with the needs of undergraduate students and their lecturers in mind. As such 1) and 3) have been emphasized. Postgraduate (and research) students are more independent and engage on individual programmes. This means their requirements are different. They require technology that can support collaboration and the mutual construction of understanding (Brown, 1989). In proposing eStudio, we hoped it might be used to create a knowledge community (Bruffee, 1999) or a community of practice (Lave and Wenger, 1991), *i.e.* a tightly knit group of students engaged in common practices, who communicated, negotiated and shared their best practice with one another. The need for such a community had been expressed by the students and staff many times.

... ...and take up of technology

For CSCW (Computer Supported Co-operative Working), of which VLEs are an instance, Rogers (1995) identified five attributes of innovations that correlate with their adoption

- Relative advantage – the degree to which a new innovation surpasses current practices
- Compatibility – the degree to which the innovation is consistent with the adopters' existing values, past experiences and needs
- Complexity – perceived difficulty of the new system
- Trialability – the ease of experimenting with the new system
- Observability – the extent to which the results are easily seen and understood.

Additionally, technology uptake can be explained through the Technology Acceptance Model (*e.g.* Davis and Vankatesh, 1996) where Actual Usage is dependent on Perceived Usefulness, Perceived Ease of Use and Behavioural Intention to Use; or Task-Technology Fit Models (Goodhue and Thompson, 1995), *i.e.* how well the technology fits the requirements of a particular task; or a combination of both (Klopping and McKinney, 2004). The results will be explained in terms of these theories.

The investigation

eStudio was piloted for just over 12 months. All CSAD PhD and MA: Design and Digital Media students, their supervisors and research support staff were invited to participate. Woodcock and Slater set up the initial file structure; Slater provided overall technical support and system management, Woodcock continued to populate the system with material relevant to postgraduate study (such as course notes and reference material). Training was provided by Slater and De and participants encouraged to use the facilities of eStudio for posting up work, conference announcements etc. The aims of the trial were to firstly to determine whether eStudio provided a useful environment for research students, and secondly, to discover factors which may affect its adoption by staff and students.

These were measured through system usage, usability questionnaires, focus groups, expert walkthroughs and semi structured interviews with 10 representatives from the stakeholder groups (students – both MA and PhD (part time and full time, UK and overseas), supervisors, administrator and initiators). Given the criticism of technology-focussed distance education evaluation (*e.g.* Hara and Kling, 1999), this study sought primarily to understand the student experience of the VLE.

Results

Although all students, and some staff, were enthusiastic about the concept and recognised the potential benefits of such a system, participation rates were low with the user access log showing between 9 and 43 users a month. Of these, most were MA:DDM students required to participate in the trial as part of a research training module. Staff were especially poorly represented although there was an acknowledged need to make their research activities more accessible. The results from the separate studies have been pooled together.

Functionality

The calendar, task tracker and chat facilities were rated as least useful. The calendar should have incorporated module, university and outside events, but was not used sufficiently or appropriately during the trial. The chat, discussion forum and quick note facilities were not used because of the under population of the system, it was difficult to find the person who was online at the same time, and the use of other systems. The task tracker was relatively easy to use and appeared to have the potential to be of great use to students and their supervisors in scheduling work, and setting milestones. However, it required levels of commitment and project management awareness not possessed by the participants.

Usability and aesthetics

The main usability issue was the slow connection speed (even over Broadband) which meant that some users were unable to access eStudio, and uploading files was problematic. Once users experienced such problems they were unlikely to make repeat visits, especially when the "studio" was under populated. The cumbersome directory structure meant that most areas were empty but there was no way of knowing this before opening a folder. The "corporate design", layout and icons were also offputting. Many of the problems could be traced back to the lack of administrative and technical support. As "innovators" the authors were largely self reliant, and took on these duties in addition to their own work.

Support for collaborative working

Despite the problems everyone saw the potential of the VLE and if improvements were made would use it. However, some reticence was expressed regarding going into other people's directories to look at work (even if this was granted public access), and uploading work for others to share. Some of these have been overcome with a new directory structure. The archive facilities would provide a means of looking at previous work and building up a group history and identity. Also, uploading something into a "neutral system" was seen as good practice in a computer culture where attachments become disassociated from emails, email addresses change regularly, and work is "lost"!

Appraisal in terms of cognitive diversity

Clearly an environment VLEs have the potential to aid those with cognitive processing disorders such as dyspraxia and Aspergers syndrome who could benefit from facilities such as event reminders and the task tracker. However, more attention needs to be given to areas such as page layout and task design to avoid mental fatigue.

Recommendations

A number of specific and general recommendations were made that were fed to the system developers. These ranged from "jazzing up the icons", through to increasing the space of file names, providing a thesaurus and spell checker, more choice for reminders of events and an intelligent reception area that could filter events for relevance based on user profile.

Conclusions and Discussion

Low uptake and satisfaction could be due to the use of other systems with which the users were happy and familiar, such as WebCT, email and on-line chat; lack of enforcement – we provided a supplementary service, which was not as easy to use as those the students were accustomed to; lack of technical and administrative support – the system was implemented at a local level and maintained by the innovators when they had free time; usability issues – in particular slow speed and inappropriate file structure; lack of a core group who consistently used the system and recruited others; lack of interest in participating in the wider community. However, the need for such a system was evidenced in the focus groups where students admitted that it was only through eStudio that they had started to talk to some of their classmates and discussed how useful a network of eStudios would be where students from other universities could share their work, best practices and assist each other.

The lack of uptake of eStudio may be explained in terms of innovation diffusion theory (Rogers, 1995), the system failed in terms of relative advantage and trialability. These were influenced by the poor directory structure and system speed, both of which discouraged exploration and more extensive usage. For example, administrators stopped importing material into the workspace and reception areas because it took five separate operations, whereas email required one. Also, given that the items were not read by a high proportion of the students, this material had to be emailed to them as well. A directory structure had been created with every student and member of staff having private and public folders. However, no feedback was provided on which folders were active without opening each one. This lack of transparency and slowness, together with the high proportion of empty folders, discouraged browsing essential to maintaining awareness of group activity.

Although most agreed that the functionality was something they desired individually and as a community, few were prepared to spend the time and effort needed to make this work. Although it is not possible to formally assess the pilot in terms of TAM and TTF, our results indicate that perceived ease of use (when compared to existing systems) and perceived usefulness may both have been responsible for the lack of uptake.

The overall pattern of uptake may follow the typical diffusion model (Rogers and Scott, 1997). Our pilot relied on the innovators, and early adopters. It was only when eStudio was being withdrawn that we discovered it was being championed by a postgraduate lecturer, who was enforcing its use with the current MA year group. The system now has information more relevant to the cohort and a revised directory structure. This seems to imply that it is acting in part as a replacement to WebCT and offers the functionality required by postgraduate students who need to collaborate and share information.

References

Brown, S. 1989, Towards a new epistemology for learning. In C. Frasson and J. Gauthier (Eds.), *Intelligent Tutoring Systems at the Crossroads of AI and Education* (pp. 266–282), Norwood, NJ; Ablex

Bruffee, K.A. 1999, Collaborative Learning: Higher Education, Interdependence and the Authority of Knowledge (2nd ed.). Baltimore, MD; John Hopkins University Press

Davis, F.D. and Vankatesh, V. 1996, A critical assessment of potential measurement biases in the technology acceptance model: Three experiments. *IJHCS*, 45, 1, 19–45

Goodhue, D.L. and Thompson, R.L. 1995, Task-technology fit and individual performance, MIS Quarterly, 19, 2, 213–236

Hara, N. and Kling, R. 1999, Students' Frustrations with a Web-Based Distance Education Course, *First Monday*, 4, 12. Accessed 23rd Nov '05, url: http://firstmonday.org/issues/issue4_12/hara/index.html

Klopping, I.M. and McKinney, E. 2004, Extending the Technology Acceptance Model and the Task Technology Fit Model to Consumer E-Commerce, *Information Technology, Learning and Performance Journal*, 22, 1, 24–47

Lave, J. and Wenger, E. 1991, *Situated Learning: Legitimate Peripheral Participation*. Cambridge, MA; Cambridge University Press

Rogers, E. 1993, *Diffusion of Innovations*, The Free Press; New York

Rogers, E. and Scott, K. 1997, The Diffusion of Innovations Model and Outreach from the National Network of Libraries of Medicine to Native American Communities, Accessed on 23rd Nov '05, url: http://nnlm.gov/pnr/eval/rogers.html

HCI – INTERFACES

A HAPTIC FISH TANK VIRTUAL REALITY SYSTEM FOR INTERACTION WITH SCIENTIFIC DATA

Wen Qi

Department of Industrial Design, Eindhoven University of Technology, Den Dolech 2, 5600 MB Eindhoven, The Netherlands

In this paper we describe a haptic representation of volumetric data in a Fish Tank virtual reality (VR) system for interactive visualization. Volume rendering is a powerful tool for visualizing scientific data. However, only visual representation of volumetric data on a traditional 3D desktop system or even an immersive VR system is still hard to comprehend the inside structure because of data complexity, occlusion and lack of rich depth cues. The haptic Fish Tank VR system that we present uses a pair of active stereoscopic shutter glasses to provide the 3D displays, a head tracker to provide the motion parallax. It also allows a user to feel inside the volumetric data by interacting with a haptic device (a robot arm) through a stylus in real time. The haptic device provides active force feedback that enhances the 3D interaction by coupling itself with visual feedback.

Introduction

What is virtual reality about? Fred Brooks defines a virtual reality experience as "any in which the user is effectively immersed in a responsive virtual world. This implies user dynamic control of viewpoint" (Brooks, 1999). We categorize a VR system according to its display technology:

- Projection based VR system, for example CAVE (Cruz et al, 1993), Workbench (Kreuger et al, 1995).
- Head mounted display (HMD) based VR system (Sutherland, 1968).
- Monitor based desktop VR system, such as Fish Tank (Ware et al, 1993).

Colin Ware firstly introduced the Fish Tank VR system in his 1993 CHI paper (Ware et al, 1993). The term "Fish Tank" is intended to contrast with "Immersion" and suggests a small-localized workspace. The idea is to use a conventional monitor, a head tracker to measure head position and derive use eye position from this. Hence it provides a correct perspective view of a small virtual environment. With stereo viewing a high quality monitor can provide a more vivid, high quality VR experience than immersion systems such as a CAVE or HMD.

The recent progress on visualization research has lead to many effective algorithms (surface rendering and volume rendering) to visualize different kinds of scientific data (scalar, vector, tensor field data) in a large amount (Hansen and Johnson, 2005). Therefore, building up a useful VR system for visualization application becomes possible and has been proposed as a promising method to observe and interact with scientific data (Bryson, 1996). In Van dam's report, he described the recent progress on immersive VR for scientific visualization and pointed out some of the potentials (Van Dam et al, 2000).

The idea of multimodal interaction in Human Computer Interaction has been shown as a important approach to improve user performance for a variety of tasks. An ideal

VR system tries to make use of all sensory modalities to create a realistic environment as well. As one of crucial sensorial modalities, haptics means both force feedback (simulating object hardness, weight, and inertia) and tactile feedback (simulating surface contact geometry, smoothness, slippage, and temperature). Providing such sensorial data requires desktop or portable special-purpose hardware, usually a robot, called haptic interfaces. For example, the PHANTOM^R from SensAble Technologies is a widely used haptic interface that involves a robotic armature (Massie and Slisbury, 1994).

In this paper, we present a Fish Tank VR system with force feedback for visualizing and interacting with scientific data (3D scalar field data). We discuss the current research on this topic and present our solution. Several results will be discussed in order to show the effectiveness of our system and the relation to the scientific question.

System

Based on the concept of Fish Tank VR system, we set up our haptic VR system as a modular and multiple threads, networking transparent environment. Within the system, it is very important that the haptic is be calibrated correctly with visual rendering. The accurate collation and registration between visual and force feedback could provide faithful interaction feedback in a VR application, especially in the application with scientific data.

Hardware components

The central computing platform is a Dell Precision Workstation 530 (Dual 2.4-GHz Xeon, 1GB RDRAM) with the following components:

- 17′ DELL monitor with Nvidia Quadro 4900 XGL graphics card, an infrared emitter with shutter stereo glasses from StereoGraphics Inc.
- A PHANTOM Desktop™ haptic device has been chosen and put beside the monitor. It provides an affordable desktop solution for haptics and is ideal for the users who work in small size of physical space and perform certain types of haptic interaction (such as the observation of inside structure of a volumetric data set).
- The DynaSight 3D Optical tracker is a compact, self-contained sensor that measures the instantaneous 3D position of a passive target. The tracker uses embedded signal processing to automatically acquire and track the target in uncontrolled environments. To cue the sensor, two adhesive-backed disposable targets (two circles that are made of reflected materials) are mounted in front of the stereo glasses so that the central application involves tracking the human head for use.

These hardware components have been organized in such a way that accurate and easy calibration could be reached. The stereo emitter and the transmitted box have been synchronized through a cable between these two. There is another cable from the emitter to the graphic card. They are also fixed on a plate together that is supported by a metal arm. The height of these two components guarantees the detection of the tracking and stereo signals continuously (see Figure 1).

Visual rendering with isosurface

The scientific data the system interacts with are volumetric data (3D scalar field data). A volume is represented as a 3D rectilinear array of volume elements, or voxels, each specifying a set of scalar properties at a discrete grid location. An interpolation function could be used to produce a continuous scalar field for each property. This is critical for producing smooth volume and haptic rendering. The Visualization Toolkit (VTK)

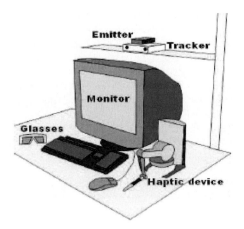

Figure 1.   The system diagram.

is an open source library that provides several different algorithms to visualize volumetric data (ray-casting, isosurface and 2D texture mapping). Considering the rendering efficiency and the real-time interaction, we chose the Marching Cube as the main algorithm for rendering the volumetric data in the form of isosurfaces. However the inherent structure of VTK does not provide any good mechanism to integrate with input device for virtual reality application. We have adopted the VTK library into our code so that the visualization capability from VTK could combine with 3D interaction devices.

Haptic volume rendering

Bringing the PHANTOM into a monitor screen based virtual environment presents several challenges. The PHANTOM $16 \times 12 \times 12$ cm physical workspace is smaller than a 15' monitor screen environment volume. Moreover, typical force update rates generated by force feedback system are 1 KHz that is higher than update rates of the graphics loop. Therefore, a software module is necessary to integrate the operation of the PHANTOM with the software controlling the virtual environment and any other simulation software.

The algorithm of force calculation is based on two principal goals (Lundin et al, 2002). First, the active forces could be calculated and presented fast enough to the user in real time for interactive purpose. Second, the forces applied to the user should be consistent with the visual rendering of the volumetric object. We simplify the force calculation to a point contact. The general equation we used for feeling an object inside a volume using a point contact model is:

$$\vec{F}(x, y, z) = \vec{F}_A + F_R(\vec{v}) + F_S(\vec{N}) \tag{1}$$

The overall force $\vec{F}$ supplied to the user located at position $P(x, y, z)$ and moving in direction $\vec{v}$ is equal to the vector sum of an ambient force $\vec{F}_A$, a motion retarding force $F_R(\vec{v})$, and a stiffness force normal to the object $F_S(\vec{N})$. The ambient force is the sum of all forces acting on the tool that are independent of the volumetric data itself. The calculation of the retarding and stiffness force functions for haptic volume rendering is dependent on the material opacity, which is estimated by both the density and the magnitude of the density gradient at that location.

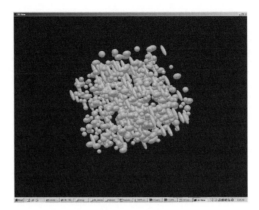

Figure 2. An example of simulated volumetric data.

Results

In order to evaluate the validity of the system, several simulated volumetric data sets have been created. Each volumetric data set is organized in such way that different shapes of objects with specific properties could be easily added into. The volumetric data are usually $256 \times 256 \times 256$ big. Inside each data set, four kinds shape of objects (sphere, ellipsoid, cylinder, curved tube) are generated with random size, amount, and position (see Figure 2). These objects might overlap with each other. The user could feel inside the volume, identify the structure of the object through the stylus while the virtual proxy of the haptic device touches the object. Usually, the more the gradient is, the stronger force the user could feel.

Conclusion

We have presented a Fish Tank haptic VR system for visualizing scientific data (volumetric data). The system provides real time volume rendering for 3D scalar field data. Haptic rendering has been implemented through a PHANTOM device. Combing the Fish Tank VR with the power of volume visualization and the addition of the haptic feedback make the multimodal interaction possible. A future user study has been planned to investigate the effectiveness of the force feedback on the properties judgement of the objects in a more quantitative level.

Acknowledgements

This project is part of the IOP-MMI program (Innovative Research Program on Man-Machine Interaction), funded by SenterNovem under the Dutch Ministry of Economic Affairs.

References

Brooks, J.F.P., 1999, What's real about virtual reality? *IEEE Computer Graphics and Applications*, 19: 16–27

Bryson, S., 1996, Virtual reality in scientific visualization. *Comm. ACM*, 39: 62–71

Cruz-Neira, C., Sandin, D. and DeFanti, T., 1993, Surround-screen projection-based virtual reality: The design and implementation of the CAVE. *ACM Computer Graphics*, **27**(2): 135–142

Hansen, C.D. and Johnson, C.R., 2004, *The Visualization Handbook*. Elsevier Butterworth Heinemann, Massachusetts

Kreuger, W., Bohn, C., Froehlich, B., Schueth, H., Strauss, W. and Wesche, G., 1995, The responsive workbench: A virtual work environment. *Computer*, **28**(7): 42–48

Lundin, K., Ynnerman, A. and Gudmundsson, B., 2002, Proxy-based haptic feedback from volumetric density data. In *Proc. of Eurohaptics 2002*, 104–109

Massie, T. and Salisbury, J., 1994, The phantom haptic interface: A device for probing virtual objects. In *Proceedings of the ASME Dynamic Systems and Control Division*, 295–301

Minsky, M., Ming, O., Steele, F., Brook, F. and Behensky, M., 1990, Feeling and seeing: issues in force display. In *Proceedings of the symposium on 3D Real-Time Interactive Graphics*, 235–243

Sutherland, I., 1968, A head-mounted three-dimensional display. In Proceeding of the Fall Joint Computer Conference. *AFIPS Conference Proceedings*, vol 33. AFIPS, Arlington, VA, 757–764

Van Dam, A., Forsberg, A., Laidlaw, D., LaViola, J. J. and Simpson, R., 2000, Immersive VR for scientific visualization: A progress report. *IEEE ComputerGraphics and Applications*, **20**: 26–52

Ware, C., Arthur, K. and Booth, K., 1993, Fish tank virtual reality. In *Proceedings of CHI 93*, 37–42

ROLES FOR THE ERGONOMIST IN THE DEVELOPMENT OF HUMAN–COMPUTER INTERFACES

Hugh David

R+D Hastings, 7 High Wickham, Hastings, TN35 5PB, UK

Individual ergonomists can make significant strategic contributions to the design of systems, in addition to the traditional "knobs and buttons" work. Ergonomists should be aware of the ways in which "the human element" or its neglect can derail systems. They can warn of unrealistic assumptions (the "paperless office", for one example). They can warn of the dangers of "horseless carriage" systems, mimicking traditional displays while losing saliency and failing to exploit the flexibility of the computer. They can emphasise the need to integrate systems and system displays. On deeper levels, they can advise on how systems can accommodate the conflicting requirements of speed, efficiency, reliability, flexibility and safety by making proper use of human operators. They should reflect on the social costs of computerising systems (to the designers, managers, users and clients), and the potential dangers of highly integrated systems.

Introduction

Individual ergonomists can make significant contributions to the design of human-computer interfaces. In general, interface designers tend to be computer literate but lacking in knowledge of human capacities and limitations. Ergonomists may be able to contribute knowledge or at least awareness of human abilities and weaknesses to the processes of planning, design and implementation of interfaces at virtually all stages.

Planning

Planning is often the weakest point in the production of a human-computer interface. It is easy to accept ill-defined specifications in the hope that they can be clarified "in the light of events". This is usually disastrous.

Initial specifications

Most projects begin with a specification. A significant contribution can be made by the ergonomist who insists that this specification should be written in operational terms, rather than in vague "public relations" language. It is also necessary, from time to time, to accept that no specification, however rigorously determined, ever covers all the points that arise in the course of implementation. Some means of solving emergent problems, without permitting endless "improvements" can usually be defined. (It is also often the case that there are disagreements within the initiating organization – "management" and "shop-floor" views may be very different.)

Ergonomists should reflect on the social costs of computerising systems (to the designers, managers, users and clients), and the potential dangers of highly integrated systems.

Perrow's classic "Normal Accidents" (1984) provides a good starting point. Its second chapter is titled: "Nuclear power as a high-risk system – why we have not had more TMIs (Three Mile Islands) but soon will". The Chernobyl incident occurred two years *after* Perrow's book was published. Sometimes the best advice will be "Don't do it!" Unfortunately this advice is rarely heeded or remembered.

Task analysis

Task analysis is a necessarily time-consuming but essential part of the design process. Ergonomists should be familiar with, at least, Hierarchical Task Analysis (Shepherd, 2001). Kirwan and Ainsworth (1992) provide a wider range of techniques, in less detail. Ergonomists should also be aware of the limitations of logical analysis and the danger of over-simplifying the processes, particularly if the operator is treated as if he or she is a logical device. It is often true that the rule book represents an accumulation of reactions to past incidents that taken together make an impossible operating method. In the recent past, "working to rule" has been an effective way of causing the complete collapse of large systems. (For example, a rule requiring the guard to check every door of a twelve-coach "slam-door" train at every station required a fifteen-minute stop in each station, converting a ninety minute journey to four hours, resulting in total chaos.)

Many contemporary systems work only because of the "little black books" in which operators store their experience of short-cuts and "work-arounds" without which the system is inoperable. It is this sort of information that the ergonomist should be aware of, take care to find out about, and bring to the attention of the system designers, as tactfully as possible.

Design

Task allocation

The allocation of tasks to human or computer operation is a critical part of the design of a new system. It is easy either to assign tasks exactly as in the previous system, or to automate every function that is technically possible – both potentially fatal mistakes.

The ergonomist, assuming that the decision to involve (or retain) human operators has been made, must assign tasks to provide satisfying workloads, with adequate flexibility and necessary safeguards to provide skill development, while also providing computer-based backup procedures to ensure safety.

Interface strategy

Ergonomists should be aware of the ways in which "the human element" or its neglect can derail systems. They can warn of unrealistic assumptions (the "paperless office", for one example). They can emphasise the need to integrate systems and system displays. On deeper levels, they can advise on how systems can accommodate the conflicting requirements of speed, efficiency, reliability, flexibility and safety by making proper use of human operators' skills, while avoiding their weaknesses.

Interface tactics

Once the allocation of tasks has been decided and the overall design strategy determined, the actual design of the interface can be determined. ("Strategy" is what does not change over the years, "tactics" change as equipment changes). There are many standardised "off-the-peg" display solutions. A well grounded ergonomist should be able to indicate how these displays can be used to provide the information the user needs in forms they will find natural. They can warn of the dangers of "horseless carriage" systems, mimicking traditional displays while losing saliency and failing to exploit the flexibility of the computer. The traditional "knobs and buttons" ergonomics has a role to play at this stage. For example, it may be necessary to demonstrate that displays that are legible by 27-year-olds at midday on a bright day in a well-lit office may not be legible to 70-year-olds on a garage forecourt in the small hours of a winter morning.

Interface testing

Most ergonomists acquire a practical knowledge of experimental design and methodology during their training. Electronic engineers generally are unaware of the need for testing, or the procedures for evaluating potential interfaces in relevant conditions, and with the variation in capacity observed in the "real world". Knowledge of the process of learning an unfamiliar interface may help avoiding trials confounded by "order effects". Equally, a good ergonomist will be aware of the limitations of "subjective" responses, particularly from unrepresentative "subjects". Ergonomists know the differences between a questionnaire and a list of questions, or an interview and an argument. Ergonomists have, of course, read the revised edition of Oppenheimer (1996).

Implementation

Application

Once an interface has been defined, designed and tested, it leaves the hands of its designers. The ergonomist has still several roles to play. Depending on the circumstances, he may be able to affect the way that the interface is put into operation. There are many case histories of failed systems where it is clear that the confidence of the operators was lost, often unnecessarily. Time and effort spent in persuading users to use systems is rarely wasted. Although there is no easily accessible text on this topic, experience and common sense, combined with some management techniques, and the expenditure of thought and imagination to anticipate and ease the psychological problems of adopting a new system will greatly help.

Monitoring

No system is perfect, even when ergonomists are employed throughout its development. Blunders, mistakes and errors in operation are inevitable. Some degree of automatic monitoring may be acceptable, although care must be taken to avoid the development of "blame" culture. Recent developments in the law of accident liability have added significantly to the difficulties of obtaining agreement to monitor working systems. Although aircrew have accepted the presence of a recorder of cockpit speech, with some reservations, at least one railway network decided not to install such recorders in their locomotive drivers working positions. The reasoning was interesting. Drivers on long, monotonous journeys keep alert, and check each other, by talking. Such conversation tends to cover subjects outside the immediate, deeply boring scene. Sports, entertainment and inter personal relations are discussed, often in highly politically incorrect terms. If this conversation were to be recorded, and an accident occurred, the recording could be subpoenaed in subsequent liability litigation, with embarrassing and emotionally disturbing consequences for victims' families. Accordingly, the opportunity to find out what was happening just before an accident has been sacrificed.

Voluntary, confidential recording of incidents is the norm in some advanced fields, such as commercial aviation and air traffic control. In such cases, it is vital that information offered in confidence shall be kept confidential and be seen to be so. The facilitation of such a system might be a valuable contribution from an ergonomist.

Rectification

While it is clearly valuable to know what is going wrong with a system, it is far more useful to be able to put it right. Techniques for the detection and correction of potential errors exist, and ergonomists should be aware of these and of the psychological and social difficulties involved in getting them implemented.

Conclusion

No single ergonomist can aspire to all the roles suggested here. Most ergonomists should be aware of the techniques available for solving these problems. All ergonomists should be aware of their existence.

References

Kirwan, B. and Ainsworth, L.K., (Eds.) 1992 *A Guide to Task Analysis* (Taylor and Francis, London)
Oppenheimer, A.N., 1996 *Questionnaire Design, Interviewing and Attitude Measurement,* Revised Edition (Cassel, London)
Perrow, C., 1984 *Normal Accidents* (Basic Books, New York)
Shepherd, A., 2005 *Hierarchical Task Analysis* (Taylor and Francis, London)

FUNDAMENTAL EXAMINATION OF HCI GUIDELINES

Yasamin Dadashi & Sarah Sharples

Human Factors Research Group, School of Mechanical, Materials and Manufacturing Engineering, University of Nottingham, University Park, Nottingham NG7 2RD

HCI guidelines are intended to assist designers, smooth the progress of designing and improve the quality and consistency of interface components. Although these guidelines, standards and style guides are widely accepted by user interface designers, users often still experience problems while interacting with an interface. Therefore it can be concluded that guidelines and standards do not always provide suitable guidance on different features of interface design. The vast majority of the available HCI guidelines are typically based on either past experience or best practice and the experimental evidence is rare. In this study, an attempt has been made to empirically verify some of the available recommendations on use of colour, text style and text size in interfaces. Three simple interfaces were designed to subjectively analyze the recommended effective colour combinations, optimum number of colours used in a page, text size and text style. The result of the experiment shows that some of the "recommended" guidelines might cause usability issues for the users and could be enhanced.

Introduction

The design of user interfaces is not only expensive and time consuming, but it is also critical for effective system performance (Smith and Mosier, 1997). The user interface design can be characterized as an art rather than a science, depending more upon individual judgment than systematic application of knowledge (Nielsen, 2000). However, there are not enough trained HCI experts to take part in every design team. Even if there were enough trained and experienced human factors and HCI experts, they could not possibly guide every step of the design. Therefore, a form of explicit design guideline is needed as a way of embodying the expert knowledge and transferring it to interface developers and designers (Smith and Mosier, 1997).

Although guidelines are not the only support available for designing usable and efficient interfaces, they are intended to assist designers, smooth the process of designing and improve the quality and consistency of interface components (Stewart and Travis, 2003). Human computer interface guidelines attempt to provide a "platform-independent and comprehensive set of guidelines" on different aspects of human computer interaction (Henninger, 2000). Even taking into account the early efforts of publishing standards to address the ergonomic problems of visual display terminals (for instance DIN 66–234), until recently there were very few standards and guidelines which could be applied directly to the user interface design.

Guidelines provide a "means of common understanding" among the experts and they also prioritize user interface issues, making it almost impossible for designers to avoid

---

Tel: +44 (0) 115 951 4196, Email: epxyd2@nottingham.ac.uk, Email: Sarah.Sharples@nottingham.ac.uk

them (Stewart and Travis, 2003). Numerous guidelines are available for user interface designers in form of books, formal standards and different style guides provided by professional software companies such as Microsoft and Apple Mac. Although it seems that such guidelines are widely accepted by user interface designers, users still experience various problems while interacting with an interface. This makes it clear that on one hand, as mentioned earlier, applying guidelines is not sufficient, and on the other hand the guidelines themselves might not provide proper guidance on different features of interface design. In other words, even an interface which is designed based on the current HCI guidelines might not be as usable as the guideline promises.

This study is an empirical examination of a set of guidelines that currently exist, and attempts to identify whether in fact these existing guidelines do meet the aims outlined above.

Method

Experimental design

In order to empirically verify the reviewed HCI guidelines, three experiments were conducted. For each experiment a simple interface was designed and each time one of the design features (i.e. background-foreground colour combinations, number of colours used in an interface, text size and text style) was manipulated. The aims of the three experiments were:

Experiment 1: Background-foreground colour combination

Predictions from published guidelines: In general the guidelines suggest that bright foregrounds (e.g., white text on green, cyan text on blue, yellow text on blue and black backgrounds) are preferred and dark on light polarity, blue and cyan backgrounds are more suitable (Lalomia and Happ, 1987; Pastoor, 1990). In this section of the experiment a set of 29 recommended background-foreground colour combinations were tested.

Experiment 2: Optimum number of colours used

Predictions from published guidelines: Most guidelines recommend a maximum of seven colours used simultaneously in an interface (ISO 9241-12; Murch, 1984; Galitz, 2002). An interface was developed with a colour palette ranging from 2 colours (Black and White) to 9 colours.

Experiment 3: Text size and text style

Predictions from published guidelines: Most Guidelines suggest a simple font style with an average of 12 to 14 point size (Galitz, 2002; Ivory, 2003). Four text styles (plain, bold, italic and underlined) and font size ranged from 11–14 points were compared.

Participants

24 male and female participants took part in this experiment. The average age of participants was 24 years old and all of them had prior experience of using different web-based interfaces and intranets.

Apparatus

The experiment was conducted using a laptop with Radon 7500 graphic card and a 15 inch LCD and a resolution of 1024 by 768 pixels. Interfaces were designed using Microsoft Office Publisher 2003.

Questionnaire design

In order to assess the effectiveness of the recommendations, each participant was asked to subjectively rate the interface by rating a set of statements. A four point likert rating scale was used where four represented the highest rating (strongly agree) and one the lowest (strongly disagree). The value of 2.5 was identified as the cut-off point for acceptability of the interface elements. In other words, the recommended guidelines which were examined in these interfaces were considered to be effective.

Results

Background-foreground colour combinations

The result of this experiment complies with Lalomia and Happ's (1987) conclusion that bright foregrounds (e.g., white text on green, cyan text on blue, yellow text on blue and black backgrounds) are preferred and also with Pastoor's (1990) recommendation that dark on light polarity, blue and cyan backgrounds are more suitable. Of course, it seemed that users tend to relate the effectiveness of the colour combination with the context of the website and also with their own persona; colour preferences. Figure 1 illustrates participants' ratings for effectiveness of the recommended background-foreground combinations. It can be seen that 15 colour combinations (magenta on yellow, yellow on black, cyan on black, yellow on red, red on black, magenta on cyan, cyan on white, cyan on blue, green on red, magenta on blue, cyan on red, blue on green, blue on red, red on green and cyan on yellow) have ratings of less than 2.5 and therefore are considered to be less appropriate.

The optimal number of colours

There are several guidelines on efficient colour coding and it seems that these guidelines are more effective than guidelines that merely give recommendations on the number of colours which should be used in an interface. As illustrated in figure 2, in response to the question "there are too many colours used", only two colour sets received a rating that was worse than the cut off point of 2.5. Since the statement is negative, any rating of 2.5 or higher is considered unacceptable. The interface with 8 colours received the highest (and hence the worst) ratings. This is surprising considering that the interface with 9 colours was deemed to be acceptable. But in this interface the extra colour was used to provide a border for the text, which is likely to have made the appearance of the interface more acceptable.

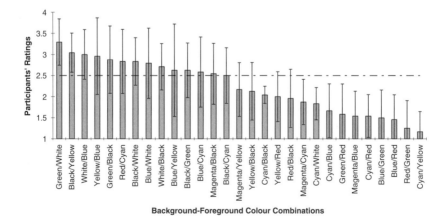

Figure 1. Participants' ratings: "The background-foreground colour combination makes reading easier".

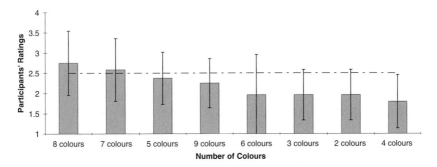

Figure 2. Participants' ratings: "There are too many colours used".

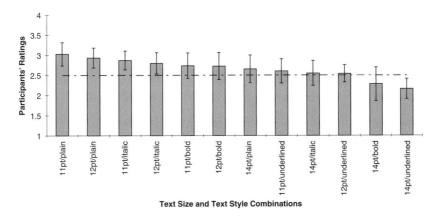

Figure 3. Participants' ratings: "Average ratings for all statements"

Text size and style

The result of this experiment revealed that the plain text style is the most effective and underlined is the least effective text style. It was also concluded that some of the text styles work better with smaller fonts. For instance, italic and bold text styles have higher ratings when combined with a text size of 11 pt, whereas bold text style combined with a text size of 14 pt has the lowest rating. This result is illustrated in figure 3.

Discussion

The results of the experiment on the effective background-foreground colour combinations as well as the points mentioned in the literature indicate that many factors affect the effectiveness of a colour combination. It is not possible to address some of these factors in the guidelines, such as individual users' personal preferences. Nevertheless, the recommendations for colour combinations need to be specified in RGB or other coding formats. General guidance such as "dark on light background" has only limited usefulness. Furthermore, the results of the experiment identify a set of combinations that are unacceptable.

The results of the experiment investigating the optimal number of colours used indicate the importance of considering the context of use. In addition, it seems that use of extra colours can be justified if they serve a purpose. It should be mentioned that in this experiment, the use of shading was not considered. The impact of personal differences is apparent from the large standard deviations of participants' ratings for experiments one and two.

The third experiment which tested the effectiveness of different text size and style combinations reveals that the guidelines on the text size are over generalized. Of course, appearance of text size on a PC screen is dependent on screen resolution. For instance a font of size 12 appears differently on different screen sizes and resolutions. Therefore, all text size regulations should be specified in order to take into account the actual size of the text as displayed on the screen.

Of course, it should be mentioned that the results of the experiments conducted in this study depend heavily on the design of the interfaces and the sensitivity of the statements used for subjective measurement of the attributes which were being studied. In addition, even above the cut-off of 2.5, there was a large range of average ratings which may indicate a variability in user acceptance of the different formats.

The results of these experiments confirm the notion that despite the usefulness and effectiveness of the HCI guidelines, it is necessary to verify and validate the available recommendations in an empirical context. Furthermore, it seems that it is not possible to provide precise guidelines for every aspect of interface design.

References

Galitz, W. O., 2002, The Essential Guide to User Interface Design, 2nd Ed., (John Wiley and Sons, New York).

Henninger, S., 2000, A methodology and tools for applying context-specific usability guidelines to interface design. Interacting with Computers, **12**, No.3, pp.225–243.

Ivory, M.Y., 2003, Characteristics of Web Site Design: Reality vs. Recommendations, In Proceedings of the 10th International Conference on Human Computer Interaction (HCII'03), June 22–27, Crete, Greece.

Lalomia, M.J., Happ, A.J., 1987, The Effective Use of Colour for Text on the IBM 5153 Colour Display, Proceedings of the Human Factors Society 31st Annual Meeting, Santa Monica, CA: Human Factors Society.

Murch, G. M., 1984, Physiological Principles for the Effective Use of Colour, IEEE Computer Graphics and Application, **4**, No.11, pp. 49–54.

Nielsen, J., 2000, Designing Web Usability: The Practice of Simplicity (New Riders, Indiana).

Pastoor, S., 1990, Legibility and Subjective Preference for Colour Combinations in text, Human Factors, **32**, No. 2, pp 157–171.

Smith, S.L., Mosier, J.N., 1997, Smith and Mosier HCI Guideline, Available at: http://www.deakin.edu.au/~malcolmc/hci/hci_contents.html [2-12-2005].

Stewart, T., Travis, D., 2003, Guidelines, Standards, and style guides, In: Jacko, J.A., Sears, A., Ed., The Human Computer Interaction Handbook, (Lawrence Erlbaum Associates, Mahwah, NJ), pp. 991–1005.

VIRTUAL AND REAL WORLD 3D REPRESENTATIONS FOR TASK-BASED PROBLEM SOLVING

Ian Ashdown & Sarah Sharples

Human Factors Research Group, School of Mechanical, Materials and Manufacturing Engineering, University of Nottingham, University Park, Nottingham NG7 2RD

This paper describes a study investigating the differences between using virtual and real world 3-dimensional representations of data. The representations were created from 3D models and outputted using VRML and rapid prototyping to ensure informational equivalence. One of the aims of this comparison was to identify the potential impact of, or barrier presented by, the computer interface on performance of analytical tasks. The type of representation design was also a consideration. The virtually displayed representations suffered from navigational difficulties. Generally the more familiar, square graph format yielded faster responses but less accurate results. Users also felt that tangible, real world models would potentially stimulate discussion and collaboration.

Introduction

The interpretation of data which has been visualised graphically shares its external cognition principles with making notes or sketching a diagram to aid solving a problem. Information visualisation therefore, carries with it the benefits of computational offloading (Scaife and Rogers, 1996), allowing quick, perceptual judgements to be used for problem solving. The design characteristics of visualisations representing the data have their advantages and disadvantages and this study will consider these while investigating alternatives. Also, using multiple axes for representing information permits a large dataset to be presented at one time and a three axis representation leads to being displayed in 3D space. The display of a representation as a virtual model then is evaluated, with respect to the interface that supports interaction. When using a 2D mouse as an input device, there are however navigational issues with 3D virtual models which can hinder manipulation. The quest for a transparent interface is progressed through printing out the models in three dimensions and is evaluated to see if this provides a better solution.

This paper presents a study that aims to identify usability issues associated with the interface of a virtual model and explore the potential benefits of an alternative real world display format. This allows consideration of design of representations and their effect on computational offloading when re-represented in an alternative format. The experiment presented follows a methodology developed by Hicks (2005).

Method

Computer Aided Design (CAD) software was used to model two designs of graphical representation of a selection of 2005 UK general election data. The designs were then

---

Tel: +44 (0)115 951 4196, Email: ianashdown@hotmail.com, Email: Sarah.Sharples@nottingham.ac.uk

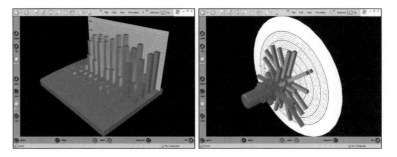

Figure 1. Virtual square and spindle representations.

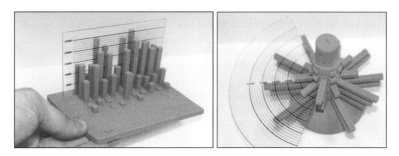

Figure 2. Real world square and spindle representations.

either displayed via VR software (Cortona VRML97 client) or produced as a rapid prototyped solid model (see Figure 1 and Figure 2).

Experimental design

12 participants aged between 18 and 34 completed the experiment. The experiment was a within subjects design and to ensure that there were no learning effects were presented were balanced using a Latin-square design. Participants were drawn from the University of Nottingham campus and Queens Medical Centre and were a mixture of students and administrative staff.

Experiment description

The independent variables were the type of representation, each of which had two levels, display type and design type. Participants were required to respond to analytical task questions that were themed into categories of A, B and C and required a variety of strategies to solve them. A previously designed usability questionnaire (Hicks et al, 2004) was applied to measure the participant's attitude towards the representation's performance, user control, affective experience and usability. The dependent variables were therefore subjective responses for each dimension, measured on a Likert scale which would then yield interval data, and time taken to complete the analytical task, measured in seconds, which would then provide ratio data. Qualitative analysis of videos of interaction with the models was also performed using theme-based content analysis (Neale and Nichols, 2001). Due to space restrictions this data is not reported here.

Results

Task response times

For the theme A questions, multivariate tests found there were significant differences for the display factor (F = 81.181; df = 1, 11; p <0.000), where responses for the real world representations were faster than for the virtual format. Significant differences were also found for the design factor (F = 7.401; df = 1, 11; p <0.020), where responses for the square format were faster than for the spindle format. For the theme B questions. Significant differences were however found for the design factor (F = 34.546; df = 1, 11; p <0.000), where responses for the square format were faster than for the spindle format. For the theme C questions, multivariate tests found there were significant differences for the display factor (F = 5.255; df = 1, 11; p <0.043), where responses for the real world representations were faster than for the virtual format. Table 1 summarises the formats that yielded the fastest response time for each question type on the basis of the ANOVA and post-hoc analyses.

Usability questionnaire

The descriptive statistics for the usability questionnaires show that across the 4 dimensions tested there was a consistent pattern in terms of mean values. For the four dimensions – Performance, User Control, Affective Experience and Usability – the most positive results were consistently achieved by the real world square representation, this was followed by the real world spindle format, the virtual square format and lastly the virtual spindle format.

For the performance dimension, multivariate tests found there were significant differences for the display factor (F = 23.189; df = 1, 11; p <0.001), where attitudes towards performance for the real world representations were more positive than for the virtual format. For the user control dimension, multivariate tests found there were significant differences for the display factor (F = 11.068; df = 1, 11; p <0.007), where attitudes towards user control for the real world representations were more positive than for the virtual format. Significant differences were also found for the design factor (F = 7.619; df = 1, 11; p <0.019), where attitudes towards user control for the square

Table 1. Summary of the preferred formats that achieved significance for response time to each question theme.

	Theme A	Theme B	Theme C
Display	Real World	–	Real World (Square)
Design	Square (Virtual)	Square	–

Table 2. Summary of the preferred formats achieving significance for each usability dimension.

	Performance	User Control	Affective Experience	Usability
Display	Real World	Real World	Real World	Real World
Design	–	Square (Real World)	Square (Real World)	–

design were more positive than for the spindle. For the affective experience dimension, multivariate tests found there were significant differences for the display factor ($F = 13.215$; $df = 1, 11$; $p < 0.004$), where attitudes towards affective control for the real world representations were more positive than for the virtual format. Significant differences were also found for the design factor ($F = 8.622$; $df = 1, 11$; $p < 0.014$), where attitudes towards affective experience for the square design were more positive than for the spindle. For the usability dimension, multivariate tests found there were significant differences for the display factor ($F = 27.729$; $df = 1, 11$; $p < 0.001$), where attitudes towards usability for the real world representations were more positive than for the virtual format. Table 2 summarises the formats that yielded the fastest response time for each question type on the basis of the ANOVA and post-hoc analyses.

Conclusions

The virtually displayed representations suffered from navigational difficulties when compared to the more direct manipulation of the real world models. Also, generally the more recognisable form of square graph design yielded faster responses but less accurate results. The results for perceived effectiveness further confirm that the real world form of presentation is a popular means by which to represent data and the square form would be the recommended design. The re-representation of equivalent data in square form carries benefits for computational offloading through allowing quicker perceptual inferences to be made. This is achieved through having all of the information aligned in the same plane and contained in an ordered grid structure allowing spatial constraining of the data and restricting the number of inferences that can be made. This spatial constraining would have previously been graphical constraining when in two dimensions. When rendered in 3D the representations have volume and spatial features so are therefore subject to spatial constraining (Larkin, 1989).

The performance data suggest that a slower response provides increased accuracy. If a particular design type was capable of stimulating more accurate responses then a trade-off against speed would probably be acceptable. Also, considering the additional information gained from the qualitative theme-based content analysis, participants spent much more time in exploration and reorientation when using the spindle designs. This is in contrast to the square format where users were confident to start investigation straight away. Such confidence would seem to be misplaced and the time spent exploring would seem to be well spent after all since it led to increased accuracy. The real world formats also recorded fairly positive opinion responses for the performance dimension of the usability questionnaire.

There were significant issues with the navigation of the interface when using the virtual models and this would support the theory that there is a period of learning required when using virtual interfaces. One point to consider was whether the central axis of the spindle when rendered in VR would be seen as an axis around which users could rotate, as would be the case for the real world form. This did not appear to be the case, and this finding supports Norman's (1999) distinction between perceived and real affordances, where perceived affordances are based on learned conventions and feedback. Since the onscreen representation of the input device is an icon, the action no longer naturally involves using fingers and thumbs. The icon was actually a pointer, 3-dimensionally curved around an axis, which on the face of it would seem to be an appropriate metaphor, the axis would then change depending on the direction of the mouse. Since these axes were screen-based rather than object-based, they do not however naturally map onto the form of an object. In total, using the navigational tools (representing pitch, roll and yaw), there were three rotational degrees of freedom possible, as opposed to one afforded by the real world spindle (although in reality, infinite degrees of freedom would be possible.)

It might be more successful if it were possible to restrict the rotation by directly selecting axial object features to allow back and forth rotation around the correct axis. This interface barrier then, would seem to hinder manipulation, which is unfortunate since, as noted above, exploration yielded more accurate responses.

It would be useful to consider whether prior experience (i.e. proficient users of virtual environments or CAD) has an effect on faster or more accurate perception of information displayed in three dimensions. This would relate particularly to virtual manipulation since it is known that there a learning issue associated with the navigation. More time should also be spent investigating the response accuracy and how it could be improved. Given the benefit of hindsight, if focussing on speed leads to misinterpretation of the results by rewarding representations that provide incorrect responses, then perhaps the overall performance of the representation should be combined with a measure for accuracy.

References

Hicks, M. O'Malley, C. Nichols, S. and Anderson, B. 2001, Comparison of 2D and 3D representations for visualising telecommunication usage. *Behaviour & Information Technology,* **22** (3), 185–201.

Hicks, M.J. 2005, *Designing human centred visualisations to support collaboration.* Ph.D thesis, University of Nottingham.

Larkin, J.H. 1989, Display-Based Problem Solving. *In:* Klahr, D and Kotovsky, (eds). *Complex information processing: The impact of Herbert A. Simon.* Hillsdale, HJ: Lawrence Erlbaum Associates, 319–341.

Neale, H. and Nichols, S. 2001, Theme-based content analysis: a flexible method for virtual environment evaluation. *International Journal of Human-Computer Studies.* **55**, 167–189.

Norman, D.A. 1999, Affordance, Conventions and Design. *Interactions.* **6** (3), 38–43.

Scaife, M and Rogers, Y. 1996, External Cognition: How do graphical representations work? *International Journal of Human-Computer Studies.* **45**, 185–213.

HERE'S LOOKING AT YOU: A REVIEW OF THE NONVERBAL LIMITATIONS OF VMC

Sarah J. Davis, Chris Fullwood, Orsolina I. Martino, Nicola M. Derrer & Neil Morris

University of Wolverhampton, School of Applied Sciences, Psychology, Millennium Building, Wolverhampton WV1 1SB, UK

Video-mediated communication (VMC) constrains our ability to use nonverbal cues and this is often a consequence of video system set-up. For example, camera placement means that the normal modulations of eye-gazing behaviour are impossible. Considering research suggesting that nonverbal signals are important in human communication, it is proposed that any form of communication that restricts or degrades these signals will be less effective. The following review will consider the different ways in which VMC attenuates nonverbal signals and the manner in which this effects communication under a number of different contexts. The review will also suggest methods for compensating for these restrictions.

Introduction

Video-mediated communication (VMC) is widely used for applications such as conferencing and distance learning, allowing geographically dispersed individuals the opportunity to interact. Part of the appeal of VMC is that, unlike the telephone, access to visual information is also possible. Because visual information is available, VMC should replicate face-to-face communication more closely than the telephone. However, there are fundamental differences between face-to-face and VMC that impact upon how successful interactions are. Current VMC systems are often of low quality, resulting in delays in transmission (for a review see Martino et al, this volume). It is also difficult to get a clear picture of a conversational partner's face if the transmission quality is low. These technological factors can impact upon the manner in which nonverbal cues are received. However, even when systems are of high quality VMC often distorts or attenuates visual information, which can result in nonverbal behaviours being ignored or lacking in communicative purpose (Heath and Luff, 1991). This review discusses the effects that technological constraints have on nonverbal behaviour and how this impacts upon the message being conveyed.

Hand gestures and posture

Many video-mediated systems are set up so only the head and shoulders are showing (Monk and Gale, 2002). This is because expanding the scope of the image would mean compromising the clarity of the facial image. As a result of this set-up some hand gestures may remain unseen and postural information may also be obscured. Posture can be used to signal the degree of interest or engagement of a conversational participant (Bull, 1978). Restricting postural information may therefore limit these cues. This may help explain Armstrong-Stassen et al's (1998) conclusion that videoconferencing participants often find it difficult to relate to the image on the screen and are also less likely to feel part of the conversational exchange. Posture and gesture also act as "turn-yielding cues", and therefore restricting this information will impact on effective turn-taking (indeed this has been indicated by Doherty-Sneddon et al, 1997).

Facial expressions

Facial expressions are also an important element of nonverbal communication, particularly as they can provide the conversational partner with feedback as well being a general indicator of emotional state (Whittaker and O'Connaill, 1997). However, when video-mediated visual quality is poor, many facial expressions may be missed or misinterpreted. Misinterpretation may be one factor that helps to explain why it is more difficult to promote trust in VMC. Trust can nevertheless be increased by allowing users to meet up face-to-face beforehand (Rocco, 1998). Moreover, initial meetings can positively enhance impressions formed (Derrer et al, this volume).

Eye gaze

A major problem with VMC is that many systems are not set up to allow mutual eye gaze to occur. This is due to the positioning of the camera in relation to the monitor. The camera is normally placed above the monitor, meaning that if someone looks at the image of the person they are talking to, it will appear to the other as if they are looking in a downward direction. A degree of gazing behaviour can be replicated by getting participants to look directly into the camera, giving the impression of mutual gaze (Fullwood and Doherty-Sneddon, in press). Monk and Gale (2002) also found that due to only the head and shoulders showing in most videoconferencing set ups, it is difficult for the listener to see what someone is looking at in the environment around them. Research has shown that if the head and eye levels are adjusted to a desired position, a lower number of turns and words are used in order to complete a task (Monk and Gale, 2002). Thus, although gaze awareness cannot be utilised as effectively as it can in a face-to-face setting, making technological alterations goes some way to improving the situation.

Because eye gaze is often used in impression formation, Fullwood and Doherty-Sneddon (in press) suggest that the perception that a person is avoiding eye contact may result in negative attitudes; this is an obvious problem for a range of different situations. For example, Daly-Jones et al (1998) showed that people find it more difficult to monitor whether a conversational participant is paying attention, which in turn may effect how the person is perceived. Fullwood and Doherty-Sneddon (in press) found that more information was recalled over video when presenters looked into a camera, therefore, replicating face-to-face findings (Fry and Smith, 1975). This is further evidence that it is possible to adapt to the pitfalls of the technology. Furthermore, this can have implications across a number of different contexts such as video-mediated distance learning, which requires students to attend to, retain and recall educational material.

Conclusion and Recommendations

The nature of university teaching and the multinational business climate means VMC is and will continue to be an important part of communication. The technological issues mean that it is unlikely to ever replicate face-to-face communication fully but if VMC is the most practical way to communicate there are a few points worth considering.

1) Although it is useful to increase the scope of the image so non-facial nonverbal information can be conveyed, it is also important that the quality of facial expressions is not compromised.
2) Giving individuals the opportunity to make face-to-face contact prior to using the equipment may help increase levels of trust and have a positive impact on impression formation (this would be useful in interview scenarios and business negotiations).
3) The speaker should be encouraged to look into the camera, giving the impression of eye gaze (this is particularly useful when information needs remembering).

References

Armstrong-Stassen, M., Landstrom, M., & Lumpkin, R. (1998). Students' Reactions to the Introduction of Videoconferencing for Classroom Instruction. *The Information Society*, **14**, 153–164.

Bull, P. (1978). The Interpretation of Posture Through an Alternative Method to Role Play. *British Journal of Social and Clinical Psychology*, **17**, 1–6.

Daly-Jones, O., Monk, A., & Watts, L. (1998). Some Advantages of Video Conferencing Over High-quality Audio Conferencing: Fluency and Awareness of Attentional Focus. *Journal of Human-Computer Studies*, **49**, 21–58.

Doherty-Sneddon, G., Anderson, A.H., O'Malley, C., Langton, S., Garrod, S., & Bruce, V. (1997). Face-to-face and Video-mediated Communication: A Comparison of Dialogue Structure and Task Performance. *Journal of Experimental Psychology*, **3**, 105–125.

Fry, R., & Smith, G.F. (1975). The Effects of Feedback and Eye Contact on Performance of a Digit-Coding Task. *Journal of Social Psychology*, **96**, 145–146.

Fullwood, C., & Doherty-Sneddon, G. (In Press). Effect of Gazing at the Camera During a Video Link on Recall, *Applied Ergonomics*.

Heath, C., & Luff, P. (1991). Disembodied Conduct: Communication through Video in a Multimedia Environment. *Proceedings of CHI '91 Human Factors in Computing Systems* (pp 99–103), New York: ACM.

Martino, O.I., Fullwood, C., Davis, S.J., Derrer, N.M., & Morris. N. (this volume). Maximising video-mediated communication: A review of the effects of system quality and set-up on communicative success.

Monk, A.F., & Gale, C. (2002). A Look is Worth a Thousand Words: Full Gaze Awareness in Video-Mediated Conversation. *Discourse Processes*, **33**, 257–278.

Rocco, E. (1998). Trust Breaks Down in Electronic Contexts but Can be Repaired by Some Initial Face to Face Contacts. *Proceedings of CHI 1998*, pp 496–502.

Whittaker, S., & O'Connaill, B. (1997). *The Role of Vision in Face-to-Face and Mediated Communication*. In K.E. Finn, A.J. Sellen & S.B. Wilbur (Eds). *Video Mediated Communication*. Lawrence Erlbaum Associates, New Jersey.

THE EFFECT OF AN ICEBREAKER ON COLLABORATIVE PERFORMANCE ACROSS A VIDEO LINK

Chris Fullwood, Nicola M. Derrer, Orsolina I. Martino,
Sarah J. Davis & Neil Morris

University of Wolverhampton, School of Applied Sciences, Psychology, Millennium Building, Wolverhampton WV1 1SB, UK

This study investigated the effects of an icebreaker on collaborative task performance across a video link. Half of the participants took part in a "getting to know you" style task before completing a map reading task, and the other half completed the map task without the icebreaker. Analyses indicate that when the icebreaker took place, participants completed the task significantly faster, in significantly fewer words and negotiated turns during conversation more effectively. One explanation for these findings is that the initial communication task allowed for the development of common ground, which lead to more efficient communication during the collaborative task.

Introduction

There is growing body of evidence suggesting proximity benefits group interaction. It is thus expected that any form of distance collaboration, for example video-mediated communication (VMC), is going to be less successful than face-to-face collaboration (Kiesler and Cummings, 2002). One explanation for this is a loss of social co-presence, a likely consequence of physical remoteness and the attenuation of visual cues. Therefore, although VMC creates the illusion of closeness participants still experience feelings of distance (Abbott et al, 1993). There seems to be something tangible about being in the same location as someone that makes collaboration easier. Indeed, Handy (1995) suggests that remote teams are less effective than face-to-face counterparts because "trust needs touch." Accordingly, it would be beneficial to create a method of promoting collaboration in geographically distributed teams. One suggestion is to use initial warm-up sessions with an emphasis on informal interaction to create feelings of togetherness, for example an icebreaker. Sciutto (1995) proposes that icebreakers help to reduce anxiety and increase interest levels. The current study investigated the effect of an initial icebreaker on collaborative task performance across V.M.C. It was expected that performance would be enhanced when an icebreaker was used.

Method

Participants

The sample comprised of 48 participants from a large U.K University. All participants were split into pairs. Half the participants were male, and the other half were female. Participants were randomly split into two groups: the icebreaker condition and the control condition. Twelve pairs were allocated to each condition. All participants were unfamiliar with their partner, and all participants had no prior experience with video-mediated technologies. Written informed consent was gained from each participant. This was so the interactions could be recorded for later analysis.

Materials

In room 1, a colour monitor (JVC TM-14EK(B)) was mounted in a wooden box, with a video camcorder (Sony CCD-TR2200EPAL) placed directly above. A microphone was placed to the right of the monitor, and video and audio quality were as high as achievable in the laboratory. The monitor and camcorder in room 1 were connected to room 2, adjacent to room 1, and with the exact same set-up. Monitors in both rooms were 14 inches in size. Each participant was distanced approximately one metre from the monitor and the scope included the participant's face and upper body.

Participants completed a "collaborative map-reading task," which involves both participants having to plan a route together (on a map of a town centre with a number of shops), picking up five items from a shopping list along the way. The participants, however, have two different priorities: participant one must complete as short a route as possible, whereas participant two must complete the route spending as little money as possible. The map was constructed in such a manner that participants would need to collaborate in order to find a route that suits both of their needs as best as possible.

Procedure

Participants in the icebreaker condition were given 10 minutes to complete a "getting to know you" style task, in which they were asked to find out the name of their partner and find a word for each letter of their partner's name that appropriately described them. After completing this task participants were given as long as they required to complete the map-reading task. Participants in the control condition completed the map-reading task without taking part in an icebreaker.

Results

Dialogues were transcribed and assessments were made on the following: 1) Time taken to complete task (measured in seconds), 2) Total word count and 3) Total number of turns. A turn began at the moment a participant started speaking and was completed at the point at which the next participant began to speak.

Using one-tailed independent measures t-tests, results indicate that participants in the icebreaker condition completed the task in a significantly quicker time ($t(22) = -2.92$; $p < 0.01$), in significantly fewer words ($t(22) = -2.27$; $p < 0.05$) and using significantly fewer turns ($t(22) = -1.91$; $p < 0.05$) compared to the control condition.

Table 1. Mean scores for time to completion, total word count and total turns for icebreaker and control conditions (standard deviations in parentheses).

	Icebreaker condition	Control condition
Time to completion	686.4 (266.5)	1099.1 (410.8)
Total word count	1007.8 (430.2)	1344.6 (280.9)
Total turns	117 (46.6)	153.5 (47.2)

Discussion

When an icebreaker took place, the map-reading task was carried out quicker, in fewer words and in fewer turns, and therefore more efficiently. Although this can partly be credited to the increased number of introductions that took place in the control condition, this alone does not explain the effect. Participants in the control condition also spent longer periods of time attempting to establish what the task was about, and what each of their priorities were. In other words, there was a need to establish common ground or mutual understanding. One example of this behaviour from the transcribed dialogues in the control condition is as follows:

Participant 1: "Good. Right so what's your priority for your shopping list, what have you got to do?"
Participant 2: "Right, I'm on my lunch break just now and I have to try and get all these items, the bread, steak, wine light bulbs and dog food as quickly as possible"
Participant 1: "As quickly as possible?"
Participant 2: "As quickly as possible."

Participants in the control condition spent on average 8.9% of the dialogue engaging in such activities, compared to 5.7% in the icebreaker condition. This result is unusual considering that participants in both conditions were given the same standardised instructions on how to complete the task. Perhaps this finding can be explained in terms of familiarity. According to Clark (1996) familiar people find it easier to establish mutual understanding. Therefore it may be the case that when we get to know someone better we feel more certain of being on the same wavelength, and therefore there is less of need to check mutual understanding.

Recommendations

Overall it would seem that knowing someone better helps to reduce psychological distance and promotes more effective collaboration. Therefore, it would be recommended that whenever meetings take place at a distance, an initial warm-up session might go some way to reduce feelings of distance between participants and improve collaboration.

References

Abbott, L., Dallat, J., Livingston, R., & Robinson, A. (1993). The Application of Videoconferencing to the Advancement of Independent Group Learning for Professional Development. *ETTI*, **31**.
Clark, H.H. (1996). *Using Language*. Cambridge: CUP.
Handy, C. (1995). Trust and the Virtual Organisation, *Harvard Business Review*, **73**, 40–50.
Kiesler, S., & Cummings, J.N. (2002). What do we know about proximity and distance in work groups? A legacy of research. In P. Hinds (Ed) *Distributed Work*, 57–80.
Sciutto, M.J. (1995). Student-centred methods for decreasing anxiety and increasing interest level in undergraduate statistics courses, *Journal of Instructional Psychology*, **22**, 277–280.

AN INITIAL FACE-TO-FACE MEETING IMPROVES PERSON PERCEPTIONS OF INTERVIEWEES ACROSS VMC

Nicola M. Derrer, Chris Fullwood, Sarah J. Davis, Orsolina I. Martino & Neil Morris

University of Wolverhampton, School of Applied Sciences, Psychology, Millennium Building, Wolverhampton WV1 1SB, UK

This study investigated the effects of initial meeting context on future video-mediated impression formation. Participants met with an individual being interviewed for a bar staff position. This initial meeting either took place over a video link or face-to-face. After the initial meeting had taken place all participants then formed part of an interview panel for a video-mediated interview. Consequently, participants were asked to rate the interviewee on aspects of personality and employability. Participants who had met the interviewee face-to-face prior to the interview rated him significantly more favourably on a number of measures (friendliness, honesty, job suitability and employability) compared to when the initial meeting took place via a video link. Initial meeting context therefore impacted on person perceptions.

Introduction

Video-mediated communication (VMC) is increasingly used to support interviewing at a distance. Although on the one hand this means that individuals can be interviewed for jobs without travelling to the interview destination, research suggests that the presentation of an individual over a video link may result in less favourable impression formation. For instance, Chapman and Webster (2001) compared candidates in face-to-face and video-mediated interviews. Results showed that face-to-face candidates were perceived in a more favourable light compared to video-mediated candidates. For example, they were judged as being better at conveying verbal and non-verbal cues. Whilst it is possible to access nonverbal information over a videoconference, Chen (2003) proposes that VMC results in these behaviours being distorted. Heath et al (1995) state that because of the use of 2D equipment (monitor), specific movement and expression are sometimes lost within the individual's general conduct. Essentially this suggests that the bodily activity generated by one person across a link may be different from what the person on the other side of the link actually sees on the monitor.

There is a wealth of evidence indicating that nonverbal communication plays a significant role in impression formation, therefore it is easy to see how the distortion of such cues could impact on an individual's perceptions of a person. The current study aims to test if negative perceptions of individuals across video links can be negated by the introduction of an initial face-to-face meeting. Therefore, highlighting whether an initial face-to-face meeting will impact significantly on impression formation in a subsequent video-mediated interview.

Method

Participants

An opportunity sample of 32 University students was used with 15 males and 17 females. Two interviewees also took part in the study; both were male and aged 21 and 24.

Materials

In room 1, a colour monitor was mounted in a wooden box, with a video camcorder placed directly above the monitor. A microphone was placed to the right of the monitor, and video and audio quality were as high as was practically achievable. Room 1 was connected to an adjacent room (room 2), which had the same set-up. Each participant (and interviewee) was distanced approximately one metre from the monitor and the scope of the image included the face and upper body.

Participants completed a questionnaire assessing their perceptions of the interviewee. The questionnaire consisted of six questions and used a rating scale from 1 to 5 (ratings ranging from "not at all" to "very"). The questions addressed perceptions of friendliness, communicative ability, intelligence, honesty, job suitability and employability.

Procedure

Participants took part in a pre-interview meeting with either one of two interviewees. Half of the participants took part in this meeting face-to-face and half via a video link. During the meeting participants quizzed the interviewee on their knowledge of drink prices. After the meeting all participants viewed the interview over a video link. Finally, participants were required to fill out a questionnaire rating the interviewee on their performance.

Results

Interviewees were scored on participant ratings (out of 5) of friendliness, communicative ability, level of intelligence, honesty, job suitability, and employability (table 1). Using two-tailed independent measures t-tests, results indicate that participants rated interviewees as being significantly friendlier (t (30) = 2.13; $p < 0.05$); significantly more honest (t (30) = 2.88; $p < 0.01$); significantly more suitable for the job (t (30) = 2.57; $p < 0.05$) and significantly more employable (t (30) = 2.95; $p < 0.01$) when they had met them face-to-face before the interview. No significant differences were found between conditions on measures of communicative ability and level of intelligence.

Table 1. Questionnaire data: video meeting and face-to-face meeting conditions, with mean scores and standard deviations in brackets.

Question	Mean Score (S.D)	Question	Mean Score (S.D)
1) Friendliness	Face 4.63 (.50) Video 4.07 (.93)	4) Honesty	Face 4.56 (.63) Video 3.88 (.72)
2) Communicative ability	Face 3.88 (.62) Video 4.13 (.96)	5) Job suitability	Face 4.44 (.81) Video 3.56 (1.09)
3) Level of intelligence	Face 4.31 (.60) Video 3.88 (.89)	6) Employability	Face 4.38 (.50) Video 3.75 (.68)

Conclusions and recommendations

The results from this study suggest that face-to-face communication has a significant influence on certain aspects of impression formation; namely friendliness, honesty, job suitability and employability. Previous research findings indicate that "trust needs touch" (Handy, 1995). This might be because it is more difficult to "read" someone when nonverbal cues are missing or distorted. Indeed, there is a body of evidence indicating that nonverbal communication plays a crucial role in impression formation. The fact that "trust needs touch" may explain why impressions of honesty were less favourable after a VMC meeting. Equilibrium theory (Argyle & Dean, 1965) also offers some explanation for the findings in this study. It is argued that in face-to-face communication intimacy is communicated using a number of cues, most of which are communicated nonverbally (e.g. proximity and eye gaze). In situations where these cues are diminished it is common for people to compensate by increasing other available cues. For example, when individuals stand far apart they increase gaze to maintain a comfortable level of intimacy. During video mediated interactions we are unable to express many of these intimacy cues, and therefore it is difficult to maintain a comfortable equilibrium. The consequence of this would be that people would appear "cold" when presenting themselves over VMC. This is likely to result in negative impressions generally.

From the results of the present study and other related research (Chapman & Webster, 2001) it is suggested that video-mediated interviewees would be disadvantaged if being compared directly to those who are interviewed face-to-face. However, even with this in mind it is still conceivable that VMC can actually be advantageous in certain situations. For example, when a child is presenting evidence in a courtroom, it has been shown to be beneficial if that individual remains distanced (Doherty-Sneddon & McCauley, 2000). To conclude, when it is not possible to attend an interview in person, from the research findings of this study, some level of face-to-face interaction prior to the video-mediated interview would be of benefit to the interviewee.

References

Argyle, M. & Dean, J. (1965). Eye contact, Distance, and Affiliation. *Sociometry*, **28**, 289–304.

Chapman, D.S. & Webster, J. (2001). Rater correction processes in applicant selection using videoconference technology: The role of attributions. *Journal of Applied Social Psychology*, **31**, 2518–2537.

Chen, M.R. (2003). Conveying conversational cues through video. *Dissertation Abstracts International*: Section B: the Sciences & Engineering, **64**, 2261, US: Univ Microfilms International.

Doherty-Sneddon, G. & McCauley, S. (2000). Influence of Video-mediation on Adult-Child Interviews: Implications for the Use of the Live Link with Child Witnesses. *Applied Cognitive Psychology*, **14**, 379–392.

Handy, C. (1995). Trust and the Virtual Organisation, *Harvard Business Review*, **73**, 40–50.

Heath, C., Luff, P. & Sellen, A. (1995). From video-mediated communication to technologies for collaboration: re-configuring media space. In S.J. Emmott (ed). *Information Superhighways: Multimedia Users and Futures*, Academic Press Ltd, London.

MINDSPACE

Hugh David

*R+D Hastings, 7 High Wickham, Hastings,
TN35 5PB, UK*

The headlong development of computer-based systems has far outstripped the human resources necessary to guarantee the oversight of human-Computer Interfaces (HCI) by professionally qualified ergonomists. It is very necessary to put relevant information into the hands of system designers before costly and irretrievable errors are made. Most ergonomics literature is aimed at the scientific community, devoting most of its content to highly technical descriptions of experimental methods, subject pools, and statistical analyses. If the information is to reach the target audience, it must be made palatable, without oversimplification or condescension. The late Stephen Pheasant's "Bodyspace" provides an example of what can be done.

Introduction

Ergonomists have long been interested and actively engaged in the design of human-computer interfaces. This interest has produced a considerable volume of academic papers, but the impact on the designers of human-computer interfaces for the "real world" has been less than might be hoped. A considerable body of information is available on the limitations of human perception and information processing. Some of this research effort has produced general guidelines, which are useful if they are expressed in terms that can be understood by designers. Unfortunately, much of the research knowledge and empirical experience is not available to the software engineers who design interfaces, or is available only in unfamiliar and forbidding terms.

Ideally, designers of human-computer interfaces should either be ergonomists or have ergonomic experience available on demand. In practice, many "designers" or "system engineers" are reluctant to involve more participants in the design process, particularly where they are under cost or time pressure. Where consultancy is available, few consultants are aware of the headlong speed of product development. Stock answers, however approximate, are preferable to "research studies". To quote a necessarily anonymous source: "We don't want to find we've been sponsoring some bloody student's Ph. D.".

Many studies have shown that good ergonomics is economically rewarding, in increased productivity, reduced accident cost and user satisfaction. How can the information we have be conveyed to the people who need to know it?

In spite of technical developments, the most widespread and cheapest means for distributing information remains the printed book. This may be supplemented by CD's, particularly where bulky archives support the information distilled into the book, and by websites where details of rapidly developing fields can be kept up to date. Unfortunately, both these methods of transferring information require equipment that is not always available and, practically, reach only the already committed.

The "Bodyspace" Paradigm

Ergonomists should be able to make their experience and background knowledge more widely available and in more attractive and compelling formats. The late Steven Pheasant's justly renowned "Bodyspace" (Pheasant 1986/98) suggests a complementary volume on "Mindspace".

Pheasant's book falls into two sections. The first part contains, after a brief but engaging introduction, chapters on the practice and principles of anthropometrics, on workspace design, on sitting and seating, on hands and handles, on ergonomics in the office, on ergonomics in the home and, in the latest edition, a chapter on health and safety at work. The second part contains tabulated data in an easily understandable form, prefaced by a discussion of human diversity. "Mindspace" could adopt a similar plan.

Ergonomics, Mental Capacity and Interface Design

It would be hard to find an author with the clarity and verve of Stephen to introduce the topic. The principles and practice of human-computer action would be an equally tough nut to crack, but could be as rewarding as Stephen's equivalent. Workspace design can be translated into display design. The equivalent of sitting and seating is difficult to envisage, however. Hands and Handles may be translated into computer-based controls for both dynamic and static systems. Ergonomics in the office might translate into the interaction of computer-based controls and displays. The ergonomics of home computing (or of computer-based interfaces in the home) might correspond to ergonomics in the home. The final chapter on health and safety at work has a direct correspondence in the computer context, where many system designers still do not know of their obligations under health and safety legislation. The statutory requirements for disabled users and their practical implications should be described.

Mental Capacity and Mental Difference

The second half of "Bodyspace" contains a chapter on human diversity, including sex and ethnic differences, class, and ageing. There is a good deal of corresponding material in the HCI field. The variety of users of systems should be explained and demonstrated. The differences, not only in physiological performance, but also in psychological attitudes, between old and young, technophobe and technophile, occasional and habitual user, partially or completely unsighted or deaf users, and other significantly different groups should be shown with directions to web sites and human consultants. Finally, "Bodyspace" concludes with tables of estimates for body dimensions standardised and interpolated by Steven from the best available data. It is hardly possible to provide a mental equivalent of these tables, but a compilation of guidelines might serve as a rather inferior substitute.

Although they are not adequate in themselves, references to other sources, such as websites and directories of consultants would form a useful addition to the paradigm.

Reference

Pheasant, S, 1986/98 *Bodyspace – Anthropometry, Ergonomics and the Design of Work* (Second (Revised) Edition 1998) (Taylor and Francis, London)

HOSPITAL ERGONOMICS

HOSPITAL BED SPACES: PATIENT EXPERIENCES AND EXPECTATIONS

Karen Taylor & Sue Hignett

*Healthcare Ergonomics and Patient Safety research Unit,
Department of Human Sciences, Loughborough University,
Loughborough, Leicestershire LE11 3TU*

Over the next 10 years, 100 new hospitals are being planned and built across the UK. This presents an opportunity for design practitioners to deliver solutions that enhance the patient's experiences of the healthcare industry. With design practitioners moving towards evidence-based design, it is important that those involved in the design process have an understanding of the end users priorities for a healthcare bed space. This paper describes a review of the literature. There is evidence to show that patients receiving care in healthcare spaces that; (1) eliminate environmental stressors, (2) provide positive distractions, (3) protect privacy and (4) accommodate social interactions will recover faster, with shorter stays, and require less analgesia. However no specific research has been carried out to investigate the functionality of healthcare bed spaces from the end users perspective.

Introduction

In-patient accommodation is the most repeated volume of space within a hospital facility. The main driving force for design has been the need to provide the most cost effective layout that will meet clinical functions within a financial constraint. Paradoxically, hospitals have become uncomfortable, stressful environments and can have a negative effect on patient outcomes (Taylor, 2005).

The patient is treated and cared for within a space that has a complex mix of stakeholders, all having a say on how the hospital environment is built, designed and operated. As the healthcare industry has a strong culture of proof through clinical trials, it seems appropriate that these principles are applied to design to provide the evidence needed to support design decisions (Hamilton, 2003).

Literature review

This narrative review identifies the broad range of issues surrounding the many factors involved in the therapeutic environment. By developing the concept model (figure 1), with the patient at the centre of the dynamic system, various factors were identified as being inter-relating and inter-dependant. Further investigation of the individual factors generated a list of underlying issues that can influence the patient. Some of the ways in which the environment of the healthcare space may impact on a patient can be seen in figure 2.

Using search engines, the internet provided a gateway to access such sites as University Libraries, Academic journals, Government sites (CHI and NPSA), and many other private

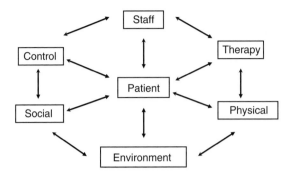

Figure 1. Conceptual Framework.

Environment

Building design	Room Layouts	Smells	Lighting
Infection control	Equipment	Comfort	Temperature
Privacy	Length of stay	Windows	Occupancy
Noise	Functionality	Gardens	Safety

Figure 2. The Environmental Stressors of a Healthcare Space.

and public organisations involved in the design of healthcare spaces. As well as incorporating the factors from the model, specific searches included:

- Therapeutic/Healthcare Environments
- Environmental stressors
- Hospital Design
- Patient Privacy and Dignity
- Patient Safety
- Evidence-based design

A wide range of studies and reports, not always specifically focused on hospital design, were reviewed. Unfortunately although a lot of work has been carried out in the design of healthcare spaces very little has been published in peer reviewed journals. The majority of the work has been carried out in the USA where the systems that deliver healthcare are different from UK. The NHS Estates have been pro-active in funding many studies; unfortunately these are rarely published in peer reviewed journals and are not readily available in the public domain, making it very difficult to critically appraise the research. Conferences have been excellent vehicles for information collection and dissemination, especially when they have focused on the Built Environment. Stakeholders, designers and clinical practitioners have all expressed the need for a system whereby they can all gain knowledge and understanding of the design process. In the next section the literature for the above six factors is summarised.

Therapeutic/Healthcare environments

Within the last 40 years health geographers have been working to understand the complex relationship between people and their environments, paying particular attention to the relationships between physical, biological and cultural features of the surroundings. Studies have shown that the environments in which people work have a direct influence on their ability to perform (CABE 2004). The term therapeutic environment is often taken to be the place in which therapy occurs, if this is the case then the environment itself must be considered part of the therapeutic process.

Environmental stressors

Noise

Studies have shown that noise generated by staff, other patients and equipment is considered an annoyance and intrusive. Despite guideline values for background noise in hospital patient rooms of 35 dB, Hodge and Thompson (1990) found that loud intermittent noises of up to 108 dB were emitted from certain equipment. A lack of sleep and increased stress due to noise can have a detrimental effect on patient outcomes.

Lighting

Detached from their normal routines and familiar surroundings, the patient is often confined to accommodation isolated from the outside world with little natural lighting. Beauchemin and Hays (1996) compared the length of stay in two groups of patients suffering from depression and showed that patients accommodated in naturally sunny rooms had shorter hospital stays than those staying in the dull rooms.

The patient cared for in a single room has more control over their environment; something as simple as being able to open a window to regulate temperature will assist the patient to adapt the room to their needs. Studies have shown that having a view of natural landscape will improve patient outcomes. Ulrich (1984) looked at recovery rates in patients recovering from Cholecystectomy from 1972 to 1981. Patients who had views of natural scenes from their windows, required less analgesia and had shorter hospital stays compared with those in rooms overlooking other buildings.

Patient safety

Good communication is vital to the safe and effective transfer of information concerning a patient's condition. This is particularly important at the changeover of working shifts and the transfer of patients between departments. It is suggested that designing patient rooms that are capable of accommodating the required diagnostic and treatment equipment within the bed space may not only help control infection, and improve patient control of the environment but may also reduce the need to move patients around (Hendrich et al, 2004).

In the design of healthcare facilities it is important to understand how social integration and the role of the family impact on patient well-being. Providing family zones in patient rooms, where comfortable and moveable furniture can allow family members to stay, may increase the social interaction and support for the patient. There is some suggestion that patients accommodated in single rooms may have less falls, perhaps because more space encourages the presence of family and friends (Hendrich et al, 2004).

Hospital design

Taylor (2005) states that; "the provision of a minimum clear space around the bed is essential in achieving an efficient and effective environment that complies with current legislation". Design practitioners can assign definitive values to the space required for equipment, and prescribe how much space staff need to deliver care, but no studies were found that considered what a patient needed to do or wanted to within a bed space.

Infection control

In 2004 NHS Estates changed the recommendations for the space around the bed based on the need to address Healthcare Associated Infections (HAI's). A National Audit Office report found that at any one time, 9% of hospital patients have an infection caught in hospital (NAO 2000). Most HAI's are associated with cross-infection from the hands of staff. In a study looking at compliance with hand washing in a teaching hospital, Pittet et al (1999) showed that the average compliance with hand washing was 48% and that this varied significantly among professional healthcare workers and was lower in surgical and intensive care workers than in other areas.

There are several models available from the USA such as the Planetree, that make sound arguments for the use of single rooms as opposed to multiple bed bays when it comes to controlling HAI's (Schweitzer et al, 2004). From a financial perspective having the flexibility to manage bed availability and patient placement would also be beneficial to the administration cost.

Patient privacy and dignity

Patient treatment areas and bed spaces have traditionally been segregated by curtains so single rooms, especially those with en-suite facilities, offer the patient increased privacy and dignity. The patients' perception of privacy and confidentiality has been investigated by Olsen and Sabin (2000). By comparing two types of rooms, walled and curtained in an emergency department, the results suggested that while both types had confidentiality issues, patients felt more comfortable disclosing personal information in the walled rooms.

Evidence-based design

Hamilton (2003) offers a definition of evidence-based design as: *"the natural parallel and analog to evidence-based medicine. It is the deliberate attempt to base design decisions on the best available research evidence"*. He believes that evidence-based healthcare design should result in demonstrated improvements not only in productivity, economic performance and customer satisfaction but perhaps more importantly in clinical outcomes.

Future research

Much of the research on healthcare facilities design has used quantitative methodologies to look at physical and psychosocial aspects. With healthcare service providers adopting a more consumerist view of the patient, and the changing ways that care will be delivered in the future, it seems appropriate to use ergonomic principles and qualitative methodologies to investigate patient needs, experiences and expectations of a healthcare bed space.

Phase 2 of this project is on-going using focus groups and repertory grid methods to collect more detailed data on the functionality, needs and expectations for patients receiving care and treatment in a healthcare bed space of the future. The outputs of this research will include:

- An end user/patient list of functional needs of health care bed spaces generated by their in-patient experience of various layouts such as; single rooms, multi-bed bays, en-suite and shared bathroom facilities and lengths of stay
- A validated, prioritised list of functional needs of a healthcare bed space for the future development of healthcare facilities.

References

Beauchemin, K.M. and Hays, P. 1996, Sunny hospital rooms expedite recovery from severe and refractory depressions. *Journal of affective disorders*, **40**, 49–51.

CABE 2004, *The Role of Hospital Design in the Recruitment, Retention and Performance Nurses in England*. CABE. London.

Hamilton, D.K. 2003, The Four Levels of Evidence-Based Practice. *Healthcare Design*, November, 18–26.

Hendrich, A., Fay, J. and Sorrells, A. 2004, Effects of Acuity adaptable rooms on flow of patients and delivery of care. *American Journal of Critical Care*, **13**(1).

Hodge, B. and Thompson, J.F. 1990, Noise Pollution in the Operating Theatre. *The Lancet*, **335**(8694), 891–894.

National Audit Office Report (HC 230,1999–00): *The Management and Control of Hospital Acquired Infection in NHS Acute Trusts in England.*

Olsen, J.C. and Sabin, B.R. 2003, Emergency Department Patient Perceptions of Privacy and Confidentiality. *Journal of Emergency Medicine*, **25**(3), 329–333.

Pittet, D., Mourouga, P., Perneger, T.V. and the Members of the Infection Control Program. 1999, Compliance with Handwashing in a Teaching Hospital. *Annals of Internal Medicine*, **130**(2), 126–129.

Schweitzer, M., Gilpin, L. and Frampton, S. 2004, Healing Spaces: Elements of Environmental Design That Make an Impact on Health. *The Journal of Alternative and Complimentary Medicine*, **10**(1), S-71–S-83.

Taylor, S. 2005, *Ward Layouts with Single Rooms and Space for Flexibility*: Discussion Document. NHS Estates. Norwich: TSO (The Stationery Office).

Ulrich, R.S. 1984, View Through A Window May Influence Recovery From Surgery. *Science*, **224**, 420–421.

Ulrich, R.S. 2002, *Health Benefits of Gardens in Hospitals*. The Centre for Health Design Texas. MD.

POSTURAL ANALYSIS OF LOADING AND UNLOADING TASKS FOR EMERGENCY AMBULANCE STRETCHER-LOADING SYSTEMS

A. Jones & S. Hignett

Healthcare Ergonomics and Patient Safety research Unit, Department of Human Sciences, Loughborough University, Loughborough LE11 3TU

This paper summarises the results of the postural analysis of loading and unloading tasks conducted using ambulance stretcher loading equipment. In recent years the ambulance services have made efforts to reduce the level of manual handling activities through the introduction of mechanical aids to assist in loading tasks. This has led to the phasing out of easi-loader stretchers and the introduction of the ramp and winch and tail lift. The new equipment has reduced lifting but the postural risk posed by the three systems has not been comparatively assessed. This analysis allowed the safest system to be identified informing future purchasing decisions made by the ambulance service. The systems analysed include the modular tail lift, the easi-loader stretcher and the hydraulic ramp and winch. The tail lift is the preferred system based on the postural analysis results.

Introduction

Since the arrival of CEN (The European Committee for Standardisation, 2000) 1789 compliancy, manufacturers and the ambulance services have had to consider new methods of loading and unloading patients from accident and emergency ambulances. While the concern for patient well-being remains high, there is increasing concern for staff welfare and the need to reduce the level of patient handling.

Vehicle manufacturers are now placing a high priority on ambulance design and safety to protect both patients and staff (Overton, 2001), and ambulance services have begun purchasing new types of equipment. This has led to a gradual phasing out of older systems such as easi-loader stretchers and the introduction of new systems such as the ramp and winch and the tail lift. These have been received with mixed emotions amongst ambulance staff.

Interviews conducted in a field study during phase one of this research (Jones and Hignett, 2005) highlighted staff concerns regarding the postures adopted during the loading activities. As yet no analysis has been carried out to assess the postural risk posed by stretcher loading systems used in the UK ambulance services and no comparison has been made to determine a preferred system.

Lifting and transporting patients is an integral part of an ambulance attendant's role. Manual handling is therefore an inherent factor within the profession making emergency rescue workers highly susceptible to musculoskeletal injuries (Lavender et al, 2000). Ambulance workers are a vulnerable occupational group and should therefore be specifically targeted by the department of health for manual handling risks (Unison, 2002).

The design of stretcher loading systems is an area in ambulance design, which has seen many advances in recent years (Boocock et al, 2000). It has been recognised that

Figure 1. Three loading systems analysed (ramp & winch, easi-loader and tail lift).

in order to reduce manual handling, systems should be mechanised as far as is reasonably practicable (Watts & Dickson, 1999) and through mechanising stretcher loading equipment ambulance services are attempting to do this however to date there has been no assessment of the postural risks posed by the three UK loading systems. With the emergence of clinical governance in the National Health Service (Scally and Donaldson, 1998) and the need to provide evidence based practice, it is important to determine these risks, providing scientific criteria on which to base future purchasing decisions.

The preliminary field study was carried out in phase on of this research, analysing the performance of the stretcher loading systems (Jones and Hignett, 2005). This identified 14 usability issues affecting the loading and unloading task. Ambulance staff across the UK ranked these issues into order of importance resulting in staff/patient safety and manual handling being the highest priority. The field study results found the tail lift to be the better system as it almost eliminates manual handling from the task.

The postural analysis supports this study by providing an objective dataset to determine the level of postural risk when operating the equipment. The three systems were comparatively assessed to determine the safest method of loading and unloading stretchers from Accident and Emergency vehicles (Figure 1).

Collaborating services

Three ambulance services collaborated in this study; East Midlands Ambulance Service NHS Trust, East Anglian Ambulance NHS Trust and Two Shires Ambulance NHS Trust. Each service uses one of the three systems, providing the range of equipment necessary to conduct the evaluation.

Methods

The postural analysis was carried out using simulations. Realistic call and transfer scenarios were developed based on the emergency calls observed during the field study as shown in figure 2. At each service the scenarios were simulated by two accident and emergency staff members using the most recent generation of loading device provided by the Trust. A pilot study was completed to ensure the data could be analysed. During this pilot, crews were asked to edit the call scenario sheet to reflect what happens in the field.

The ramp and winch was assessed both with and without the use of the winch. During the field observations staff failed to use the winch so for this reason it was necessary to assess the risks with both loading/unloading methods.

To collect the data, researchers filmed the simulations using multidirectional filming. This allowed the postures to be viewed from different angles allowing a thorough analysis to be conducted. The video footage was downloaded and photographic stills

Task	Task completed when....	Patient profile
To load patient on stretcher	In ambulance, secured on stretcher ready for transportation	Angina patient complaining of chest pains. Male aged 54, weighs approximately 20 stone
To unload patient on stretcher	Stretcher is out of ambulance ready for transfer into hospital	

Figure 2.   Example task scenarios.

Figure 3.   Multi-directional postural stills of tail lift loading task.

showing the postures were collected every 2 seconds of the loading and unloading tasks. The postures observed in each photograph were analysed by conducting Rapid Entire Body Assessment (REBA, Hignett & McAtamney, 2000).

REBA is a whole body assessment tool designed to be used in the field or with photographic stills. It gives the result as a 5-point action category score ranging from 0, no action required to 4, action required immediately (Hignett and McAtamney, 2000).

For each system, the postures adopted by both the head end and the foot end operator were analysed during the tasks. The mean REBA score was calculated for the overall task for each system, showing the overall postural risk.

Prior to coding the postures forty postural stills were randomly selected for an inter-rater reliability assessment. Two researchers coded these postures, confirming the reliability of the results.

Participants

Two participants were recruited from each ambulance service. Those selected were qualified medical technicians or paramedics, who use the equipment on a daily basis. The inclusion criteria required staff to be fit and able to carry out accident and emergency work with no signs of lower back pain. A third participant was recruited to act as the patient, providing a realistic scenario for the ambulance staff to simulate.

A risk advisor or occupational health representative working within the each service was present at the simulations to ensure the tasks were completed safely.

Results and discussion

Figure 2 shows the mean REBA score for all systems analysed. The hydraulic ramp is represented on when used on its own and with the winch.

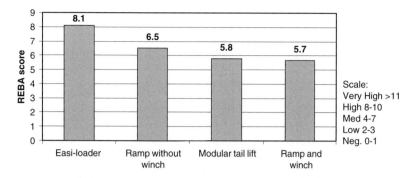

Graph 1. Postural analysis results – mean REBA scores.

The analysis showed that there was little difference between the three systems in terms of postural risk. The easi-loader stretcher poses the greatest risk to operators with a mean REBA score of 8.1 (action category 3), posing a high-risk to staff so action must be taken soon. The other three systems fell under the medium risk (action category 2) where action needs to be taken but not urgently.

The ramp and winch and the tail lift pose the lowest postural risk but there is slightly higher risk if the ramp is used without the winch.

In support of these results the findings from the preliminary study show that reduced manual handling is a high priority for the users. Staff reported manual handling risks in all the systems but more problems were identified with the easi-loader and the ramp and winch. Many of the problems associated with the tail lift system were linked to the stretcher rather than the tail lift.

The preliminary study also found that the winch is rarely used in the field (Jones & Hignett, 2005), so the risk associated with the ramp and winch is therefore greater, further supporting the recommendation that the tail lift is the preferred system.

However, the results show there is still a medium risk associated with the tail lift and action is necessary. The assessment highlighted that poor postures are adopted when raising the side protection barriers on the tail lift. Automating the raising of these barriers would reduce the risk to staff further. The Ferno Falcon 6 stretcher used with the tail lift vehicle weighs 64 kg excluding the mattress. A lighter weight stretcher would improve the postures adopted by staff and reduce the associated risk.

References

Boocock, M., Smith, C. and Morris, L. (2000). An ergonomic evaluation of loading operations using the Ferno 35A trolley cot. *Health and Safety Laboratory*, 1–20.

British Standards institution. (2000). *Medical Vehicles and their Equipment – Road Ambulances*. London: BSI BS EN 1789:1999.

East Midlands Ambulance Service NHS Trust. (2004). East Midlands Ambulance Service – NHS Trust Homepage: URL: http://www.emas.nhs.uk, taken on 29/01/05.

HSC 1999/065 *Clinical Governance: Quality in the new NHS,* Health Services Circular, NHS Executive. Department of Health. http://tap.ccta.gov.uk/doh/coin4.nsf

Jones, A. and Hignett, S. (2005) A comparative analysis of stretcher loading systems. In P. D. Bust and P.T. McCabe (eds.) *Contemporary Ergonomics*. London: Taylor and Francis. 261–265.

Lavender, S.A., Conrad, K.M., Reichelt, P.A., Johnson, P.W and Meyer, F.T. (2000). Biomechanical Analysis of Paramedics Simulating Frequently Performed Strenuous Work tasks, *Applied Ergonomics,* April. **31**(2). 167–177.

Overton, J. (2001). Ambulance Design and Safety, *Journal of Prehospital and Disaster Medicine,* 2001, **16**, (3).

Hignett, S. and McAtamney, Rapid Entire Body Assessment (REBA); Applied Ergonomics. 31:201–205, 2000.

Unison. (2002). Ambulance worker accepts £140,000 compensation for back injury, press release, November 19. 2002, URL: http://www.unison.org.uk/asppresspack/pressrelease_view.asp?id = 214, accessed on 19/01/04.

Watt, B. and Dickson. (1999). The Scottish Ambulance Service Board Manual Handling Project; Evaluation of Ambulance Loading Systems – Update, ASI International, September 1999. (5).

Scally, G. and Donaldson, L.J. (1998) Clinical governance and the drive for quality improvement in the new NHS; British Medical Journal, 1999, URL: http://bmjjournals.com/cgi/content/full/317/7150/61.

MANAGEMENT OF MANUAL HANDLING RISK IN WELSH CARE HOMES

A.D.J. Pinder, P. Marlow & V. Gould

Health and Safety Laboratory, Harpur Hill, Buxton SK17 9JN

This study benchmarked the management of manual handling risk in Welsh care homes. Questionnaires were completed by 241 managers and 860 workers on perceptions of management systems, prevalence of musculoskeletal trouble and psychosocial factors. High proportions of staff were routinely involved in manual handling. Both managers and workers had very positive perceptions of how manual handling risk was controlled. Mechanical handling aids were in general and frequent use. Prevalence rates of musculoskeletal trouble were low. Worker perceptions of psychosocial factors were significantly more positive than other UK surveys. Both management and staff believed that manual handling risks could be further reduced. Therefore, care homes should be encouraged to maintain and improve on the present high standards.

Introduction

The control of manual handling operations is part of the broader risk management that occurs in a workplace and therefore of the corporate "safety climate". The evidence about manual handling safety climate is sparse but recent evidence (Johnson and Hall, 2005) is that perceived behavioural control is directly associated with safety related behaviour in lifting. As workplace norms are malleable and vary between workplaces, and as cultural influences affect risk-taking behaviour at work, the hypothesis is that modifying the culture should improve the standard of safety behaviour. Hence, safety climate measures can be used as a means of measuring organisational performance. Therefore, this study aimed to benchmark manual handling risk management in the care home sector in Wales through cross-sectional measures.

Methods

HSE and Local Authorities (LAs) carried out joint Wales-wide inspections of care homes concentrating on manual handling issues, with a target of ten care homes in each authority area. As part of the visit they left a questionnaire for completion by managers. Workforce questionnaires were distributed later by a market research company, with a target of ten questionnaires completed at each home. Versions of both questionnaires had been previously been used in the chemicals and offshore industries to examine the perceived status of safety management systems for manual handling operations.

The management questionnaire was designed to measure attitudes on aspects of the constructs of "Corporate Safety Climate" (CSC); "Management Commitment" (MC); "Cultural Profile" (CP) and "Risk Management Systems" (RMS). Each question had a five-point scale, with each point anchored to a statement about a particular aspect of an organisation's manual handling regime.

The original workforce question set was titled "Organisational control of manual handling risks". Responses were on five point scales, with anchor points of "Strongly Agree", "Agree", "Uncertain", "Disagree", and "Strongly Disagree". The earlier work had identified three scales in the question set: "Perceptions of Management Commitment and Background Climate" (PMC); "Rules, Procedures and Monitoring" (RPM); and "Staff Training and Involvement" (STI). Lower total scores on a scale indicated positive perceptions that manual handling issues were being effectively addressed in their place of work. This questionnaire was combined with a modified version (Pinder, 2004) of the Nordic Musculoskeletal Questionnaire (NMQ) and a psychosocial questionnaire.

The section on musculoskeletal disorders included questions about time off work due to problems caused or made worse by manual handling as well as the NMQ. This asked about musculoskeletal "trouble" experienced in nine body areas in the previous three months and in the previous seven days. It assessed severity by asking if trouble suffered in the previous three months had affected normal activities and asked about work-relatedness of the trouble suffered in the previous three months.

The psychosocial question set from Johansson and Rubenowitz (1994) was used to measure "Work Characteristics" factors of "Influence on and control over work", "Supervisor climate", "Stimulus from the work itself", "Relations with fellow workers", "Psychological work load". To extend the scope of the question set, a sixth factor, "Management commitment to health and safety" was added.

Management survey results

Questionnaires were returned by 241 managers. Almost all (83%) care homes were privately owned with the remainder run by LAs or charities. Most (84%) employed 11–40 care workers, and most (75%) had 11–40 clients. Residential care was the primary type of care provided by 80% of homes. In 82% of homes over 80% of staff were routinely involved in manual handling activity and in 70% of homes more than 80% of such staff had received training in safe handling techniques. Formal programmes for conducting manual handling risk assessments were reported by 91% of managers. HSE guidance (HSE, 2004) makes it clear that it is expected that a suitable and sufficient risk assessment of patient handling must be formal and written. This implies that the 8% of homes that reported not having formal written risk assessments were not compliant with the MHOR 1992 and the visiting HSE and LA inspectors will have sought to rectify this.

The homes where high proportions of staff were performing manual handling tasks were the ones that had high proportions trained in safe handling. Where 81–100% of staff performed manual handling tasks, 78% had been trained in safe handling.

Of 216 care homes, 22 had a total of 33 manual handling incidents resulting in time off work reported in their accident books within the previous twelve months.

The four CSC subscales indicated that managers perceived high levels of awareness, priority, allocation of resources and implementation of initiatives for controlling manual handling risks in their care homes. A t-test showed that sites providing primarily residential care scored significantly higher than those providing primarily nursing care ($p < 0.05$). An ANOVA showed significantly higher scores ($p < 0.01$) at sites where 81–100% of staff had been trained when compared to sites where 0–20% of staff were trained. The four MC subscales indicated that managers believed that senior management had high levels of insight and understanding, saw control of manual handling risks as a high priority, and had played a leading role in setting up risk assessment systems in health and safety initiatives. The two CP subscales showed that care supervisors placed a high priority on control of manual handling risks and that care staff exhibited significant concern about the control of these risks. The RMS subscales indicated that managers reported that manual handling assessments, health monitoring and audit and review were all performed at high levels, and that there were ongoing programmes for training in risk assessment and in involving staff in the risk assessments. T-tests showed that sites with formal programmes for conducting risk assessments scored significantly higher ($p < 0.01$) on the CSC, MC and RMS scales than those without formal programmes.

Workforce questionnaire results

Usable workforce questionnaires were returned by 860 individuals employed in 84 care homes. Private care homes employed 80% of respondents, with LAs employing 9% and charities 11%. Almost all (90%) worked in care homes providing residential care and most respondents (75%) were care assistants.

The mean (SD) experience working in health care (N = 667) was 11.0 (10.6) years. That this workforce had a very high turnover rate with many new entrants was shown by 34% of workers having less than five years experience in health care, and 67% and 24% respectively having less than five and one year's experience in their current job.

Training in manual handling was widespread as 89% had received it at some point and 74% had received it in the previous 12 months. That some organisations were not up to date with training was shown by 8% of respondents having been employed for more than one year in health care, but never having been trained in patient handling.

Of the 860 respondents, 83% reported the availability of at least one type of hoist, with 80% reporting free-standing hoists, but only 11% reporting ceiling mounted/track hoists. Slide sheets were available to 75% and 52% reported both belts and stand aids. Other unspecified aids were available to 27%. Three or more types were reported by 68%. It is clear that handling aids are in general and widespread use as only 8% failed to report any aids. The percentages that reported that available aids were used "Whenever possible" or "Often" ranged from 61% for belts to 80% for free-standing hoists.

Only 10% of 853 respondents reported ever having had time off work due to problems caused or made worse by manual handling. Only 2% reported having had time off in the previous three months. Of these 18 individuals, 13 gave durations, with one reporting five months absence and eight with three or fewer days absence. That five (nearly 40%) did not report durations, illustrates the difficulties of collecting data about work absence especially when the cause is episodic and therefore prone to recall failure.

Workforce attitudes to organisational control of manual handling risks

The majority of responses across the sample as a whole were positive (~43%) or very positive (~41%), indicating that the workforce perceived a very high standard of manual handling safety management within their care homes. There were no clear differences between the PMC, RPM and STI scales. Only one item received significantly fewer positive responses ($p < 0.01$) than the remainder: "There aren't really any ways in which you could reduce manual handling risks any further at this care home". Only 31% believed that there were no further ways to reduce risks from manual handling.

Reports of musculoskeletal trouble

The low back had the highest prevalences of "trouble", with rates of 28% and 13% for three month and seven day trouble. Rates of "disability" and work-relatedness in the three months were 7% and 11%. This workforce had lower rates lower than a varied group of industrial workers (Pinder, 2004) with a three month prevalence of 46%, and a group of podiatrists, (71% three month prevalence). Seven day trouble rates in care staff varied between 46% and 55% of three month trouble. Mean disability across body parts was 23%. The ratio of whether trouble was caused or made worse by work varied from 0.17 in the ankles/feet to 1.00 in the upper back.

The low back seven day prevalence rate was divided by the 14 day prevalence for nurses' aides in Norway (Eriksen, 2003), giving a rate ratio (RR) of only 0.24 (95% CI 0.23–0.38). Comparisons with seven day prevalences in two other HSL studies gave an RR of 0.30 (95% CI 0.23–0.28) for a group of 148 podiatrists and one of 0.61 (95% CI 0.48–0.77) for 500 UK industrial workers (Pinder, 2004).

Workforce attitudes to work characteristics

Table 1 gives the mean and standard deviations of the factor scores for the six psychosocial factors of the Work Characteristics section of the questionnaire. For WCF1, "Influence on and control over work", and WCF5, "Psychological work load", the mean was equivalent to a score between 3 and 4, i.e., between neutral and positive. For WCF2, "Supervisor climate", WCF3, "Stimulus from the work itself", WCF4, "Relations with fellow workers", and WCF6, "Management commitment to health and safety", the mean was equivalent to a score between 4 and 5, between positive and strongly positive.

The responses from the care staff are significantly ($p < 0.05$) more positive than the responses from the groups of industrial workers and podiatrists, both of which had mean scores in the neutral to positive regions. The scores for these two groups were only significantly different ($p < 0.05$) on WCF2, WCF3, and WCF5.

Discussion

The results of this study have consistently shown that manual handling problems appear to be well controlled in the care home sector in Wales. High levels of senior management commitment were reported in the management questionnaire. Handling aids were widely available and in widespread use in the sector. That training levels were high (90% of staff) and recent (75% of staff) are other indications of management commitment. The workforce saw the management of care homes as mostly having good control of the risks of manual handling. The outcome of this good control is shown firstly by the low reported rates of absence due to problems associated with manual handling, and the short durations of most of the absences reported. It was indicated secondly by the remarkably low prevalence rates of musculoskeletal trouble obtained with the NMQ. These rates were significantly lower than in other UK surveys of different working populations and Eriksen's (2003) survey of a similar workforce in Norway.

Table 1. Comparison of Work Characteristics factor scores from HSL studies.

Factor	Care staff Mean (SD)	Industrial workers Mean (95% CI)	Podiatrists Mean (95% CI)
WCF1	18.2 (17.8–18.5)	14.7 (15.1–14.3)	15.7 (16.3–15.0)
WCF2	21.2 (20.9–21.6)	16.2 (16.7–15.8)	17.5 (18.3–16.7)
WCF3	21.2 (20.9–21.6)	15.0 (15.4–14.5)	18.4 (19.1–17.7)
WCF4	21.8 (21.5–22.1)	18.7 (19.1–18.3)	18.1 (18.8–17.4)
WCF5	18.2 (17.8–18.6)	16.0 (16.4–15.6)	14.9 (15.5–14.2)
WCF6	21.1 (20.7–21.4)	17.2 (17.6–16.7)	17.6 (18.2–17.0)

The cross-sectional nature of this survey precludes drawing conclusions as to causation, but other studies indicate that the provision of manual handling aids reduces the risks of injury and disability from manual handling (Chhokar et al, 2005).

Organisations within the care homes sector should be encouraged to maintain, and even raise, the present high standard of manual handling risk management systems. Ideally, a longitudinal study should be conducted with a fixed group of care homes and care staff. This would permit the measurement of changes over time in the profile of manual handling risk management due to further management interventions to reduce risk. However, as such studies are difficult, time consuming, and expensive, serial cross-sectional surveys may offer a more practical approach.

References

Chhokar, R., Engst, C., Miller, A., Robinson, D., Tate, R.B. and Yassi, A. (2005). The three-year economic benefits of a ceiling lift intervention aimed to reduce healthcare worker injuries. *Applied Ergonomics*, **36**, (2), 223–229.

Eriksen, W. (2003). The prevalence of musculoskeletal pain in Norwegian nurses' aides. *International Archives of Occupational and Environmental Health*, **76**, (8), 625–630.

Health And Safety Executive (2004). *Manual Handling: Manual Handling Operations Regulations 1992 (as amended). Guidance on Regulations*. L23. (Sudbury, Suffolk: HSE Books), Third Edition.

Johansson, J.A. and Rubenowitz, S. (1994). Risk indicators in the psychosocial and physical work environment for work-related neck, shoulder and low back symptoms: A study among blue- and white-collar workers in eight companies. *Scandinavian Journal of Rehabilitation Medicine*, **26**, (3), 131–142.

Johnson, S.E. and Hall, A. (2005). The prediction of safe lifting behavior: An application of the theory of planned behavior. *Journal of Safety Research*, **36**, (1), 63–73.

Pinder, A.D.J. (2004). Work relatedness of MSDs in industrial workers. In: *Premus 2004: Fifth International Scientific Conference on Prevention of Work-Related Musculoskeletal Disorders* (Zurich: Institute of Hygiene and Applied Physiology, Swiss Federal Institute of Technology), Volume 2, pp. 617–618.

KEEPING ABREAST OF THE TIMES

Hazel J. Scott & Alastair G. Gale

Applied Vision Research Centre, Loughborough University, Garendon Wing, Holywell Way, Loughborough LE11 3TU, UK

> For the past 15 years we have run a national scheme (PERFORMS – personal performance in mammographic screening) which gathers data concerning how breast screening radiologists interpret sets of difficult screening cases. The scheme functions as an educational exercise and is funded by the NHS Breast Screening Programme as part of their quality assurance. Initially participants viewed sets of difficult mammograms on radiographic multi-viewers and recorded their opinions on paper and subsequently these data were input into spreadsheets for analysis. The next development was to utilise a hand-held PDA, coupled with a bar code reader, to simplify the data recording and provide the potential for some immediate feedback on an individual's performance. In the latest incarnation a tablet PC is now used with a GUI which allows the radiologists both to record their decisions quickly and easily and also to provide very detailed performance feedback possibilities. The evolution of the approach is described.

PERFORMS is a nationwide self-assessment scheme that over 400 medical specialists in the UK NHS Breast Screening Programme complete on a bi-annual basis (described in detail in Gale, 2003) . The objective of the scheme is for imaging scientists to examine a series of difficult breast screening mammograms and report their various decisions. Once all participants have taken part then anonymous regional and national data are accrued and fed back to each individual.

Paper-based approach

Initially the scheme functioned using a paper-based method where each participant was sent a booklet to complete, together with the series of X-ray mammogram cases to examine (Figure 1). Once the cases had been interpreted then the booklets were returned to us and data entered into a spreadsheet for analysis. The system was simple and friendly for the user but the data transcription was tedious, open to potential error and there was no opportunity to provide any feedback on the decisions made by a participant.

PDA and barcodes

The next development (Figure 2) was to introduce a hand-held PDA with a bar coding system. This allowed the individual to read each case and quickly enter decisions by using a handheld barcode reader and barcode sheets (which displayed the range of possible decisions on any given case, mimicking the paper-based option) into the PDA. Once all cases had been read by a participant then the PDA was able to analyse their responses and provide some immediate feedback on their performance.

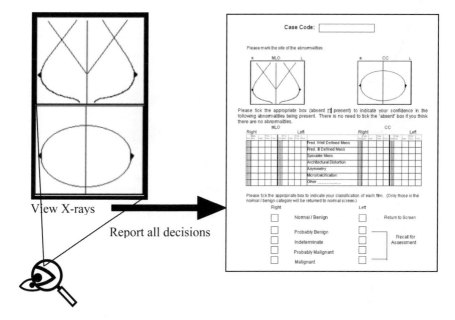

Figure 1. Original two-view mammograms reported using a booklet.

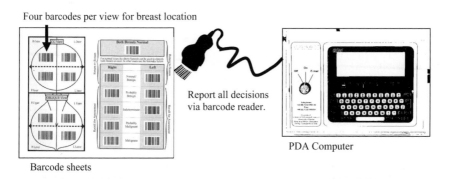

Figure 2. PDA, barcode reader and barcode sheet.

In addition to a summary statistics participants where afforded the option to review, in detail, any number of X-ray cases. This system was in use for over 5 years and was highly reliable, although somewhat restricted in the amount and nature of radiological feedback to the participant (being purely text based). In addition the system was limited in several ways, firstly, in terms of response options, particularly for the identification of breast location as logistically only four barcodes could be allocated per X-ray view (see Figure 2), and secondly, the PDA required at least 20 minutes of training time before the system could be used comfortably by a new user. Participant feedback also highlighted the need for a more responsive system than barcode readers which were at times slow to register a user's input.

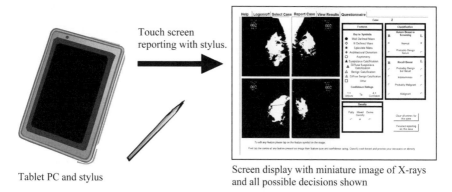

Figure 3. Pen-based PC Tablet.

Tablet PC

The latest approach is to utilise a tablet PC with pen-based GUI input. This functions in a similar way to the former PDA but the individual first identifies a particular case to the tablet whereupon a small image of the case being examined is shown on the tablet. The user can then graphically indicate the exact location of any areas of interest as well as record their various decisions about the case. This system was designed to be intuitive so far less training time is required. After all data are entered then immediate feedback on performance is possible, this time with detailed graphical feedback on every aspect of each case.

Discussion

Evolution of the methods has made it possible to contend with a medical environment with ever restricted time schedules. A system which is far more responsive (faster) and requires less training also enables the scheme to cope with a larger influx of new participants.

Acknowledgements

This work is supported by the NHS Breast Screening Programme.

Reference

Gale, A.G. 2003: PERFORMS – a self assessment scheme for radiologists in breast screening. In *Seminars in Breast Disease: Improving and monitoring mammographic interpretative skills*, **6(3)**, 148–152.

FATIGUE EXPERIENCED BY CYTOLOGY SCREENERS READING CONVENTIONAL AND LIQUID BASED SLIDES

Jayne Cole & Alastair G. Gale

Applied Vision Research Centre, Loughborough University, Garendon Building, Holywell Park, Loughborough LE11 3TU

The "Pap" smear test has traditionally been used to screen women for abnormal changes in the cells of the cervix. This method is now being replaced by Liquid Based Cytology (LBC) techniques, which when piloted resulted in increased reports of fatigue by cytology screeners. The nature and extent of this discomfort on the larger population of cytology screeners is as yet, undetermined. This working group is exposed to a higher risk of visual and physical fatigue due to the static and repetitive nature of their work and also the high level of required visual concentration. These issues may need to be addressed to maintain the high quality and reliability of the Cervical Screening Programme. Proposed research to compare the visual and physical fatigue of cytology screeners' when screening LBC and conventional slides is currently undergoing ethical approval.

Introduction

In 2000, there were 2,424 new cases of invasive cervical cancer in England (NHSCSP, 2000). It is the eleventh most common cause of cancer deaths in women in UK with some 2–3% of all women over 40 years of age developing cervical cancer. Mortality rates increase with age and the highest number of deaths occurs in the 75–79 age group (Cancer Research UK, 2003).

The NHS Cervical Screening Programme (NHSCSP) was established in 1988 to reduce the incidence and mortality of women with cervical cancer by proactively screening those most at risk. It does this by identifying pre-cancerous changes to the cells of the cervix. Since its introduction, death rates from cervical cancer have fallen significantly. Mortality rates in 2000 were 60 per cent lower (3.3 per 100,000 women) than they were 30 years earlier (8.3 per 100,000) in 1971 (Cancer Research UK, 2003). The latest relative survival figures for England show that an average of 84 per cent of women diagnosed with cervical cancer between 1993 and 1995 were alive one year later and 66 per cent were alive five years later (Cancer Research UK, 2003). It is estimated that cervical screening now saves approximately 4,500 lives per year in England (Peto et al, 2004).

The NHSCSP sends out an invitation to women between 25 and 64 years of age to be screened every 3–5 years. Cervical cells are collected from each woman using a spatula and transferred on to a slide which is then sent to a cytology laboratory. There the cells are stained and examined under a microscope by a cytology primary screener. In the period of 2003–04 over three and a half million women were screened by the NHSCSP, and four million samples were examined by cytology laboratories (Department of Health, 2004).

The Importance of Ergonomic Factors in Cytology Screening

During a typical working day cytology primary screeners visually examine approximately 32 slides, with most screeners working at a microscope for four or more hours. This task causes the screener to adopt a relatively fixed posture for periods of time in order to view the slide and operate the microscope controls. Additionally, the screener is required continuously to operate the microscope stage controls, often with their wrists and arms in non-neutral postures. The prolonged use of a microscope is a recognised cause of visual and muscular problems.

Visual problems

Druault (1946) referred to "operational microscope myopia" and claimed that near sightedness (myopia) and double vision in physicians was the result of undue accommodation and convergence caused by the use of microscopes. Occupations which have intensive microscope usage have found 75–80% of users report visual problems (Emanual and Glonek, 1975; Soderberg, 1983) and that the prevalence of symptoms increases with time spent using a microscope (Kalavar and Hunting, 1996). Frenette and Desnoyers (1986) conducted tests of visual fatigue on cyto-technicians both before and after they started work and compared the results with haemotologists who used a microscope less frequently. Some 31% of cyto-technicians showed symptoms of blurred vision at the end of the day compared with just 3% of haemotologists. Within the UK some 73% of cytologists have subjectively reported eye strain while screening conventionally prepared slides (Hopper et al, 1997). Factors which might contribute to this fatigue in cyto-screeners include; long periods of accommodation, ophthalmic factors (e.g. long sightedness or astigmatism), microscope illumination, and environmental conditions (Burrells, 1977, Ostberg and Moss, 1984).

To prevent the onset of fatigue and discomfort and also to maintain a high level of concentration by the screener it is therefore important that the screening task is well designed. This has led to the development of ergonomic standards in cytology screening (Medical Devices Agency, 2002).

Physical problems

As early as 1942, Simmons et al found that sustained contractions of the ocular and neck muscles when using microscopes can cause headaches and stiffness of the neck. In an industrial setting, Soderberg (1978) found that approximately 45% of microscope workers suffered from muscular ailments, and Kreugar, et al (1989), found that 50% of microscope users experienced daily neck pain. In the medical field, Kalavar and Hunting (1996), found that 70% of cytology technologists had a shoulder, neck or upper back problem. Specifically within cytology in the UK, Hopper et al, (1997) found that within a 12 month period 78% of cytology screeners reported experiencing musculoskeletal problems. Twenty nine percent of these respondents' reported experiencing musculoskeletal pain every day, while 54% experienced discomfort sometime during the week. Thirty four percent of respondents reported the intensity of their discomfort (on a five point scale) as moderate, and 24% reported usually feeling the discomfort after working for one hour. This discomfort was found to adversely affect the level of concentration and job satisfaction reported by the cytology screener (Hopper et al, 1997).

When compared to other health service professions (Figure 1) it can be seen that cytology screeners have predominantly more discomfort in the neck, shoulders, arms and wrists. Other studies have found the most prevalent discomfort in the neck, shoulders, wrists and back (Rizzo et al, 1992; Hopper et al, 1997).

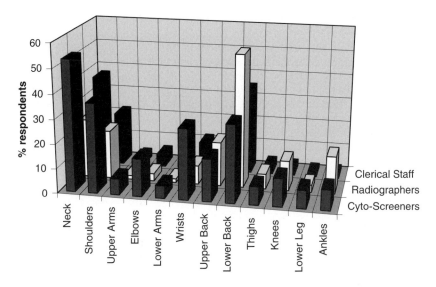

Figure 1. Prevalence of discomfort in three occupational groups.

Important Changes to the NHSCSP

Traditionally the NHSCSP has used the "Pap" smear test to screen for abnormal changes in the cells of the cervix. This test involves collecting cells from the cervix onto a spatula and transferring them to a glass slide which is sent to a hospital laboratory for analysis by a cytology screener. There are problems associated with this technique due to errors made during the acquisition of the smear and unwanted material being present, which results in a high level (9.1%) of "inadequate" slides (Fahey et al, 1995). Such smears need to be repeated, causing anxiety for the woman and extra cost for the NHSCSP.

Liquid Based Cytology (LBC) has been shown to address some of the problems of the "Pap" test (McGoogan and Reith, 1996, Moss et al, 2002). The cervical cells are still collected in a similar way, but the part of the spatula containing the cells is broken off in a tube of preservative fluid. This is then sent to the laboratory where it is spun to reduce unwanted material. A single layer of cell suspension is then deposited onto the slide and automatically stained. A cytology primary screener then examines the slide.

In 2001, the NHSCSP carried out a pilot project at three sites in England to evaluate the costs and practical implications of introducing LBC into the screening programme. Resulting from a survey of these laboratories in 2003, NICE recommended that "LBC is used as the primary means of processing samples" within the screening programme (NICE, 2003). LBC will therefore be implemented nationally across England during the next five years.

Implications of change

The way in which cervical cells are presented on the slide differs between the use of LBC methods and the traditional "Pap" test. This will potentially affect the way in which each slide is screened in terms of microscope stage movement, control operation, hand movements, scanning techniques used and eye movement patterns. Presently no work has been done to specifically investigate the implications of these changes for cytology staff. Some laboratories have informally reported that LBC slides are more tiring to read (McGoogan and Reith, 1996), and our own informal discussions with cytology screeners confirm these reports. This is reflected in the results from LBC feasibility studies (Moss et al, 2002), where 22 out of 38 screeners said they felt that the number of breaks should be increased when reading LBC slides.

Previous work undertaken by us for the NHSCSP has demonstrated that cytology screeners do in fact become more fatigued during the day when performing conventional screening (Horberry et al, 2000; May et al, 2001). This research has led to recommendations concerning break frequency and duration, and the type of work performed (NHSCSP, 2003). If for some reason LBC slides are indeed more tiring to screen, then the NHSCSP need to take steps to address this in order to ensure that the high quality and reliability of the screening programme is maintained.

Conclusions

It is therefore proposed that LBC technology will have ergonomic implications for cytology primary screeners and due to them being susceptible to fatigue because of the established problems of repetition and static postures in the nature of their work. In order to ensure the high quality of the screening programme these changes need to be fully investigated. A study proposal to investigate these factors is currently under development.

References

Burrells, W., 1977, Microscope Technique. *A Comprehensive Handbook for General and Applied Microscopy.* (New York, NY: John Wiley and Sons).

Cancer Research UK, 2003, *CancerStats*, London.

Department of Health, 2004, Cervical Screening Programme, England 2002–2003, *Statistical Bulletin* 2003/04, ISBN 1 84182 906 4.

Druault, A., 1946, Visual Problems following Microscope Use. *Annals Oculistique*, 138–142.

Emanuel, J.T., and Glonek, R.J., 1976, Ergonomic approach to productivity improvement for microscope work, *Proceedings of the AIIE Systems Engineering Conference, Institute of Industrial Engineers*, Norcross, G.A.

Fahay, M.T., Irwig, L., and Macaskill, P., 1995, Meta-analysis of Pap test accuracy. *American Journal of Epidemiology*, **141**, 680–689.

Frenette, B., and Desnoyers, L., 1986, A study of the effects of microscope work on the visual system. *Proceedings of the 19th meeting of the Human Factors Association of Canada,* Richmond (Vancover), August 22–23.

Hopper, J.A., May, J.L., and Gale, A.G., 1997, Screening for Cervical Cancer: The Role of Ergonomics. In S.A. Robertson (ed.) *Contempory Ergonomics*, 38–43.

Horberry, T., Cowley, H., Miles, J., and Gale, A.G., 2000, *Cervical Screeners and their Working Hours,* Report to the NHSCSP.

Kalavar, S.S., and Hunting, K.L., 1996, Musculoskeletal symptoms among cytotechnologists. Laboratory Medicine, **27** (11), 765–769.

NHSCSP (2003), Laboratory Organisation: A guide for laboratories participating in the NHS Cervical Screening Programme, **No 14**, ISBN 187 1997 59 3.

Krueger, H., Conrady, P., and Zulch, J., 1989, Work with magnifying glasses, *Ergonomics*, **32**, No7, 785–794.

McGoogan, E., and Reith, A., 1996, Would mono-layers provide more representative samples and improved preparations for cytology screening ? Overview and evaluation of systems available. *Acta Cytologica*, **40**, 107–119.

May, J.L., Cowley, H., and Gale, A.G., 2003. Variations in the discomfort and visual fatigue of cytology screeners, In:- P. McCabe (ed.), *Contempory Ergonomics*, 2003. Taylor and Francis, London.

Medical Devices Agency, 2002, *Minimum Ergonomic Working Standards for Personnel Engaged in the Preparation, Scanning and Reporting of Cervical Screening Slides*, MDA/021004.

Moss, S.M., Gray, A., Legood, R., and Henstock, E., 2002, *Evaluation of HPV/LBC, Cervical Screening Pilot Studies,* First Report to the Department of Health on Evaluation of LBC.

NHSCSP, 2000, National Statistics. Cancer registrations in England, 2000.

NICE, 2003, Guidance on the use of liquid based cytology for cervical screening, NHS Technology Appraisal 69, ISBN 1 84257-368-3.

Ostberg, and Moss, C.E., 1984, Microscope Work – Ergonomic Problems and Remedies, *Proceedings of the 1984 International Conference on Occupational Ergonomics,* Rexdale, Ontario, Canada, Human Factors Association of Canada, 402–406.

Peto, J., Gilham, C., Fletcher, O., and Matthews, F., 2004, The cervical cancer epidemic that screening has prevented in the UK, *The Lancet*, **364**, Issue 9430, pp 249–56.

Rizzo, P., Rossignol, M., Gauvin, J.P., 1992, A One Week Incidence Study of the Symptoms Experienced by Hospital Microscopists: The Economics of Ergonomics. *Proceedings of the 25th Annual Conference of the Human Factors Association of Canada,* Hamilton, Ontario, Canada, October 25–28.

Soderberg, I., 1978, Microscope Work II. *An Ergonomic Study of Microscope Work at an Electronic Plant*. Report No. 40, National Board of Occupational Safety and Health, Sweden.

Simons, D., Day, E., Goodell, H., 1942, Experimental studies on Headache. *Research Publications Association*, **23**: 226–224.

INCLUSIVE DESIGN –
IN THE BUILT
ENVIRONMENT
SYMPOSIUM

DECENT HOMES AS STANDARD, BUT ARE THEY INCLUSIVE?

Marcus Ormerod, Rita Newton & Pam Thomas

SURFACE Inclusive Design Research Centre, BuHu Research Institute, The University of Salford, Maxwell Building, Salford M5 4WT

This paper explores the implications of the UK Government's Decent Homes Standard (DHS), which sets targets for Registered Social Landlords (RSLs) to improve their housing stock. These targets fail to include issues of accessibility within them, which may lead to a missed opportunity for an inclusive design approach within refurbishments.

A study of six larger RSLs was undertaken by the authors into the how the DHS translates into bricks & mortar and the impact this has on accessibility of social housing. The results of the research show a lack of an inclusive design approach in refurbishment projects, and instead a "special needs" mindset towards access issues.

Introduction

The research study assessed the approach to accessibility of stock transfer properties, undergoing improvements to bring them in line with the Decent Homes Standard (DHS), within six Registered Social Landlords (RSLs) in England. The DHS gives targets that a decent home should meet but it does not include accessibility.

In the UK the Government aims to bring all social housing into a decent condition by 2010, with most improvement taking place in deprived areas. This has led to the stock transfer of a large quantity of social housing properties from Local Authorities to RSLs. There are four main criteria that the DHS uses to determine if a home is in a decent condition (ODPM 2004):

1. It meets the current statutory minimum standard for housing;
2. It is in a reasonable state of repair;
3. It has reasonably modern facilities and services;
4. It provides a reasonable degree of thermal comfort.

Within the DHS there is no specific mention that a home should be accessible both for occupiers and their visitors, but there are certain requirements and advice within DHS that relate to accessibility by implication. Guidance on implementing the DHS recommends when considering refurbishment landlords should consult with current Building Regulations and other relevant technical publications for the standard of work to be carried out. Since accessibility is part of the revised Building Regulations (Part M in England and Wales) and its Approved Document (AD M), there is an inferred requirement in achieving DHS.

The English Housing Conditions Survey (EHCS) is used to indicate the changing state of homes in the UK and is a well established set of statistics (ODPM 2003, 2005). The EHCS takes account of four Part M accessibility features – level access to dwelling; flush thresholds; 750 mm clear width door openings; and bathroom/WC at entrance level.

The Joseph Rowntree Foundation has for sometime been advocating the concept of Lifetime Homes (JRF 2005 Habinteg 2005) as a way to incorporate access features into new housing and to apply in refurbishment of existing dwellings. Lifetime Homes is better practice guidance and moves towards an inclusive design approach, with 16 access related features. Inclusive design is a way of designing products and environments so that they are usable and appealing to everyone regardless of age, ability or circumstance by working with users to remove barriers in the social, technical, political and economic processes underpinning building and design (Ormerod 2005).

Using available better practice guidance along with the authors experiences of access problems in housing a set of 13 issues were identified as indicators of approach to accessibility, to use in interviews and case studies with RSLs involved in undergoing improvements to stock transfer properties to bring them to the DHS.

Case studies

6 RSLs in England agreed to take part in the study and provide interviews with their key staff, along with site visits to DHS refurbishment projects. The RSLs chosen were Housing Associations managing large numbers of homes through stock transfers. Whilst each RSL consulted with tenants, prior to refurbishment, access was not raised as a collective issue, but was raised individually by tenants.

To provide cross case study analysis a set of 13 issues were identified by the authors as indicators of the level of access within refurbishment works of social housing –

1. There is an Access Statement, which, is related to an Access Strategy:
   a) is used in the decent home refurbishment scheme;
   b) incorporated into tender documents, specifications, and bills of quantities, which are used in engaging the contractor.
2. Where there is the opportunity to improve the external approach:
   a) steps are replaced with a shallow ramp (1:20);
   b) the entrance path is made wide enough for a parked car and a pram, buggy or wheelchair user to pass;
   c) Doors and or gates are made distinguishable, for example through colour or textures, and lighting.
3. Where entrance doors are being installed, level thresholds are incorporated of no more than 15 mm height.
4. Where there is opportunity provide a storage area for a bike, pram or wheelchair inside the entrance.
5. Where new doors are installed they will be at least 750 mm wide.
6. Where electricity supply is renewed:
   a) all switches, sockets and control panels are put in a place that is accessible (i.e. between 450 and 1200 mm from the floor);
   b) and a new fuse panel is being installed, at least one spare spur or circuit is included. (This may be for any purpose, but may be required for equipment or a recharging station).
7. Where a ground floor toilet needs to have a wash hand basin:
   a) the toilet is kept at ground floor entrance level, and a hand basin installed;
   b) new taps have lever or cross head handles, and are consistent with Hot on the left and Cold on the right;
   c) where a new toilet is to be installed it has a large handled flush;
   d) if possible the toilet room it is made large enough for a wheelchair user.
8. Where new windows are being installed they are easy to open, and operate, with low handles. (Where possible living room window glazing begins at 800 mm or lower).
9. Where new kitchen spaces are being designed, enough space is allowed for a wheelchair user to enter and circulate.

Decent homes as standard, but are they inclusive? 331

10. Where new fixtures and fittings are being installed in any room, these are tonal/colour contrasted to assist people with low vision.
11. Where new wall tiles are being installed these are matt, rather than gloss (to prevent glare) and contrast with fixtures or fittings.
12. Where stairs are wide enough two handrails are installed. Either in individual units or in communal areas.
13. In accommodation with communal areas if work is to be done to improve lifts, this will include improving accessibility for people with physical or sensory impairments.

Findings

From the case studies it emerged that there is a "special needs" mind set amongst RSLs with every effort being made to adapt properties for individual tenants once they have been identified as a disabled person. The mainstreaming of access issues to all properties was not considered by the RSLs studied as they felt that "disruption would be too much for tenants" only improving access when asked for adjustments by individual tenants. The exception to this was common areas serving a large number of tenants in multi-storey housing where access was taken into account. The same level of access was not applied to similar areas leading to individual homes.

There was an initial concern by the authors that high thresholds were being fitted due to installation of uPVC doors in refurbishments, but this proved to be unfounded. The RSLs in this study were fitting timber doors for ecological reasons and the threshold strip was generally low level. Often, however, the installation of the threshold is onto the existing doorstep with no attempt made to either achieve level access, or to increase the size of the area outside the door to make a safe platform to stand on.

There was an emphasis on giving tenant choice where possible on fixtures and fittings, especially in terms of colour and kitchen design. Whilst this increased freedom of choice is beneficial for the existing tenant, it may be detrimental for accessibility of a future tenant. Tonal/colour contrast is important for visually impaired people to determine location of items and to use them safely. Choice was being given to tenants without any explanation of the limitations their choices may have. Similarly a kitchen design that reduces circulation space may create a physical barrier to a future tenant(s) who are, or who become, a wheelchair or walking aid user.

Many RSL properties already have a ground floor toilet and even if this is in a limited space, it is a start towards accessibility and visitability. However, where an existing tenant had removed the ground floor toilet, a RSL would not insist on reinstating the toilet as part of the refurbishment programme.

Simple access related considerations were not being included in refurbishment works such as fitting second handrails to staircases where there is sufficient width to do so; selecting matt wall tiles instead of gloss to reduce glare in bathrooms and kitchens; ensuring that there is spare capacity in the electrical system for ancillary equipment such as charging equipment and stairlifts; specifically contrasting fittings from their backgrounds such as washbasin and toilet.

Access statements and Access Strategies were not being developed for refurbishment projects, but the RSLs involved in the research felt that these would be beneficial. Comprehensive findings are given in Ormerod and Thomas (2005).

Conclusions

At a planning level on the refurbishment projects studied there were no access strategy documents that detailed accessibility issues of the properties, identifying current barriers and how they could be removed. This means an absence of action to ensure that access is improved for the changing needs of the current tenant, or increasing accessibility for future tenants. The lack of explicit statements on improving accessibility as part of the DHS means that general access improvements are not being undertaken.

Overall the findings of this research demonstrate that there is often a "special needs" mind set within RSLs leading to a special adaptations, rather than a mainstream, view of accessibility. The 6 RSLs studied are prepared to do everything they can for tenants once they have been identified as requiring "special needs", but this is then on an individual basis.

Tenants are being given increased levels of choice, wherever possible, within refurbishment works. However, tenant choice of colours and fittings may have an adverse effect on accessibility. When choosing items, tenants are not informed if that choice reduces access, such as tonal/colour contrast and kitchen layout/design. In satisfying current tenant choice the changing needs of the tenant, or the future tenants accessibility requirements are not considered. Additionally simple access issues are not being offered by RSLs such as matt wall tiles instead of high gloss, which would reduce glare. Others access issues appear to be only partially addressed, such as lower door thresholds being fitted but on steps, rather than trying to create level access, or a level area outside the door.

The introduction of DHS without any inclusive design consideration would appear a missed opportunity to improve accessibility in stock transfer properties. RSLs are starting to consider how they can improve access in future refurbishment projects. This still leaves many homes that have already been refurbished in need of further work to increase accessibility to a level where a disabled tenant can even visit neighbours in the safe knowledge that they can get into the home and use the facilities.

Acknowledgements

The authors wish to acknowledge the support from The Joseph Rowntree Foundation to undertake this research as part of its research and innovative development projects (Ormerod & Thomas 2005). The facts and views expressed, however, are those of the authors and not necessarily those of the Foundation.

References

Habinteg. (2005). **Lifetime Homes.** http://www.lifetimehomes.org.uk
JRF. (2005). **Lifetime homes.** http://www.jrf.org.uk/housingandcare/lifetimehomes/
ODPM 2005. **Housing in England 2003/04.** London: ODPM.
ODPM. 2004. **A Decent Home.** London: ODPM.
ODPM 2003. **English House Condition Survey Key Findings for 2003 Decent Homes and Decent Places.** London: ODPM.
Ormerod, M.G. 2005. Undertaking access audits and appraisals: An inclusive design approach. **Building Appraisal.** Vol. 1, no. 2, June, pp. 140–152.
Ormerod, M.G. and Thomas, P. 2005. **Decent Homes Standards: Opportunity to improve accessibility in existing housing stock.** York: JRF.

ACCESSIBLE HOUSING DESIGN FOR PEOPLE WITH SIGHT LOSS

Caroline Lewis[1], Janet John[2] & Tanita Hill[3]

[1] *JMU Access Partnership – Cymru, Trident Court, East Moors Road, Cardiff CF24 5TD*
[2] *Royal National Institute of the Blind – Cymru*
[3] *JMU Access Partnership, 105 Judd Street, London WC1H 9NE*

The Welsh Assembly Government has adopted Lifetime Homes principles for all social housing in Wales. They have subsequently funded a project undertaken by RNIB Cymru, supported by technical consultancy from JMU Access Partnership, to provide valuable complementary guidance to benefit people with sight loss.

The project has several tangible outcomes:

1. Housing Sight – a design guide for new build houses
2. Adapting Homes – a design guide for adapting existing homes
3. In Sight of Home – good practice guidelines for local authorities
4. The design of a training package for relevant professionals such as Occupational Therapists, Grants Officers and Architects.

This paper will outline each outcome in turn.

Housing Sight

Housing Sight is the first in the series of innovative guides to be published by RNIB Cymru and JMU Access Partnership which provides practical advice on building houses which are accessible to people who are blind and partially sighted. It is the first guide of its kind in the UK for the building of accessible homes for people with sight problems.

Housing Sight is based on information provided by current regulations, guidelines and good practice. An extensive literature search was carried out and, most importantly, the project was informed and shaped by views of people with sight loss. To gain a wide viewpoint from both urban and rural areas, several focus groups were set up across Wales. The groups consisted of people with sight problems and professionals working in the field of sight loss, including children aged 11+ and parents of younger children. "It's great that people are actually asking us what WE want for a change… they don't normally bother to ask." Rosie, 60, who is partially sighted.

Housing Sight contains four sections; introduction, background, design principles and supplementary information.

Introduction

Housing Sight goes towards achieving RNIB Cymru's objective: "To have housing in Wales that will enable people with sight problems to live safely and independently and enjoy the right to privacy, comfort and security in their own home."

Housing Sight is intended for all those involved in design, development, management or other housing related disciplines. It will provide the necessary knowledge to allow them to build a fully accessible home for people with sight problems.

In 1998 The Chartered Institute of Housing in Wales published "Lifetime Homes in Wales" which assessed accessible homes for all in Wales at that time. Lifetime Homes describes 16 design principles for accessible housing. It sets out physical design standards and Housing Sight aims to complement this with guidelines on design features that will benefit people with sight problems.

Background

The approach taken in Housing Sight is that disabled people, including those with sight problems, should be able to participate fully in society safely, independently and without undue restriction. Nowhere is this a more fundamental right than in a person's home.

To be able to create good design for people with sight problems, it is important to know how sight loss affects individuals in different ways. What is right for one person with a sight problem is not necessarily right for another.

Design principles

The design principles within "Housing Sight" explore the needs of people with sight problems in new housing, setting design standards across nine principal areas of the approach to the house, entrance and hallway, entrance level space, kitchen, circulation, bathroom, windows and doors, utilities and communal access homes.

Each area contains recommendations in bold and additional information related to good practice in designing homes for people with sight loss. Photographs and quotes are also used to illustrate specific points.

For example, a recommendation in the toilet section is that all fixtures and fittings used within the WC should be highlighted from the wall/floor against which they are positioned. To illustrate this point, a quote was added: "Our bathroom suite and walls are champagne, so he can't see where the toilet ends and the wall starts. Now he is standing up to go to the toilet it goes everywhere." Leanne, mother of James 3, who is partially sighted.

Supplementary information

In addition to the design principles supplementary information on colour and tonal contrast, lighting and the external environment is given. Lots of recommendations are given with quotes to show certain points. For example, it is recommended that the maximum size of drainage grills, covers and gratings in the external environment should be 13 mm. The quote added to this section is: "My friend who is blind caught his long cane in a drain and it snapped off. It was a good job he was with someone or he would not have been able to get back home." Sarah 46, who is blind.

Adapting Homes

Adapting Homes is based on Housing Sight described above. Additional information was gathered for Adapting Homes from Occupational Therapists, Rehabilitation Workers, Local Authority Housing Officers, Housing Associations and Care & Repair Cymru. Many of the people consulted had years of experience and their knowledge was essential to Adapting Homes.

Introduction

The Office of Population Censuses and Surveys have identified that the number of people with sensory loss of all types exceeds the number of people with mobility impairments. However, historically, within the evolution of access guidance and standards, there has been a strong focus on the needs of people with mobility impairments. This is probably due to the lack of knowledge or understanding about how people with sight loss use their environment. Adapting Homes raises awareness and provides practical information, suggestions and solutions to help solve common problems that professionals can face when attempting to adapt an existing home for people with sight loss.

Adapting Homes is intended for anyone who has a duty or wish to ensure that people with sight loss are able to live safely and independently in their own homes. This includes local authority staff in Social Services and Housing Departments, but also surveyors and architects who undertake assessments and provide adaptations to homes for disabled people. It will also be useful for voluntary organisations working with older or disabled people in their homes, or for those providing care in residential and nursing homes.

During the research, people often said that many of the best solutions are not rocket science; Adapting Homes aims to demonstrate how simple it can be to provide solutions that will help people with sight loss.

"I asked my eye consultant what I could do because I was having problems cleaning my teeth – the toothpaste was white, the bristles were white, my sink was white and I couldn't see to get the toothpaste onto the brush. Half the time it ended up in the sink, on the floor or over me. He told me to use striped toothpaste because of the colour contrast. But do you know, I never even thought of applying the same principle to my home. I think it's a marvellous idea." Janet 54, who is partially sighted.

Colour contrast and lighting

Colour contrast and lighting are important design elements for people with sight loss. Colour and contrast, particularly when used in conjunction with good lighting, can make a huge difference to a person's ability to understand their environment, find their way, make choices and predict what is happening. For persons with sight loss to find smaller items, such as a toilet roll holder, the item needs a stronger colour differentiation from its background. Some people with low vision also require extra light to support their functioning and in many cases need additional task lighting for specific tasks, for example, under-unit lighting above kitchen work surfaces.

Adaptation principles

The adaptation principles within Adapting Homes details issues and solutions to common problems when adapting homes for people with sight loss. An assessment checklist is included with the guide, which can be used as a prompt. Examples of questions asked in the checklist are:

- Is any foliage cut back and well maintained?
- Can the occupier identify their home easily from both entrances?
- Is there a conveniently located light switch in the hall?
- Are stair carpets free from visually confusing patterns?
- Do the bath and shower have a non-slip base?

If the answer to any question is "no" then page references for further information and recommendations on that subject area are supplied.

In Sight of Home

In Sight of Home is a guide for improving housing services for people with sight problems. This guide sets standards of good practice that are capable of being copied across the Welsh social housing sector. It is for all unitary authorities in Wales who want to improve their housing services for people with sight loss, including policy makers and practitioners, such as Grants Officers and Occupational Therapists. It is also aimed at Commissioning Officers and bodies responsible for policy setting, such as the Social Services Inspectorate for Wales (SSIW) and will be useful for healthcare professionals working in hospital and community settings.

Based on an extensive literature search, suggestions from professionals in the field and the collation of questionnaires and personal anecdotes the guide makes a number of key recommendations. These focus on:

- Making improvements to referral processes.
- Improving joint working across professional boundaries.
- Altering the allocation procedures for properties to make them equitable.
- Improving communication.
- Consultation and tenant participation.
- The need for training.

Services for people who have lost their sight do not reach most of the people who are in need of help. By following the recommendations in this publication improvements can be made to housing services which will prevent people with sight loss from "slipping through the net".

Training package

The last stage of the project was the development of a training package on the three guidelines; Housing Sight, Adapting Homes and In Sight of Home. The training package was designed for amongst others, architects, designers, occupational therapists, housing providers and disabled grants officers. Upon completion of the course, staff will have an increased awareness of appropriate housing design, and know how to work effectively with and support those clients who have sight loss. The course will develop an understanding of effective communication, design issues and will consider policies and procedure.

The course has been trialed in two authorities and has been thoroughly evaluated and modified in line with comments made. The revised course offers 5 modules:

1. A basic awareness raising self-training package targeted at those who can refer clients for housing adaptations.
2. An awareness raising module suitable for gateway staff who deal face to face with clients who may have sight loss.
3. A module suitable for those involved in making recommendations regarding adaptations in people's homes, focussing on assessment and design recommendations in Adapting Homes.
4. A more advanced design module for those involved in the design of new builds and based on recommendations in Housing Sight.
5. A module focusing on improving policy and procedures through seminar style facilitation and based on the recommendations in In Sight of Home.

The training programme has been formally recognised by the College of Occupational Therapists as pertinent for occupational therapy post qualifying development. The Chartered Institute of Housing also supports the programme.

Conclusion

As a result of the considerable success of the project, RNIB Cymru worked in partnership with the Wales & West Housing Association to construct a pilot home built to the specifications in Housing Sight. This provided a wonderful opportunity to cost and evaluate the recommendations.

The Assembly Government included key recommendations from Housing Sight in their Development Quality Requirements. This means that all social housing built in Wales in the future will have to incorporate RNIB Cymru's recommendations.

A third outcome of the project was that the Welsh Assembly funded RNIB Cymru to raise awareness, encourage good practice, and demonstrate principles of design features and adaptations which will benefit people with sight loss amongst relevant professionals and individuals across Wales. A Housing Training and Development Officer has been employed to run training sessions. The programme is funded, which enables RNIB Cymru to offer them free over a two and a half year period.

References

Rees L. and Lewis C. 2003, *Housing sight: A guide to building accessible homes for people with sight problems,* RNIB and Welsh Assembly Government

Rees L. and Lewis C. 2004, *Adapting homes: A guide to adapting existing homes for people with sight loss*, Welsh Assembly Government (HMSO and the Queen's Printer for Scotland)

Rees L. 2004, *In sight of home: Meeting the needs of people with sight problems,* Welsh Assembly Government and RNIB Cymru

INCLUSIVE PRODUCT DESIGN: INDUSTRIAL CASE STUDIES FROM THE UK AND SWEDEN

Hua Dong[1], Olle Bobjer[2], Peter McBride[3] & P. John Clarkson[1]

[1]*Engineering Design Centre, Department of Engineering,
University of Cambridge, Trumpington Street, Cambridge CB2 1PZ, UK*
[2]*Ergonomidesign, Box 1400, 167 14 Bromma Stockholm, Sweden*
[3]*Ergonomics Specialist, Nortel Networks, Doagh Road,
Newtownabbey, Co. Antrim, N. Ireland*

Two case studies of inclusive product design are presented in this paper: the Selectronic Shower of a UK-based manufacturer and the Bahco Rx Cutter by Ergonomidesign, Sweden. The initiatives of the two projects and their design process were investigated, and the business case discussed. Based on evidence, we concluded that inclusive design is cost-effective.

Introduction

To promote inclusive design more effectively, we need evidence of the business case (Underwood and Metz, 2003). The selection of the two case studies for this paper is based on this criterion. They are:

1. The Selectronic Shower (Figure 1) developed by the Applied Energy Products Ltd., a UK-based manufacturer of water, heating, ventilation products. The Shower has incorporated advanced technology yet has a simple to use interface. It is endorsed by the Royal National Institute of the Blind (RNIB) and is ideal for children, the elderly and less-able people.
2. The Bahco Rx cutters (Figure 2) designed by Ergonomidesign, one of Scandinavia's largest and best-known industrial design consultancies based in Sweden. They have adjustable handle width and spring force – designed to include women users.

Some information about the Selectronic Shower project can be found from Stabler and van den Heuvel (2003). Bobjer et al (1995) and McBride (1995) also published papers discussing hand tool development, using Rx cutters as examples. However, existing literature does not focus on the business case aspect of inclusive design. Dong followed-up the Selectronic Shower project by visiting the manufacturer and interviewing the New Product Introduction Manager who co-ordinated the project. Bobjer and McBride were ergonomics specialists and they conducted user trial studies of a range of cutters, including the Bahco Rx range. So in this paper, we are able to draw the two case studies together with the same focus on the business case of inclusive design.

Project initiative

Since the adoption of inclusive design is not yet wide spread in industry, it is interesting to know what motivated or initiated the two projects.

The Selectronic Shower was developed with the Disabled Discrimination Act (DDA) in mind. The design brief was to produce a high specification shower with operational features suitable for the National Health Service (NHS) and Care Sector market, which

Inclusive product design: Industrial case studies from the UK and Sweden 339

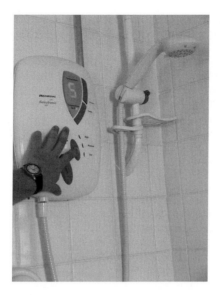

Figure 1.   The Selectronic Shower.

Figure 2.   The Bahco Rx cutters.

complied with current NHS safety guidelines relating to safe showering temperatures. The design of Rx cutters was an initiative of the Bahco Group, one of the world largest handtools producer. Cutters and pliers are intensively used in the manufacture and assembly of printed circuit boards; the manual use of such tools can be 10,000 times per day or more. A number of risk factors are associated with using cutters, such as repetition, posture, force and contact stress; they contribute to cumulative trauma disorders of the operators. In recognition that the design of high quality cutters has been technology driven rather than operator oriented, the Bahco Group developed a series of Rx cutters based on its evaluation of ergonomic hand tools. The Rx Cutter user trial was motivated by Nortel Networks, a major producer of electronic hardware. The ambition of Nortel Networks was to suggest worldwide recommendations on the purchase and use of tools best suited to the user, the production and the quality demands.

The design process

Both the Selectronic Shower and the Rx Cutter projects were developed with input from ergonomics specialists and end-users.

The Selectronic Shower was developed in partnership with the RNIB. At an early stage, the manufacturer approached the RNIB with an initial design model of the shower. The RNIB ergonomists helped validate the design, drawing upon their knowledge and experience of product design for older and visually impaired people. Based on the RNIB ergonomists' early assessment and suggestions, the manufacturer was able to improve the ergonomic features of the model. End-users were involved later: six focus groups (over 50 participants, with the majority of them over 60 years of age) were organised in residential homes and local associations for visually impaired people. They were invited to test the mock-up prototype with fixed buttons, together with an existing shower to provide an impression of button pressure and audible feedback. The RNIB ergonomists observed the process and conducted discussions using a structured questionnaire; they then presented the findings to the manufacturer. According to the New Product Introduction Manager, the in-house design team of the manufacturer had got the design 85 per cent right with the expertise of the RNIB ergonomists before it went out to user testing. Long-term user trials were conducted after the fully functional showers were manufactured. The RNIB ergonomists also helped the manufacturer in making the user instructions inclusive, and training sales people and installers.

The Rx cutters were designed in eleven steps by Ergonomidesign in association with the Bahco Group, these steps are (see more detail from the web site: http://www.lindstromtools.com/ergonomics_step.htm):

1. Preliminary specifications: decide which tool to develop.
2. Market analysis: study existing tools and their market strategy.
3. Background research: research on work-related injuries caused by hand tools.
4. Prototype design: develop the first working prototypes.
5. User test 1: test prototypes with many different kinds of hands.
6. Prototype evaluation and modification: modify the best prototypes.
7. User test 2: ask more users in more countries to test modified prototypes.
8. Final design recommendations: specify detail for a true-to-life prototype.
9. Product specifications: manufacture a small run of "finished" tools.
10. User test 3: validate the final solution and prepare for launch.
11. Follow-up.

These eleven steps have become a corporate design procedure for the Bahco Group. By involving end-users in the design process, the designers were no longer using their own hand dimension as the norm. Also, by introducing ergonomics specialists into the design process, a new priority and focus was given to operator demands and physical requirements. Hence the size, shape and function of the tool and the dynamics of the tool in actual use were better taken into account.

The business case

The Applied Energy Products Ltd. launched the Selectronic Shower in early 2003. Although its price is about a quarter higher than most of the existing showers, it still sold well. In the first year, its sales increased over a hundred per cent, so the manufacturer doubled its volume. The second year, sales continuously grew at a very good level.

The Selectronic Shower is more expensive because it has an integral motorised flow control valve and needs to meet extra requirements for power control and temperature control. In addition, they are treated with Microban, an anti-bacterial addictive which prevents the growth of fungi and bacteria for the life of the shower. Having been approved by the British Electrotechnical Approvals Board (BEAB), it became the first product to meet new industry standard for thermostatic electric showers. It has passed a rigorous testing schedule and complies with the additional safety requirements of the care industry, and was awarded BEAB Care Mark.

Although the Selectronic range has been primarily sold to the Care Market at the moment, the New Product Development Manager believes that it has a potential to reach a much broader market. If the sales volume is increased significantly, the unit price will reduce.

For the Rx cutters, after a 28-month period of testing, Peter McBride, the ergonomics specialist from Nortel Networks strongly believed that all operators who have a need to use wire cutters should be supplied with the Rx range. Although the cost of the Rx cutter is higher than traditional cutters, it can be justified as follows:

- During the user trial period, the average cost of existing cutters was $32.40. In general, these cutters have to be replaced at least four times a year, which equates to a total cost of $129.60 per operator per year, or $10.80 per cutter per month.
- The Rx cutter cost $64.80 and is expected to last at least one year (with constant usage). Of the 26 pairs of the Rx cutters issued, 22 pairs were still in good condition after 28 months of use. So the cost of these Rx cutters is only $2.30 per cutter per month; the 22 pairs in 28 months save up to $5236 in total, which equates to $103.50 per operator per year.
- The traditional range of cutters causes considerable pain and discomfort, while the Rx range is comfortable to use.
- Nortel Networks health care providers reported fewer calls concerning upper limb complaints/discomfort since the Rx cutters had been introduced as the company's regular cutter.
- For traditional cutters, if the spring breaks, the whole cutter becomes inoperative, while the spring in the Rx is a separate component, which can be replaced by the operator.
- Electro Static Discharge (ESD) is a very important production safety measure within the electronics industry, but all traditional tools are not ESD approved. The Rx tools are ESD safe.

Discussion and conclusions

The Selectronic Shower was initiated by the DDA and the Rx cutters were concerned with risk factors associated with using hand tools. They both achieved inclusive design which becomes premier products for all. With advanced technology yet accessible interfaces, the Selectronic Shower is good for less-able people, elderly, children and indeed all who want a high quality shower. The Rx cutters not only accommodate a wide range of user hand size (including women users) but also make reaching and gripping tools much easier – Bobjer et al (1995) reported that users "liked" the Rx tools at the beginning of the evaluation study, and they did not want to give them up by the end of the evaluation period simply because Rx range was so much better.

The involvement of end-users and ergonomics specialists in the design process was critical for these two inclusive design projects. The ergonomics specialists helped to change the focus of design from technology-driven into human-centred as well as from product-focused into context-centred, while end-users helped evaluate prototypes and validate solutions.

There is a need to understand the business case of inclusive design in a broader sense: not only in terms of profit, but also in terms of competitive edge. The Selectronic Shower project has gone beyond inclusive design to inclusive performance, and the manufacturer has been able to jointly set a new standard BEAB for the industry. Bahco's unique corporate design procedure, i.e. the eleven steps, has made Bahco tools premium. On the web site of Lindstrom – a trade name of the Bahco Group – it says: "Some competitors have been able to implement one facet or another of the Lindstrom manufacturing process. Others have tried to copy the form, appearance and even the actual part numbers of Lindstrom tools. However, none has been able to successfully blend all the elements required to achieve the level of performance recognized as a true Lindstrom tool." (http://www.lindstromtools.com/about.htm).

To conclude, there is a business case for inclusive design because:

- It increases long-term profits.
- It improves companies' competitive edge.
- It produces better products for all end-users.

Acknowledgements

We would like to thank Professor Maria Benktzon of the Ergonomidesign for helping identify the Swedish case study, and Barry Tanner for helping with the interviews.

References

Bobjer, O., Bergkvist, H., Jansson, H., and Lohmiller, W.R., 1995, *Development of ergonomically designed pliers for the electronics industry, Sandvik-Lindstrom Rx*, (Ergonomidesign AB, Sweden).

McBride, P., 1995, *Using Ergonomics to Improve Working Conditions and Reduce Costs*, In *Ergonomics Society of Gt. Britain Conference 1995*, (Leicester University, England).

Stabler, K. and van den Heuvel, S., 2003, *The Selectronic Shower: an inclusive design case study*. In *Include 2003*, (the Helen Hamlyn Research Centre at the Royal College of Art, London). 8:296–8:301.

Underwood, M. J. and Metz, D., 2003, *Seven business drivers of inclusive design*. In *Include 2003*, (the Helen Hamlyn Research Centre at the Royal College of Art, London). 1:39–1:44.

SMART HOME TECHNOLOGY IN MUNICIPAL SERVICES; STATE OF THE ART IN NORWAY

Toril Laberg

The Delta Centre, Directorate for Health and Social Affairs, PO Box 7000, St. Olavs plass, N-0130 Oslo

Most of us are going to live in our own homes when we become older, and we will want to feel safe and to be independent. Building institutions with the aim of housing people is outdated. Smart home technology as part of the municipal services has proved to contribute to independent living and safety in homes. The Norwegian experiences with smart home technology started some 10 years ago. Several municipalities have installed smart home technology in residential homes, aiming at the resident's safety and independence. The prospective barriers are linked to the planning, implementation and management of the technology, to the utilisation of the existing technology in private housing and to the building up of services to support the inhabitants of such housing when the need for municipal service arise.

The Norwegian case

The main trend in Norway is to support disabled and elderly people in their own home by giving healthcare, practical assistance and support from the municipality at home. The services are multidisciplinary teams, where nurses, occupational therapists, physio-therapists and home helpers forms a core. The type and extent of services are decided individually. Some services are free of charge, some services the user pays a part of. The use of institutions with the aim of housing people has been reduced, and the institutions are more often used for short term stays aiming at the relief of the family or for rehabilitation.

The trend of building institutions for people in need of medical, practical or emotional care changed during the 1990's, supported by political incentives like cheap loans and grants supporting the building of residential homes. The residential homes consists of flats grouped together, in groups of from 4–5 and upwards. The profit for the municipality is having several users of their services gathered geographically, and saving the cost of running institutions. The profit for the inhabitants is to have the opportunity to live in a flat which is a private home, instead of having to move to an institution when their need for care increases. The laws and regulations controlling the economic and public goods depend on the type of ownership to the dwelling. In a flat, including the new-built residential flats, the resident buys a share or pays the rent. Living in a home entitles the person to apply for free assistive technology from the state. In an institution the clients pay for "bed and breakfast", and the owner of the institution supplies the necessary assistive technology. Apart from these technical exercises, the most important effect of living in a flat is the impact it has on a persons self-esteem and dignity, as there is a great difference in having the role of a premise provider or the role of a client/ patient. Knowledge of empowerment has given us arguments that the role you play might influence your health, providing we define health as wider than absence of illness.

E-mail: toril.laberg@shdir.no

Several local authorities took advantage of modern technology as part of the municipal services, when planning and building the new residential homes. There is no exact register as to how many of the flats are installed with smart home technology, but some 20 000 flats were built. Two surveys have been conducted to collate experience on smart home technology as part of home care services. We found the technology only in the newly built residential flats. This finding inspired us to further investigate the implications of installing smart home technology in existing houses and flats. We performed a retrofitting in the flat of a disabled man, following the ordinary procedures, applying for financial support for assistive technology. The aims were to test out available and suitable technology for retrofitting, the cost and available financial support and the functionality of smart home technology versus environmental control systems. The project is documented in a film called "Smarthus".

Definition of smart home technology

Smart Home Technology is a collective term for information and communication-technology in homes, where the components are communicating through a local network. The technology may be used for monitoring, alarming and executing actions, according to the programmed criteria. The smart aspect is the integrated communication between the devices, and the possibility to generate automatic actions. An automatic action often used is the generation of alarms when something abnormal occurs, or when a normal action fails to appear.

The local network communicates with the external world by telephone or through the Internet, sending messages or alarms to one or more recipients. In a smart home one may integrate:

- Safety; for example alarms
- Environmental control systems, for example remote control or programmed control of doors, windows and lights
- Energy-control-systems, for example adjusting the heating at all hours
- Communication, linked to the telephone or the Internet
- Entertainment, such as television, film and music

Benefits of using smart home technology

As the residential homes were built in clusters, it made it possible to take advantage of modern technology as part of the home services. The focus when installing smart home technology has been on safe and independent living for the residents. Safety is the biggest advantage of smart home technology. The resident, their family or their carers can feel safe due to the versatility of the alarm systems. The benefits for the municipalities are improved quality of services, improved working conditions for employees and to some extent financial benefits.

Safety

There are a number of ways to program the system to trigger alarms, according to the need of the individual person. The alarm may be triggered by the user, like an ordinary call alarm. Alarms may also be triggered when a pre-determined criteria is met, without the person having to take personal action. The alarm may be activated when something irregular occurs, or when an normal action fails to happen. This is called passive alarms, and is used to avoid dangerous or harmful situations. The use of such alarms must be carefully assessed ethically and legally. Alarms are not bells or red lamps associated with alarms in the health and caring sector. Most often the alarms are transferred via mobile phones, as a SMS (short message sender) to the employers.

Incidents like fire, a fall of unfortunate consequences of nightly wandering can be prevented. Sensors to register the weight of the bed can activate the lighting of the route to the toilet, when the bed is left. The technology can also register the time lapse before the person has returned to bed. If too long a time lapse might be connected with a dangerous situation, for instance a fall, an alarm might be triggered.

Sensors in doors and windows register if these are open or closed. For example may the heating decrease when the door is opened during daytime, and an alarm can be triggered if the door is opened during the night. The last instance could be a sign of burglary or a resident on his way out. Ordinary burglar alarms can also be connected to the system, as can with smoke detectors.

Safety can include some negative aspects, especially if the border of surveillance is trespassed. For some people, in particular for the older part of the population, safety based on technology may seem frightening. The use of technology can be experienced as alien and cold, while the regular care from employees may be experienced as "real" care. Others will happily use technology exactly to limit the number of calls, and thus the number of persons visiting them. The main point is that individual appreciations and adaptations must be made.

Independent living

The first and foremost impact on independent living is the principal opportunity to live in your own home, and not in an institution. A typical example of increased independence is the automation of several actions, for example by using a "day-switch", a "night-switch" and a "not-at-home-switch". These switches are programmed so that, at the touch of one switch, the house is set in the desired state. The automation may give increased independence to a person with reduced mobility, as he does not have to move around the whole house in order to turn off several appliances - lamps, computer, coffee-machine - check doors or activate alarms. For a person who tends to forget important tasks, like turning gadgets off or locking doors, the automation may serve as a memory jogger.

Impact on health

There is a lack of systematic research on the use of smart home technology and its impact on health. One could claim, however, that independent living and empowerment are two basic requirements that have an influence upon health. A user said:

> "The technology has made my flat more modern. It is not full of technical aids, and still it has become more convenient to me. It does not make me feel disabled!"
> "I have less pain in the neck and shoulders, because many of my routine activities, like turning off several lamps and open the balcony door has become automatic."

The staff say that the biggest profit is on the qualities of the services they produce:

> "The alarm systems make me more relaxed when I am assisting a person. I know that if something irregular happens with another resident, I will receive a SMS on the mobile phone."
> "The smart home technology results in more quiet and peaceful surroundings, no bells, no checking in on the residents from time to time. This tranquillity makes the demented residents more calm, too. They can be quite agitated if the surroundings are noisy and busy."

The caretaker in a building of residential flats said:

> *"It makes me a more important person, in a way. I have much more responsibility at work now."*
> *"I have even received a diploma for finishing a course in running the system!"*

The next of kin of the residents have reported that they feel confident because of the safety. They are often involved in decisions regarding the use of passive alarms.

The economy

The discussion on economy is closely connected to the quality of the services, and economic advantages have so far not been focused upon. Smart home technology has been introduced as a support, not as a compensation to the caring human hands. Some figures from Norway can highlight the economic aspects.

The first residential home with smart home technology was built for 5 persons with dementia, in 1995. They planned to staff the flats with 2 awake staff at night, based on the experience from other lodgings. On opening the place, they realised that 1 awake staff was sufficient, due to the smart home technology. The suppliers of the technologies estimates the cost of the technology per flat to €2 000 when installed in several flats during building.

The yearly cost of running a place in a nursing home is between €60 000 and €100 000 per person. A study carried out in a municipality documents that the price of giving care in a nursing home is twice the price of assisting people in their own home. The study compared a nursing home with a residential home with residents with equal score on an ADL-index (activities of daily living).

The retrofitting of smart home technology amounted to €30 000. Some half of the sum, the resident was granted from the state, as assistive technology.

Smart home technology may support independent living out of institution, and it seems that this might be good economy for the society.

Prospective barriers

The barriers are more often linked to the infrastructure around the user technology, not to the technology itself.

Planning, implementation and management

The success criteria are closely linked to the user involvement and a multidisciplinary approach. In municipalities where inhabitants, their relatives, employees and the planners are working together, there seem to be a higher grade of satisfaction with the technology. As the technology represents advanced systems with lots of potential, it requires much more involvement than enthusiasm in parts of the municipality. From the level of the local politician, via the responsible leaders of the departments of technology and of home-care to the homehelper, it is important with a joint understanding of the implication of using the technology. Education at all levels forms a base in understanding the potential of the technology.

The utilisation of the existing technology

Smart home technology is frequently offered to all kinds of buyers, installed in apartments with relatively high standards. An increasing number of apartments with a so-called "senior profile" are financed and built privately, incorporating smart home technology. In addition providers and electricians inform that they more and more often are installing smart home technology in private homes. The purpose of these private installations are safety, comfort and home-amusement.

There is a challenge on the technological to develop compatible systems, so that systems installed for entertainment and comfort, can be utilised for safety and independent living, if the inhabitant should need this with age.

Building up of services

The enormous increase in the number of elderly in the years to come, force the responsible planners to look to alternatives of today's organising and staffing of the municipal services. We have to look to how new technology may support the helping, caring human hands. We have for example already seen how smart home technology contributes to more efficient use of staff, by utilising the same staff at nursing homes, residential flats and staff serving the residents in the community at night.

In the future there will be a need to build up home care services to people living in homes with smart home technology, which not are part of a residential cluster. The key area is the building up of wire-less communication between our homes and the home-care-team.

Conclusion

Norway has gained quite some experience from practical use of smart home technology. Smart home technology is adding to a safe and comfortable living environment for everybody, and is becoming increasingly more common to ordinary consumers. For the elderly or disabled person, the benefits are even greater. There is a responsibility resting on governmental bodies and professionals to involve in the future planning of community care, to ensure the optimal use of technology to assist the human helpers that always will be the basis in all care.

LOOK IN, TURN ON, DROP OUT

Alastair G. Gale & Fangmin Shi

Applied Vision Research Centre, Loughborough University, Loughborough LE11 3TU, UK

The ART (Attention-Responsive Technology) research project is developing a system to enable mobility-impaired individuals to access technology efficiently. The system monitors both the individual and any ICT devices in his/her environment. It then uses the individual's gaze direction to determine to which ICT device, if any, they are potentially attending. This information is relayed to a user-configurable control panel, which then displays only those controls that are appropriate, both to the user and to the particular device in question. The user can then choose to operate the device if s/he wishes. The initial ergonomic challenges in the development of the ART system are described.

Introduction

The title re-iterates Timothy Leary's 1962 exhortation to individuals to forego their normal working routines and follow his psychedelic lifestyle approach. Here the quotation is used with similar sincerity but in a markedly different fashion – to describe the operation of a control system (ART – Attention Responsive Technology) under development which is designed primarily for use by disabled users (Gale, 2005a).

The system selects a device for control by means of the user simply gazing at an identifiable electronic device (an ICT device or "object") that they want to operate ("look in") so enabling it for potential use. Then the user can operate that device if they so wish ("turn on"). One of the key issues for the approach is the question of both selection and operation accuracy and the potential for errors ("drop out"). It is therefore this last issue which forms the key interest considered here.

What are the potential ergonomic issues which will impact on the actual deployment of such a control system and that will give rise to possible errors? An "object" here is taken to be any ICT device or related electrical piece of equipment in the user's environment and which could be operated by the ART methodology. An error is seen as the selection by the ART system of the wrong object, non-selection of an intended object or incorrect operation of a correctly selected object.

ART

The ART system and its operation have been described in outline previously (Gale, 2005b; Shi et al., in press; Shi et al., 2006). A user's eye gaze behaviour is monitored automatically and when this is measured to fall on a particular ICT object then the user is offered a dedicated interface simply for that object. Thus, where the user looks governs the selection of an ICT object for possible operation but not its actual operation, hence overcoming inadvertent operation of objects (c.f. Jacob, 1990). If s/he wishes to operate the selected object then they can do so by various interface methods tailored to the specific needs of the individual user.

What then are the potential sources of error in the use of such a control system? This paper describes the main issues which are foreseen as being important.

In parallel, a survey is currently under way of potential end users to glean information about what they would like such a system to offer them and also the issues that they foresee in utilising the ART system.

Challenges

In the initial stages of development it is assumed that our target user is:-

- Mobility restricted in some way, although possibly moving about the environment – for instance, in an electric wheelchair.
- In terms of physical movement – potentially only capable of moving their eyes but could have some other restricted physical movement (e.g. limited hand movement but with no real physical strength with which to operate interfaces).
- Wishes to operate and control various systems autonomously without the need for another person's intervention.

For the ART system the difficulties anticipated relate to:-

- The potential user
- The environment
- ART system accuracy
- ICT objects

Potential users

Users of the ART system are targeted as being anyone with mobility problems or a mobile person in an environment where, for some reason, they are restricted in their physical movements. However, the key driver for the development is individuals with severe mobility issues.

The extreme of such conditions are those persons with Locked in Syndrome (LIS), as first identified by Plum and Posner, 1966. Such quadriplegic patients are very rare and have no mobility, except for the voluntary use of eye movements and blinking; however the patient remains alert, generally with a good prognosis for long term (>5 year) survival after the first year (Hemsley, 2001). Murthy et al. (2005) report a recent case which indicates the potential rapid development of the syndrome.

There are different kinds of LIS: "classical" locked in syndrome where patients can only move their eyes vertically as well as their upper eyelids, "incomplete" (where other eye and eyelid movements are possible) and "complete" paralysis (no movements). Patterson et al. (1986) reviewed 139 cases of LIS and emphasised the need for an effective communication system for such patients, an issue well addressed by COGAIN (2004).

The ART system would potentially then be of use to a wide cross section of individuals, including other patient groups (e.g. spina bifida patients) where the person has some degree of restricted mobility but, importantly for the project, has voluntary control over their eye movements.

User environment

A key question is in what environments would ART be used by potential users? Two eye movement monitoring systems are used (Gale, 2005b) in the development of the ART system so that the product will be of use to a great number of users and in a wide range of situations. Both methods utilise the principle of directing incident infra-red

(IR) illumination at one or both eyes of the user and then monitoring reflections of this light from the front and other ocular surfaces of the eye/s. The level of IR used in such commercial eye movement systems is very low and well within acceptable standard safe limits. However, it is easy for such an approach based on IR to be overwhelmed when used externally, for instance on a bright sunny day, by the IR naturally present in sunlight which can cause the system to fail to monitor the user's eye movements appropriately so giving rise to potential error. Hence such systems, whilst operable externally, are most easily used indoors.

Consequently, it is initially envisaged that three environments are probably important, namely; living, working, and the kitchen. Therefore, the ART approach would have to deal adequately with any/all ICT objects in these environments that a particular user may wish to control independently.

ART accuracy

The ART eye movement system must be capable of accurately identifying where the user is looking so that the appropriate object is selected. The key question here is just how accurately must the system measure eye gaze behaviour?

All eye movement recording systems produce some output indicating the point of gaze of a user with some degree of accuracy. There is always some inherent error in measuring eye gaze depending upon a number of factors including the particular recording technique used. Equally, all systems require calibration before use by an individual. To do this the user looks at a matrix of points in space (typically arranged in a planar 2D square about 20–25° visual angle in size) and their gaze locations are recorded. Software then compares the known locations of the calibration points to the system's recorded gaze locations and performs a transformation between these two matrices.

After calibration then subsequently all recorded raw gaze locations are first transformed by the software according to the outputs from the calibration trial. For accurate eye gaze measurement then good calibration is a necessity. However, good calibration usually requires some degree of compliance by the user and this may not always be possible for the deployed ART system. For instance it is always difficult to know when an individual is actually fixating on a calibration point when instructed to do so and this is particularly the case if the user has communication difficulties.

Consequently, can poor calibration or even no calibration be used in the ART approach and if so is the resultant accuracy level of user gaze location acceptable? One of the advantages of the ART system is that objects for the user to select between can be spatially distributed about the environment, therefore in principle it is possible to array objects around so that relatively poor estimated measurements of where someone is looking could be used to select one object from many.

The potential for error therefore depends upon the number of objects, their size and spatial arrangement. In order to be able to distinguish between a user attending to one of several overlapping objects, or objects in close proximity, will require quite accurate eye gaze recording.

ICT objects

Clearly it is essential to know which ICT objects are important to the specific user as well as knowing which ICT objects potential users currently use and which they would like to additionally use through implementing the ART system. The frequency of operation of each object in a given time period is also important and could lead to error. For instance, electrically operated curtains may only be opened and closed once a day by the user, therefore not necessarily requiring a fast responding control mechanism to the user's command.

The speed of selection and operation of a particular object may be important for some and not for other objects. The current speed of selection and operation of objects by a user's current communication/control system would be taken as a metric against which to measure the ART system's performance. The failsafe operation of an object is also an important consideration. ICT objects may also need to be operated by others apart from the target end users and so other additional interfaces may be required for their helpers or family members.

An advantage of the ART approach is that an object can be selected for operation by such an eye gaze contingent system when the object is at some considerable distance from the user, as long as the measured gaze location – including an allowance for overall measurement error (essentially the object's "operational zone") – does not overlap the operational zone of an adjacent object. For some objects this will be eminently suitable (e.g. opening/closing curtains or operating room lighting levels). For other objects, whilst these could be selected at some distance, the actual useful operation of them would require that these be near to the user or else errors will result.

Conclusions

The ART system is just in its second year of a three year development process and progress to date has concentrated upon finding state of the art technical solutions to the overall research problem. Potential user opinions on the utility of such a system have been canvassed and these will be presented elsewhere. Here, we have highlighted the issues which are foreseen as particular challenges to be overcome. Whether the user opinions simply concur with these or add yet further significant issues which need to be addressed is an issue for the future.

Acknowledgement

This research is supported by the ESRC PACCIT (People At the Centre of Communication and Information Technology) Programme.

References

COGAIN Communication by Gaze Interaction (2004) EU IST Network of Excellence http://www.cogain.org.
Gale A.G. 2005a. The Ergonomics of Attention Responsive Technology. In: D. de Waard, K. Brookhuis, R. van Egmond, and T. Boersema (Eds.), Human Factors in Design, Safety, and Management, Shaker Publishing, Aachen.
Gale A.G. 2005b. Attention responsive technology and ergonomics. In P.D. Bust & P.T. McCabe (Ed.) Contemporary Ergonomics 2005. Taylor & Francis, London.
Hemsley Z. 2001. Locked-in syndrome: A review. *CME Journal Geriatric Medicine* **3(3)**:114–117.
Jacob R.J. 1990. What You Look at is What You Get: Eye Movement-Based Interaction Techniques. In Human Factors in Computing Systems, (CHI '90 Conference Proceedings, ACM Press), 11–18.
Murthy T.V.S.P., Pratyush Gupta, Prabhakar T., and Mukherjee J.D. 2005. Locked in syndrome – a case report. *Indian J Anaesth*. **49(2):** 143–145.
Patterson J.R., and Grabois M. 1986. Locked-in Syndrome: A review of 139 cases. *Stroke* **17(4):**758–764.

Plum F., and Posner J.B. 1966. The diagnosis of stupor and coma. Blackwell Scientific Publications, Oxford, 92–3.
Shi F., Gale A., and Purdy K. 2006. Sift Approach Matching Constraints for Real-Time Attention Responsive System. In Proceedings of the 5th Asia Pacific International Symposium on Information Technology Conference (in press).
Shi F., Gale A., and Purdy K. 2006. Eye-Centric ICT Control. In P.D. Bust & P.T. McCabe (Ed.) Contemporary Ergonomics 2006. Taylor & Francis, London.

LIVE FOR TOMORROW – FUTURE-PROOF YOUR HOME

Alison Wright

*Easy Living Home Ltd, The Hermitage Design Studio,
Holmes Chapel, Cheshire CW4 8DP*

Home improvement is a high priority for many retired people, but older consumers rarely feature in the magazine and TV programmes that specialise in the subject. Without good advice and examples we all find it difficult to envisage what could be done in terms of design to make our homes an attractive and enjoyable base for an active retirement. Traditionally home adaptation has tended to be a reactive solution to diminishing independence based on a medical model. The results are often unattractive and stigmatising to the user merely highlighting their disabilities and loss of independence. In this paper it is argued that the traditional medical model of home adaptation for older people is no longer appropriate and a fresh approach is required where everyone, but particularly older people, are self-empowered to be smart consumers and live for tomorrow by "future-proofing" their homes.

Home improvement for independent living

This paper is based on the results of a two year research project "Home Improvements for Independent Living", carried out through the Helen Hamlyn Research Centre, based at the Royal College of Art and jointly sponsored by the Centre and the Lifespan Trust. Inspired by the original Lifetime Homes project, built by Habinteg Housing Association in 1994 and sponsored by the Joseph Rowntree Foundation, the research explored how older people adapt their homes to suit their changing needs. The Research focussed mainly on kitchen and bathroom design, rooms that focus groups of older people had identified as problem areas and involved visiting homes around the UK. The research methods used were:

1. Interview – home owners/occupiers were interviewed about their lifestyles, any problems experienced with the design of their homes and any changes they had made to their homes to alleviate problems or take into account their changing needs.
2. Observation – home owners/occupiers were observed preparing and cooking in their kitchens. It was interesting that what the householder perceived themselves as doing and what was observed were often two different things. For example one candidate who purportedly had no problems with the design of her kitchen was observed as kneeling on the floor in order to empty her oven.
3. Survey – each kitchen and bathroom was measured and design inadequacies or defects were noted.
4. A review was carried out on existing literature available to consumers on home design for the 50+ age group and to identify "inclusively" designed home products.

The aim of the research was to understand where current home design fails older people and to identify what changes older people make for themselves as coping strategies in order to remain independent. Common areas of difficulty and concern identified from the bathrooms and kitchens were:

Safety
Bending and stretching
Wrist and hand dexterity

From the review of consumer literature it was identified that:

1. Best sources of information were outlets advising people with disabilities i.e. Disabled Living Foundation.
2. Older people do not go to disability outlets as a primary resource, for help with the design of their homes because they do not consider themselves disabled.
3. Despite a wealth of product information in the disability outlets, inspirational interior design images were hard to find.
4. Older people look for inspiration, from friends and family first, then in mainstream DIY stores and home improvement magazines or television programmes.

With these factors in mind a theoretical re-design of each of the kitchens and bathrooms visited took place. The data collected was used to guide a set of proposals for interior design solutions by combining carefully selected mainstream products, which would help to alleviate as many as possible of the original problems identified by the householders. The aim was to design-in accessibility without compromising style or affordability, directly challenging the traditional prescriptive "medical model" in which an Occupational Therapist prescribes purely functional solutions, often overlooking aesthetic consideration in favour of Local Authority budgetary constraints. The new "consumer model" aims to offer a pro-active solution, whereby the householder remains in control of the design process and the solution is a balance of function with aesthetic. The new designs were presented to the householders for feedback with interesting results:

Denial

A proportion of householders denied that the issues raised during observation were a problem. For example a particularly tall lady was observed dropping to her knees to take hot dishes out of her oven, an action she considered a normal compromise for being tall, rather than being a fault in the design of her kitchen. Several candidates were in denial about their own ageing, refusing to acknowledge that the research findings could benefit their own lifestyles, although they did see the relevance for older friends or relatives.

Peer group interest

All the research subjects were interested in the findings of other households, especially the "before" and "after" images of the re-designs. They equated the "before" and "after" format of the feedback with the entertainment-based media formats of magazines and television shows.

Age related passivity

Candidates felt that some of the theoretical design suggestions made would be helpful to them however, when questioned on whether they would go ahead and implement the changes in their own homes, there was a marked age related difference in the response, unrelated to affordability. Candidates aged between 50 and 70 years old said they would consider implementing some or all of the design ideas, whereas candidates aged 70+ considered it unlikely that they would implement the designs, despite confirming that they envisaged the changes would benefit their daily lives. The reasons they gave are broadly summarised as follows:

Spending priorities

On reaching 70 the emphasis of people's spending appeared to change from spending on themselves, their home and lifestyles, towards spending on children/grandchildren, saving towards an inheritance nest-egg or saving to offset future care costs.

"Hassle" factor

This term was identified during the research as a perceived fear of, or inability to cope with, major disruption in daily routine and the perceived disruption of the physical home environment caused should building works be undertaken in the home. An inability to cope with the financial and organisational decision making required to carry out such a project, a lack of design inspiration from which to envisage the results and not knowing who to engage to do the work combined with a general fear of "being ripped-off" by builders.

Story telling

The householders were at initially curious about the designs presented to them but they were unanimously defensive of any observed behaviour identified during the research. For example, the safety hazard of an electric kettle being stored and operated from a metal sink draining board. The retort to this observation was "I've been doing it that way for 20 years and I'm still here!" However, when offered examples of other people's "before" and "after" interior design observations and design scenarios, the candidates were immediately transformed into pro-active consumers, immediately identifying with many, if not all of the observed problems and the solutions.

The feed-back sessions made it clear that older people will reject what they perceive to be a "prescriptive" approach to problem solving. It was also clear that if the design solutions were to be effective in helping to keeping people independent at home, then the way the design ideas were to be delivered, the vehicle developed would also have to be informed by the new "consumer" model. In other words, a design template, or tool would have to be designed for use by older consumers themselves and their families, as proactive consumers at one end of the scale and by design professionals and housing providers at the other end of the spectrum of users. The language used for this next stage of the research would crucial because it would have to be understood by consumers and professionals alike and, if successfully identified, could potentially be used as a design communication tool between client and contractor. The term Future-proofing emerged from the research and was carried forward as a description of the design concept.

Brief summary of research conclusions

1. People recognise and accept their own needs more readily through images of design examples implemented by their peers.
2. Changes to the home need to be implemented before the age of 70, preferably no later than at retirement, while their decision-making powers remain active.
3. Encouraging 50+ consumers to be proactive in this area is a question of marketing, not prescription using upbeat terminology such as stress-free, smart consumer, wise investor, stylish.
4. The consumer offer has to be less medical, more inspirational, offering aesthetic and budgetary choice. It has to be promoted through mainstream outlets e.g. DIY retailers, magazines, television.

BBC2 "Home Front"

The theoretical designs from the research findings were transformed into practical outcomes when in 1998, BBC2's first interior design programme, "Home Front", commissioned the redesign of a kitchen for a retired couple using the Future-Proof concept. The results were broadcast to an audience of 4.6 million viewers with a "Facts Sheet" supporting the programme in "BBC Good Homes" magazine. Following this success a further commission was forthcoming to redesign the bathroom for a young family with relatives aged from 4 to 84. The format of the television make-over show proved to be a popular vehicle for delivering the Future-proofing message, with the educational element of designing for the future being delivered as palatable entertainment. The impact on the audience was clear with viewers contacting the show to enquire about where to purchase the products used. The power of television was also clear in that the designs attracted good press coverage and a number of newspaper and magazine articles followed, including a regular "Easy Living" column in a home and garden magazine, identifying mainstream home products using Future-proof criteria as identified in the original research.

However, the project was short lived due to changes in the format of the "Home Front" show. Further approaches to the broadcasting industry were rejected on the grounds that the viewing public was not interested in anything to do with age, compounded by the misconception that older viewers were in the minority. The problem was further compounded by copyright agreements in broadcasting which have prevented the Future-proofing "Home Front" material from being reproduced on any level, including printed material, so the images of the "Home Front" designs have been archived and the Future-proofing message is prevented from reaching the screens of consumers by ageist "gate-keepers", most of whom were in their late forties or early fifties, within the broadcasting industry.

Creating a bridge between older consumers and design professionals

Dissemination of the research findings was influenced by the "Home Front" experience which confirmed the appropriate "language" to be that of images, rather than words. In response, the research findings were developed into the website www.futureproofhome.co.uk incorporating:

Inspiration – to stimulate the user through a visual medium, enabling them to identify their own future needs and aspirations through an interactive, approach.

Information – to support the user's design choices by collating product information in an accessible format, highlighting and promoting existing products, considered to have inclusive elements, identified through the research findings.

The result was a visual design tool which empowered older people, their relatives and friends, to interpret their own requirements, in terms of making changes to their homes on a DIY basis and as a visual "language", or "bridge" to interpret and share their design ideas and requirements with design/construction professionals who will implement the changes for them. The research highlighted the common faults of home design, which needed to be addressed. The web site addressed these areas with design inspiration and product suggestions showing people how best to match living environments to their changing needs and lifestyles, as they reach retirement age.

Collaboration with industry

One outcome of the research has been collaboration with a leading Housing Association on two "showcase" Lifetime Home properties in Bradford, in which the theoretical outcomes of the research are once again being translated into practical interior design solutions (due for completion April 2006). The aim being to demonstrate to design and construction industry professionals in this sector, that function need not compromise style and affordability. This project is the natural conclusion for the research, because it aims to build upon the Lifetime Homes project from 1994, which was the original inspiration behind the research.

A further outcome has been collaboration with a leading bathroom manufacturer on the design and development of inclusively designed bathroom products, to directly address the issues raised in the research. This collaboration was the catalyst for the first private partnership between the design and occupational therapy professions. The partnership aims to build on the findings of the initial research, by introducing inclusive design principles into the training of Occupational Therapists who specialise in housing adaptation. The original website is being upgraded and supplemented by a catalogue of inclusive interior design ideas and products, to meet the needs of non-web literate consumers.

In order to achieve these aims a hybrid business model has been developed to act as a bridge between the existing medical model, in preparation for changes in legislation around home adaptation, and the newly emerging consumer model. The business aims to improve the outcomes for older people who require the support of professionals to meet their changing needs at home, whilst creating a vehicle to match pro-active older consumers with inclusively designed home products.

TAKING THE TABLETS HOME: DESIGNING COMMUNICATION SOFTWARE FOR ISOLATED OLDER PEOPLE AT HOME

Guy Dewsbury, Peter Bagnall, Ian Sommerville,
Victor Onditi & Mark Rouncefield

*Computing Department, Lancaster University, Infolab 21,
South Drive, Lancaster LA1 4WA*

Designing new technology for older people requires the ability to translate specific requirements from a heterogeneous group of people into a suitable application framework. Currently technology for the home is limited to sensor based systems that are supposed to support activities of daily living but omit emotional needs from the design requirements (Down 2005). This paper describes the work of the authors in developing an informal communication tablet computer based tool that is designed to decrease isolation and loneliness. We describe the person-centred design process of the technology and overview the process of developing CATCH a checklist to support the design of assistive technology systems. The paper proposes the link between the design process and the use of a dependability assessment as an aid to translating design requirements into real world designs.

Introduction

As the proportion of older people continues to increase worldwide, we are faced with a number of challenges to enable older people to remain at home and still have a quality of life. Current focus in this area tends to centred on technologies that support alert systems which can assist in preventative medicine by preventing falls or major illnesses. Technology use is being encouraged by the Government as a means of keeping older people in their own homes and out of hospitals or care institutions (DOH 2005). These technologies tend to be based on medical rationales and tend to not be user-centred in design or implementation.

This paper charts the work partially funded in the UK by the Engineering and Physical Sciences Research Council as part of the DIRC Interdisciplinary Research Collaboration (www.dirc.org.uk) and partially funded by Microsoft Research as part of their "Create, Play and Learn" collaborative research programme in which we have developed software that runs on a tablet computer which is designed promotes communication between older people. On this project, we worked with a number of older people in the community to design a technology system that would be suitable for them. In the process of this investigation, our original conceptions of designing some form of high tech system was found to be misguided by the participants whose concerns were mundane, relating to undertaking everyday tasks. Our investigations found that, even in housing specifically designed for elders with available social facilities, older people felt lonely and isolated. They used the artefacts such as the television as a means of avoiding the consequence of their isolated experience, the television taking the place of the lost partner, in a silent static form.

As a response to the isolation that many of the participants were experiencing, we started to investigate how we could make use of computer-mediated communications as a way to facilitate non-intrusive, informal communications between older people. Critical to this was the development of a system that was both usable by people suffering from many of the normal infirmities of old age and which did not look like a conventional computer with all of the preconceptions that this entailed.

Design issues

Our intention was to design some form of technology to support older people but through out work with older people we realised that the heterogeneity of this population is considerable, hence any single design would automatically fail to meet the requirements as many people would need to accommodate themselves to use the technology. We intended to make a technology that was useable, useful and intuitive and thereby avoid the issues of technology abandonment (Pape et al 2002). To this end, we employed a number of field research methods to enable the older persons to participate and for their views to be heard. These included an adapted form of cultural probes, observations, interviews, questionnaires and technology tours (Dewsbury et al 2006). Through these research tools we built up a picture of each participants current lifestyle and activity patterns. The qualitative research demonstrated that the concerns of the participants, who ranged in ages from 54–96 years old, were often centred on mundane activities such as washing curtains, sweeping the yard etc. The findings also highlighted the ergonomic issues related to designing for the participants'. The participants wanted some form of tool to promote communication and assist in decreasing loneliness.

Whilst designing a computer based communications tool, there are several competing considerations (Fiske et al 2004). Many older people are technophobic, unfamiliar with modern computers or somewhat reticent about learning to use them, so ergonomic design is essential to ensure usability (Dickinson et al 2003). Some older people have limited space in their homes and they might find difficulty in reading from a small typeface and thus require a larger screen. The options available currently are desktop PCs, Laptops, Tablet PCs, and handheld devices such as the Palm. Desktop PCs can occupy too much space and be importable, Laptops can be awkward to use, and some older people might find them too heavy to move around comfortably. Palm handheld sized devices can be too small for some older people to read easily. This leaves Tablet computers as the best option, specifically slate style tablet PCs. The one we actually being used weighs 1 kg, which we have found to be generally acceptable.

Tablet PCs have few hard buttons; they have been disabled so that they do not produce confusion with the users. Input is via a stylus, which is somewhat familiar to older people since it looks and behaves in a similar way to a simple pen. Tablets also generally include 802.11 wireless networking, which makes connecting to the Internet physically much easier than having to deal with wires and allows the user to walk around with the Tablet without losing connectivity.

For the software platform, the most obvious choice might be to build applications on a standard OS such as Windows XP. However, these contain numerous interaction options, which can make them hard to learn and could overwhelm some users. It was decided to build a system which sits on top of Windows XP, but provides a simpler user experience, hiding the complex Windows GUI. The software design that emerged over a period of a few months was inspired partly by instant messenger systems. The ability to see quickly who else was available on the system was seen as an essential to allow discrete communication. An important aspect of instant messenger is the ability to advertise "status". Typically systems allow "Online", "Busy" and a small number of other status messages. However, it seemed unnecessary to restrict older people to such a small set of status messages. Instead a simpler and richer option was to provide a small area where they could write a message which would be seen by all their friends and family. Messages can be changed just by wiping the area clean by tapping a button and writing the new message. Early trials of this feature suggested this was both simple enough to be readily understood, and enjoyed. This design meant that most of the screen would be taken up with an "address book", but since the device is focussed on communication, and helping people to see who else is available to potentially communicate with, this has proven a successful option.

The other feature present on the first screen (known as the Chooser) is a list of activities. Currently we have only tested only one, namely Chat, which allows synchronous written communications. To start an activity the users taps on the activity's button and on the other participants, and then taps "Start". The start button allows users to change their minds if the accidentally tap the wrong thing, and also makes the order in which choices are made (people first or activity first) more flexible. The Chat activity design is aimed at being simple but powerful and presents an interface which looks much like a shared whiteboard. The various participants can write and draw, and see each other's writing and drawing almost instantly. This design helped to avoid a problem with instant messenger style applications, namely threading. In instant messenger it is common (Isaacs et al 2002) to see more than one conversational thread. For younger people who are familiar with the technology this is generally acceptable, but even for them is can become confusing and hard to follow. The shared whiteboard style design avoids this since responses are likely to be made near to the original utterance, so the spatial layout can be used to alleviate the confusion threading might cause. Once the whiteboard fills up the design relies on a paging system. This was instead of a scrolling system, which can be physically awkward with a stylus. Making sure the software was designed in a way that was appropriate for the hardware was a major consideration of the design.

The software we have developed also comprises a game of Freecard, which is a non-rules based card game software. This allows the users to chat about the rules and how the game should be played. Surprisingly or at least surprising to anyone who spends any time observing the daily life of older people, little attention has been paid to the role of games in the lives of older adults. Our fieldwork has highlighted how games – crosswords, card games, Scrabble etc. – were an important and valued part of their individual and social lives. Games appear important for a number of reasons. Firstly, as a source of entertainment and enjoyment; secondly as a mechanism to develop new social contacts and reinforce existing friendships; and finally as therapy, slowing down memory loss and maintaining motor skills as well as improving self-esteem and independence. We see particular benefits arising where games facilitate social interaction. To support both individual and social game playing requires research to help understand the most effective means of interaction with computing platforms and how we can support, for example, people talking during games, player awareness, coordination, etc. Freecard and the Chat applications provide a vehicle to stimulate conversations and interactions virtually, which might lessen the isolation experience by many of the older participants. These systems are currently being installed in people's homes for a full evaluation. The applications Chat and Freecard will be followed by other applications determined by the participants in due course.

CATCH

The dependability model for assistive technology (AT) systems developed by Dewsbury et al (2003) has been developed as a basis for discussing and analysing home technology systems such as the Tablet software. However, we believe that models should be of practical use and so we have used the dependability model (See Dewsbury et al 2003) as a basis for a specification method, used by professionals who are specifying and designing AT installations for disabled people. The dependability model breaks down technical systems into four main components namely, Fitness for purpose, Trustworthiness, Acceptability, Adaptability and each component of these can be broken down into further elements as well as considering things from a human and system perspective.

A product of the fieldwork with the older people was the development of the CATCH method (www.smartthinking.ukideas.com/CATCH.html) which highlights dependability issues of AT systems that specifiers should consider when proposing a technology for an older or disabled people. The basic process of AT provision that we assume is that the system is specified by social care professionals and they work with technical advisors and installers who chose the particular devices that make up the system. As vast arrays of devices are potentially available, it was not our intention to recommend what model of device should be chosen.

Our aim was to design a systematic approach whereby a professional could combine their knowledge of the needs for technology with our dependability model thus leading to more dependable installations. The need for dependability was evident from a number of studies carried out by one of us (Dewsbury et al 2004) where it was found that a very high percentage of installed AT systems were either non-operational for most of the time or were never used. Phillips and Zhao (1993) suggest that the non-use of AT is due to one of four reasons:

- lack of consideration of user opinion and selection,
- simplistic device procurement,
- poor device performance,
- changes in users needs or priorities.

As our method was intended for use by professionals working with disabled users and their relatives, we could not suggest a specification method that required detailed technical knowledge or the use of specialised notations. We therefore decided to adopt a checklist-based approach and this led to the development of the CATCH (Collection of Assistive Technology Checklists for the Home) method. CATCH is presented to users in a handbook and the introduction in that handbook sums up our overall goal:

"The CATCH method is a support tool for assessing whether a person can benefit from an AT installation. It is a way of highlighting some of the key issues that might be forgotten in an AT system. Most importantly it is a way of producing a better, more dependable system for the user that meets explicit and implicit needs."

CATCH includes a general discussion of AT dependability issues, design guidelines and structured checklists, which collect information about a user and their setting. The process of using the checklists assumes that a professional has identified a requirement for technology and has made some initial suggestions for the support required. The initial checklist is used first to help the designer decide if this is really the best solution and then to identify the relevant dependability attributes followed by the system assessment checklists. These allow the requirements specifier to assess whether or not the proposed solution is likely to be dependable in the particular environment where it will be installed. If not, the method prompts the designer to propose alternative solutions. On completion of this process, the designer is then able to write a equipment and installation specification that can be used for equipment procurement and deployment.

The evaluations of CATCH have demonstrated that there is a distinct need for an instrument that can determine the efficacy of technology (Down 2005). A further aspect of CATCH is that it demonstrated that the dependability model can be adapted in the real world to provide a useful instrument.

Conclusion

We have presented two forms of user-centred design approaches that complement each other and were developed under the same projects. CATCH was initially developed to support technology systems but was also used in the development of the Tablet software. For our participants, the tablets might be a useful method of promoting communication and a virtual method of alleviating feelings of isolation. Throughout both of these projects an ergonomic approach was used ensuring a person-fit to the intended technologies. By considering the needs of our participants and directly reflecting them in the design process has led to a more efficient design which is clearly useable by the participants.

References

Dewsbury G, Sommerville I, Bagnall P, Rouncefield M, and Onditi V, 2006, Software Co-Design with Older People, in *Proceedings of Cambridge Workshop on Universal Access and Assistive Technology* (CWUAAT) 2006.

Dewsbury G, Clarke K, Rouncefield M, and Sommerville I, 2004, Depending on Digital Design: Extending Inclusivity, *Housing Studies* Vol 19, No 5, September 2004.

Dewsbury G, Sommerville I, Clarke K, and Rouncefield M, 2003, "A Dependability Model of Domestic Systems", in Anderson S, Felici M and Littlewood B (Eds), *Computer Safety, Reliability and Security: 22nd International Conference, Safecomp 2003, Proceedings, Lecture Notes In Computer Science,* 2788, Springer Verlag, Heidelberg, pp103–115.

Dickinson A, Goodman J, Syme A, Eisma R, Tiwari L, Mival O and Newell A, 2003, Domesticating Technology: In-home requirements gathering with frail older people In Stephanidis C. (ed) *10th International Conference on Human – Computer Interaction HCI* (22–27 June, Crete, Greece 2003) 4, pp.827–831

DOH (Department of Health, Older People and Disability Division), 2005, *Building Telecare in England,* DH Publications

Down K, 2005, Anxiety grows over lack of critical debate, Society Guardian, 19 Oct 2005.

Fiske A, Rogers A, Charness N, Czaja S, and Sharit J, 2004, *Designing for older adults: Principles and creative human factors approaches*. Boca Raton, Florida, CRC Press.

Isaacs E, Walendowski A, Whittaker S, Schiano D J, and Candace K, 2002, The Character, functions, and styles of Instant Messaging in the Workplace, *CSCW 2002,* Nov 16–20, 2002.

Pape, L-B, Kim J, and Weiner B, 2002, The shaping of individual meanings assigned to assistive technology: A review of personal factors. *Disability and Rehabilitation, 24* (1/3), 5–20.

Phillips B, and Zhao H, 1993, Predictors of Assistive Technology Abandonment, *Assistive Technology*, 5(1): pp36–45.

THE BENEFITS OF ADAPTING THE HOMES OF OLDER PEOPLE

Peter Lansley

The University of Reading, Whiteknights, Reading RG6 6AW

If more older people are to enjoy the quality of life which comes from living in their existing homes, then housing, care and social services providers have to be convinced that investment in adaptations and assistive technology (AT) will be more economical than the provision of traditional care services. The adaptability of properties varies according to many design factors and the needs of occupiers, so do the requirements for care services. The study showed that adaptations and AT could substitute and supplement formal care, and in most cases the initial investment would be recouped through subsequently lower care costs within the average life expectancy of a "user", and for half of the users within reduced life expectancy due to ill-health. The findings add considerable weight to current initiatives aimed at the more extensive provision of AT.

Introduction

The adaptation of the home and the provision of Assistive Technology (AT) are attracting much attention as a means of helping older people maintain their independence. The hope is that new developments will be widely adopted and will lead to savings in the cost of health and social care by enabling the use of AT in place of direct care services, such as home nursing, domestic care, shopping and meal on wheels (Royal Commission on Long Term Care, 1999; Audit Commission, 2000; Office of the Deputy Prime Minister, 2003).

The study, undertaken with the financial support of EPSRC, considered the adaptability of a wide range of properties in the social housing sector, typical of those occupied by older people, and the cost of providing adaptations and AT, as well as an investigation of the acceptability of a wide variety of these to a sample of older people. In particular it included an examination of the extent to which adaptations and AT could substitute for and supplement formal care in terms of cost and enhanced quality of life. The study is reported in Tinker (2003) which covers some of the implications for policy; Lansley et al (2004) which considers the technical aspects of assessing care needs and the adaptability of homes, the associated costs and the implications for housing providers and the methodology for combining the cost of care, adaptations and AT; and McCreadie and Tinker (2005) which considers the experiences and views of older people of adaptations and AT. The research was carried out by a team from King's College London and the University of Reading, comprising specialists in social gerontology, surveying, occupational therapy, rehabilitation and finance. They followed the definition of AT used by the Royal Commission on Long Term Care – *"an umbrella term for any device or system that allows an individual to perform a task which they would otherwise be unable to do or increases the ease and safety with which the task can be performed"*. This covers building adaptations and items of AT which have to be installed and fixed to a building (collectively termed adaptations) and portable AT, items of personal equipment used by an individual (termed AT).

Table 1. The health and functional impairments of the users.

User	Age	Initial Condition and Condition Five Years Later	Disability Scores	
A	78	Arthritis, mild sight loss then Further deterioration	1.3	6.3
B	75	Diabetes, mild sight loss, occasional fainting and falling, reduced strength, some forgetfulness then Further deterioration, angina, occasional falls	1.5	8.6
C	70	Parkinson's disease, occasional falls, hearing loss then Further falls	4.9	9.5
D	75	Chronic obstructive pulmonary disorder, reduced strength in left arm and hand then Decreased lung function, large weight increase, arthritis	6.9	8.8
E	78	Diabetes, loss of sensation, weakness in limbs then Further deterioration, amputation of one leg below knee	6.3	12.6
F	80	Stroke (not severe), poor eyesight, hearing loss, multiple impairments then Further stroke, complications following broken hip, deterioration in sight, hearing	5.1	13.1
G	70	MS since 50 then Further deterioration	18.5	19.1

Costs

Drawing on OPCS data on disability prevalence in older people (e.g. Grundy et al, 1999) profiles of seven standard users were developed, Table 1. Each detailed the user's functional and sensory capacity and covered a full spectrum of needs, excluding cognitive impairment. Each profile was considered at two points in time, five years apart, with the functional abilities of the user deteriorating between the current and future time periods. On the basis of various assumptions about the input of informal care, each was attributed a "package" of adaptations and AT to address their needs at both points in time. These profiles were used to benchmark the adaptability of the properties.

Detailed surveys of 82 homes were carried out. Three quarters of these were conventional mainstream properties, of which a third had community alarms and grab rails but no other adaptations, the rest were conventional sheltered homes. The implications of the range of adaptations required, the equipment needing to be installed, and the AT to be accommodated, was assessed for each property in relation to each user's needs at both points in time. Where adaptation was possible, designs and specifications of the necessary building work and equipment were drawn up and the costs estimated. The number of properties capable of adaptation and the average cost of undertaking this work are shown in Figure 1.

Further assumptions categorised the adaptations and AT into three types for each user profile; *Basic* (essential, regardless of the type of care available to the user in order for them to remain in their own home); *Care Reducing* (further adaptations and AT hypothesised as having the potential to substitute for formal care); *Good Practice* (additional adaptations and AT suggested as supplementary to improve the quality of life of the user). The adaptations and AT were then combined with the input of human care to create a "care package". The receipt of formal care services (home care, day care, community nursing and meals, allowing for care management) is heavily contingent upon the availability of informal care. Three assumptions were made about this, that the individual had: no informal care; a non-resident carer (helping with domestic tasks such as cleaning and shopping, but not personal care); or, that they had a co-resident carer (helping with both personal care and domestic tasks). The likely realistic input of formal

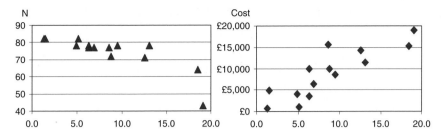

Figure 1. Number of adaptable properties and average cost of adaptation by OPCS ordinal disability score.

Table 2. Total cost of packages of care, adaptations and AT (no informal care).

Period	5 years			10 years		
	Traditional (£)	Augmented (£)	Maximum (£)	Traditional (£)	Augmented (£)	Maximum (£)
A	130	351	1090	20772	9100	15637
B	3522	664	5701	38717	17331	31393
C	4029	4306	9046	43898	24578	32474
D	10180	6239	11811	34600	23794	31377
E	6133	7367	9874	69342	41131	45732
F	1240	–	2583	80614	64289	69789
G	107765	–	145412	198501	–	241508

care services was then considered in relation to the different levels of AT and informal care. This produced the following packages sufficient to enable individuals to remain in their own homes: *Traditional* (basic adaptations and AT and full formal care); *Augmented* (basic and care reducing adaptations and AT and reduced formal care); *Maximum* (basic, care reducing and good practice adaptations and AT and reduced formal care).

The combinations of informal care, adaptations and AT and formal care were considered for each user profile at both points in time. Formal care costs were combined with the estimated costs of the adaptations and AT taking into account, installation, maintenance, replacement and recovery costs and applying standard investment appraisal methods, on an annual basis for 15 years. Table 2 provides estimates of the cost of the packages for all of the "no informal carer" scenarios for five and ten year periods. The period after which the cost of both augmented and maximum packages would be equal to or lower than the cost of the traditional package, that is, the break-even period (which has been measured from the time of the initial investment even though further investments are made after 5 years). This is different from the time taken to recoup an investment in adaptations and AT.

Findings

Taking the assumption of no informal care, with the exception of User G, the additional investment in the first five years, during which most users have mild or modest levels of disability, for the augmented and maximum packages is relatively modest and in most cases involves additional expenditure over the traditional package. In the most extreme case the additional expenditure is just over £5,000 but generally it is much less. More interesting is a ten-year time span, since after five years the health of each user deteriorates and additional adaptations, AT and care are required. The result is that the cost of all three packages rises. However, the cost of the additional investment in adaptations and AT is quickly recouped because of reduced care costs. The breakeven periods for the augmented packages are very short, all within 1.1 years after the fifth year, and for all of the maximum packages the period is between 1.6 and 3.1 years after the fifth year, that is excluding User G.

Consideration of both the average life expectations (OPCS, 2002) and reduced life expectations because of ill health (Lansley et al, 2003), suggests that for the augmented packages adaptation and AT costs are usually recouped within life expectancy periods. For the maximum packages costs are usually recouped within the average life expectancy and for about half the packages within the reduced life expectancy. These findings apply to all users where there is no informal care, a non-resident informal carer, or a co-resident carer, except for User G, who is very disabled throughout the ten year period, when there is no carer. For this user the traditional and augmented packages are not feasible and have to be replaced with residential care.

Although these findings are sensitive to a number of assumptions, they do suggest strongly that introducing adaptations and AT that result in reduced formal care can lead to savings compared with traditional packages, and in some cases significant savings can be achieved. In terms of practice, the successful achievement of this desirable outcome would depend on the sensitive specification of care, adaptations and AT requirements arising from a user's needs and ensuring that these were appropriately matched to the user's home and their preferences.

Costs vary greatly in relation to the type of disability and the characteristics of a user's housing. When these factors are taken into account a number of important relationships emerge. Locomotion disabilities have a major impact on the feasibility of adaptation and on cost. Two storey properties become progressively more difficult and eventually impossible to adapt with increasing locomotion impairment. There are very strong correlations between scores of overall disability and cost, with the locomotion disabilities, and hence mobility requirements, having the most important impact. Despite large variations in costs within groups of similar properties statistically significant differences were identified between groups. As might be expected, the properties that provided the most scope for adaptation were ground floor flats and bungalows, and properties with at least two bedrooms. Often they were characterised by combinations of: accommodation on one level – no vertical circulation; spacious layout with rooms separately approached from hall or landing; internal stud partitions and timber floors; large bathrooms or space to enlarge an existing bathroom; and large walk in cupboards. Sheltered bungalows, however, were generally smaller and more difficult to adapt than conventional bungalows for user profiles requiring wheelchair movement within the property. Flats in converted houses, maisonettes and flats with either no lift or an inadequate lift, and one bedroom properties and bedsits provided the least scope for adaptation. Often they had: changes in floor level within the same floor; restricted accommodation layout; small bathrooms and no scope for enlargement; and restricted spaces around the property – limiting space for ramps, scooter stores and extensions.

Conclusions

The paper has drawn on many other analyses and sources of information and employs a range of assumptions. However, the approach used to generate the user profiles, the specification of formal care requirements, adaptations and AT, and the costing of these, has been strongly based on information and experiences which are generally acceptable to professionals in the health care, social services and housing sectors. The analyses support the view that given careful selection of adaptations and AT to match the needs of a user and their preferences these can not only enhance quality of life but can do this in a cost effective way. The analyses suggest that across a wide range of cases the provision of good practice adaptations and AT can be funded through savings in formal care provision achieved through the adoption of care reducing adaptations and AT, and in some cases there will still be savings in total costs.

The work has shown that there are important relationships between the functional abilities of an older person, the costs of adapting their homes and providing AT to meet their needs, and the potential savings in the costs of care which would otherwise be required. It highlights the value of the ergonomic analyses which need to be undertaken when assessing the individual and then specifying the adaptations and AT from which they will benefit.

References

Audit Commission 2000, *Fully equipped. The provision of equipment to older or disabled people by the NHS and social services in England and Wales*, (Audit Commission, London)

Grundy, E., Ahlburg, D., Ali, M., Breeze, E., and Sloggett, A. 1999, *Disability in Great Britain*, (HMSO, London)

Lansley, P. R., McCreadie, C., Tinker, A., Flanagan, S., Goodacre, K., and Turner-Smith, A. 2004, Adapting the homes of older people: A UK study of costs and savings, *Building Research and Information*, 32, 6, 468–483

McCreadie, C., and Tinker, A. 2005, The acceptability of assistive technology to older people, *Ageing and Society*, **25**, 1, 91–110

Office of National Statistics 2002, *Annual Abstract of Statistics 2002.* Table 5.22. (The Stationery Office, London)

Office of the Deputy Prime Minister 2003, *Delivering adaptations: responding to the need for adaptation*, Consultation paper, (ODPM, London)

Royal Commission on Long Term Care 1999, *With Respect to Old Age, Main Report*, (The Stationery Office, London)

Tinker, A. 2003, Assistive technology and its role in housing policies for older people. *Quality in Ageing.* **4**, 2, 4–12

INDUSTRY'S RESPONSE TO INCLUSIVE DESIGN: A SURVEY OF CURRENT AWARENESS AND PERCEPTIONS

Joy Goodman, Hua Dong, Pat M. Langdon & P. John Clarkson

Engineering Design Centre, Department of Engineering, University of Cambridge, Trumpington Street, Cambridge CB21PZ, UK

Although there are great potential benefits to companies of widening their target market to include older and disabled users, many companies still target a narrow range of users. In the context of research into inclusive design, we are addressing this by conducting a survey of UK companies to examine their awareness of inclusive design, the barriers to it in their companies and the drivers that would encourage them to design more inclusively. In this paper, we summarise results from the first 87 companies and discuss their implications for the development of inclusive design promotion and support tools.

Introduction

The number of older and disabled people in developed countries like the UK is rapidly increasing (U.S. Census Bureau, 2005), making this a large and influential consumer group with considerable spending power (Coleman, 2001). Widening one's target market to include such users makes sense both commercially and ethically but many companies still target young, able-bodied people (Clarkson et al, 2003). To overcome this hurdle, we need to understand what causes it: to gain a more complete picture of the situation and the reasons that companies themselves give for not designing more inclusively.

Related work

Previous studies have started to develop this understanding. In the US, the Universal Design Research Project interviewed 26 consumer product manufacturers (Vanderheiden and Tobias, 2000) and, in Japan, a survey examined 307 companies (Unpublished, 2000). Both studies identified a range of barriers and drivers for universal (or inclusive) design, covering areas such as regulation, training, data, demand and interest.

More recently, in the UK, Dong et al (2004) surveyed a range of manufacturers, retailers and design consultancies to identify their perceived drivers and barriers to inclusive design. They describe how barriers fall into three main categories: (1) those due to how inclusive design is perceived; (2) those relating to practical difficulties; and (3) those relating to the culture or nature of the organisation. They also show that companies are motivated by a range of drivers for inclusive design.

These results indicate some of the key barriers and drivers to inclusive design but there is still a long way to go in our understanding. We need to find out more about what companies think inclusive design is and what their current response to it is, as well as probing further their response to the barriers and drivers.

To this end, we are conducting a survey as part of a project sponsored by the UK Department of Trade and Industry and the EPSRC funded i~design project. The survey

---

{jag76, hd233, pml24, pjcl0}@cam.ac.uk

builds on the UK study described above and seeks to address some of its limitations by expanding the sample size and investigating companies' current position and understanding of inclusive design in more detail. The results of the survey will feed into the development of approaches for overcoming industry barriers to inclusive design and of tools to support designers in carrying it out. Some of the survey results are reported elsewhere (Goodman et al, 2006), focusing on the factors involved in a company's response to inclusive design and the differences between different types of company.

Method

Questionnaire design

The first half of the questionnaire sought to understand the respondent's context, examining the company's demographics and its current understanding of and response to inclusive design. Respondents were then presented with a set of drivers and barriers to inclusive design, derived from those identified in the previous survey (Dong et al, 2004; Dong, 2004). They were asked to indicate how much they saw these as drivers or barriers for their organisation. A final section elicited responses to some possible approaches for encouraging inclusive design, specifically related to the interests of the research teams undertaking the survey.

Sampling and questionnaire distribution

This paper presents intermediate results from the ongoing survey, describing the responses from the first 87 companies. Responses were elicited from UK companies in design, manufacturing and retail. Many of them were obtained through industry contacts, while others were recruited by phoning contacts obtained through a web search of companies in these sectors. A degree of self-selection means that companies are more likely to take part if they have a prior interest in the topic. Thus the sample is likely to represent higher levels of interest in inclusive design than UK companies as a whole.

Results

Awareness of inclusive design

Survey respondents were asked if they had heard of the terms "inclusive design", "universal design" and "design for all". A majority of the companies (78%) had heard of inclusive design, with fewer having heard of the other terms. However, fourteen companies (16%) admitted to not having heard of any of them.

Respondents were then given a standard definition of inclusive design against which to gauge their companies' current levels of awareness, effort and interest in the concept of inclusive design and the current levels of inclusivity of their products and services. Inclusive design was defined as: "a process whereby designers, manufacturers and service providers ensure that their products and services address the needs of the widest possible audience, irrespective of age or ability" (derived from the DTI's Foresight Programme). Responses were given on a 4-point scale and are shown in Figure 1.

The current situation is not as positive as might be expected from a sample with a degree of self-selection. Over half the respondents rated their companies as low or very low on current inclusivity of products or services (51%) and on the current level of effort utilised to make these inclusive (54%). Levels of awareness were also low with 43% of respondents considering there to be little or no awareness of inclusive design in their organisation. There was a higher level of interest in making products or services

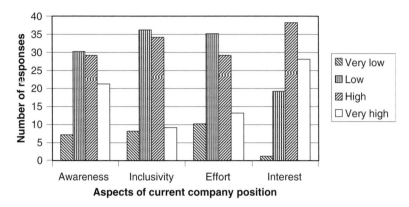

Figure 1. Current company position.

more inclusive, but, even so, a significant proportion of respondents (23%) indicated low or very low interest. Given the self-selection in the survey, the real levels of awareness, inclusivity, effort and interest are likely to be even lower than those reported here.

Drivers for inclusive design

To overcome these hurdles, we need to ascertain what factors are likely to encourage companies to design more inclusively. Respondents were, therefore, asked to indicate how much they agreed with a set of five statements about drivers for inclusive design (shown on the left of the bar in Figure 2), and to say how effective they thought inclusive design could be in helping them to achieve a range of commercial benefits (on the right of the bar in Figure 2). The results indicate that key drivers for companies as a whole are demographic and consumer trends and brand enhancement. The commercial benefits highlighted included increasing customer satisfaction and being a source of innovation and differentiation.

The drivers identified by respondents indicate areas that it may be useful to focus on in efforts to encourage companies to take up inclusive design. If companies value, for example, inclusive design's capability to increase customer satisfaction, then it may be useful to explain more carefully how it can be used to achieve this and to show more clearly how customer satisfaction is linked to financial benefits.

Conversely, drivers that were not identified by companies, such as legislation, may indicate areas where such efforts are currently failing. This should cause us to re-evaluate these efforts, examine the reasons for their failure and possibly redirect them in new directions.

Barriers to inclusive design

If we are to encourage inclusive design effectively, we also need to know what companies perceive as the main barriers to its use. Respondents were therefore asked whether or not they agreed with a series of statements on such barriers. Their responses are shown in Figure 3.

The most commonly acknowledged barrier to inclusive design was the lack of time and budget to support it. Other significant barriers were a lack of knowledge and tools to practise it and the feeling that inclusive design was not a perceived need of the end users.

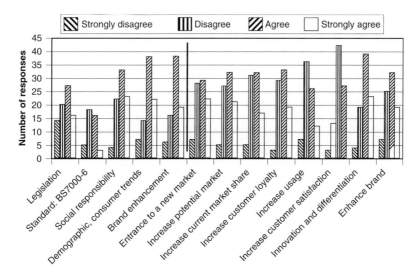

Figure 2. Responses to drivers for inclusive design. Agreement with listed drivers is shown on the left of the bar and perceived effectiveness on the right.

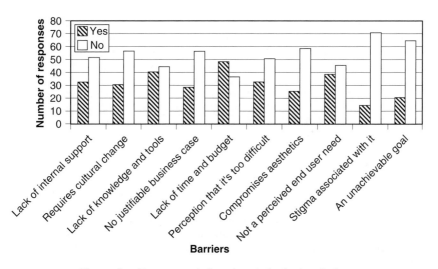

Figure 3. Responses to barriers to inclusive design.

To address these barriers, we need to develop tools to support the practice of inclusive design, taking into account concerns over time and budget. Tools are needed that make the process of inclusive design faster and cheaper and that do not themselves require a lot of time or money, either in training or in practice. More education on the nature of inclusive design is also needed, demonstrating why it is important and how it can be useful for companies whose traditional end users are not older or disabled.

Conclusions

If we are going to have products in the marketplace that are truly usable by the wider population, then we must encourage companies to put inclusive design into practice and equip them to do so. This paper has identified low levels of awareness and response to inclusive design among UK companies and has highlighted key barriers to its uptake and drivers that could encourage its use. To address these, we need to develop a more tailored business case and develop methods and tools to support designers in carrying out inclusive design, taking into account concerns over time and budget.

References

Clarkson, P.J. and Keates, S., 2003, Inclusive design – a balance between product demands and user capabilities. In *ASME '03*, USA

Coleman, R., 2001, Designing for our future selves. In W.F.E. Preiser and E. Ostroff (eds.) *Universal Design Handbook*, (MacGraw-Hill, New York), 4.1–4.25

Dong, H., Clarkson, P.J. and Keates, S., 2004, Requirements capture for inclusive design resources and tools. In *ESDA 2004*, 7th Biennial ASME Conference Engineering Systems Design and Analysis, Manchester, UK

Dong, H., 2004, *Barriers to inclusive design in the UK*, Ph.D. thesis, Department of Engineering, University of Cambridge

Goodman, J., Dong, H., Langdon, P.M. and Clarkson P.J., 2006, Factors involved in industry's response to inclusive design. In *Designing Accessible Technology: Proceedings of CWUAAT 2006*, (Springer)

Unpublished report, 2000, Kyoyo-hin (Universal Design) in Japan. Available from the i~ design collection of the Helen Hamlyn Research Centre, Royal College of Art, UK

U.S. Census Bureau, 2005, International Data Base (IDB). Website: http://www.census.gov/ftp/pub/ipc/www/idbnew.html. Accessed: 9 Dec 2005

Vanderheiden, G. and Tobias, J., 2000, Universal design of consumer products: current industry practice and perceptions. Available at: http://trace.wisc.edu/docs/ ud_consumer_products_hfes2000/index.htm. Accessed: 8 Dec 2005

EXPLORING USER CAPABILITIES AND HEALTH: A POPULATION PERSPECTIVE

Umesh Persad, Patrick M. Langdon & P. John Clarkson

Cambridge Engineering Design Centre, Department Of Engineering, University of Cambridge, Trumpington Street, Cambridge CB21PZ, UK

This paper presents an analysis of the structure of the prevalence of disability and medical conditions in a representative sample of over 5000 people in the Great Britain 16+ adult population. Using data from 1996/97 Great Britain Disability Follow-Up Survey, exploratory data analysis revealed the general increase in loss of ability in the older population. Summary characteristics are presented for five age groups and design implications are drawn for the practice of inclusive design.

Introduction

Inclusive design is an approach to design that aims to include the needs of users with various physical, cognitive and sensory ability limitations in the design process. As such, an understanding of the structure of user capability at the population level is vital (Carlsson, Iwarsson, & Sthål, 2002). This paper presents an analysis of the structure and distribution of capabilities and medical conditions in a representative sample of over 5000 British people in the Great Britain 16+ adult population.

A derivative data analysis was performed on the 1996/97 Great Britain Disability Follow-Up Survey (DFS) compiled by the UK Office of National Statistics (ONS) (Grundy, Ahlburg, Ali, Breeze, & Sloggett, 1999). Thirteen scales measuring levels of disability and fifteen variables making up the International Statistical Classification of Diseases and Related Health Problems (ICD) were used in the analysis. The scales measured locomotion, reach and stretch, dexterity, seeing, hearing, continence, fits (or consciousness), communication, behaviour, intellectual functioning, digestion, scarring (or disfigurement) and independence. The health categories were Infectious, Neoplasm, Endocrine, Blood, Mental, Nervous, Eye, Ear, Circulatory, Respiratory, Digestive, Genito-Urinary, Skin, Musculo-Skeletal, and Other. The prevalence of health conditions and ability loss were analysed in five age groups of 16–20, 21–40, 41–60, 61–80, and 81–100 years using SPSS 12.01 software for Windows.

Results and Analysis

Figures 1 and 2 show proportions of people in the Great Britain population at the time of the survey with various ability limitations and health conditions in 5 age groups.

Compared to other age groups, the 16–20 age group showed the highest prevalence in fits, communication ability loss and intellectual functioning loss with the most prevalent health conditions being infectious, mental, respiratory and skin diseases. The 21–40 age group showed the highest prevalence in behaviour problems and digestion ability loss compared to other groups, with the most prevalent condition being nervous disorders. The 41–60 age group showed the highest prevalence in dexterity ability loss,

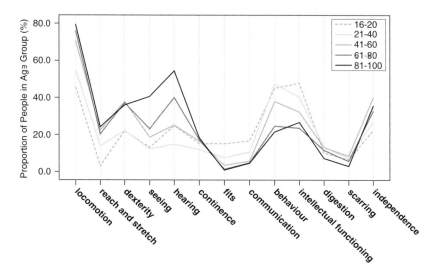

Figure 1. Proportions of people with ability limitations in the GB population.

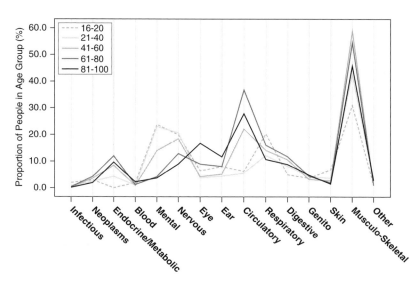

Figure 2. Proportions of people with a health condition in the GB population.

scarring and loss of independence compared to other groups, with Musculo-Skeletal disorders being the most prevalent condition. The 61–80 age group showed the highest prevalence of Neoplasms, Endocrine, Metabolic, Circulatory and Digestive diseases. Finally the 81–100 age group showed the highest prevalence in locomotion, reach and stretch, seeing, hearing and in continence compared to other groups. The most prevalent medical conditions in this group were Blood, Eye, Ear, Genito-Urinary and Other diseases.

Table 1. Description of ability and medical conditions in different age groups.

Age	3 most prevalent ability loss	4 most prevalent health conditions
16–20	(1) Intellectual functioning, (2) locomotion, (3) hearing	Musculo-skeletal, Mental, Respiratory and Nervous
21–40	(1) Locomotion, (2) intellectual functioning, (3) dexterity	Musculo-Skeletal, Mental, Nervous and Respiratory
41–60	(1) Locomotion, (2) dexterity, (3) intellectual functioning	Musculo-Skeletal, Circulatory, Nervous, Respiratory
61–80	(1) Locomotion, (2) hearing, (3) dexterity	Musculo-Skeletal, Circulatory, Respiratory and Nervous
81–100	(1) Locomotion, (2) hearing, (3) seeing	Musculo-Skeletal, Circulatory, Eye and Ear
All ages	Locomotion, dexterity, hearing, intellectual functioning, seeing, reach and stretch, communication	Musculo-Skeletal, Circulatory, Nervous Respiratory, Digestive, Mental, Endocrine and Metabolic, Eye, Ear, Genito-Urinary, Neoplasms, Skin, Other, Blood, and Infectious

Table 1 shows the three most prevalent ability losses (from locomotion, reach and stretch, dexterity, seeing, hearing, communication and intellectual function) that impact on the design of mainstream products and environments, together with the four most prevalent health conditions in each group. The last row of the table lists 7 ability losses and 15 health conditions for all ages in order of decreasing prevalence.

The prevalence of motor ability loss such as locomotion and dexterity and the high prevalence of Musculo-Skeletal conditions are evident across all ages. It also shows that cognitive ability loss is more prevalent in the younger age groups of 16–20 and 21–40 with associated high prevalence in mental conditions. The older age groups of 61–80 and 81–100 show high prevalence of sensory ability loss (hearing and seeing) likely due to ageing and the increased prevalence of Eye and Ear conditions.

The data structure was further analysed by utilising hierarchical agglomerative clustering methods to explore the groupings of ability loss and health conditions. Figure 3 shows clustering results for disability scales on the left and the same variables converted to categorical ability variables (i.e whether there is a loss or not) on the right. For the disability scales, a Euclidean distance measure was used for similarity, and Ward's method was found to be the most effective clustering algorithm (Everitt, Landau, & Leese, 2001). For the converted categorical ability variables, a Jaccard measure of similarity was used, and Complete Linkage was used as the clustering algorithm (Everitt et al, 2001). The clustering results for ability variables are shown in Figure 3.

The cluster diagram on the left of Figure 3 shows that seeing, communication and hearing ability group together. Intellectual functioning ability and reaching/stretching ability is then joined to this group. Locomotion and dexterity abilities form their own group at a larger distance. The distance measure calculates the similarity between the scores of any two abilities, thus seeing and communication ability scores tend to occur at the same level. There are larger variations between levels of locomotion, dexterity and the other abilities indicated by the larger distances at which they group.

The right-hand cluster diagram in Figure 3 shows groupings for the variables representing similarity in terms of co-occurrence. It can be seen that locomotion,

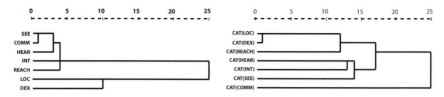

Figure 3. Cluster diagrams of ability loss.

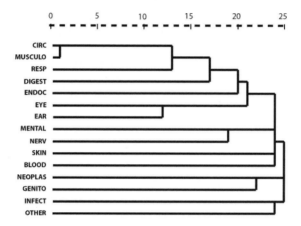

Figure 4. Cluster diagram of health conditions.

dexterity and reach form a physical ability group. Hearing, intellectual functioning and seeing also group together forming a sensory/cognitive group. Communication is last to group with the previous two groups due to lower levels of co-occurrence with the other variables.

Figure 4 shows groupings of medical conditions. Circulatory and Musculo-skeletal conditions group at a very small distance, indicating that of all the medical conditions, these two tend to co-occur most frequently. Respiratory conditions tend to co-occur with Circulatory and Musculo-Skeletal diseases as well. Other co-occurring groups include Eye and Ear conditions, Mental and Nervous conditions and Neoplasms and Genito-Urinary conditions.

Design Implications and Conclusions

Understanding the abilities of the disabled population and how it is structured with age is an important first step to designing inclusively. Only after understanding the loss of ability that occurs with age and/or disabling conditions can designers begin to adjust products and environments that cater to ability loss. Though it is very difficult to predict the specific level of ability loss of an individual based on the presence of different combinations of medical conditions, it is important to understand these major causes of ability loss to better understand lifestyle issues and the origin of multiple ability loss.

The high prevalence of Musculo-skeletal conditions (such as all forms of arthritis that occur mostly with increasing age) and the loss of locomotion and dexterity ability suggest the design of products with reduced physical demand. The forces required to open packaging, operate various controls and physically manipulate products need to be reduced. The high prevalence of sensory ability loss in the older population requires that products provide stronger and clearer sensory signals (e.g. larger text, increased contrast, better lighting and adjustable volume levels). The prevalence of intellectual ability loss was shown to be greater in the younger age groups (Figure 1), possibly due to differential mortality in the older age groups (i.e. people with various conditions such as dementia, do not survive, leaving the 81–100 population with only survivors). In any event, loss of fluid cognitive ability is known to decline with increasing age (Fisk et al. 2004), and product design must accommodate this. Simplicity in product design is needed, with an understanding of users' prior experience and mental models. Complex interaction sequences that place large demands on executive function, working memory, learning and visual cognition should be avoided in favour of less demanding sequences of use.

Inclusive design requires the simultaneous consideration of motor, sensory and cognitive ability loss, as these losses tend to co-occur in the older population. In addition, a better understanding of combinations of motor, sensory and cognitive ability loss in relation to interface demands is needed for the successful design of more accessible products.

In conclusion, it was observed that the loss of ability to interact with everyday products generally tends to increase with increasing age. This is likely to be due to the effects of the ageing process and higher incidence of medical conditions that cause impairment leading to loss of ability. Designers are faced with meeting the challenge of designing inclusive products, services and environments that place minimum demands on people's sensory, cognitive and motor abilities. Designing inclusively is becoming more and more important in an ageing world, bringing about an increase in the quality of life for older and disabled people.

References

Carlsson, G., Iwarsson, S., and Sthål, A. 2002, The Personal Component of Accessibility at Group Level: Exploring the Complexity of Functional Capacity. *Scandinavian Journal of Occupational Therapy, 9*, 100–108.

Everitt, B., Landau, S., and Leese, M. 2001, *Cluster Analysis* (4th Edition ed.). London: Arnold.

Fisk, A. D., Rogers, W. A., Charness, N., Czaja, S. J., and Sharit, J. 2004, *Designing for older adults : Principles and Creative Human Factors Approaches*: CRC Press.

Grundy, E., Ahlburg, D., Ali, M., Breeze, E., and Sloggett, A. 1999, *Disability in Great Britian: Results from the 1996/97 disability follow-up to the family resources survey*: UK Department of Social Security.

"SLEEPING WITH THE ENEMY!" – A SURVIVAL KIT FOR USER AND PROVIDER COLLABORATORS

John Mitchell[1], Michelle Turner[2], Simon Ovenden[3], Julie Smethurst[4] & Tracy Allatt[5]

[1]*Ergonova*
[2]*Sheffield Access Liaison Group*
[3]*Access Officer, Sheffield City Council*
[4]*Sheffield Transport4All Groups*
[5]*Transport Planner, Sheffield City Council*

The purpose of this paper is to provide a survival toolkit for building good working relationships between disabled people and providers of the environment we live and work in. Sheffield has a number of well-established disabled user/provider collaboration groups. These include Sheffield City Council's Access Liaison, Transport4All and Housing Scrutiny Groups. The groups are advisory and their remits are to use the expertise of disabled users, councillors and council officers to improve the quality and inclusivity of council services and facilities. In view of this Symposium's focus on Inclusive Homes and Neighbourhoods, the work of the Housing Scrutiny Group is covered in more detail elsewhere (See Wright, 2006). This paper is concerned with the development of working relationships between disabled user groups and Sheffield City Council. It reviews their, evolution, successes and failures, natural history and dynamics and ground-rules for effective collaboration.

Introduction

"Sleeping with the enemy" is a pejorative expression first used to describe the survival-driven collusion of disempowered victims with their oppressors during wartime, and has often been used synonymously with the word "collaboration". Thus the word "collaboration" can now have overtones of compromise and betrayal. Fortunately however, the interests of those who use and those who provide local services need not be in direct opposition and they can often be highly compatible. The quality of homes and neighbourhoods is often decisive in determining whether or not disabled people can live their lives fully, productively and inclusively (Mitchell, 2004). In the past, many authorities were unaware of local barriers, of the restrictions they impose on people with impairments and of the cost to the local economy in terms of reduced productivity and increased support costs. Since non-inclusive homes and neighbourhoods can have such direct effects on the activities of local authorities and their disabled ratepayers, it is clearly of mutual interest that they should work together to get rid of avoidable restrictions to achieve full social and economic participation. Changes to legislation and policy since the 1990s have improved the way local authorities work and consult with user groups, including disabled people. The following are changes that have taken place in Sheffield as a result of such collaboration.

The "disability" voice in Sheffield

During the 1970s, disabled Sheffielders lacked:

- An agreed agenda for inclusion
- An autonomous group to represent their interests
- Methods of communication, negotiation and collaboration with providers.

The Sheffield Co-ordinating Committee for the Disabled

From the 1970s until 1986, the Council for Voluntary Services hosted the "Sheffield Co-ordinating Committee for the Disabled". The focus of this group was on meeting disabled peoples' need for support from health and social services rather than the removal of barriers. A few members of the committee had impairments and represented groups such as the Polio Fellowship and the Central Deaf Club. Its Chair and the majority of committee members were non-disabled representatives of established local and national charities, such as the MS Society, RADAR, the Red Cross, the Sheffield Royal Society for the Blind and the Spastics Society. When issues were raised by disabled members, the Chair would speak on their behalf to the officers and departments concerned. This "third party" approach was only rarely successful. Perhaps its most notable achievement was the establishment of the first Sheffield Shopmobility scheme during the early 1980s. In 1981, the International Year of Disabled People with its mission of integration and participation strengthened the desire of disabled members to speak up for themselves. When the first disabled chair was elected in 1983, "everyday" issues, such as the usability of Sheffield's pavements, assumed a higher profile (Mitchell, 1985a and b).

The forum of people with disabilities

Although direct interaction between disabled people and council officers had borne some fruit, disabled people decided to set up an organisation that they could control and manage themselves. Accordingly, the Co-ordinating Committee agreed to disband itself and the Forum of People with Disabilities was established in 1986 to take its place. Until that time, people with different impairments had tended to join groups representing people with the same impairment. Some feared that a "pan-impairment" group might supplant and undermine the interests of these groups. Nevertheless, the new group thrived and established a series of Action Groups to represent disabled peoples' interests in mainstream areas such as housing, transport, education and access in order to tackle the main barriers against inclusion. A major obstacle to progress was that councillors, officers and the general public seemed to be unaware of these barriers. The Forum therefore put considerable effort into exposing these barriers, explaining them, publishing the results and involving the local media in the subsequent debates. For example, after providing councillors and officers with blindfolds and long canes, or putting them in wheelchairs, they were asked to use their pavements to get back to the Town Hall under the gaze of local newspaper, TV and radio reporters. The same approach was used to expose barriers in Sheffield's libraries (3), World Student Games facilities (4, 5, 6 & 7), its new Supertram system (8) as well as its places for eating out (9). As links between providers and disabled people improved, the need for campaigning diminished and the Forum was wound up in 2004. Disabled people now work with providers mainly as individuals and as members of inclusive and impairment groups.

Developing communication, negotiation and collaboration with providers

The City Council's desire to use the expertise of disabled people was tempered by the fear that their participation in decision-making committees would remove power from locally elected representatives. As a result, in its earlier disability consultative committees, disabled people were prevented from speaking and their attendance was not recorded in the minutes. The Disability Discrimination Act, the Disabled Persons Transport Advisory Committee (DIPTAC) and the Social Inclusion Mobility Unit have raised national awareness politically. Similarly, Sheffield's second Local Transport Plan has reinforced the need for providing inclusive transport and neighbourhoods by having four shared priorities, one of which is accessibility. A key development came in 1992 with the formation of the Access Liaison Group. Its role was advisory and its members included councillors, officers and disabled people. Its function was to promote informed decision-making at operational and policy levels.

The range of impairments among disabled members was particularly useful in converting apparently unconnected, "one-off" feedback from members of the public into a more systematic picture of inclusive and non-inclusive practice. The combination of site visits and the development of "good practice" guidance for planners and developers was used to explain non-inclusive practice and show how it could be avoided in future. The group was rapidly accepted as the "authoritative local body" for access and transport. Subsequently, the South Yorkshire Passenger Transport Authority established a sub-regional group, and adopted an Inclusive Transport Policy. The focus of transport users in Sheffield then turned to the advantages of "low-floor" buses which led to the establishment of the Transport4All group in 2004. The group's terms of reference have had an important influence on maintaining their focus without degenerating into forums for complaints.

This has encouraged the Council and the Passenger Transport Executive to continue their support, the same also applies to the disabled community who want to see action and momentum rather than empty words.

The success of the groups can be judged by:

- The gradual improvement of access, particularly physical access in Sheffield.
- The willingness of officers to seek and use the expertise of disabled people.
- The desire of other collaborative groups to learn from the Sheffield experience.

Access successes include:

- A Mayoral Reception to mark its 10th Anniversary.
- Refurbishment of the 19th century Town Hall with distinctive ramped access, an inclusive lift and automatic doors.

Transport successes include:

- Inclusive standards for bus mini-interchanges.
- Improved passenger information systems.
- The PTA's Inclusive Transport Policy.
- Improvements to Sheffield's Midland Station regeneration proposals and the.
- Transport Interchange.

However, there were some major schemes where disabled users felt that unsatisfactory compromises had been made:

- The inclusion of steps in the Supertram System.
- The choice of more aesthetic rather than inclusive paving sets.
- Inclusion in the newly built Millennium Gallery.
- Step-free access in Sheffield's Peace Gardens.
- Non-inclusive way finding information in the Connect Sheffield System.

A "Survival Kit" for collaborators

Establishing the mutual benefits of collaboration

There is little point in collaboration unless it benefits all those involved. Officers may initially feel that they have little to gain by collaborating with users who have no professional qualifications, training or administrative experience. Users must therefore demonstrate to them that:

– Barriers are real and costly.
– Those who experience them can help eradicate them.
– Users can evaluate and expose providers' performance.
– Users can help providers avoid pitfalls.
– User and provider interests often coincide.

For their part, disabled people are well aware that some officers may prefer to follow procedures rather than to achieve inclusion. Officers therefore need to demonstrate to users that they:

– Value user inputs.
– Recognise inclusion as a measure of quality-desire and will promote inclusion.
– Recognise their duties under the DDA but are not limited by them.

In themselves, mutual respect and collaboration do not guarantee progress towards inclusion. A number of unexpected barriers have emerged that reduce its effectiveness and sustainability. These include:

– Non-inclusive methods of communication and consultation.
– Difficulty in retaining skilled collaborators once they have become expert.
– Solutions remain "soft-wired" high profile exceptions instead of standard practice.

References

1) Mitchell, J, 1985a, *"The Causey Campaign – Pavements in Sheffield"*, Report for the Sheffield Co-ordinating Committee for the Disabled
2) Mitchell, J, 1985b, *"The Magic Carpet Scheme – Better Pavements for Sheffield"*, Report for the Sheffield Co-ordinating Committee for the Disabled
3) Stubbs, J, 1990, *"Sheffield's Libraries"*, Forum Access Report No 1, Forum of People with Disabilities, PAVIC Publications, ISBN 086339 3241, Sheffield
4) Mitchell, J, 1991, *"The Waltheof Centre"*, Forum Access Report No 2, Forum of People with Disabilities, PAVIC Publications, ISBN 086339 3241, Sheffield
5) Mitchell, J, Harker, L, Maclean, L and Southgate, C, 1991, *"The Sheffield Arena"*, Forum Access Report No 3, Forum of People with Disabilities, PAVIC Publications, ISBN 086339 325X
6) Mitchell, J and Maclean, L, 1991, *"The Don Valley Stadium"*, Forum Access Report No 4, Forum of People with Disabilities, PAVIC Publications, ISBN 086339 3268
7) Mitchell, J and Maclean, L, 1992, *"The Hillsborough Leisure Centre"*, Forum Access Report No 5, Forum of People with Disabilities, PAVIC Publications, Sheffield Hallam University, ISBN 086339 3276
8) Mitchell, J, Ravetz, A and Southgate, C, 1994, "Sheffield's Supertram: *Transport for Everybody – or just another broken dream?*", Forum of People with Disabilities, Access Report No 7, PAVIC Publications, ISBN 0 86339 4299
9) Disabled peoples' Forum, 2003, *"All Sheffielders Eat – a guide to some local venues in Sheffield"*

WHAT'S IN IT FOR ME? ALTRUISM OR SELF INTEREST AS THE "DRIVER" FOR INCLUSION

John Middleton[1], John Mitchell[2] & Robert Chesters[3]

[1]*Director of Public Health, Sandwell Primary Care Trust,* [2]*Director of Ergonova,* [3]*Inclusive Design manager for Medilink West Midlands and Sandwell, Primary Care Trust*

> Disabled service users have suffered from the failure of designers and commercial markets to deliver affordable, aesthetically pleasing mass produced non-stigmatizing "aids" and "adaptations for daily living" or what would better be called *"products for everyday life"*. The general public has also lost out in this. The particular insights of design with disabled users in mind benefits all users. Bringing the spending power of over 5 million disabled people into the mainstream market would also serve to keep down prices. The health service and other public services and the benefits system has lost out because the potential of disabled people has not been unlocked, people have been kept dependent for the want of inexpensive technology from design solutions which are taken for granted in mainstream consumer markets. This paper describes some of the problems to be confronted and overcome if consumer markets are to embrace inclusive design and deliver for all of us, the "inclusion dividend".

Mainstream and assistive markets

At present, barriers in commodities (places, products and processes) prevent people with impairments from accessing them through the "mainstream" market. Mitchell (2003) illustrates such barriers in homes and neighbourhoods.

Instead of being able to choose what they want and obtain it at the same price as other people; barriers compel them to pay more for things that they might prefer not to purchase at all.

As the Disability Rights Commission (2003) points out, "Design is a social issue because it is poor design which turns impairment into a disability." If it was ever justified to divide the market into mainstream and assistive sections, the grounds for its continuation are rapidly being undermined.

At a time when the understanding of barriers against people with impairments is growing, technology is also becoming more flexible and therefore able to accommodate a much wider range of user characteristics than was possible in the past (Mitchell, 2004). This flexibility could allow the components and functions of modern commodities to be reconfigured to get rid of their earlier barriers. If the barriers are eliminated in this way, people with impairments will begin to make unrestricted use of mainstream commodities instead of being segregated into the assistive market with its low volumes, its poor quality and choice and high prices. The distinction between impaired and unimpaired users has always been rather tenuous and arbitrary. For example, in terms of their functions, powered golf carts and wheelchairs are virtually indistinguishable. Why should a golf cart be seen as a "niche" of the mainstream market while a wheelchair is seen as belonging to the assistive market?

The inclusion of people with impairments in the mainstream market would increase the volume of sales for manufacturers by ensuring that no potential customers are excluded by inherent barriers. Many mainstream providers also have to divert their resources to provide support to enabled people with impairments to cope with the barriers they have built into their commodities. For example, support staff who are needed to help customers negotiate steps are no longer required in a "step-free" setting.

Investment for mainstream and assistive innovation

Traditionally, there has been a sharp dichotomy between the ways in which "mainstream" innovation and "assistive" innovation have been funded.

Large-scale developments of both mainstream and assistive products are generally "altruism-free" and supported by companies and their shareholders in the expectation that new products will thrive in the market, become economically self-sustaining and contribute to the company's profitability and trading position. The decision to invest often depends on a careful analysis of the competition, the costs of development, production and marketing, the likely demand for their new initiative and their forecast for its overall profitability. Major initiatives can involve considerable analysis and investment in view of the potential size of the resulting profits or losses. Smaller assistive innovations occur in a world that appears to be completely outside the marketplace. Initiatives are often funded by small grants from charities, voluntary groups or research sponsors supplemented by unpaid time from the development team. There may be no expectation that the initiative will ever sustain itself in the marketplace without continual subsidy. Altruism does not seem to be an adequate basis for inclusion.

Integrating the market for homes

Dividing the home market into mainstream and assistive components has pernicious consequences for home users and providers alike. The Helen Hamlyn Foundation (1990) presented this exclusion starkly when they asked their delegates "Are you male, fit and aged between 18 and 40, not very tall or very short? Do you have good sight and good hearing and are you right handed? If you are, you are part of the 18% of the population for whom British houses are designed".

As Chesters (2005) pointed out, barriers in homes and home markets affect disabled home users by:

– Restricting their lives and activities.
– Making them less productive and more dependent.
– Restricting their choice and value for money.

They affect those offering homes for sale or rental by:

– Excluding a large proportion of their customers from the market.

They affect those providing support for home users by:

– Increasing the need for support services.
– Increasing the demand for financial support.
– Reducing social and economic activity.

They affect hospital and rehabilitation services by:

- Increasing the complexity and expense of returning newly impaired people to their homes.

The division of the market has been in place for so long that it is now rarely questioned, let alone challenged. There is a need to shift public presumptions away from expecting a divided, non-inclusive market to accepting an integrated, inclusive market as the "norm".

Similar shifts in public and political expectations occurred during the 1970s and 1980s when widespread support was engendered for disarmament and for the reduced use of nicotine by promoting the "Peace" and "health" dividends. (Middleton, Routley and Lowe, 1987 Middleton, 1996, Middleton, 1997 and Coote, 2002) The time is now ripe for promoting the concept of the *"Inclusion Dividend"* by establishing the extent of the unnecessary restrictions and losses that are caused by non-inclusive homes and domestic commodities and by contrasting these with the gains in social and economic life that can accompany inclusion.

The Sandwell Primary Care Trust is therefore leading a campaign to engage the interest and support of local and national stakeholders in developing a model for inclusive home refurbishment and rebuild. (Sandwell Inclusive Design Partnership, 2004) Disabled home Users, local industry seeking regeneration, researchers and providers in the West Midlands, North Staffordshire and Sheffield are therefore engaging with providers to obtain mainstream funding for inclusive research, development, evaluation and implementation.

Sandwell PCT is engaged in a major health and social care redesign programme called "Towards 2010" covering Sandwell and western Birmingham. The project serves over 500,000 people with high levels of vulnerability and poor housing. We are planning for a substantial investment in inclusive housing through mainstream health service investment and through partnership efforts with a housing market renewal area and a regeneration zone.

There is a high level of aspiration to create independence and appropriate care. The PCT is therefore probably unique in the country in employing a designer to lead inclusive design. The Inclusive Design manager within PCT works directly with other NHS trusts and local industry on projects where end user input is essential. Characteristically these design projects range from clinical devices to assistive technologies. The ethos of this initiative is to tackle new innovations and conduct user centred testing and evaluation before the product goes to market. This is a step often over looked by industry, two recent innovations include an innovative walking stick and a new wheelchair cushion.

The walking stick project illustrates an excellent example of where inclusive design can expand the market potential of a product. A local manufacturer came forward with a prototype for a new high visibility walking stick with a mind solely on the elderly mobility market. Design support by the Inclusive Design manager expanded the scope of the original innovation and through user research identified that high visibility was indeed important, but the biggest problem with current walking sticks is the handle design. Through basic block modelling a new handle was developed which when returned to the user group was widely accepted as a comfortable and hugely versatile design solution. The Inclusive Design process also steered the company away from simply looking at the elderly mobility market. The design also has huge benefits to walkers and hikers who also require high visibility and a versatile handle which can be gripped in several ways depending on the terrain.

The end product is an innovative, stylish and versatile walking stick suitable for any user. The *inclusion dividends* on this product for the company will be a larger share of the market and for end users a stick which is better suited to there need.

The wheelchair cushion project illustrates a slightly different approach. A local manufacturer came forward with an idea to take the companies existing seating technology into the mobility market. The original design has been developed for the aerospace, rail, automotive and military sectors to improve comfort and reduce the risk of deep vein thrombosis (DVT). The Inclusive Design manager took the company to talk with a group of wheelchair users and conduct some initial tests. This consultation exposed some of the key issues with existing wheelchair cushions (High cost and poor performance), and showed that the new prototype design had some encouraging potential. This product has moved on significantly and is entering a clinical trial stage. However what is important to remember about this project is it's another example of *self interest* as a driver. The product is being developed by a company looking for financial return and significant market share; to achieve this they are developing a high volume cushion at a significantly lower price than other market leaders. Before they *tool up* for production the company is listening to and conducting trials with wheelchair users to ensure the design is right and clinically sound before launching into the market.

In contrast many of the existing market leaders were designed specifically for the mobility market only and have no other mainstream market placements.

Stakeholders anticipate that inclusion will benefit them by the restoration of economic activity, the enlargement of markets and the reduction of support costs in the ways summarised below:

- Health care – Early, effective return home following illness or impairment
- Social services – smaller, more focused user support packages
- Housing providers – inclusion of elderly and disabled customers in market
- Local authorities – sustainable regeneration of derelict housing
- Manufacturing – markets for new materials and components
- Revenue agencies – higher contributions from greater economic activity
- Disabled users – greater autonomy, social and economic participation.

Disabled users will of course be the principle beneficiaries of the mainstreaming of inclusive design, but the general public, health and public services, manufacturers, housing developers and the rest of the private sector all stand to gain from the liberation and independence of disabled people brought about through good design. We will all benefit from the *"inclusion dividend"*.

References

Coote, A. ed. *"Claiming the health dividend: unlocking the benefits of NHS spending"*, London: Kings Fund, 2002.
Disability Rights Commission, 2003, *"Inclusive Design – products that are easy for everybody to use"*, Disability Rights Commission, London.
Helen Hamlyn Foundation, 1990, conference folder, *Lifetime Homes Seminar*, Helen Hamlyn foundation, London.
Mitchell, J, Chadwick, A and Hodges, B, 2004, *"Why can't we have a Home with Nothing Wrong in It?* – Disabled peoples' aspirations and barriers in inclusive living at home", Study undertaken by Ergonova for Inclusive Living Sheffield and the Sheffield housing Department.
Chesters, R, McClenahan, J, Middleton, J and Mitchell, J, 2005, *"Models for Inclusion: Evidence for an Inclusive Home Marketplace"*, Contemporary ergonomics, The Ergonomics Society, Loughborough, ISBN 0-415-37448-0, pp 302–305.
Middleton, J D, Routley J and Lowe B, *"Swords to Ploughshares: improving the world's health" Medicine and War* 1987; 3:93–100.

Middleton J, "*Converting Sandwell to a healthier economy*", In Bruce, N, Springett, J, eds. Research and change in urban community health, Liverpool and London: Arrow publications and Liverpool university, 1996.

Middleton, J. "*Public health, security and sustainable development*", Health and hygiene 1997;18: 149–154.

Mitchell, J and Chesters, R, 2004, "*Designing out the cell of Exclusion*", Contemporary Ergonomics 2004, Ed Paul T McCabe, General Ergonomics, pp 445–449, ISBN 0 8493 2342 8.

Sandwell Inclusive Design Partnership. *Smart Inclusive Housing Manifesto*. West Bromwich: Rowley Regis and Tipton Primary care trust, 2004.

"CHOICES NOT BARRIERS" HOUSING STRATEGY – LEARNING FROM DISABLED PEOPLE

Clair Wright

Sheffield Homes New Bank House, PO Box 1918, Sheffield S1 2XX, Sue Stones, Joint Chair Monitoring and Scrutiny Group

In September 2004 Sheffield Homes and Sheffield City Council Neighbourhood Directorate launched "Choices Not Barriers – a housing strategy for people with physical and sensory impairments, mental health problems and learning disabilities" This was the first such strategy for a major city in the country, and set out issues, priorities and planned actions to address the housing needs of disabled people in all tenures. It represented the culmination of a 12 month process which had involved disabled people, partner organisations and a wide range of officers from Sheffield Homes and the Council. Most importantly, the process had been an opportunity for the two organisations to work with disabled people, and to develop a positive, constitutive and learning relationship, which has continued into the strategy's implementation phase.

Who are sheffield homes?

Sheffield Homes is an Arms Length Management Organisation, which was formed in April 2004. It is a not for profit company, wholly owned by Sheffield City Council, with a board of directors and local boards which include tenants, council representatives and independent people. Sheffield Homes manages the Council's housing stock (52,000 properties), delivering a range of services such as lettings, repairs, rent collection and arrears management, through a number of customer access points across the city. In addition, Sheffield Homes is responsible for delivering the Council's Decent Homes programme, a potential spend of £456 million, to bring the housing stock up to the government's Decent Homes Standard by 2010.

Sheffield Homes maintains strong links with the Council, and is a key partner in delivering many of the Council's priorities. The Council has maintained the provision of services to private sector tenants and owner occupiers, homeless people and asylum seekers. In addition, the Council has a strong strategic role in relation to housing, working with developers and housing associations to meet the housing requirements of the city, including regenerating failing housing markets.

The purpose of "choices not barriers"

The need for a housing strategy for disabled people, to look at access to housing and related services, was identified as part of a Best Value Review of Services to people with physical and sensory impairments across the Council. The strategy was to include clear priorities and targets. It aimed to address the requirements of disabled people in accessing mainstream housing services, as well as the need for specific housing and support services for those with more acute needs. In particular, the strategy sought to ensure that disabled people would benefit from the large scale investment in housing taking place in the city through the Decent Homes Programme and Housing Market Renewal regeneration funding. The Government's Decent Homes Standard currently does not include any requirements to improve the accessibility of the housing stock, focussing instead on basic measures such as thermal comfort, age of bathroom and kitchen fittings. However the investment which would be taking place under the Decent Homes Programme offered an opportunity to make some improvements to access, within the restrictions of the available resources.

Consultation with disabled people

Sheffield City Council Housing Services had for a number of years attempted to consult with disabled people. It was acknowledged that our main mechanisms for consulting with our customers, our extensive network of tenant and resident associations, did not represent disabled people. A group called the "Housing Disability Consultative Group" had been formed in the late 1990s, however by the time the strategy was in development, in 2003, this group was evidently not effective. This was for a number of reasons. Officers from within the organisation were not experienced in consulting disabled people, and so did not tailor their approaches to meet their diverse requirements nor were they able to progress many of recommendations made by the group. Some of the members were not disabled people themselves, but worked in organisations which provided services to disabled people, and so were less able to comment from direct personal experience. The group lacked a clear focus or remit, and members felt frustrated that their investment of time and commitment in attending was not resulting in any tangible results. Numbers attending the group dwindled.

In developing the strategy, it was clear that the views of disabled people needed to be at the heart of it; however experience had demonstrated that officers from within the organisation did not have the skills to be able to undertake effective consultation. A decision was taken to commission Inclusive Living Sheffield (ILS), a relatively new organisation of disabled people working with and for disabled people, to undertake the consultation on behalf of the Council. This was a developing area of work for ILS and a significant amount of officer time was invested in drawing up the scope of the project as well as regular liaison with ILS throughout its implementation. ILS commissioned Ergonova, a consultancy run by one of their members, to carry out the project. A short framework document was developed by Council officers which set out the issues that the strategy would address, with some key questions. This would form the basis of the consultation which ILS developed using their expertise in working with disabled people. ILS conducted eight focus groups with people from Sheffield with different impairments, and with mental health problems and learning disabilities. There was representation from a range of ages and ethnic backgrounds. ILS were able to use their local knowledge and networks to identify groups where people met together in a supportive and open atmosphere, where discussion could take place effectively. By drawing on their own personal experience of the barriers experienced by disabled people within the home and in using housing services, ILS were able to facilitate and encourage disabled people to share their own experiences, and also to identify potential solutions. ILS produced a comprehensive report of their findings, which included direct quotations from the individuals within the focus groups. This identified a number of key issues which were incorporated into the strategy, and informed decisions about priorities and targets to address the barriers identified. More importantly, however, the work undertaken by ILS brought the experience of disabled people of housing and housing services to the attention of officers and managers within the organisation in a direct and challenging but constructive way.

Impact

In February 2003 a Partner Consultation Event was held as a further stage in the development of the strategy. This was attended by officers from across the Council, Mental Health services, Primary Care Trusts, Housing Associations (Registered Social Landlords), providers of tenancy support, and voluntary sector organisations. The purpose of the event was to share the findings of the ILS consultation project, and to gather views on the direction of the strategy. ILS co-hosted the event, which was chaired by ILS's Chairperson, Christine Barton MBE. ILS members delivered a presentation of their findings, which, like the report, included direct quotations from the focus groups. This proved to be extremely powerful and had a major impact on those attending. Even managers who had previously considered themselves to be well versed in issues around disability, and who worked in related fields, were challenged by the presentation of the consultation project in this direct way. The presentation brought managers face to face with the barriers inherent in their own services which were preventing disabled people from accessing them effectively.

Following the presentations, workshops took place in which members of ILS participated. This provided an open and honest forum where the issues raised could be debated and potential solutions identified. The ILS consultation project demonstrated that the organisation, its managers and officers, needed to effectively engage with disabled people in order to learn from them. Where previously consultation with disabled people had often been a superficial process, more concerned with informing than with listening, it was clear that a new model was needed which would be a more open, mutual learning environment.

The monitoring and scrutiny group

One objective of the ILS consultation project was to identify a group of disabled people who would be interested in being involved in a long standing group to work with the Council on the development and implementation of the strategy, to monitor and scrutinise progress against the targets and priorities it had set. ILS identified fifteen people with a range of physical or sensory impairments, or with learning disabilities or users of mental health system, who were interested in such a role. The group initially met in the autumn of 2003, with Council officers and ILS members to facilitate. It was evident from this initial meeting that many of the disabled people present had a low opinion of the Council and particularly of housing services, and that there was a lack of trust that the group were to be effective. There were concerns that this was a "tick-box" exercise and that officers were not willing to listen to their views and to act on them. This further underlined the need for ILS's involvement as they were able to offer some assurances and to provide some credibility to the process. The group now meets every two months, and has grown in size through advertising in the tenants' newsletter, and through word of mouth. The group has agreed terms of reference, and is jointly chaired by two of the group members. A programme of topics for meetings for the year ahead is set in January, which reflects priorities within the strategy. Some members of the group are skilled in attending and participating in meetings and at identifying and challenging the barriers faced by disabled people, while for others this has been a new experience.

Capacity building

In the public sector where consultation with service users is seen as a mainstream and critical activity, there is much debate about the need for capacity building within the community to enable individuals to effectively engage with services. However, in order for the monitoring and scrutiny group to operate effectively, capacity building within the organisation, amongst managers and officers, has been necessary. This has involved detailed pre-meetings and careful briefing, to ensure that managers coming to the group to discuss an area of their service are equipped and prepared to do so effectively. For example, managers need to be prepared to deliver a briefing that is accessible to all group members, that does not rely on visual materials, or use complex language or jargon, and presents the key issues in a way which is immediate and practical rather than theoretical. Managers need to be ready to listen more than talk, so briefings are generally only ten to fifteen minutes long, and include some questions to prompt discussion. A shift in attitude has also been necessary; there may sometimes be an assumption in consulting with the public that managers will already be aware of the issues raised. In consulting with disabled people, this is not the case.

As a non-disabled people, without personal experience of disabled barriers, managers need to be ready to listen and to learn, and to avoid a position of defending established practices. As a result of these measures, the group has developed into a constructive, honest and open forum in which issues are debated and solutions can be identified. Managers attending the group find it to be a genuine learning experience, and are keen to maintain and develop the relationship.

Outcomes

The consultation with disabled people undertaken by ILS, and the work of the monitoring and scrutiny group, has brought about some significant improvements in housing services in Sheffield to the benefit of disabled people. Examples include:

- Decent Homes Plus Standard: Sheffield has developed an enhanced standard which includes where possible measures to improve accessibility such as lever taps, showers over baths, hard standings for accessible parking. In addition, surveyors identifying works needed to bring properties up to the Decent Homes standard are trained to identify whether the tenant has any needs for adaptations, and these are then completed as part of the improvement work. So far 780 adaptations have been carried out, a significant proportion represent unidentified need.
- Planning Powers: As part of the strategy the Council is working with developers to agree an accessible property standard, to be applied to new developments in the city, in particular, those in regeneration areas where the Council owns the land. This will create more new accessible housing for disabled people.
- Choice Based Lettings: Sheffield uses a choice based lettings system to let available Council properties. Properties are advertised in the press, on a website and in an estate-agent style "Property Shop", and applicants place "bids". Improvements have been made to the wording of advertisements to provide clearer information on accessibility and any adaptations in the properties. Properties are also advertised on the Accessible Property Register website.
- In person access: members of the monitoring and scrutiny group visited the Property Shop, and made recommendations to improve access to the reception area and to the computers used to access the website. These recommendations proved very useful and were also implemented in other in-person access reception points.
- Sheffield Homes website: The monitoring and scrutiny group made recommendations to improve access to information, particularly through the website. Members recommended the use of text to voice software which can be installed on a server and accessed by users without specialist equipment. This recommendation will be implemented early in 2006, giving greater accessibility to people with visual impairments and people with literacy problems.
- Accessible Meetings Checklist: The monitoring and scrutiny group suggested compiling checklists to ensure that meetings are accessible to all the disabled people who wish to attend both for TARAS and a more detailed list for officers. These distinguish between what is important, what is desirable and what can be got around with assistance plus a Meetings Charter.

References

Sheffield City Council and Sheffield Homes Jan 2005 *"Choices not Barriers: A Housing Strategy for People with Physical and Sensory Impairments, Mental Health Problems and Learning Disabilities"*.

Office of the Deputy Prime Minister Feb 2004 "A Decent Home: The Definition and Guide for Implementation".

The Disability Rights Commission, Habinteg Housing Association and the Joseph Rowntree Foundation Dec 2003 "Decent Homes: Evidence from The Disability Rights Commission, Habinteg Housing Association and The Joseph Rowntree Foundation".

BARRIERS AGAINST PEOPLE WITH DIFFERENT IMPAIRMENTS IN THEIR HOMES AND NEIGHBOURHOODS

Arnold Beevor[1], Paul Mortby[2], Carol Townsend[3], John Mitchell[4], Maureen Sanders[5] & Roy Waller[6]

[1]*Mental Health Action Group, Sheffield*
[2]*Transport4All*
[3]*Speaking Up For Action, Sheffield*
[4]*Ergonova*
[5]*Central Deaf Club, Sheffield*
[6]*Sheffield Royal Society for the Blind*

At the outset of this workshop, non-disabled participants will be asked to predict the most formidable barriers that they would have to overcome in their homes and neighbourhoods if they were to acquire an unexpected impairment. While the responses are being analysed, people with impaired sight, hearing, mobility and dexterity, understanding and mental health will set out the barriers that make it difficult for them to find, reach, understand, relate to and control the things they wish to use in and around their homes (the "Sheffield Formula for Inclusion"). The Formula exposes specific barriers against people with different impairments as well as common barriers that affect people with any impairment. Specific barriers include those which prevent people with impaired sight from finding things, with impaired mobility from reaching them and with impaired hearing from understanding what is being said via non-inclusive communication. Generic barriers include the lack of clear, plain language or "plain picture" guidance and instructions that impedes people with and without impairments. After these presentations, the workshop will, discuss the barriers put forward, choose a single barrier for detailed consideration, and, suggest how it could be eradicated to achieve full inclusion.

Are barriers optional?

Living in an "untamed" natural environment might allow its users very little scope for modifying its demands and challenges. Those who were strong, agile and resourceful enough to meet these demands would survive but others would be likely to perish. In these conditions, people with impaired sight, hearing, mobility dexterity, understanding or health would be highly vulnerable.

However, in 21st century urban Britain, very few of the places we use remain unchanged. A whole series of places, products and processes have been developed that modify our natural world and make it easier, safer, more reliable and more convenient to use.

When they first emerged, many new technologies contained inherent barriers which excluded some of their potential users. For example, printed text demands the ability to see and read and it excludes those who cannot meet these demands. More recently, technologies have acquired greater flexibility and can therefore, potentially, cater for all their would-be users. The critical distinction between the demands of the primitive and the "man-made" world is that, in the first, the demands are inherited from an unforgiving natural world. In the second, the barriers are "man-made". As such, they are potentially optional. Their eradication depends on developing technology and on developers' decisions to require inclusive outputs.

"Designing-out" and "choosing-out" barriers

At present, people with impairments have no systematic ways of alerting design-developers to the barriers that they face. Their feedback and ideas are not sought, collected, analysed or used to evaluate existing designs or to develop inclusive successors. Also, while some good quality evidence is available about barriers against people with impaired mobility, little is available on those against people with impaired sensation, cognition or health. As a result, those who wish to eradicate barriers must find out what their targets are from those who experience them at first hand (Mitchell, 2004a).

Barriers in homes and neighbourhoods

The Sheffield Housing Department had become aware that barriers in their homes restricted their users and their markets while considerably increasing their support and adaptation costs. They therefore commissioned a study to expose these barriers and to suggest methods of eradicating them (Mitchell et al, 2004b).

A considerable range of barriers was reported by people with impaired sight, hearing, mobility, use of their hands, understanding and mental health. It is often believed that such barriers are complex, contradictory, difficult to comprehend and, as a result, difficult to address (Goldsmith, 1963). Nevertheless, the study found that the barriers exposed fitted readily into the implicit sequence of functions that are required for using places, products and processes. Under this sequence the "Sheffield Formula for Inclusion", (Chesters, 2005) suggests that before anyone can use anything, they must first find it, then reach it, understand how to use it, relate to it without being intimidated and, finally, control it. The following profiles of barriers against people with different impairments emerged from the study.

Examples of barriers encountered in neighbourhoods

Various barriers made it difficult for people with impairments to use their local shops, pavements, facilities, transport, parking and gardens. For people with impaired sight and reading, the main barrier was the lack of a simple, inclusive method of way finding. For people with impaired hearing, the main barrier was the lack of a simple, portable, inclusive method of communicating with the people they meet in their neighbourhood. For

Table 1. Barriers against people with different impairments.

Barriers in non-inclusive places, products and processes made it difficult for people with impaired:
- Sight to find, understand and control things
- Hearing to understand, relate to and control things
- Mobility & dexterity to reach and control things
- Learning to understand, relate to and control things
- Mental health to relate to, understand and control things

people with impaired mobility and dexterity, the main barrier is the lack of safe, easy, "step-free" access between their homes and gardens and the places they wish to use in their neighbourhoods. The main barriers reported by people with learning difficulties and by people with impaired mental health are the lack of activity, passers-by, lighting, visibility and welcome in environments that therefore appear to be threatening and intimidating.

Examples of barriers in domestic utilities and systems

Many common problems were reported in domestic utilities and systems. These affected all the participants even when their function was disrupted at different points by different barriers.

Every participant reported difficulties in using their domestic utilities and systems. For example, the locations of controls such as water stop-cocks, smoke alarms and central heating systems in unfamiliar, inaccessible and poorly lit places impeded all participants. Many operating procedures are complex and poorly standardised. This together with the lack of plain language or plain picture instructions and sometimes of any instructions at all prevented people with learning difficulties and many others from using them.

References

"*Designing out the cell of Exclusion*", Contemporary Ergonomics 2004, Ed Paul T McCabe, General Ergonomics, pp. 445–449, ISBN 0 8493 2342 8
Mitchell, J, Chadwick, A and Hodges, B, 2004, "*Why can't we have a Home with Nothing Wrong in It? – disabled peoples' aspirations and barriers in inclusive living at home*", Study undertaken by Ergonova for Inclusive Living Sheffield and the Sheffield housing Department
Chesters, R, McClenahan, J, Middleton Mitchell, J and Chesters, R, 2005, "*The Sheffield Formula for Inclusion*", Include, the Helen Hamlyn Foundation, Royal College of Art, London, ISBN 1-905000-10-3
Goldsmith, S, 1963, "Designing for the Disabled", Architects Press, London

INNOVATION AND COLLABORATION BETWEEN USERS AND PROVIDERS

Liz Birchley[1], Garin Davies[2], Sean Gamage[3], Conrad Hodgkinson[4], John Mitchell[5] & Julie Smethurst[6]

[1]K Barriers Ltd
[2]MacDonalds Hotels
[3]South Yorkshire Passenger Transport Executive
[4]The Accessible Property Register
[5]Ergonova
[6]Sheffield Royal Society for the Blind

Many of the places, products and processes that people must use in their daily lives contain barriers that exclude anyone who has impairment. Unfortunately, while these barriers are very familiar to those who are excluded by them, other people may be quite unaware of their existence, let alone of their effects. To improve the quality and inclusivity of the places, products and processes that they offer, it is important that innovators and providers are fully aware of these barriers, how they could be overcome and of the benefits that they and their users could gain from inclusion. Disabled users in Sheffield have been providing developers with these direct insights as a means of ensuring that their mutual interests will lead to better, more inclusive homes and neighbourhoods. The workshop will presents examples of the collaboration that is being undertaken to develop and provide more inclusive hotel bedrooms, museums and their exhibits, barriers on country walks, information for home rental and purchase and "step-free" access to local transport vehicles. These examples will enable workshop participants to discuss the processes of exposing and eradicating barriers and of bringing more inclusive outcomes into mainstream, economically viable use.

Introduction

Barriers prevent people with impairments from taking advantage of the choices, opportunities and activities that non-impaired people enjoy. The exclusion of disabled clients and customers inevitably reduces the size of market for providers by limiting the number of people who can use the commodities on offer.

Inclusive, barrier-free commodities which could be used by the full spectrum of the population would maximise use, productivity and profitability and benefit both providers and users alike. Unfortunately, although these barriers are very familiar to those they exclude, other people may be quite unaware of their existence, their causes, their effects or their potential solutions (Mitchell, 2004).

It is important that innovators and providers are fully aware of these factors, as the starting-point for developing better, more inclusive and more productive commodities.

There are currently no routine systematic methods for seeking, collecting and analysing feedback from disabled users about the barriers they face in their homes and neighbourhoods. Innovators and providers must therefore seek direct inputs from users on a "one-off", "ad-hoc" basis to find out what the barriers are and how they can be overcome.

Collaboration between users and innovators has an illustrious history. For example, the Everest and Jennings folding wheelchair was developed by a wheelchair user and an engineer working together. The POSSUM environmental control system was developed by an innovator seeking to provide autonomy and control for paralysed users and the King's Fund healthcare bed was developed by a design team working in conjunction with bed users. Each of the following examples represents efforts to combine the skills and expertise of disabled people, innovators and providers in Sheffield.

Innovations and collaboration between users and providers

Inclusive hotel bedrooms

In the development of the Four Star MacDonald St. Paul's Hotel, Sheffield and other hotels in the MacDonald group, the MacDonald development team had envisaged that they would require three distinct types of bedroom. These were planned to accommodate guests with no impairments, guests with physical impairments and guests with sensory impairments.

The challenge is always for design to go hand in hand with practicality and this is particularly the case when designing facilities for guests with impairments. This process has with out doubt evolved during the building of this hotel. After discussion with local users in Sheffield's Access Liaison Group they are now questioning the need for these three distinct types of bedroom and the limitations it imposes on the booking procedures. The hotel is now considering whether these three types of room could be integrated to provide a single, inclusive bedroom that can accommodate the needs of all its potential users.

Features of a room designed for guests with physical impairment would include wet area shower with built in seat, holding rails adjacent to the WC, wider doorways and repositioned furniture to allow for the turning of a wheel chair etc. Features of a room designed for guests with a sensory impairment would include light systems for fire alarm and doorbell, signage and printed material in Braille and a vibrating pillow alarm etc.

In reality all three types of rooms (standard rooms and modified rooms) are all very similar. Standards of luxury have not been compromised in the modified rooms at all and these rooms are highly suitable for all guests.

Over the months following the opening of the hotel it is hoped that we can establish a framework for a future where all hotel bedrooms are designed in a way that they are accessible for all guests. The bedrooms are just one aspect of a hotel that through close collaboration with the Access Liaison Group has many features to enable access for all. As many of the hotel's access features do not have a clear precedence or guideline within the industry; the hotel is taking the post opening stage of the hotel as an opportunity to continue to work closely with the Access Liaison Group. Hopefully. This will show continual improvement and collaboration, which can be of benefit to future projects.

The Weston Park Museum Project

The museum needs to attract everyone by being fully inclusive. This means that the building needs to be fully accessible and the services it offers – exhibitions, educational opportunities, workshops, lectures, need to be able to be appreciated by as many people as possible from all sections of society.

Sheffield Museums and Galleries recognized that in order to make their building and services accessible to disabled people they needed to consult with experts in the field of disability – disabled people who would be interested in enjoying their facilities provided they were enabled to do so. These experts needed to be able to advise at all stages of the museum refurbishment project, from meeting with the architects on a site visit, to visiting examples of good and bad practice, and to investigating innovative and creative ways of making the final product meet the criterion of full inclusivity.

This approach ensured that:

- Mistakes could be avoided
- Issues were not overlooked due to ignorance or lack of information
- Previous good practice could be drawn upon, thus avoiding the re-inventing of wheels
- Access solutions were realistic and workable the overall aim would be achieved.

Inclusive barriers on country walks

How do you design a barrier that isn't a barrier?

This was the question Keith Barraclough, retired sheet metal worker, conservation volunteer and designer, asked himself when it became apparent that his first, very successful restrictive access barrier wasn't as "user friendly" or effective as he would have liked. The task was to design a new structure that gave access to trails, footpaths and open spaces for all legitimate users whatever their needs, but excluded motorcycles and cars. Having had experience of consulting disabled users with his early design Keith was able to produce the K Barrier, however it was vital that the people the K Barrier was designed to include were actually involved in its fine tuning. Sheffield Access Liaison Group provided invaluable advice and experience and were able to demonstrate the practical requirements of people with a range of impairments as the design needed to be accessible not only to people who use a wheelchair, scooter or walking aid but also to people with visual impairments.

By having a series of consultations and visits it was possible to make the design as inclusive as possible, while maintaining a high degree of effectiveness against illegitimate use.

The Accessible Property Register

The Accessible Property Register (APR) works with Sheffield Homes to identify and promote accessible and adapted property. APR specialises exclusively in accessible and adapted property for sale and rent. Two of the three directors are disabled people. Property adverts (free for private individuals) are accepted from all sources and detailed information is provided about access features and adaptations.

APR has developed "access criteria" based on Building Regulations Part M. Criteria relate to parking, approach to the property, access, entrance level bathroom or WC, and access to entrance level rooms. All properties managed by Sheffield Homes are now inspected and assessed against these criteria when they fall vacant. Any adaptations are recorded at the same time. Ultimately therefore, a full record of accessible and adapted housing stock will be created.

The information is used by marketing arm, Sheffield Property Shop, to promote qualifying properties in one of three categories: Accessible, Adapted, or Accessible and Adapted. Under choice based letting procedures, these properties are advertised on a weekly basis on the APR website. Details can then be viewed alongside adverts from other sources, private individuals, estate agents etc. Potential tenants/purchasers therefore have the opportunity to view the full range of accessible and adapted property available in the area at any given time.

The next stage will be for Property Shop to provide the same information in all print media advertising and on their website. There are also plans to include a search facility on the website, allowing visitors to search specifically for accessible and adapted property.

Transport 4 All

Transport 4 All in Sheffield is a collaborative group of disabled public transport users, officers of both Sheffield City Council and South Yorkshire Passenger Transport Executive, and managers from some of the public transport operators in Sheffield.

The group is an evolutionary development of a "Low Floor Bus Group" that was established when "accessible" buses were being introduced and intended to create a link between disabled bus users and bus operators. The remit of the original group was too narrow, and changing legislation such as the PSV Accessibility Regulations (2000), and PSV Conduct of Drivers, Inspectors, Conductors and Passengers, Amendment Regulations (2002) meant that the bus industry was being forced into changing the ways that it operated anyway.

In the meantime, there was growing concern about the accessibility of theoretically accessible transport (like hackney carriages and the Supertram network) and new projects such as the redevelopment of Sheffield Rail Station and the building of new PTE infrastructure that meant that the group needed to broaden it's outlook and refresh it's approach. To this end, a revised terms of reference for the group was developed, and elected member support was gained, giving the group a more "official" standing, reporting via the Sheffield Access Liaison Group to Cabinet.

One of the early achievements of the group was agreeing the layout and type of tactile paving to be used at bus stops in Sheffield, compromising between the needs of wheelchair users and those with visual impairments. Ongoing work includes design standards for SYPTE facilities such as interchanges and travel information centres, as well as work with First to improve training for drivers to help them meet the needs of passengers with hearing impairments. Transport 4 All is chaired by a disabled member of the group and is facilitated and serviced by Sheffield City Council.

References

The Accessible Property Register www.accessible-property.org.uk

Mitchell, J and Chesters, R, 2004, "*Designing out the cell of exclusion*, Contemporary Ergonomics 2004, Ed Paul T McCabe, General Ergonomics, pp. 445–449, ISBN 0 8493 2342

TACTILE COMMUNICATION IN THE HOME ENVIRONMENT

Elizabeth Ball & Colette Nicolle

Ergonomics and Safety Research Institute (ESRI), Loughborough University, Holywell Building, Holywell Way, Loughborough, Leicestershire LE11 3UZ, UK

Communication is a crucial part of home-life. It includes interactions, both face-to-face and remotely, with others living in the household, friends/relatives and service providers. Effective communication in these, and other situations, is important to quality of life.

Deafblindness has a profound impact on communication. This paper considers some of the issues that arise when using English-based tactile communication and the implications for the home environment.

Deafblindness

Deafblindness is a combination of vision and hearing impairment, which causes difficulties with mobility, communication and access to information (Department of Health, 2001). The majority of deafblind people have some residual vision and/or hearing, which is sufficient to enable them to communicate aurally using speech or visually using sign languages. A minority of deafblind people have such severe vision and hearing impairments that they rely on tactile communication. This paper focuses on methods of tactile communication that are based on spoken languages.

There are two dominant approaches to attempting to address the communication needs of deafblind people, both of which take a content-focused, information processing view of communication. The rehabilitation approach internalises the problem to the deafblind person and attempts to treat and train the individual. The access approach externalises the problem in situational barriers and attempts to remove these. However, in many situations, especially within the home environment, communication is about more than the transmission of content – it is about relationships. Thus relational aspects of communication must be considered when evaluating tactile communication.

Communication in the home environment

Effective communication is crucial to quality of life. It forms the foundations of healthy family relationships and friendships, which are important to well being. Relationships may break down completely if communication is disrupted (Aguayo, 1999).

Communication is also important for interactions between service users and providers and for educational and vocational achievement and fulfilment. With an increasing number of services being offered to users in their own homes (e.g. home delivery of shopping) and increased opportunities for home-based learning and employment, effective communication in the home environment may become even more important and influential on an individual's overall quality of life.

English-based tactile communication

The two most widely used methods of English-based tactile communication are Braille and fingerspelling.

Braille is a form of tactile writing. Refreshable Braille displays enable deafblind people to access computers. Thus Braille can be used for distance communication, such as email or text-telephony. It can also be used for face-to-face communication if the speaker, or a language support professional, types onto a computer keyboard and the deafblind person reads from a refreshable Braille display – a form of speech-to-Braille reporting. Potentially, Braille could be used with automatic speech recognition.

In fingerspelling, each letter of the alphabet has a sign, which is felt against the hand. The speaker, or language support professional, spells out each word, letter by letter, onto the deafblind person's hand. In the United Kingdom, the most widely used form of tactile fingerspelling is the deafblind manual alphabet. Fingerbraille is a form of fingerspelling, which is rarely used in this country, but which warrants mentioning because of its potential for overcoming a major drawback of other forms of fingerspelling. The speaker or language support professional taps the deafblind person's fingertips as though they were the keys of a brailler. This enables all Braille contractions and abbreviations to be used. Thus, it increases brevity, maintains clarity and increases speed, relative to other forms of fingerspelling.

Tactile communication: is it effective?

Many issues arise when using tactile communication. Here we focus on four areas: attitudes, concurrent tasks, speed and nonverbal communication.

Attitudes

The British are well known for our inhibitions about touching. For some people, discomfort with using touch acts as an attitudinal barrier to the use of tactile fingerspelling. These people insist, if they have to communicate at all with a deafblind person, that it is done through a language support professional or using Braille, thus eliminating the need for physical contact.

Concurrent tasks

One of the most immediately obvious issues with tactile communication is that it occupies the hands. Thus it becomes impossible to do much else whilst communicating.

In the case of doing research interviews, a deafblind interviewer cannot take notes whilst using fingerspelling. Thus, if having notes is important, speech-to-Braille may be the better option, as the transcript can be saved. In the home environment, it becomes impossible to have conversations over meals or whilst engaged jointly in other activities. This inevitably alters the experience and may negatively impact upon the relationship.

Where there are more than two people present, including everyone equally in the conversation becomes difficult. Particularly if the deafblind person has their hands otherwise occupied, others may talk amongst themselves leaving the deafblind person out. If the deafblind person is included in the conversation, the person communicating to him/her must simultaneously voice the words for the benefit of others. Whilst some people find this easy to do, even helpful, others find it difficult.

Speed

Compared to speech or sign languages, both fingerspelling and Braille are slow. In the case of fingerspelling, Reed, Delhorne, Durlach and Fischer (1990) showed that at speeds of up to 5 letters per second, which equates to approximately a quarter of the speed of speech, fingerspelling could be received with high accuracy of 80–100%.

However, accuracy of reception fell as speed increased or complexity of sentences increased. The exception to this is fingerbraille. In Japanese, fingerbraille can be received at approximately seven-eighths of the speed of speech (Hoiuchi and Ichikawa, 2001). In the case of Braille, Hislop (1984) showed an average paper-based Braille reading speed of 126 words per minute. Most refreshable Braille displays have larger Braille cells and dots than paper-based Braille. Enlarged Braille letters slow down reading (Millar, 1977). Additionally, as Stuckless (1994) points out using QWERTY keyboard entry, it is impossible to enter text at the speed of speech. Thus speech-to-Braille though faster than most forms of fingerspelling, is still slower than speech.

Using slower methods of communication may have a number of consequences. Amongst them is that overall less communication may take place. This may sometimes mean whole topics are avoided. Often it means that summarisation is used. Adjectives are usually the first to go. A person who responds to nouns without the adjectives loses relational ground.

A further consequence of using slower methods of communication, especially if through a language support professional, is that there are inevitable gaps between the speaker finishing and the deafblind person receiving and responding. These gaps seem to cause a great deal of discomfort. For example, in a focus group involving a deafblind researcher using speech-to-Braille and four participants, three of the participants found the gaps so disconcerting that they refused to go ahead with the discussions. Gaps disrupt the flow of communication and relational ground is lost.

In the home environment, not only may slower communication lead to less communication and the loss of relational ground, it may also lead to interactions being broken off prematurely which may not only adversely affect the relationship but also prevent a deafblind person from giving or requesting further information.

Nonverbal communication

Nonverbal communication and context are critical to full understanding. Relational ground is lost, inclusion and control suffer, and little pleasure is realised, if nuance is not picked up, if emotional overtones are missed or sarcasm is taken seriously, for example.

Braille is a purely text-based system. Thus, it fails to transmit nonverbal communication, such as gaze, gestures and tone of voice (Fox, 1999). Communication, especially socio-emotional communication, is less effective in purely text-based systems (Sproull and Kiesler, 1991). Consequently, signals needed to understand the conversation, cues for turn taking and context may be missing. Frustrations and misunderstandings abound and development of relationships is impeded (Spears and Lea, 1992).

Unlike Braille, fingerspelling is capable of transmitting at least some nonverbal communication. Some emotions, such as shock, tension or laughter can be felt. Hoiuchi and Ichikawa (2001) showed that timing structure, that is, duration of touch and gaps between characters, is used to add emphasis and meaning in fingerbraille. The same is true of the deafblind manual alphabet, in the author's experience. However there is still a need to ensure that adequate cues about words, nonverbal communication and context are transmitted (Fuglesang and Mortensen, 1997). Interestingly, in the author's experience, speakers often underestimate how much nonverbal communication can be picked up through fingerspelling.

This has important implications for communication in the home environment. The use of Braille is likely to lead to less effective communication, misunderstandings, frustrations and inhibited relationship building. Yet, Braille is sometimes the only option: a visiting service provider, for example, is unlikely to be able to fingerspell but may be able to type or, in the future, to use automatic speech recognition systems, to communicate through Braille.

Considerations for designing home environments

Both the systems of English-based tactile communication commonly used in the United Kingdom, Braille and fingerspelling, have strengths and weaknesses. As a result, they are appropriate for different types of tasks.

Braille is well suited to tasks that require large amounts of factual information to be transmitted, but poorly suited to tasks where there is a high socio-emotional content.

In contrast, fingerspelling is relatively well suited to tasks involving high socio-emotional content, but poorly suited to tasks where speed matters or where either of the communicators feels uncomfortable about physical contact.

Consider, as examples, the following pairs of tasks. Within each pair, the topic and situations are similar. However, the first task (a) involves primarily factual information, therefore, most likely to be suitable for using Braille. The second task (b) involves a higher socio-emotional content and is, therefore, most likely to be appropriate for using fingerspelling.

1. a) Discussing instructions for cooking a recipe.
   b) Discussing how much a meal was enjoyed.
2. a) Accessing a broadcast news bulletin.
   b) Accessing a broadcast drama.
3. a) Talking to a social worker about the community care assessment process.
   b) Talking to a social worker about social care needs during a community care assessment.
4. a) A conversation with a partner about details of the travel arrangements for a holiday.
   b) A conversation with a partner about where you both want to go on holiday.

Currently, the most accessible home environment, with respect to tactile communication, would be one which enables communication partners to choose which method of communication is most appropriate to the task. Both systems, it seems, have their strengths and weaknesses, so must be used in combination for the best overall result. However, this is often not practical and is rarely achieved.

Often, communication is viewed simply as the transmission of information and the relational aspects are overlooked. This, combined with the facts that technology can support Braille and that strangers can usually type but can rarely fingerspell, has led to a tendency for Braille to be used more and more. Due to its limitations in relational communication, we must, it seems, be cautious not to overuse Braille-based technological solutions to deafblind communication.

However, technology may have much to offer in the future. To be most useful, technologies would need to not only interpret verbal information (speech recognition), but also nonverbal information. Both the verbal and nonverbal information would need to be transmitted in a tactile code that can be received at high speeds. Both of these issues need substantial work before technologies of this type could be of significant benefit to deafblind people.

Future work is needed to try to develop a method of English-based tactile communication, high and/or low tech, which combines the strengths of both Braille and fingerspelling and overcomes the weaknesses of both. The ideal system would be usable by strangers, socially acceptable and fast. It would provide quality verbal and nonverbal information and serve both the informational and relational purposes of communication.

References

Aguayo, M. 1999, *Rehabilitation of deafened adults: a puzzle with missing pieces.* Unpublished MSW thesis. Waterloo ON Canada: Wilfrid Laurier University

Department of Health 2001, *Social Care for Deafblind Children and Adults*. LAC, **8**
Fox, B. 1999, Directions in research: language and the body. *Research on Language and Social Interaction*, **32**, 51–59
Fuglessang, L. and Mortensen, O.E. 1997, Communicative Strategy – Including transfer to tactile mode. Plenary presentation at the *4th European Conference on Deafblindness*, Madrid, Spain, July 1997
Hislop, D.W. 1984, *Characteristics of tactual reading by blind optacon and Braille readers*. Doctoral dissertation, University of Illinois at Chicago. Dissertation Abstracts International, **45**(02B), 0799
Hoiuchi, Y. and Ichikawa, A. 2001, Teletext receiver by fingerbraille for deafblind. In *Proceedings of CSUN 2001 Conference on Technology and Persons with Disabilities*, California State University, Northridge
Millar, S. 1977, Tactual and name matching by blind children. *British Journal of Psychology*, **68**, 377–87
Reed, C.M., Delhorne, L.A., Durlach, N.I. and Fischer, S.D. 1990, A study of tactual and visual reception of fingerspelling. *Journal of Speech and Hearing Research*, 786–797, USA
Spears, R. and Lea, M. 1992, Social influence and the influence of the "social" in computer-mediated communication. In M. Lea (ed.), *Contexts of computer-mediated communication* (Harvester Wheatsheaf, New York), 30–65
Sproull, L. and Kiesler, S. 1991, *Connections: New ways of working in the networked organization*, (MIT Press, Cambridge, MA)
Stuckless, E.R. 1994, Developments in real-time speech-to-text communication for people with impaired hearing. In M. Ross (ed.) *Communication Access for Persons with Hearing Loss*, (York Press, Baltimore MD), 197–226

DEVELOPING NEW HEURISTICS FOR EVALUATING UNIVERSAL DESIGN STANDARDS AND GUIDELINES

Chris M. Law[1], Julie A. Jacko[2], Ji Soo Yi[3] & Young Sang Choi[3]

*[1]UMBC (University of Maryland, Baltimore County),
Department of Information Systems, 1000 Hilltop Circle,
Baltimore, MD 21250 USA
[2]Georgia Institute of Technology, Wallace H. Coulter
Department of Biomedical Engineering, 313 Ferst Drive,
Atlanta, GA 30332 USA
[3]Georgia Institute of Technology, School of Industrial and
Systems Engineering, 765 Ferst Drive,
NW, Atlanta, GA 30332 USA*

Inclusive/Universal Design (UD) standards and guidelines are widely promoted to aid in the design of electronic devices for use by people with disabilities. For the success of such UD Resources (UDRs), the needs of designers (the target readers) have to be met. To determine this, a systematic analysis using heuristic evaluation methods was conducted. Ten new heuristics based on design psychology, design methodology, technical writing, document design, and UD were developed. Five expert evaluators analyzed eight UDRs. This paper presents the process of creating the new heuristics, and highlights of the results, which revealed shortcomings in the content and presentation styles in a number of UDRs. The results are being utilized to frame the design of further studies of UD in practice. The authors briefly discuss implications of this set of heuristics beyond the realm of UD.

Introduction

Inclusive/Universal Design (UD) standards, guidelines and other resources are widely promoted to aid in the design of electronic devices for home and office use for use by people with disabilities. The needs of end-users with disabilities are typically addressed by such resources as the content focus; but the needs of the designers, who are the actual users and recipients of UD Resources (UDRs), also have to be met in order for successful utilization of UDRs and product development as a final goal.

Some investigations of design guidelines have previously been undertaken in the Human Factors/Ergonomics (HF/E) field. For example, Burns et al (1997) conducted experiments on the usability, usefulness, and viability of HF/E handbooks, reporting that up to that point, few studies had been carried out on the utilization of HF/E guidance; and criticizing the HF/E community by suggesting that "practicing ergonomists [...] may wonder how the handbooks are designed." Burns et al (1997) recommended that future HF/E guidance should be based on the explicit analysis of the needs of industry. In keeping with this recommendation, we sought to understand the use of UD guidance from the position of one of the primary "customers", of UDRs i.e., designers. This analysis was the first step in a multi-year project looking into Universal Design in Practice (UDiP). A heuristic analysis methodology was developed and utilized to examine eight different universal design resources.

The process of developing the new heuristics

Heuristic analysis is a method that has traditionally involved judgment of user interfaces according to recognized usability principles or "rules of thumb" (collectively "heuristics") (Nielsen, 1994). In the heuristic analysis method, a number of evaluators evaluate target designs according to pre-established design principles. Most published works in the human factors literature which employ heuristics, utilize the principles developed by Nielsen for the design for computer applications. These heuristics had to be modified for use with print resources. Purho (2000) created a set of heuristics for documentation, but an extensive literature survey did not reveal any past studies that created heuristics for analyzing documentation, guidelines, and similar design resources from the point of view of a reader who is a product designer. Therefore, a set of heuristics specifically for this purpose had to be created.

Literature from diverse fields such as design psychology (e.g., Petroski, 1996; Lawson, 1997), HF/E (e.g., Norman, 1990; Nielsen, 1994; Buchanan, 1995), instructional design (e.g., Van der Meij and Carrol, 1998), and technical writing (e.g., Shriver, 1997; Melenbacher, 2003) were consulted for creating the new set of heuristics.

From a systematic analysis of the rules, guidelines and recommendations in each of the various literature resources, a large set of principles and heuristics were proposed. The research team then ran pilot evaluations of the heuristics with selected UDRs, revised some of the heuristics, and narrowed down the list of principles and heuristics to what was thought to be manageable for a typical heuristic evaluation process. The resulting three principles and ten heuristics are briefly described below.

Principles and heuristics employed in the study

PRINCIPLE 1. Address the pertinent product design aspects

HEURISTIC 1.1. *A coherent vision, purpose and central idea should be provided*: The "central idea" can be regarded as the main purpose of the resource. The principle motivations, methods to be employed, and reasoning behind the creation of the resource should be clearly stated. If the central idea is not conveyed clearly, then it can be easy to misinterpret the aims of the authors.

HEURISTIC 1.2. *End-user and product goals should be addressed*: The purpose of the product existence for the end user is to help them achieve specific goals. It is sometimes easy to overlook the main tasks and instead focus on the components of products and this can make it difficult for readers/designers to be creative in problem solving.

HEURISTIC 1.3. *Potential end-user errors and failure scenarios should be addressed*: Some end-users will inevitably make mistakes and will need help with any interface. The guidance in the resource should address errors that might be made by users and suggest methods to appropriately deal with those errors.

HEURISTIC 1.4. *The factors in product development that are beyond the domain of the designer should be considered*: The reader/designer will be only one part of the whole team that is responsible for producing any given product. Resources should include advice on how to incorporate other pertinent product development factors into the overall design of the product.

PRINCIPLE 2. Support the design process and design psychology

HEURISTIC 2.1. *Action oriented approaches should be supported and encouraged for readers*: The design process is seldom straightforward and eludes reduction, and no one design approach or process can yield the complete answer. However, what is clear about designers is that they are action oriented, and this type of approach should be supported in the way the resource is designed.

HEURISTIC 2.2. *Inevitable trade-off decision-making should be supported*: In any design project, there will be design trade-off decisions needed due to limitations. Trouble typically occurs when one (design) factor dominates all of the others (for example, aesthetics, build quality, cost, ease of use, or accessibility). A consideration of balance of various factors is necessary, and this should be explicitly stated/addressed in the guidance.

HEURISTIC 2.3. *Designers should be allowed to bring past experience into a prescriptive problem-solving activity*: Scientists normally *describe* how things work whereas designers normally *prescribe* how things will work. This is an important distinction. The purpose of the resource should not be to only relay specifications (e.g., say what size the buttons need to be), but also to persuade and influence designers to carry out their prescriptive design activities in a manner that incorporates access for people with disabilities.

HEURISTIC 2.4. *Commonly employed self-referential viewpoints of readers should be actively countered*: A common mistake that all people make is thinking that other people are mostly like themselves. With designers, this self-referential approach can be dangerous and lead to interfaces that are usable for only the user who can behave like a designer. If this view is allowed, the fundamentals of interaction become lost on the typical user as the number of functions and modes increases, and greater likelihood of user error is also expected. This common behavior should be *actively* discouraged.

PRINCIPLE 3. Design the document effectively

HEURISTIC 3.1. *The design of the document should be clear and appealing*: Designing end-user manuals and documentation requires properly addressing a number of criteria, such as style, grammar, use of white space, illustrations, and so on.

HEURISTIC 3.2. *Different types of readers, and different usage over time should be supported*: Guidelines and standards documents should be readable for first-time learners and also for readers who are coming back to them for reference. The time-lag for returning readers from the initial learning may be extensive, so it should not be assumed that the reader will remember all of the pertinent information that is *not* contained within any given subsection of the resource. The design of the document, with proper indexing and cross-referencing, should facilitate use as both an introduction and a reference tool.

Results highlights

The subjective evaluations of five evaluators were aggregated at a full-day session in which all of the results for all eight resources were discussed. The individual evaluations are intrinsically subjective, but when summed across the evaluators, an objective measure can be achieved. Using scores, evaluators identified which of the resources were the most problematic. Major problems were found in 7 of the 8 resources. Fully detailed results will be submitted for publication elsewhere (see the project website given at the end of this paper for details). The following gives a summary of some of the interesting results that emerged from the analysis:

A coherent vision is rare: The resources were chosen for the study because they included guidance that was universal in nature (i.e., design of standard or mainstream Information and Communication Technology (ICT) products to accommodate functional limitations of users, including users with disabilities). However, one of the most immediate and striking results was that it was hard to find clear objective statements in the resources stating what the authors were trying to achieve with the resources (the "central idea"). In this field, there are a variety of approaches including, but not limited to, universal/inclusive design, design for disability, design for compatibility with assistive technology, and design of assistive technology. For the readers (designers) this presents a situation whereby it is incumbent on them to find out what the authors of the guidance wished to achieve. This could represent additional work for the readers and potentially lead to misinterpretation of the basic design idea.

Principle 2 (Design process and design psychology) was the most problematic: The process and psychology of design was the principle with which the most number of problems was found across the board. Only within this principle, we found instances of what were considered "catastrophic" problems (things that would be imperative to fix before release). The most instances of "major" problems (things that would be important to fix) were also found under this principle. In practice, this implies that the practical actions common in the design process are not supported by the majority of the resources. Readers would have to take the time and effort to re-interpret the guidance in the context of their work.

Supporting trade-offs in design decision making was a common shortcoming: In writing heuristic 2.2, the word "inevitable" was included ("Inevitable trade-off decision-making should be supported"). In all design texts and design psychology references, decision-making and trade-offs are addressed as an integral part of any process. This heuristic was linked to major problems in the majority of the resources. What was commonly seen was an all-or-nothing approach. Long lists of design requirements were typically presented with no comparative value or priority system attached to them.

Discussion and conclusions

A number of problems were found with the resources examined in this study, from the point of view of designers who are charged with implementing inclusive/universal design requirements. These problems were examined only in terms of meeting the common design aspects (principle 1), design process and psychology (principle 2), and document design (principle 3). The appropriateness of the content of the resource (for meeting the needs of end users of the eventual products) was not included in this study. However, even with the most appropriate content, if the delivery mechanism contains mismatches between the way that designers think/operate, and how the authors of the guidelines want to see their recommendations implemented, problems are likely to occur in practice. In subsequent phases of research, examining whether there are such mismatches exhibited in industry-based practice will be a major focus.

The heuristics used in this study could be applied (and modified if necessary) to the analyses of other design resources beyond the universal/inclusive design fields. It would also be possible to use these heuristics to analyze non-ICT arenas of universal/inclusive design, such as architecture, website design, and the design of non-electronic consumer products. By understanding the limitations, shortcomings (and positive, non-problematic) elements of existing guidance from the point of view of its recipients, the researchers hope to gain knowledge that will be beneficial to the production of future resources in the inclusive/universal design field.

The development of a set of heuristics to support the evaluation of design resources in the manner described in the study has proven to be useful in designing subsequent research methods for field studies of design in practice. The knowledge that there are shortcomings in the design of the majority of resources (that we examined) will be useful in framing the approaches to field studies including interviews with designers who are charged with using them to meet the needs of customers who have disabilities and/or who are aging, or who have functional limitations for any reason.

Acknowledgment and project website

This work is part of the *Universal Design in Practice (UDiP) Project* in the Wallace H. Coulter Department of Biomedical Engineering at Georgia Institute of Technology funded by the National Institute on Disability and Rehabilitation Research (NIDRR), US Department of Education (grant number: H133G040151). The opinions and content are those of the grantees. They do not necessarily represent the policy of the Department of Education.

More information about the *UDiP Project* can be found on the project website at http://www.isye.gatech.edu/lhci/udip.

References

Buchanan, R. 1995, Wicked problems in design thinking. In: V Margolin & R Buchanan (eds.), *The idea of Design*, (MIT Press), 3–20

Burns, C., Vicente, K., Christoffersen, K., and Pawalk, W. 1997, Towards viable, useful and usable human factors design guidance, *Applied Ergonomics*, 28(5/6), 311–322

Lawson, B. 1997, *How designers think: the design process demystified.* (3rd ed). (Architectural press)

Mehlenbacher, B. 2003, Documentation: not yet implemented, but coming soon! In: JA Jacko & A Sears (eds.), *The Human-Computer Interaction Handbook: Fundamentals, Evolving Technologies and Emerging Applications.* (Lawrence Erlbaum Associates)

Nielsen, J. 1994, Heuristic Evaluation. In: Nielsen & Mack (eds.), *Usability inspection Methods* (John Wiley & Sons)

Norman, D. 1990, *The design of everyday things.* (1st Doubleday/Currency ed.), (Doubleday)

Petroski, H. 1996, *Invention by design: how designers get from thought to thing,* (Harvard University Press)

Purho, V. 2000, Heuristic inspection for documentation – 10 recommended documentation heuristics. STC Usability SIG Newsletter, 6(4), http://www.stcsig.org/usability/newsletter/0004-docsheuristicshtml

Schriver, K.A. 1997, *Dynamics in document design*, (John Wiley & Sons, Inc)

Van der Meij, H. and Carrol, J.M. 1998, Principles and heuristics for designing minimalist instruction. In: JM Carrol (ed.), *Minimalism beyond the Nurnberg Funnel*, (MIT Press)

DESIGN, USABILITY AND UNSAFE BEHAVIOUR IN THE HOME

H.J. McDermott[1], R.A. Haslam[1] & A.G.F. Gibb[2]

[1] Health & Safety Ergonomics Unit, Department of Human Sciences,
Loughborough University, Loughborough, Leicestershire LE11 3TU
[2] Department of Civil and Building Engineering, Loughborough University,
Loughborough, Leicestershire LE11 3TU

Each year, almost 4000 deaths occur within the UK as a result of a home accident and 2.8 million domestic accidents result in the casualty requiring hospital treatment. New homes include many safety features to protect occupiers from injury, however the effectiveness of these can be dependent upon user behaviour. This research examined how behaviour interacts with design and how this may lead to an increased risk of injury. Forty, in-depth, semi-structured interviews were conducted with individuals inhabiting a new home. Various behaviours were reported in relation to building features including fire doors, pipes and loft access. The accounts demonstrate that designers need to give greater consideration to the interaction between occupier behaviour and building features so that improvements in design and occupier education may lead to improved health and safety.

Introduction

Unintentional home injuries are a serious public heath and safety problem worldwide. In the United Kingdom, almost 4000 deaths occur annually as the result of a home accident (DTI, 2001) and approximately 2.8 million home accidents occur which result in the casualty requiring hospital treatment (DTI, 2003).

House fires also present a cause for concern. Every year, more than 400 people are killed and over 10,000 injured in house fires within the United Kingdom (ODPM, 2004). The effects of fire can be devastating due to the disruption of domestic life and the loss of personal belongings. The average cost of damage caused by a house fire in the UK is £21,500 (ODPM, 2004).

The careful design of dwellings can help minimise the risk of unintentional injury. Despite this being widely recognised within governments and advisory groups, injury in the home remains common. In 2002, 20.1% of all UK home accidents were associated with a construction feature within the home (DTI, 2003). The features involved included stairs, banisters, stair posts, walls, windows, doors and door frames. According to Bonnefoy et al (2004), human behaviour and dwelling design are important contributory factors in home accidents, structural features can present physical dangers (steps, stairs and balconies) and occupiers themselves can create additional hazards through their behaviour. Haslam et al (2001) for example, identified a number of unsafe behaviours in relation to stair use amongst older people, including hurrying and the carrying of bulky or heavy items. They also identified patterns of behaviour that changed the nature of the environment itself, such as the leaving of obstacles on the stairs.

Legislation aims to protect the health and safety of individuals within their home through the application of national standards, codes and Building Regulations. Heimplaetzer & Goossens (1991) argued that whilst many "solutions" to health and safety problems have been translated directly into building codes or Regulations, these are then interpreted by architects and designers as a guarantee that maximum safety is provided. Architects and designers do not see the need to go any further than this. Heimplaetzer & Goossens claim many accidents within buildings continue to be "architecturally triggered", arising from an interaction between occupant behaviour and design. For example, falls may occur due to the need to climb on things to reach windows or cupboards, and impact injuries occur due to the positioning of doors, windows and low ceilings (Heimplaetzer & Goossens, 1991). Interestingly, they also claim that many solutions aimed at reducing the number of domestic accidents are chosen on the basis of partial or incomplete modelling. For example, in preventing children from falling down the stairs a closure may be fitted at the top of the stairs, but the consequences of this modification for adult occupants is ignored. In this manner, safety measures introduced to protect occupiers from one element of danger, can introduce additional hazards within the home. Pickett (2003) also highlighted this in a report on finger-trapping risks created by fire doors installed within dwellings.

The aim of this investigation therefore, was to gain an improved understanding of the ways in which people interact with their homes and to identify how behaviour interacts with design to affect health and safety. This included attention to the problems people have using home features and systems. This knowledge should be of value to those responsible for the design and construction of new homes.

Methods

Forty face-to-face, semi-structured interviews were conducted within the participant's home to collect information on the personal experiences of individuals inhabiting a new-build home. Home audits were also undertaken to identify where problems arose with design features and where modifications had been made. Each interview was conducted by the same researcher, and lasted approximately one and a half hours. The interviews were recorded, and later fully transcribed.

The qualitative data analysis followed three steps: data reduction, data display, verification and conclusion drawing (Miles and Huberman, 1994). Data reduction was achieved by coding of the data using the qualitative software NVivo and subsequent pattern coding of the initial codes (Miles and Huberman). Validation of the coding was achieved by independent review of a sample of the data and subsequent interpretation by another researcher. The pattern coding provided the basis from which the conclusions within this study have been drawn.

Results

The mean length of occupation of the properties in this study was 12.5 months (SD = 8.6). The age of participants ranged from 20 years to 65 years, (mean = 37.5 years, SD = 12.9). All participants were recruited on the basis of being the first occupiers of the property. Of the 40 properties visited, 4 were classed as detached, 3 as semi-detached, 5 as terraced, 20 as "town houses" and 8 were apartments/flats.

It was observed that 26 of the 40 properties were fitted with self-closing fire doors in line with UK Building Regulations. In all 26 of these, the occupiers had interfered with the doors in some way. Participants reported removing or jamming the self-closing mechanism itself in 9 of the properties, and in 25 of the properties, fire doors were wedged open preventing them from closing.

A 25-year-old male living in shared accommodation said:

> "I think obviously they are a good idea ... I'm sure there's another way of doing it"

A 35-year-old married female commented:

> "I understand the health and safety behind it but it drives me [mad], it worries me, they really go with a bang"

The participants provided a number of reasons for disabling their fire doors including inadequate internal lighting when the doors are shut, noise due to the doors slamming and the prevention of finger-trapping injuries.

Unsafe behaviour was also reported in conjunction with design features and systems within the home. Of the 40 properties visited 32 had been built with a loft (roof void accessible via a hatch), and a purpose built extending loft ladder had been fitted by the occupiers in only 5 of these properties. A loft-ladder had not been provided as standard by any of the house builders. In the remaining properties, access to the loft was achieved by various means including the use of general-purpose ladders, stools, furniture and fixtures. A 33-year-old male described how he had fallen from the loft access hatch when the drawers he was standing on fell:

> "So I could have fallen down the stairs quite easily. It's right next to the stairs. It fell that way fortunately, if it hadn't, I would have gone over the stairs."

Unsafe behaviour was also reported in relation to DIY tasks undertaken in the home, specifically in relation to electrical and water safety. In all the properties visited, the service cabling and piping was located within the walls. In 10 of the properties, occupiers reported that they did not consider the location of these services before drilling into the plasterboard, and described taking risks when hanging pictures or curtain rails. A further 15 properties stated they were unaware of the location of these services but did take care when drilling. However, 8 households had purchased a services detecting tool. Of these, 2 of the items only located electricity cables and in one instance, because of this, a water pipe had been damaged as a result of drilling. Only 2 of the householders recalled having been given a services map by the house builder outlining the location of electricity cables.

Participants spoke of problems in relation to a number of design features and systems within their home. Scalding occurred in 2 cases resulting from high water temperatures and 9 occupiers complained that the water temperature was too high. Although mixer taps were fitted in some bathrooms, the design did not always prevent the problem of scalding. A 21 year old female described the problem.

> "Even when you have hot and cold on at the same time it still comes out in columns, and if you put your hand under, ironic as it sounds you can still actually scald yourself"

In 2 of the 40 properties, the occupiers identified stair newel posts as dangerous features within their homes (newel posts are located at the top and bottom of a stair case and positioned at stair turns for structural support). These lead to the risk of injury as the result of head impact.

Another feature resulting in an increased risk of impact or head injury was sloped internal ceilings. These were located above the stairs and also within bathrooms on the top floor of three story properties. In 3 properties visited, occupiers complained of having struck their head due to these low ceilings.

Sloped external access to properties, in line with current UK Building Regulations, was suggested as presenting an increased risk of slips and falls during bad weather. Participants from 2 properties reported that the floor surface of the external access became slippery due to ice and water.

An installed safety feature, introducing an additional risk for falls, particularly for children, were emergency egress windows [a window provided for emergency egress purposes which should have an unobstructed openable area that is at least $0.33\,m^2$ and at least 450 mm high and 450 mm wide, (ODPM, 2004)]. An emergency egress window was fitted to windows on the first floor in 18 of the properties visited. Only 5 of these windows could be locked and only 8 were fitted with a restrictor bar which could be over ridden in the event of an emergency. In 6 of the 18 properties the windows could not be locked or restricted in any way. This led to a concern for the safety of children.

Discussion

This study has identified a number of unsafe behaviours present among occupiers of new dwellings, arising as a direct result of the occupant's interaction with the building features and systems within the home.

The study has also identified a number of problems that are experienced with specific architectural features and systems of the home. These problems may lead to a risk of injury as a result of interacting with the feature or through inadequate occupier modification.

The behaviours and problems identified in this study arose as a direct result of the occupant's interaction with the features and systems within their home, and are amenable to prevention through alternative design. This supports the suggestion by Heimplaetzer and Goossens (1991) that improved design of architectural features would reduce the potential for "architecturally triggered" accidents within buildings. The main cause of unintentional injury to children of all ages is falls. The provision of unlockable fire egress windows at height presents additional risk of falling for young children. This is a further example of what Heimplaetzer and Goossens (1991) refer to as "partial" or "incomplete modelling". In providing a safe egress route for occupants of the property, the installation of an egress window at height may have consequences for children in relation to falls. "Complete modelling" in this case, would refer to designers having taken into account all categories of users and all predictable patterns of use and misuse.

The findings also support Bonnefoy's (2004) claim that both human behaviour and dwelling design are important contributory factors in home accidents. This study has shown how structural features can present physical dangers (fire egress windows, loft access hatches) and also how behaviour can create additional hazards (tampering with fire doors, drilling into electricity cables). Complex interactions arise between the occupant and building features and it is important that architects and designers are aware of these interactions in order that preventative efforts through alternative design will be successful.

The results also support the work undertaken previously by Pickett (2003) in relation to self-closing fire doors. In the present study, it is notable that in each of the homes the occupiers had interfered with the fire-door mechanism in some way, thereby reducing the level of protection afforded through their installation. If the results of this study are indicative of behaviours practised in other homes, as suggested by Pickett (2003), the provision of internal self-closing fire doors is ineffective as a safety measure.

The findings reported in this paper are based on self-report data gathered during retrospective interviews with occupiers and the limitations of this methodology should be acknowledged. The sample of participants were self-selecting and may have held particularly strong views or had particular experiences, motivating their participation. The study called for participation at a very busy time for individuals, subsequent to moving into a new house and this may have influenced response levels.

This study has provided insight into the interactions of occupiers and their homes. The findings should be of interest to those responsible for the development of building standards, procedures and guidelines, informing them of the impact of occupier behaviour.

References

Bonnefoy, X.R. et al (2004) *Review of evidence on housing and health; Background Document,* Fourth Ministerial Conference on Environment and Health, Budapest, Hungary, 23–25 June 2004. World Health Organisation.

Department of Trade and Industry (2001) *23rd Annual Report of the Home and Leisure Accident Surveillance System – 1999 data,* Department of Trade and Industry, London.

Department of Trade and Industry (2003) *24th (Final) Report of the Home and Leisure Accident Surveillance System – 2000, 2001 and 20002 data.* Department of Trade and Industry, London.

Haslam, R.A., Sloane, J., Hill, L.D., Brooke-Wavell, K. & Howarth, P. (2001) What do older people know about safety on stairs? *Ageing and Society* 21, 759–776.

Heimplaetzer, P.V. & Goossens, L.H.J. (1991) Risks and accidents in the built environment. *Safety Science,* 14, 87–103.

Miles, M.B. & Huberman, A.M. (1994) *Qualitative Data Analysis, An Expanded Sourcebook.* Sage Publications.

Office of the Deputy Prime Minister (2004) *Fire Statistics Monitor. Issue No. 2/04* Covering period up to 30th June 2003. ODPM, London.

Pickett, J.W. (2003) *Report on the finger-trapping hazard created by fire doors installed in two types of three story houses.* Report prepared on behalf of The Royal Society for the Prevention of Accidents. ROSPA, Birmingham.

LIFESTYLES AND VALUES OF OLDER USERS – A SEGMENTATION

Patrick W. Jordan

University of Leeds, Leeds LS2 9JT

A psycho-social market analysis identified 18 separate segments distinguishing consumers in terms of factors such as demographics, social-economic status and attitudes and values. Six of these segments were either dominated by, or contained a significant proportion of, people of over 55 years of age. This paper looks at these segments in some depth, considering the design and marketing implications of appealing to each segment. When designing products and services for older people, it is important that we understand the different lifestyles and values of various people within this demographic. Thinking of "older people" as if they are a single group is not sufficient if we are to create products and services that will connect with users both practically and emotionally.

Introduction

Designing for an aging population has become established as a key area of practice and research within design and related disciplines. The contribution of ergonomics to this has been considerable. Substantial progress has been made in understanding how to design products for those who are experiencing the effects of aging. These effects can include reductions in physical and cognitive function and the onset of disabilities.

However, as well as designing products so that they fit people's physical and cognitive characteristics, it is also important to design products that connect with people emotionally (Jordan 2002, Norman 2005). A positive user-experience is not only about how easy the product is to use, but also about how people feel when they use the product. A well-designed product can make people feel good about it and about themselves.

Creating products that offer a positive user-experience requires that we understand people not only cognitively and physically, but also emotionally. We need to understand people's lifestyles, values and aspirations – their hopes, their dreams, their fears.

Contemporary Trends Institute Segmentation

A study carried out by the *Contemporary Trends Institute* (CTI) created a segmentation of users based on a number of psycho-social variables. These included demographic variables such as age, affluence and family circumstances and also attitudinal variables, capturing the way that they see life in general and their attitudes towards products and services in particular.

In total, 18 market segments were identified, 6 of which contained a significant proportion of older people. For the sake of this analysis "older" refers to people of 55 and over. This is the age that is typically used in the inclusive design literature when referring to older people – although in Western society it might be argued that 55 hardly

---

p.jordan@leeds.ac.uk

seems "old" anymore and certainly many people in this age bracket would not think of themselves in this way. Nevertheless, it is often an age at which lifestyle changes are occurring, for example, people's children are leaving home. It is also an age after which physical disabilities and deficits may be significantly more likely to occur.

In this paper, we will look at the segments in more depth. A descriptive overview is given of each and "icons" associated with each are also identified. Icons are celebrities or others in the public eye whose personality and lifestyle either captures something of the essence of what each segment represents or is in some way aspirational for the people in the segment.

Advice is also given about the sorts of design issues that are likely to be important to the people in each segment and the brands that are likely to appeal. Although the segments are global, the icons and brands that appeal are likely to vary from country to country. In this case, those listed are with reference to the UK. This does not mean to say that the icons and brands will be from the UK, but it does mean that they will be ones that are familiar to UK residents.

After the older segments have been described, brief descriptions of the others in the CTI Segmentation will also be given. The aim of doing this is to help to put the older segments into context.

Older segments

Senior professionals

Tend to be older people – although not always – who have worked their way up the corporate ladder. They are highly educated people who excel at their jobs, but who also value their home and family lives. May be highly variable in terms of social attitudes. Many will be status aware and will tend to favour traditional high-end brands which affirm their professional success.

Many in this segment will be "baby boomers", the post-war generation who had their youth in the 1960s and many of them may have values that reflect this. This might mean, for example, that they value things such as youth, excitement, freedom and rebellion and will enjoy products that speak to those values.

Harley-Davidson is an example of a brand that appeals to this group and they have become increasingly successful in the marketplace based on sales to this segment. Another brand with a strong appeal here is Porsche. In the home they may favour Sony. An icon for some in this segment would be Mick Jagger – still a rebel in his sixties and someone with the energy and dynamism to show the younger rivals in his industry a thing or two about rock and roll!

Traditional professionals

Tend to be older people in the traditional professions. They will be highly educated and will have a strong sense of values, tending to be somewhat conservative. Many will have kids who have since left home. Many will also have grown up in wealthy families and will have an understanding of traditional quality brands. They will recognize these brands and tend to be willing to pay more for them. However, they may pay little attention to newer brands.

When it comes to automotive brands, this group is likely to favour the sensible high-quality German brands such as Audi, BMW and Mercedes. Philips is a brand that we might find in their homes. Generally speaking, this group is a little less rebellious at heart than the boomers in the Senior Professionals Segment.

In terms of taste they tend to be more likely to favour understated, perhaps even slightly conservative designs and will look for good build quality, durability and reliability in the products that they buy.

Affluent retired

These people have retired with sufficient savings to be able to treat their retirement as a sort of extended vacation. They travel a lot, both domestically and abroad and enjoy eating out, shopping and other leisure activities. They have a high disposable income and few financial responsibilities. They also have plenty of time and opportunity for spending their money. As well as being attracted by traditional upscale brands they may also be discovering a number of new brands and will be comparatively open to them. Typically they are quite highly educated.

This group are not likely to see retirement as simply being a period to wind down and take it easy, but rather as an opportunity to do the things that they never had time to when they were working. Many will become active in the local community and the voluntary sector, using the skills that they have developed over their lifetimes to use in public service. Although many might have been in the Senior Professionals segment during their working lives, they will be older than the people currently in that segment and may be less likely to hold "baby boomer" values – in particular they may be less image-conscious and less concerned about appearing old.

They are affluent, but still careful in their purchase decisions and will generally prefer products that offer quality and value. Trusted automotive brands might include Toyota and Lexus, while Philips might be a brand they would trust in the home. Favourite celebrities might include Terry Wogan and Michael Parkinson.

Downscaled retired

This segment has retired and are now struggling to make ends meet. Retirement has meant a significant drop in income and this group is mainly dependent on their savings, which may not be that great. As a result they may have had to take a drop in their standard of living, although with few responsibilities they may still have a reasonable level of disposable income. This is likely to lead to a more conservative attitude towards brands with a preference for "safer" value brands. This group may come from a wide variety of different professional backgrounds and also be varied in their social and moral attitudes.

Rather than seeing retirement as opening up a world of adventure, Downscale Retired will see it as an era where they have to carefully manage their finances. However, even without a large disposable income, they may still have enough money to treat themselves modestly on a regular basis. On a day-to-day basis, this may mean things like going to the pub for a drink or enjoying other social activities such as Bingo and gaming. Some will take a holiday in a traditional UK destination or perhaps a package holiday abroad.

Favourite celebrities for this group might include people such as the game-show host Bruce Forsyth and actor Sean Connery.

Traditional skilled

This group consists largely of blue-collar skilled workers in industries where there is a strong tradition of solidarity with their co-workers. They want the best for themselves and their families and see their personal prospects as being strongly linked to the prospects for their profession as a whole. They will want to buy good quality and good value brands for themselves and their families. The segment includes a wide age-range and wide range of family circumstances.

Although this group may not be as heavily skewed towards an older age group as some of the others it does nevertheless include a significant proportion of older people. Because those in this group have spent their career in skilled manual labour many of them may have picked up musculo-skeletal or other problems associated with the job.

These might include, for example, back problems, repetitive strain injury (RSI) related impairments and hearing deficits.

This group may be quite traditional, even conservative in their values and perhaps less open to new products and services than some of the other older segments. They will also tend to have a lower level of education than some of the other groups. This is mainly because their careers have relied on skills rather than academic qualifications.

Many in this group will tend to instinctively prefer British brands. Ford and Vauxhall may appeal for example.

Assisted elderly

This group of people have age-related disabilities and need to be cared for in their old age. They may be cared for by their families, a care professional who comes to their homes, or they may be resident in accommodation with on-site care facilities. The group will experience a wide variety of financial circumstances and may have a wide variety of values and attitudes towards brands. Others – those who care for them – are likely to have a large influence over their brand choices.

Because this segment are likely to be facing physical challenges, they may be particularly aware of inclusive design issues and may also use a variety of assistive devices. The practical elements of what a product or device can do for them might therefore outweigh any brand or image considerations. However, this is not to say that they are happy about this. Many in the segment may feel stigmatised by having to use "special" devices with aesthetics that mark them out from the mainstream.

A challenge with respect to this group is to create products and services that are designed in such a way that they help them to live as independently as possible. Examples are products such as the Good Grips range from Oxo. This range includes a series of kitchen appliances that are designed so that they can be used by people with arthritis. In practice, this has meant designing them so that they are well balanced and have a handle that is designed from materials that are easy to hold and grip. However, they are not designed to look like "special" products but styled attractively. This has given them a far wider appeal and they have been successful in the mainstream market – a genuine inclusive design success.

Other Segments

Explorers. Creative, educated people who enjoy exploring new things and discovering things about themselves.
Pleasure Seekers. Their career is very important to them. Enjoy the trappings of success.
Life Balancers. Aim to strike a balance between home and work life.
Creative Entrepreneurs. Imaginative, driven people who have built successful businesses out of doing things that they love.
Urban Socialites. They want to do well at work, but tend to "work to live".
Strivers. They are hard workers with a strong work ethic.
Intellectual Liberals. Highly educated people with a strong liberal ethical agenda.
Strugglers. These people are often struggling to provide the basics for themselves and their families.
Self-Reliant. Strongly independent. Take pride in supporting their families.
New Traditionalists. Tend to be middle aged and often in the traditional professions. Strong sense of values.
Teenage Rebels. Tend to like "edgy", rebellious brands.
Young Adults. Teenagers who tend to be more mainstream in their tastes.

Conclusion

When designing products and services for people it is important not only to take into account their cognitive and physical characteristics, but also to look at their attitudes and values and understand how these might influence what they want from a product. The descriptions in the segmentation demonstrate a wide variety of attitudes and values across the six older segments. If we are to design with these people in mind, then it is essential that we are mindful of these and that we consider their implications for design.

The older groups demonstrate a diversity similar to that in the segmentation as a whole. It is no longer sufficient for us to think simply in terms of designing for "older people". Rather, we must seek to understand the lives and values of older people at the same level of detail and nuance as we would the market as a whole. In this way, we can create effective, efficient and pleasurable products and services that will be a joy for older people to own and use.

References

Jordan, P.W., 2002. How to Make Brilliant Stuff that People Love and Make Big Money Out of It. (London: Wylie)

Norman, D.A., 2005. Emotional Design. (New York: Basic Books)

GOING OUTSIDE OF THE FRONT DOOR: OLDER PEOPLE'S EXPERIENCE OF NEGOTIATING THEIR NEIGHBOURHOOD

Rita Newton, Marcus Ormerod & Vanja Garaj

SURFACE Inclusive Design Research Centre, BuHu Research Institute, The University of Salford, Maxwell Building, Salford M5 4WT

This paper reports on one aspect of the preliminary findings of a research project investigating the most effective ways of ensuring that the neighbourhood environment is designed inclusively to improve the quality of life of older people. Focus groups with older people were conducted in different locational environments. Whilst these discussions covered a broad range of issues, we report on what older people perceive as the benefits of going outdoors and where they typically go within their local neighbourhood, plus an examination of the barriers to walking. The results largely confirm anecdotal evidence in that older people have concerns about the quality and maintenance of the environment, provision (pavements, suitable crossing points, adequate lighting) and personal safety (fear of accidents with cyclists and cars, fear of falling due to pavement surface, anti-social behaviour and fear of crime).

Introduction

There is currently a drive within many countries to focus on the quality of public spaces. This is because there is an increasing body of research which suggests that going outdoors is good for us, and also the quality of public spaces is poor due to lack of funding, coherent strategies, and only pockets of good practice. People place the quality of their local environment high on the issues that concern them and most need improving, often higher than issues of healthcare and education (ODPM 2004). It is the issues that affect people's quality of life at a local level that they want most improved. MORI (2002) suggest that this is not surprising considering that the majority of people do not use schools, public transport, health or social services, but do on a daily basis use "the street outside their front door, the local neighbourhood and the environment around their workplace". Research has also shown that the external environment makes a significant contribution to the well-being and quality of life of the people who use it. CABE (2004) suggest that "open spaces are a powerful weapon in the fight against ill-health" referring to the positive effects of walking in reducing heart disease (Bird 2003) and in helping us to live longer (research in Japan: Takano et al 2002) and in reducing blood pressure and stress (Hartig et al 2003).

Additionally, designing the external environment such that people can continue to live in their own homes, particularly if they acquire an impairment or are older, is important. A study by Hok Lin Leung (1987) looking at the housing concerns of older homeowners in Ontario, Canada, found (amongst other things) that older people generally place a high value on independence and wish to remain in familiar surroundings, and in the same neighbourhood in order to maintain existing social relationships. In terms of access to and use of public transport by older people, research has shown that the physical environment has a major impact on this (Salvage & Zarb 1995, DETR 1998, DPTAC 2002).

What are the difficulties that people experience when going outdoors?

We can start to build up a picture as to what are the barriers that people face when they go outdoors, based on findings in Nasar et al (1993), Lavery et al (1996), Kitchin and Law (2001), DTLR (2002), and Brook Lyndhurst (2004) and we see issues emerge of:

1. Uneven pavements, broken paving stones;
2. Gradients that are too steep (hills);
3. Poor lighting;
4. No shelter at the bus stop;
5. No rest facilities (seating) at the bus stop;
6. Lack of dropped kerbs, and awkward kerbs;
7. Roads without crossings;
8. Environmental quality issues such as litter, graffiti and vandalism;
9. Lack of, or the poor condition of, facilities (such as public parks, public toilets);
10. Other users of the environment, including undesirable characters;
11. Concerns about dogs and dog mess;
12. Personal safety and other "psychological" issues;
13. Fear of crime.

The list above is not exhaustive since there are many reasons why people do not wish to visit a particular area, or even struggle to navigate the environment in order to get to the local bus stop. The difficulty is that not all these reasons will be documented, because as previously stated, the study of external environments has been patchy and disparate.

Seven focus groups were undertaken with forty older people. Whilst nine participants preferred not to provide personal information due to cultural reasons relating to their ethnic background, of the thirty-one who did provide information, 81% were white and 19% were non white (Asian British – Pakistani or Bangladeshi); 75% were female and 25% were male; most participants lived in their own homes (94%), the age range was 65 to 84 years of age with 52% of participants under 75 and 48% of participants 75 to 84 years old. 65% of participants had one or more impairments (eyesight sixteen participants, hearing thirteen participants, walking difficulties ten participants). Participants discussed the benefits for themselves of going outdoors, and the typical places they visited in their neighbourhood environment, and these are summarised below.

Benefits of going outdoors:

- For exercise;
- To relax;
- To socialise with people you know, and people you don't know;
- For variety;
- A feeling of being in control (even if only pottering in the garden);
- To enjoy the countryside and nature;
- To benefit from fresh air;
- To aleviate depression of being indoors all the time.

Activities being carried out outdoors:

- Walks in the park, along the canal, in the countryside;
- Walks along the streets;
- Local shopping;
- Visit the GP surgery;
- Visit family members or friends;
- Going to church;
- Maintaining the garden;
- Sitting and relaxing in the garden;

- Visiting the local restaurant or pub;
- Going to the local community centre.

Participants stressed the need for local facilities that were within easy walking distance of home (such as green space, a local shop, the doctor), particularly as car driving became

Table 1. Older people's experience as a pedestrian using their local environment

Design and maintenance issues	
Quality of the paving	Uneven and cracked paving and the fear of tripping and falling over; low levels of cleanliness *"footpaths are terrible .... dirty, chewing gum, litter"*
Provision of a pavement	Where there is an absence of a pavement particularly in rural areas, concerns for personal safety due to traffic; when pavement is a grass verge (rural areas) maintenance of the grass so that a suitable walking surface is provided
Street works	Difficulty in negotiating works when on the pavement, difficulty in crossing the road due to road works
Road crossings	Poor visibility *"sometimes when you stand at a crossing, you cannot see the traffic coming because a building blocks the line of sight"*; difficulty of crossing the road at roundabouts; traffic lights seemingly to favour motorists rather than pedestrians (length of time to wait, length of time to cross the road); absence of controlled crossing point; faulty traffic lights (urban focus group participants) making crossing roads dangerous and difficult; fear of falling on tactile paving at crossings
Street lighting	Poor light levels at night along the street, and during the day in underpasses; no street lighting (rural)
Topography	Facilities and services inaccessible because of steepness of the environment
Seating	Insufficient number; poorly maintained; not replaced when un-useable
Public toilets	Insufficient number, poorly maintained, quite often locked
Social behaviour affecting walking	
Speed of traffic	Makes crossing a road difficult, also a problem when there is no pavement (rural areas)
Cycling on pavements	Fear of collision as a pedestrian with cyclists due to the absence of cycling lanes and number of cars parked on roads which were not originally designed to accommodate parked cars and traffic
Fear of crime	During quiet times particularly out walking in the evening; during busy times (eg shopping during the day) in crowds
Anti-social behaviour	Of young people on the streets
Parked cars	On pavements – *"you have to go around them all the time"* On roads – on yellow lines such that crossing the road is difficult (particular issue for participants with poor eyesight)
Vandalism	Public toilets, benches, bus shelters and information points
Car drivers not stopping at uncontrolled crossings	*"Crossing roads is difficult and dangerous because drivers cannot be trusted to stop"*
Weather	
Snow	Left on pavements makes them slippery
Wet weather	Provision of shelter (bus shelter)

more difficult with increasing age. When asking participants about their local neighbourhood environment and what they felt hindered their interaction within the neighbourhood and enjoyment of it as a pedestrian, the issues emerged are listed in Table 1.

Conclusion

Sir Stuart Lipton (CABE 2004) makes the point that public space is "our open air living room, our outdoor leisure centre", and the planning, design and management of this space is crucial (ODPM 2004). Local neighbourhoods are no exception especially since we access them from our front door step. Our discussions with older people in different neighbourhood environments have largely confirmed anecdotal evidence in that older people have common concerns about the following:

1. Quality – cleanliness, good walking surfaces, vandalism;
2. Maintenance – street works and road works, seating, toilets;
3. Provision – pavements, suitable crossing points, adequate lighting;
4. Personal safety – fear of accidents with cyclists and cars
   – fear of falling due to pavement surface (uneven, slippery, tactile)
   – anti-social behaviour
   – fear of crime

This summary reinforces our understanding and has provided a platform for a more in-depth study comprising detailed interviews with older people and audits of neighbourhood environments. Further information available at www.idgo.ac.uk

References

Bird, W. 2003. *Nature is good for you*. ECOS. **24** (1) 29–31.
Brook Lyndhurst 2004. *Sustainable Cities and the Ageing Society: the role of older people in the urban renaissance*. Final report for ODPM.
CABE 2004. *The value of public space*. London. CABE Space.
DETR 1998. *Older people, their transport needs and requirements*. London. DETR.
DPTAC 2002. *Attitudes of disabled people to public transport*. London. MORI.
DTLR 2002. *Improving urban parks, play areas and open spaces*. London. DTLR.
Hartig, T., Evans, G., Jamner, L., Davies, D., and Garling, T. 2003. Tracking restoration in natural and urban field settings. *Journal of Environmental Psychology*. **23** 109–123.
Kitchin, R., and Law, R. 2001. The socio-spatial construction of inaccessible public toilets. *Urban Studies*. **38** (2) 287–298.
Lavery, I., Davey, S., Woodside, A., and Ewart, K. 1996. The vital role of street design and management in reducing barriers to older peoples' mobility. *Landscape and Urban Planning*. **35**. 181–192.
Leung Lin, H. 1987. Housing concerns of elderly homeowners. *Journal of Ageing Studies*. **1** (4) 379–391.
MORI 2002. *The rising prominence of liveability*. London. MORI.
Nasar, J., Fisher, B., and Grannis, M. 1993. Proximate physical cues to fear of crime. *Landscape and Urban Planning*. **26** 161–178.
ODPM 2004. *Living places: caring for quality*. London. ODPM.
Salvage, A. and Zarb, G. 1995. Measuring disablement in society – Working paper 1 Disabled people and public transport. www.leeds.ac.uk/disability-studies/archiveuk
Takano, T., Nakamura, K., and Watanabe, M. 2002. Urban residential environments and senior citizens, longevity in megacity areas: the importance of walkable green spaces. *Journal of Epidemiology and Community Health*. **56** 913–918.

AN EVALUATION OF COMMUNITY ALARM SYSTEMS

Sue Brown[1], Maxine Clift[1] & Lorraine Pinnington[2]

[1]Assistive Technology Evaluation Centre, Centre for Evidence-based Purchasing,
[2]Division of Rehabilitation and Ageing, University of Nottingham, Derby Hospitals NHS Foundation Trust, Derby City General Hospital, Uttoxeter Road, Derby DE22 3NE

A growing number of older people and those with disabilities are living in the community and often alone. It is increasingly important, therefore, for providers of support services to include telecare as part of the "care package". Evaluations of community alarm systems have been undertaken as recently as 2003. As technological changes have been rapid during the intervening period and many product ranges have increased in number or complexity, research data quickly become less relevant (Ricability, 2003). Confidence in the use of community alarms is dependent on many interrelated factors, such as ease and success of use. This evaluation will adopt an ergonomic approach to identify the features and functions of community alarm equipment. It will also assess the degree to which the devices meet the needs and abilities of different clinical populations.

Background and rationale

Assistive Technology (AT) is defined as equipment used by older people and those with disabilities to maintain independence and functional ability (Audit Commission, 2004). People who are at risk of personal injury or vulnerable to environmental hazards may still be able to live independently provided they are supplied with community services and appropriate assistive technologies. Although AT cannot fully replace the personal care that is often provided by relatives, friends or staff, it should always be considered as an option and included as an integral component of "care packages" provided to people in need of comprehensive community support.

Community alarms were first introduced in 1948 (Porteus and Brownsell, 2000). These early systems enabled individuals to call for help in an emergency. This Electronic Assistive Technology (EAT) has developed into "second generation" equipment which is now referred to as telecare (Porteus and Brownsell, 2000). Through technological advances and the increasing number of people needing support in the community additional components have been developed to monitor both the environment conditions and the individual. A basic community alarm system comprises a base unit which is connected to a telephone monitoring control centre, activated by a manually operated trigger. More recent systems can incorporate sensors and monitors which detect changes in the environment and user circumstances to alert a control centre automatically.

An ageing population and a changing focus towards independent living have resulted in an increasing demand for community support services. To meet this increasing demand the UK government has awarded an £80 million Preventive Technologies Grant (PTG) to local authorities for implementation of technology focused care provision. These funds are designed to "initiate a change in the design and delivery of health, social care and housing services and prevention strategies to enhance and maintain the well-being and independence of individuals" (DoH, 2005). In order to utilise these funds effectively, the Centre for Evidence Based Purchasing has commissioned this study to evaluate community alarms and monitors from a features based perspective. The data will enable key stakeholders to use objective evidence to inform the selection of devices. As a result, the design of care packages will incorporate equipment which is appropriate to the changing needs and abilities of the user.

Aims and objectives

The overall aim of the study is to evaluate the advantages and potential limitations of community alarms and monitors available on the UK market.

More specifically, we aim to:

- Describe and appraise the product range and their features in the context of different user needs and abilities
- Appraise the suitability of different product designs and supporting literature in relation to physical, cognitive and sensory impairment
- Identify factors which impede or enhance user compliance.

Scope and methods of the evaluation

Consultation period

The scope of the study will be partially informed through consultation with users, manufacturers and professionals prior to submission for ethical approval.

Selection of technology

Telecare is defined as "the remote or enhanced delivery of health and social care services to people in their own home by means of telecommunications and computer-based systems" (Barnes et al, 1998). The purpose of this equipment is to support independent living and safety in the home by tailoring the system to the meet the needs of the individual. In this study, the term telecare refers to community alarm components which are compatible with a basic package comprising a base unit and manual trigger system. Some sensors and monitors detect potentially hazardous changes in the environment, or the circumstances of the user, and are designed to initiate an alert automatically. Other devices are activated through the actions of the user. Once alerted control centre staff initiate contact with the user and if necessary activate an appropriate response.

Participants

Participants in the study include the following:

Professionals – Health and social care workers are involved in referral for community alarm systems. Control centre staff assess the environment, install the equipment and instruct the user and informal carers.

Informal carers – include family, friends, neighbours and colleagues; people who are in regular contact with the user and therefore need to know how the system works.

Equipment users – consideration should be taken of the age range, abilities and impairments of telecare users as well as their lifestyle and home environment.

Manufacturers – have a vested interest in the format and findings of the study and are included at various stages of the evaluation such as scoping meetings and commenting on the findings for the final report. They are also asked to provide a sample of their products for evaluation.

Evaluation stages

The evaluation will utilise a multi-method approach and will focus on the needs and abilities of the user. The methods can be structured into four stages:

a) Literature review

A search of current product, clinical and research literature will be undertaken to identify key issues and research priorities.

b) Professional appraisal

Health and social care staff and control centre staff will be asked to evaluate the features and functions of the full range of monitors and sensors on the market.

c) Ergonomic evaluation

The features and functions of products will be considered in relation to usability addressing both physical and cognitive abilities. Product literature and training material will also be evaluated.

d) User trials

Monitors and sensors which are triggered by the actions of the user will be included in the user trials. A representative sample of products on the market will be selected to include:

- triggers activated by pressing a button or pulling a cord
- sensors which register that the user has fallen over
- pressure sensors which register the presence or absence of the user
- monitors which use pixelated images to register the presence or absence of the user.

New and established users of community alarms will be interviewed in their home environment to establish:

- the extent to which the alarm system meets their needs
- usability and compliance issues.

Deliverables and dissemination activities

- a product and research literature review
- a web-based searchable database of telecare products and their features
- a full technical and summary product evaluation report
- a summary evaluation report for non-professionals.

References

Audit Commission, 2004, *Assistive technology independence and well being 4*, (National Report, Audit Commission Publications, UK)
Barnes, N.M., Edwards, N.H., Rose, D.A. and Garner, P. 1988, Lifestyle Monitoring – Technology for Supported Independence, *IEE Computing and Control Engineering Journal* 9 (**4**): 169–174
Department of Health, 2005, *Building Telecare in England*, (Department of Health, Policy Document, Older People and Disability Division, London) p.8
Porteus, J. and Brownsell, S. 2000. *Using telecare: exploring technologies for independent living for older people*, (Anchor Trust, Pavilion publishing, Brighton)
Ricability, 2003, *Calling for help. A guide to community alarms*, (Ricability, London)

A STRATEGIC SPATIAL PLANNING APPROACH TO PUBLIC TOILET PROVISION IN BRITAIN

Clara Greed

Faculty of the Built Environment, University of the West of England, Bristol BS16 1QY

Everybody needs to go to the toilet. Public toilets should be seen as an integral component in urban design and city planning, and not as a money wasting, add-on provided with little consideration of the spatial, social or design aspects of the city. Emphasis is put upon seeing policies for downtown toilet provision within the context of establishing a hierarchy of toilet provision across the city as a whole, looking at city-wide "macro" level, the "meso" district level and last, but not least, the local "micro" level of toilet block siting. Public toilets need to be mainstreamed into strategic urban policy, transportation policy and urban design considerations. Emphasis is put upon survey, analysis and plan-making processes and upon urban design principles in making effective toilet provision.

A toilet distribution hierarchy

A great deal of attention has been given to the internal design of toilets (ODPM, 2004), but there is not commensurate emphasis on toilet location and distribution. But, availability is of greater importance, for example, to incontinent users, than the specifics of toilet cubicle fixtures and fittings (Hanson et al, 2004). Many people are severely constrained by "the bladder's leash" (Bichard ct al, 2004). Women and the elderly have to plan their journeys carefully, or give up going out altogether, as a result of toilet closure. Over 40% of public toilets have closed over the last 10 years. Some local authorities have no public provision anymore. Everyone needs "away from home toilets" if they are out all day (BTA, 2001). They are "the missing link" essential to getting people out of their cars and back on to public transport. Yet there is no statutory requirement that public toilets must be provided or that public toilet locations should be shown on urban development plans. The still extant, 1936 Public Health Act. Section 87, sub Section 3 gives local authorities permissive powers to build and run on street "public conveniences", but does not compel them to do so (Greed, 2003:52). The lack of a requirement for mandatory toilet provision is the fundamental problem that needs to be addressed. Nevertheless, this paper seeks to move the debate forward by providing guidance on where and why toilet facilities might be provided to meet the needs of modern society.

City-wide "macro" level provision

A high-level strategic perspective on toilet distribution across the city is needed, creating a spatial hierarchy of toilet provision that is appropriate to toilet needs in each locality. An *ad hoc* approach to siting has resulted in many a good toilet being under-used and vandalised because it is in the "wrong" place, whilst under-provision exists in areas where there is a heavy footfall (BTA, 2001). Toilet "hotspots" need to be identified. For example, the city centre generates high levels of toilet need, being the focus of retail, tourist, and employment activities. Secondary district centres and local points of attraction also need to be identified where toilet demand is likely to be high, such as around sports stadia or tourist attractions.

Many local authorities have no toilet strategy or plan. They have inherited a hotchpotch of toilets, built at various times for different reasons, which they kept open as best they could against difficult odds. A minority of authorities do have a toilet strategy, but the purpose is to "rationalise" and reduce the number of toilets available rather than to plan for unmet needs. One local authority had a location policy which put toilets into three categories, strategic toilet locations (on main roads and in central areas); tourist areas; and local district facilities. In the course of its cost-cutting rationalisation process the local ones were prioritised for closure, as they were seen as the least important (Greed and Daniels, 2002).

But local facilities linked to district shopping centres are vital for the elderly and those without cars, who must shop locally. Women need local public toilets, more than men, as they are the ones who are out and about in the day time, travelling on public transport, often accompanied by children, the elderly and family members with disabilities. Yet there are less toilet facilities for women than men nationally. Men have twice the number of facilities in an average toilet block as they have urinals in addition to cubicle provision. This has been condoned by regulatory standards which are currently being updated (Barkley, 2005).

City centre toilets

Decisions about toilet location should be based on adopting traditional spatial planning techniques, based on a "survey, analysis, plan" approach, along with public participation and consultation. Central areas should be allocated high levels of provision, with the fullest range of facilities for diverse user groups. Toilet blocks should be located alongside main pedestrian routeways and foci. Railway stations, bus stations and transport termini – the gateways to the city – should be provided with toilets which are accessible for people carrying luggage. In all public toilets barrier payment systems should be avoided as they restrict access. Instead, toilet attendants should be installed, whose presence reduces vandalism and anti-social behaviour. At least some toilets should be available on a 24 hour basis. The provision of male street urinals may deal with the street fouling generated by "the evening economy", but offers no relief to women, whilst many men would be wary of using them in full public view.

District "meso" level provision

At the district and neighbourhood level toilets should be provided in conjunction with local shopping centres, car parks and transport termini. The survey process should identify and disaggregate local needs. It should not be assumed that because toilets exist, off-street in public houses or fast food outlets, that, therefore no on-street public toilets are required. Rather, on-street public toilets should also be provided that are available to all. However, applying the principles of toilet hierarchy, the number of facilities and level of provision will be less than in the city-centre, because the catchment population is smaller. Attention should be given to the surrounding suburban residential settlements as well as local centres. Bus drivers, postal workers, domestic workers, health carers, mothers with babies, the elderly and schoolchildren, all need suburban toilets. Facilities should be provided on housing estates at key road junctions and along main spine roads, and, for example, at edge-city bus termini. Residents and shopkeepers should not have public toilets imposed upon them; public participation should be encouraged, with rate reductions, and even payment for keeping an eye on the toilets.

Less built-up areas

Toilets need to be provided to serve green spaces within cities, such as in parks, playing fields, allotments and landscaped areas. Many such localities were originally equipped with public toilets but many have been closed. Restored provision, but with better supervision, for example, the return of park keepers, would lead to more people visiting these amenity areas.

In villages, at least, minimal facilities should be provided, such as a small block comprising 4 Ladies cubicles, plus two cubicles and two urinals for the Gents, with each "side" including one accessible larger "disabled" compartment, plus one unisex and multi-use toilet available 24 hours (Greed, 2003:183). This may seem excessive but many an apparently isolated rural block of toilets may be used by coach tour passengers, during local markets and sports events – resulting in queues for the Ladies. In calculating need, attention must be given to the "interval factor": when large numbers of people suddenly need to use the toilets during a limited amount of time, be it between acts in a theatre, or during a short toilet-break on a coach trip.

Some rural local authorities are keen to enter into agreements with local shops, halls and cafes to make their toilets available to the public or to maintain adjacent local toilet facilities. Other solutions include opening up the toilets at a village hall or church to passing ramblers and tourists. In remote areas, where there are no settlements, rural planners should draw up a spatial strategy of distributing public toilet blocks at key intersections of footpaths, at scenic spots, car parks, and other honey pot areas. Local farming diversification policies should include tax allowances for the provision of toilets for hikers and walkers.

At least one minimal toilet block should be provided in all settlements over 5000 population, a figure which is used in France as the threshold for the provision for various public amenities. Toilets should be provided at least every 35 miles along motorways. Toilets should be provided along cross-country cycle routes. Likewise, cycle parking facilities with toilets, good washing and drying facilities, and left luggage lockers, should be provided in town centres. Toilet facilities should be provided for every car park of over 100 cars. All this may also seem excessive but many such facilities did exist before the closures of the last ten years. Many high-rise car parks used to have customer toilets, now closed because of drug-taking or for "security reasons". High provision levels are already found in the Far East where there has been a restroom revolution. Thousands of new public toilets being built in Chinese cities (Xu, 2005). The location of the nearest toilet is shown on the lid of the litter bins placed at every street corner (Xu, 2005).

Pride of place, "micro" level

In deciding where to site a new toilet block in a particular locality, the principles of good urban design should be applied. The infrastructure beneath the city streets is a major constraint, as is the availability of a cheap plot, but these should not be the only considerations. Toilets should be located in central public thoroughfares and squares, as high-status design features. The surrounding townscape should be legible and permeable so that people can easily make out the layout and main routeways, and thus quickly identify where there is likely to be a toilet in a strange town (Roberts and Greed, 2001). Once people know they can rely on finding toilets, and feel reassured that they will be open and usable, the "missing millions" will venture out of their houses.

Toilets should not be located on steep slopes, rendered inaccessible by steps and badly-designed ramps. Ideally all toilets should be positioned, "at grade". Ramps should be provided where there is no alternative. Handrails and gentle steps should be provided for those, with arthritis, who cannot cope with ramps. The area around the toilet should be carefully managed. Putting recycling skips alongside toilets in car parks creates an unsanitary environment. The toilets should be surrounded by clear paved surfaces without obstructions. Good visibility and surveillance contribute to the reduction of criminal behaviour. Good lighting – outside and inside the toilet – is essential. Where boundary demarcation is necessary around toilets, open railings or spacer bricks are preferable to high hedges or solid walls. Entrances should face on the main street: and not be located around the back. Temporary parking should be provided outside, with a combined parking and toileting payment system, to accommodate desperate motorists.

Public toilets need an internal circulation zone, especially in the Ladies, so that users do not have to wait uneasily outside. There needs to be safe space to leave pushchairs, toddlers, dogs, bicycles, friends and suitcases whilst people go into the toilet cubicle. Therefore APCs (Automatic Public Toilets) are not recommended; even the best designs are spatially restrictive internally, and they can generate queues. Many people are wary of using them for fear of technical malfunction and because of personal safety concerns.

As to the level of facilities provision, it is recommended that "a local authority should provide no fewer than 1 cubicle per 500 women and female children and one cubicle *or* one urinal per 1100 men, and no fewer than one unisex cubicle for use for people with disabilities per 10,000 population and no fewer than one unisex nappy changing facility per 10,000 people dwelling in the area (Cunningham and Norton, 1993). Currently women have to queue owing to under provision and because, for biological reasons, take twice as long to "go" than men (Asano, 2002). The relevant "population" in question should include commuters, tourists and visitors as well as residents. All age groups should be taken into account including children, who cannot, by law, just pop into a pub to use the toilet. In comparison, Japan has introduced toilet provision ratios of 2:1, and even 3:1 in favour of women in heavily used tourist and city-centre locations (Miyanishi, 1996). Recent changes in North America have also introduced 2:1 potty parity in some states, but it is not retrospective (Kwon, 2005).

Conclusion

In the absence of legislative compulsion, toilet provision depends on the goodwill of individual authorities. Indeed well-intentioned legislation can exacerbated the situation. Since October 2004, the Disability Discrimination Act 1995 (the DDA), has placed a duty on toilets providers to make "reasonable adjustments" to improve access. Some local authorities have pre-empted this requirement by closing down their toilets altogether to save money upgrading them (Hanson et al, 2006). Disability and incontinence groups have argued that "continence" is a valid disability under the DDA (albeit less visible than other conditions) and therefore even larger groups of people are discriminated against because of toilet closure. Clearly legislative change is needed. Good public toilet provision would not only meet the needs of all citizens, but would contribute towards creating more inclusive, accessible and sustainable cities for all.

Reference

Asano, Y. 2002, *Number of Toilet Fixtures: Mathematical Models*, Faculty of Architecture and Building Engineering (Shinshu University, Nagano, Japan)

Barkley, M. (ed.) 2005, *Draft Revisions to BS6465 Part 1, Sanitary Installations*, (Chapman Taylor Architects, London)

Bichard, J., Hanson. J. and Greed, C. 2004, *Access to the Built Environment Barriers, Chains and Missing Links*, (University College London, London)

BTA, 2001, *Better Public Toilets: the provision and management of "away from home" toilets*, (British Toilet Association, Winchester)

Cunningham, S. and Norton, C. 1993, *Public Inconveniences: Suggestions for Improvements*. (All Mod Cons and the Continence Foundation, London)

Greed, C. 2003, *Inclusive Urban Design: Public Toilets* (Elsevier, Oxford)

Greed, C. and Daniels, I. 2002, *User and Provider Perspectives on Public Toilet Provision*, (University of the West of England, Bristol)

Hanson, J., Greed, C. and Bichard, J. 2004, Inclusive design of public toilets in city centres *Vivacity: Sustainable Urban Environments – Urban Sustainability for the Twenty-Four Hour City, Project GR/S18380/01* (Engineering and Physical Science Research Council, London)

Hanson, J., Greed, C. and Bichard, J. 2006, The Challenge of Designing Accessible Toilets, in City Centres *Proceedings of the Ergonomics Society Annual Conference*, April 2005

Kwon, H. 2005, *Public Toilets in New York City: A Plan Flushed with Success?* (Faculty of Urban Planning, Columbia University)

Miyanishi, Y. 1996, *Comfortable Public Toilets: Design and Maintenance Manual*, (City Planning Department, Toyama, Japan)

ODPM. 2004, *Access and Facilities for Disabled People: Approved Document M*, (Office of the Deputy Prime Minister, London)

Roberts, M. and Greed, C. 2001, *Approaching Urban Design*, (Pearson, Harlow)

Xu, C. 2005, Code of Practice for Management of Public Toilets, *World Toilet Forum Proceedings Shanghai*, (World Toilet Organisation, Singapore)

THE CHALLENGE OF DESIGNING ACCESSIBLE CITY CENTRE TOILETS

Julienne Hanson[1], Jo-Anne Bichard[1] & Clara Greed[2]

[1]*Bartlett School of Graduate Studies, University College London, Gower Street, WC1E 6BT*
[2]*Faculty of the Built Environment, University of the West of England, Bristol BS16 1QY*

For almost a generation, "accessible" toilets have been provided in British towns and cities as an alternative to mainstream provision, and they remain essential for people with disabilities seeking to participate in city life. But the distinction between "general" and "special" needs, embodying a medical approach to disability, has proved unhelpful in meeting the toilet requirements of everyone in society. The chances of a more inclusive approach to toilet provision, based on the social model of disability and supported by recent legislation, are discussed with reference to research on the role of public toilets in creating accessible city centres. Two ergonomic variables, the overall dimensions of the WC cubicle and the height of the toilet seat, are used to illustrate the complexity of ergonomic variables entailed in the design of an accessible WC compartment.

Introduction

This paper presents findings from a thirty month (September 2003–March 2005) EPSRC funded research project into the inclusive design of public toilets in city centres. From October 2004, service providers have had to make "reasonable adjustments" to the physical features of their buildings in order to overcome barriers to access. This includes making toilets for customer use accessible to disabled people, not just for wheelchair users but also for people with a whole range of "hidden" disabilities such as poor eyesight, a learning disability or incontinence.

The research, which is studying the impact of this new legislation on toilet provision in London, Sheffield and Manchester, is therefore taking place in "real time". The project involves consulting disabled people about the design features that will assist them in using the toilet, and auditing "accessible" (formerly disabled) toilets provided by Local Authorities and private businesses for public use, to evaluate precisely how accessible these are.

Personas, based on the collective biographies of disabled users, have been constructed to capture and describe a wide range of user needs. The development of a Toilet Audit Tool, based on the relevant British Standard (BS8300) and Part M of the Building Regulations (2004), has enabled 50 design variables that are crucial to the design of an accessible toilet to be monitored. This has led to a database that can compare the needs of people with a very wide range of disabilities to what has actually been provided in the "real world", and thus to identify the gaps in provision where people's needs currently are not being met.

A "contested site" for human behaviour

Toilets house an apparently mundane activity, but the public toilet is a highly "contested site". It shelters a very private activity that takes place in public space, in proximity to complete strangers. Differential provision exists for men and women, the needs of late night "binge" drinkers often compete for resources with those of daytime shoppers and toilets are colonised by unplanned uses such as vandalism, drugs or sex, that make the environment unattractive or even dangerous for ordinary members of the public. This has led to a conflict between toilet users, who require easy access, and toilet providers, who aim to defend the facilities like a fortress in order to deter unwanted uses, (Greed, 2003).

Public toilets therefore offer a remarkably clear example of how the design of the built environment can either "enable" or "exclude" individuals and groups from city centres. The objective of achieving "access for all" has its own conflicts, in respect of whether an "inclusive design" or "special needs" approach is adopted. Some disabled users prefer an enlarged cubicle in the separate sex toilets, an inclusive approach to design, whilst others require a "unisex" accessible toilet. The very existence of this "third way" as an alternative to accessible mainstream provision can be regarded as a hangover from an era when designing for special needs rather than social inclusion was taken for granted. In defence of the accessible unisex toilet, however, it is fair to say that whilst people who can use the toilet independently prefer to use an enlarged cubicle in the separate sex toilets, those who need assistance to use the toilet especially where this is given by a spouse or a carer of the opposite gender, usually prefer a "unisex" accessible toilet.

This is not the only bone of contention, however, and opinions differ on many other aspects of toilet design, such where an adult and baby room should be provided. Some advocate that this should be placed within the unisex accessible toilet whilst others assert that a separate adult and baby room should be located in the men's and the women's toilets. Wheelchair users have been known to challenge the rights of people with a "hidden" disability to use the accessible toilet.

Finally, and importantly for ergonomics, people with different medical conditions require different detailed design features within the WC cubicle itself. The most fundamental challenge to dimensional coordination is the actual size and overall dimensions of the WC compartment, since this affect people's ability to access the WC in the first place. Currently, the recommended dimensions for an accessible toilet cubicle are 1500 mm wide by 2200 mm deep, and these minima are usually treated as maximum dimensions.

Turning to a matter of detailed design, the height at which the top of the toilet seat is set, 480 mm, which was originally specified to permit easy transfer by a wheelchair user from the wheelchair to the toilet, is too high for people of short stature comfortably to use the toilet and yet it is too low for people with stiff or painful knee and hip joints to get up off the seat without discomfort.

Dimensional "dissonances" of this kind between different user groups affect just about every design feature of the accessible toilet and it is simply not possible to optimise the design to suit everyone; someone will always be inconvenienced or excluded. An interesting by-product, therefore, is that different design guides, access consultants and local access groups each have their own preferred dimensions that are adopted locally in preference to those set out in the Building Regulations. From a design perspective, the accessible toilet cubicle is not so much the "smallest room" as the "most complex building".

Mainstream and accessible toilets

Before 1979, the needs of disabled people (especially in a wheelchair) to access a toilet when away from home were not catered for. Even if the public toilet was at pavement level, most ordinary cubicles were far too small for a wheelchair to access. However, 1979 saw the inception of the Royal Association for Disability and Rehabilitation's (RADAR) key scheme, whereby a special, locked unisex WC was provided for the use of disabled people. After 1979, Local Authorities began to provide accessible toilets for the public, described by Goldsmith, (1997) as the "icon" of disability rights groups for the last twenty years. Meanwhile, the provision of an accessible toilet somewhere in the city centre meant that mainstream public toilets need not be accessible to people with disabilities.

The design of these facilities was specified in a British Standard, originally BS5810 (1979) and now BS8300 (2001). Previous research (Feeney, 2003) established the design standard for access, predominantly though not exclusively with respect to wheelchair users. The current Building Regulations (ADM, 2004) drew heavily on this work to specify the precise layout of unisex corner accessible toilets for use by disabled members of the public. The layout of the compartment is critical for wheelchair use, but at the same time whilst an estimated 20% of people in the UK are considered to be "disabled" (DRC, 2000), less than 5% of these are actually wheelchair users.

To a greater or lesser extent, when anyone is away from home they plan the excursion around the provision of ordinary, mainstream public toilets, avoiding locations where there is no provision. However, disabled people have to consciously plan their daily spatial routines to ensure that they can always access the toilet if they need it. There are about 10,000 public toilets in the UK, of which just 3,500 were originally designed to be accessible to wheelchair users. As there are fewer accessible toilets than mainstream ones, choice is severely curtailed. Consequently, the lives of disabled people are narrowly constrained to a small everyday "home range" and governed by restrictive patterns of behaviour, like not leaving home for more than about two hours at a time. The lack of accessible public toilets is therefore contributing to a "Catch 22" situation where not many disabled people use public toilets because they are badly designed, but because disabled people do not use public toilets in large numbers, providers think there is no need for them to design accessible facilities.

Public toilets provided by Local Authorities have been closing in recent years due to high maintenance costs; 40% of all public toilets have closed within the last ten years. The DDA appears to be further accelerating public toilet closures. Most were not designed to be accessible and adapting them to meet the new access standards would be very expensive for Local Authorities. Closing all public toilets does not discriminate against anyone, but it reduces provision for everyone. However, private providers are improving their toilet provision in the interests of good customer care, so responsibility for making "away from home" toilets accessible to disabled people is gradually being transferred to the private sector.

Cubicle sizes and toilet seat heights

Before the DDA came into effect in October 2004, researchers audited toilet provision for public use in sixty premises in Clerkenwell, in the south of the London Borough of Islington. Clerkenwell exemplifies an attractive, historic, mixed-use urban area that plays host to diverse populations of visitors, workers and residents and has a range of users that appeal to a wide-cross section of society, including disabled people. Islington was recently identified as the London borough with the lowest per capita provision of public toilets in the capital (London Assembly's Green Party Group, 2004), with just one facility per 58,600 people. The borough has suffered a 70% decrease in its public toilets since the 1990s and, despite its popular appeal, all the public conveniences and urinals in Clerkenwell area were closed as long ago as 1991. Visitors to the area who need to use the toilet are therefore forced to rely on private provision, or on public toilets located in the adjacent boroughs of Camden and the City of London.

The study revealed that, whilst grab rails were being used as a token gesture towards making toilets appear accessible, all of the toilets audited had major design flaws.

Clerkenwell was revisited in the summer of 2005, and this time seventy premises accessible to the general public were audited, to benchmark provision now that the DDA had come into effect. Disappointingly, out of the seventy providers thirty-eight had made no provision for disabled people, thirty-two claimed to have an accessible toilet and seventeen of these gave permission for an audit to be conducted. None of the audited examples had achieved a fully accessible toilet in respect of the design guidance in ADM. The features that were most often provided were an outward opening door (16/17), a door of the correct 800 mm clear opening width (15/17) and a sturdy horizontal 680 mm grab rail on the inside face of the door (also 15/17). However, once through the door, the user was faced with a WC cubicle that contained a motley array of sanitary ware and equipment, rather than a standard interior as specified in the ADM.

Only thirteen accessible toilet compartments were of the right width and just three were of the correct depth. The rest were smaller than these minimum dimensions and would effectively excluded wheelchair users. This is particularly worrying as research has revealed that a 1500 mm by 2200 mm cubicle is too small to accommodate the growing numbers of users of power wheelchairs. The new standards are rapidly becoming obsolete in this respect.

So far as the height of the toilet seat is concerned, just five providers had achieved this objective. The majority had not with the result that, even if all the other features of the cubicle were correct, wheelchair users would still find it difficult to accomplish a transfer. The unreliability of this dimension means that many wheelchair users are obliged to carry an adjustable seat with them, to adjust unsuitable sanitary ware to the correct height for their needs.

Design templates

The requirement for access under the terms of the Disability Discrimination Act (DDA) extends to a much wider constituency of disabled people than wheelchair users, including people with physical, sensory and cognitive impairments. The expectation is that greater social inclusion will be achieved through inclusive design of the built environment. In the case of the accessible toilet, every aspect of its design and construction, from the entry and exit to fixtures, fittings and furnishings, now needs to be reconsidered from the point of view of a very wide range of user requirements, including a new generation of users of large power wheelchairs.

Toilet audits have revealed that private providers simply do not understand the complexity of the design of an accessible toilet compartment, nor do they appreciate that the precise technical specification is critical for wheelchair users. Meanwhile, other disabled user groups are missing out altogether.

The research has also revealed that the ADM 2004 for an accessible toilet does not suit everyone's needs. The next step is for the desired design solution for each persona to be translated into a design template that will explain to providers and designers what each user group requires by way of "reasonable adjustments", and why these features are important if their entitlement to a toilet that is safe, private, clean, comfortable and dignified to use is to be realised.

References

British Standards Institution 2001, *BS8300: 2001 Code of Practice for the Design of buildings and their approaches to meet the needs of disabled people*, (BSI, London)

Disability Rights Commission 2002, *Code of Practice Rights of Access Goods, Facilities, Services and Premises*. (DRC, London)

Feeney, R. 2003, *BS8300: The Research Behind the Standard*, (International Workshop on Space Requirements for Wheeled Mobility, New York)

Goldsmith, S. 1997, *Designing for the Disabled: the new paradigm*, (Architectural Press, London)

Greed, C. 2003, *Inclusive Urban Design: Public Toilets*, (Architectural Press, London)

Green Party Group 2004, *Toilets going to waste! London's public loos are being flushed away* (The London Assembly, London)

Office of the Deputy Prime Minister 2004, *Building Regulations Approved Document M: Access to and use of buildings.* (The Stationery Office, Norwich)

ACCESSIBLE HOUSING? ONE MAN'S BATTLE TO GET A FOOT THROUGH THE DOOR

H.J. McDermott[1], R.A. Haslam[1] & A.G.F. Gibb[2]

[1]Health & Safety Ergonomics Unit, Department of Human Sciences, Loughborough University, Loughborough, Leicestershire LE11 3TU
[2]Department of Civil and Building Engineering, Loughborough University, Loughborough, Leicestershire LE11 3TU

In 2000–2001, 18% of adults in England aged 16 and over reported having some form of disability, 5% of whom reported having a serious disability. The most common type of disability reported amongst adults was loco motor disability. Current legislation on housing design, to incorporate the needs of those individuals with physical disabilities, is contained within Part M of the Building Regulations. These requirements have had a significant impact upon dwelling design. This case study documents the experiences of a disabled man occupying a brand new home. Forming part of a wider research project, this study identifies how current Building Regulations only go some way in accommodating the needs of those with a disability and that in some "parts" these regulations display a lack of systems thinking.

Introduction

In 2000–2001, 18% of adults in England aged 16 and over reported having some form of disability, 5% of whom reported having a serious disability. The most common type of disability reported amongst adults was loco motor disability (Department of Health, 2003), the issue most affected by housing (Heywood, 2004).

Up until the 1990's, the majority of housing within Britain was not accessible to wheelchair users (Barnes, 1991), and the provision of housing for the disabled was guided by policies which were firmly rooted within an individual model of disability (Stewart et al, 1999). Such a view places the "disability" within the individual.

The paradigm shift from the individual, or medical model of disability to the social model of disability informed British housing policy during the 1990's and reflected a growing acceptance of a collective responsibility to create a fully inclusive environment. Stewart et al, 1999, for example, argue that the inability to access a dwelling in a wheelchair is not the result of a disability, but due to the fact that architects have consistently failed to design dwellings which are accessible to everyone. Instead of viewing the needs of the disabled as "special needs", requiring separate provisions for housing, the social model of disability supports a collective response to a socially created problem.

In 1999, the scope of Part M of the Building Regulations relating to disabled access to public buildings, was extended to include new dwellings. These regulations have had a major impact upon dwelling design, but have been the subject of intense disapproval by builders on the grounds of impracticality and cost, and condemnation by disabled groups on the grounds that they do not go far enough (Ridout, 1997). Indeed, the overall aim of the regulations, to promote "visitabilty" housing, has been heavily criticised. Madigan and Milner (1999) argue that such a label places emphasis on the occasional visitor rather than emphasising the more important objective – to make homes adaptable for a whole range of future occupiers.

Imrie, (2004) argues that the physical design of the dwelling and home environment can have a significant impact on the health and well-being of an individual and that the design of the majority of dwellings is underpinned by values that do not relate to disabled people. Heywood, (2004) identified a number of health outcomes that were associated with unadapted or badly adapted housing which fully support Imrie's argument. These health outcomes included pain, accidents, exacerbated physiological illness and psychological illness. Such findings have major implications in respect to Part M of the Building Regulations and the health, safety and well-being of disabled people within Britain.

An improved understanding of the ways in which disabled people interact with their home would benefit those responsible for the design and construction of new homes, and would also inform those responsible for the development of building codes and Regulations.

Methods

This case study documents the experiences of a disabled man, recently having occupied a new build home. The case study is taken from a wider research project which explored the interaction between design, usability and occupier behaviour in the safety of new dwellings.

Letters inviting participation were delivered to completed properties on new build developments in the areas of Leicestershire and Nottinghamshire. Two press releases detailing the study objectives were also issued inviting interest.

Semi-structured interviews were conducted within participants homes to collect information on the personal experiences of individuals inhabiting a new dwelling. Home audits were also undertaken with the researcher accompanying occupiers around their properties to identify where problems existed with design features and where modifications had been made. Each interview lasted approximately one and a half hours. The interviews were recorded with the consent of the interviewees and later fully transcribed.

Qualitative data analysis followed three steps: data reduction, data display, verification and conclusion drawing (Miles and Huberman, 1994). Data reduction was achieved by coding the interview data using the qualitative software package Nvivo and pattern coding these units of analysis into a smaller number of themes and explanations.

Results

The participant in this case study was a 57 year old male who was dependent on a wheelchair for mobility. He had purchased a two bedroom, ground floor apartment where he lived with his ambulant wife and 32 year old son. At the time of the study, the family had been resident in this property for 9 months.

The study identified a number of problems connected to the physical design and layout of the property, affecting the participants mobility and functioning within the dwelling. Some of these design features relate to the provisions contained in Schedule 1 of the Building Regulations, whilst others were not the subject of any legislation.

Attitude of house builder

The property was purchased by the occupier "off plan" (prior to construction), and there were a number of changes that the occupier requested of the house builder during construction to assist him with his disability. These changes included alterations in the bathroom in order that a fully accessible bathroom could be fitted. It appeared that the house building company were reluctant to make any changes or deviate from the standard plan, even though the request was made very early on in the process:-

> *"We asked for changes to be made, they weren't interested. They knew they could sell them [the homes] ten times over"*

Despite such reluctance, the occupier did push the company to make a change to the bathroom:-

> *"They agreed to that, we had a lot of arm twisting to do it, they didn't want to obviously. Why go to any hassle if somebody's prepared to pay the price as it stands"*

Fire doors

The main entrance door to the accommodation block where the apartment was located was fitted with a self-closing fire door in line with the provisions contained in Part B of the Building Regulations. This fire door caused problems for the participant in gaining entry to the block:-

> *"you cannot get into this building in a manual wheelchair, I think you would struggle to get out, there is quite a strong swing on it" [the door]*

A self-closing fire door was also fitted to the front door of the property preventing the occupier from gaining easy access to his own property:-

> *"There was a self-closer, I had to take it off, have you tried opening it with a wheelchair? I wouldn't be able to get in to start the fire"*

Self-closing fire doors were also fitted to all habitable rooms within the dwelling itself and once again they caused problems for the occupier:-

> *"No, there's no way I could open the doors in a wheelchair, it's just not possible, so I've had to take them off"*

No ramped exit

Part M of the Building Regulations require all dwellings to be accessible for wheelchair users. The entrance to the accommodation described in this study was compliant with these provisions, in that there were no steps to negotiate, but the occupier was concerned regarding his safe exit from the property to his garden:-

> *"That doesn't have to be suitable for myself, as long as I can get in that's all right"*

The difficulties he had experienced in exiting to the garden had resulted in the occupier building a DIY ramp to assist him with his exit:-

> *"I've had real problems, I built that ramp out there, I've got another ramp outside as well, but if you saw me going out, you'd have a fit. I didn't like that!"*

Due to the rear exit of the property having a raised door threshold, the self-built ramp was necessary for the occupier to gain entry to his garden. However, when negotiating the ramp, the participant's wheelchair tipped, leaving him at risk of falling forwards from his wheelchair. (Figure 1)

Figure 1. Unassisted exit from property in wheelchair.

Features

There were a number of other features of the property that affected the occupier's functioning. These included the fuse box and the spy hole in the front door being at an inaccessible height and the fitted fire blanket being out of reach for a disabled occupier. The communal letter boxes were positioned in the hallway of the block, with the occupier's own box also out of his reach. He also became aware that the disabled parking space provided for the block had been allocated to a first floor apartment, to a fully ambulant occupier.

Discussion

The examples from this case study demonstrate how some dwellings in Britain continue to be designed with little or no consideration for disabled occupiers. The findings also suggest that the policies of at least one house building company remain rooted in an individual model of disability.

Many of the problems encountered by the participant in this case study (fuse box, spy-hole etc) could have been avoided through modifications during design and construction. The apparent reluctance of the house builder to make changes during the building of this dwelling, suggests a lack of understanding of the varying needs of disabled people within the house building industry. This supports the claim made by Imrie (2004).

The findings are also in line with previous claims that the provisions within Part M of the Building Regulations do not go far enough to incorporate the needs of the disabled, (Ridout, 1997). Part M requires electrical sockets and switches within new dwellings to be accessible for a wheelchair user, but these regulations do not cover other internal features within the home.

With regard to injuries in the home, Heywood (2004) identified unadapted or badly adapted housing as one source of these. Examples from this case study that illustrate how this might arise include the need for a self-built ramp and the removal of the fire door closers. This participant's independent exit from the property led to an increased risk of injury to himself and the removal of the self-closers from the internal fire doors created health and safety risks for his family.

The amendments to the Building Regulations (Part M), aimed to ensure new dwellings within Britain were, at a minimum, accessible. Part B of the Building Regulations provide the requirements for fire doors in dwellings of a particular design. This study has suggested how the requirements of Part B can inhibit the aims of Part M. The ability of wheelchair users to negotiate fire doors is perhaps an important issue that has received insufficient consideration. A systems approach to the development of Building Regulations is desirable.

Although this case study documents the experiences of only one disabled man, it has highlighted some important issues. The prevalence of these issues in wider society is worthy of future investigation.

References

Barnes, C. 1991 *Disability and Discrimination in Britain.*(Hurst and Co, London), cited in Imrie, R. 1998 Focusing on Disability and Access in the Built Environment. *Disability and Society*, Vol. 13, No 3, 1998.

Department of Health 2003 *Health Survey for England 2001,* (TSO, London).

Heywood, F. 2004 The health outcomes of housing adaptations. *Disability & Society,* Vol. 19, No. 2. March 2004.

Imrie, R. 1998 Focusing on Disability and Access in the Built Environment. *Disability and Society*, Vol. 13, No 3, 1998.

Joseph Rountree Foundation, 2000 *Lifetime Homes,* (Joseph Rountree Foundation).

Madigan, R. and Milner, J. 1999 Access for all: housing design and the Disability Discrimination Act 1995. *Critical Social Policy,* 19: (3) 396–409.

Miles, M.B. and Huberman, A.M. 1994 *Qualitative Data Analysis, An expanded sourcebook.* (Sage publications)

Ridout, G. 1997 Access Bill. *Building Homes.* No. 10, pp 28–9. Dec 1997.

Stewart, J., Harris, J. and Sapey, B. 1999 Disability and Dependency: origins and futures of "special needs" housing for disabled people. *Disability & Society,* Vol. 14, No. 1.

INCLUSIVE DESIGN – IN SOCIETY

GUIDE DOGS AND ESCALATORS: A MISMATCH IN URBAN DESIGN

Diane E. Gyi & Louise Simpson

Department of Human Sciences, Loughborough University, Leicestershire LE11 3TU

Accidents involving guide dog users on escalators can result in serious injuries to the dogs themselves and severe distress to their owners. As escalators are becoming an increasingly common feature in our urban environment, an exploratory study of the problem was conducted focussing on the potential solution areas of guide dog training, protective equipment design and escalator design. Interviews were conducted with guide dog owners in the UK to more fully understand their experiences. Existing dog protection systems were also examined and anthropometric data collected on a sample of guide dogs. The aim of the research was to provide recommendations to facilitate a more inclusive design solution.

Introduction

"Stepping onto an escalator is an act of faith. Riding the moving stairs is an adventure for the toddling young and a challenge to the tottering old! In the early days, they had to be persuaded to get on at all. A one-legged man "Bumper" Harris, was hired to ride for a whole day on the first installation (it was at Earls Court) to show how easy it was. Some people were sceptical (how had he lost his leg?) but others broke their journey just to ride up and down". (Campbell, 2002)

Many of us now take escalators for granted whilst travelling around the urban environment but for some people, they are difficult and dangerous to navigate leading to social exclusion. Whilst guide dogs bring independence to many blind people one disadvantage is that dogs are prohibited from using escalators in the UK; BS EN115 (1995) indicates that *"escalators are not suitable for use by animals"*. In fact, the advice in the UK is for guide dog owners not to use escalators when working their dogs but to use fixed stairs and lifts, or to ask for the escalators to be stopped. If these alternatives are not available, they are advised to carry the dog for the duration of the ride and are offered training in this technique.

In spite of this advice, it is known that some guide dog owners in the UK work their dogs on escalators and consequently accidents are occurring which may result in serious injuries to the dogs themselves as the dogs paws and claws become caught in the cleats. This is distressing for their owners and as escalators are becoming an increasingly common part of urban environments poses a major hazard. In some countries, for example the USA, dogs are successfully trained to walk on and off escalators with instruction. Canine Personal Protective Equipment (PPE) boots are used to protect these dogs in the early part of training. In terms of escalator design itself, very little research has been conducted, but obviously this is an important part of the long-term solution as escalators are an important means of encouraging "flow" in the urban environment. The potential of solution areas such as dog training, canine PPE and escalator

Table 1. Questions and issues raised during interviews with guide dog owners.

	Questions and issues
Escalator use	Where? Why?/Why not? Frequency? Experiences? Alternatives?
Training	Any received? Quality? Comments on policy?
Canine PPE (boots)	Experience of dog boots? Boot design e.g. fit, protection, durability, ease of putting on/off, acceptance, style, colour.
Escalator design	Physical prompts e.g. jet of warm air, for the dog.

design need to be explored with guide dog owners in the UK. The main aims of the research were:

- To gain a better understanding of the risks of taking guide dogs on escalators from the perspective of the guide dog owners.
- To provide recommendations/guidelines to reduce the risks of travelling on escalators for this user group.

Methodology

Purposive and directed sampling was used to select blind and partially sighted guide dog owners from two cities, London and Nottingham. Semi structured interviews took place in a variety of locations convenient to each individual participant. For practical reasons, six of the interviews were conducted over the telephone. A summary of the interview questions and issues addressed is given in Table 1. Participants were provided with details of the interview both verbally and in large print or Braille prior to the interviews. To facilitate discussion regarding dog boots (canine PPE), participants were shown a sample of commercially available boots.

Following advice from an orthopaedic veterinary surgeon, key landmarks, tools and techniques were identified for the collection of canine anthropometric data. The differences in limb dimensions in guide dog breeds and gender were thought likely to be small. It was therefore planned to select 50 dogs based on the proportion of breeds currently used for guide dogs in the UK i.e. Labradors crossed with Golden Retrievers (47%), Labradors (35%), Golden Retrievers (13%) and German Shepherds (5%).

Results and discussion

The sample included a wide range of ages, escalator users and non-users, guide dog experience and gender. The sample of 17 individuals, n = 9 (London) and n = 8 (Nottingham), aged between 25 and 65+ were interviewed over the summer of 2005. 35% of the participants were with their first guide dog, 35% their second or third and 30% had worked with more than four guide dogs. The interviews were recorded on audiotape and transcribed by the researcher. Key themes were identified from the information rich data. Due to the small sample size, except where indicated, data from the Nottingham and London guide dog owners were analysed together.

Experiences of guide dog owners

41% of these guide dog owners (the majority from the Nottingham sample) had never used an escalator. In contrast, there was a need to use escalators on the Underground in London. Despite a concern of possible litigation in the event of an accident, "necessity" and "convenience" were the main reasons given for guide dog owners using escalators despite current advice in the UK. Many of these participants had a guide dog for the independence it offered. The concept that a guide dog might exclude them from freedom to travel where they wished i.e. on escalators was a strongly felt concern.

Of those guide dog owners who use escalators (n = 10), 40% managed to carry their dogs for the duration of the ride. However, this was generally unpopular not just because of the physical size and weight of the dogs, but also "spoiling clothes" and "annoying other escalator users". Interestingly, the remainder of the sample adopted very different "coping strategies" i.e. lift paws on, ride, lift paws off; carry dog on, ride, carry dog off; walk dog on, ride, walk dog off; walk dog on, ride, carry dog off; and walk do on, ride, lift paws off. It is likely that a larger survey would show a variety of other self-taught techniques.

Where alternative ways of changing level (e.g. lifts, fixed stairs) were available, 70% reported that they would use that mode, although this was a more positive experience for the Nottingham participants. Negative experiences (largely the London participants) included being unable to find assistance, long waiting times and complicated journey planning. With regard to stopping the escalator, once again this was largely a negative experience for the London guide dog owners and several individuals expressed a reluctance to ever do this.

Three of the guide dog owners reported "near misses" involving their dogs on escalators which not surprisingly gave them concerns about using them again. Risks identified by owners generally were; difficulties carrying the dog (weight and size), dog unable to guide whilst being carried, no free hands to orientate themselves, anxiety in the dog, limitations in the self taught technique, and difficulties carrying bags/luggage (and a dog!). The findings also indicate that the risks are higher for guide dog owners in London, not just because of the physical challenges but also the increased emotional challenges (e.g. greater risk of distraction, the limited patience of busy commuters). Several of the London guide dog owners who use escalators on the Underground, refrain from this is shopping centres where they believed there was less necessity.

Guide dog training

Clearly the general training given to guide dogs and their owners is of a high standard, however, the majority of these guide dog owners felt that carrying their dogs on an escalator was not a practical solution. Those who walked their dogs on escalators were confident in their self taught techniques and would like to have had the option of this training from the training centres. The concept of training the dogs to jump over the combs was not thought feasible in a crowded urban environment. From discussions with guide dog trainers, a concern was raised about infrequent users of escalators and therefore how successful training could be for these dogs and owners. Training would also need to be initiated when the dogs were young; older dogs would not be amenable to being trained to walk on escalators. In addition not all guide dogs would respond well to such training and could restrict matching clients to dogs and create longer waiting times.

All of guide dog owners in London compared with only 25% of those from Nottingham disagreed with advice given in the UK. Again, this is due to the current necessity to use escalators in London as part of the public transport network.

Canine PPE

None of the guide dog owners in this sample had any first hand experience of using dog boots, it was therefore surprising that only 6% of the sample were totally negative about the idea (53% were undecided and 41% were positive). Views were split as to how they should be used i.e. just for training, just for escalator use, or all the time. Some guide dog owners felt strongly that they wanted to avoid damaging the dogs paws and could see additional benefits of dog boots in the urban environment, for example,

Table 2. Canine anthropometric for a sample of guide dogs (n = 30).

Dimension (mm)	Mean	SD	5th %ile	50th %ile	95th %ile
Weight (kg)	32.1	4.7	24.4	32.1	39.8
Height	550	53	463	550	637
Fore limb paw breadth	59	9	44	59	74
Fore limb paw length	85	10	69	85	101
Fore limb paw thickness	32	4	25	32	39
Claw-carpal length (fore)	79	10	63	79	96
Claw-forearm length	140	8	127	140	153
Pastern circumference (fore)	113	12	100	113	133
Carpus circumference (fore)	138	13	117	138	160
Rear limb paw breadth	52	7	41	52	63
Rear limb paw length	82	10	66	82	98
Rear limb paw thickness	33	4	26	33	40
Claw-carpal length (rear)	75	10	59	75	91
Claw-hock length (rear)	212	18	182	212	242
Metatarsus circumference (rear)	103	10	87	103	119
Hock circumference (rear)	115	10	99	115	131

protection from glass, ice, salt or cigarette ends. Others were more sceptical particularly about the unwanted attention a dog wearing boots could attract.

Provision of the four sets of boots, designed for a variety of applications (e.g. bandage protection, police dogs) prompted discussion about how the designs could be improved for guide dog PPE. Although, none of the boots in the sample were identified as a good solution for these guide dog owners, valuable feedback was given which will facilitate the future design/evaluation of these products for use in the UK.

Canine anthropometry

Due to time constraints and practical difficulties it was only possible to measure 30 dogs (7% Labradors, 23% Golden Retrievers, 10% German Shepherds). Although the techniques for measuring had been piloted with guide dogs, it was necessary to include some dogs that had not been fully trained; these dogs were less co-operative leading to practical difficulties with the data collection. These data are summarised in Table 2. It can be seen that there is little variation in limb and paw measurements.

Escalator design

The overall view was that any changes/modifications to escalator design should assist the human part of the team; a system to warn the dog to step over the comb such as a brush or warm jet of air were not popular. Several participants commented that an underfoot tactile system to inform them of the direction of the escalator would be helpful.

Discussions with a human factors expert on transport in London, indicated that it is likely that more lifts will be fitted in the future rather than focussing on replacing/modifying escalators. This should benefit some guide dog owners but will obviously still exclude others who would like the freedom to travel exactly where they wish.

Conclusions and recommendations

- An urban environment that works for all is essential to any living city and an important quality of life factor for guide dog owners. Escalators are becoming a more

popular means of changing level such that there is an increasing necessity to negotiate them.
- The majority of guide dog owners in this sample do not carry their dogs on escalators and were unhappy with current advice in the UK.
- Techniques for walking the dogs on escalators are successfully taught in some other countries, for example, the USA, Australia, New Zealand. A pilot study for training dogs in walk on/off techniques in the UK is urgently needed.
- The variety of different techniques for using the escalator highlights the resourceful nature of these users. However, standardising techniques (through training) will allow control measures to be put in place to reduce risk.
- The role of canine PPE (i.e. the boots) needs further exploration as a solution to protect the dogs feet. As there was little variation in limb and paw measurements of the guide dog breeds, a design solution specifically for the needs of guide dogs may be feasible.

Acknowledgments

The authors would like to thank the team at the Guide Dogs for the Blind Association, the guide dog owners and their dogs for their involvement in the research.

References

BS EN115. 1995, *Safety rules for the construction and installation of escalators and passenger conveyers. P33, Code of Practice*, (British Standards Institute, London)

Campbell P. 2002, *Why does it take so long to mend an escalator?* London Review of Books. Available at http://www.lrb.co.uk/v24/n05/print/camp01_.html (Accessed November 2005)

COST 219TER: AN EVALUATION FOR MOBILE PHONES

Edward Chandler[1], Elizabeth Dixon[1], Leonor Moniz Pereira[2], Angelika Kokkinaki[3] & Patrick Roe[4]

[1]*The Royal National Institute for the Blind, Innovation and Disability Access Services, Bakewell Road, Orton Southgate, Peterborough PE2 6XU, United Kingdom*
[2] *Faculdade de Motricidade Humana (FMH), Universidade Técnica de Lisboa, Estrada da Costa, 1495-688 Cruz Quebrada – Dafundo, Portugal*
[3]*Department of Management & Management Information Systems, Intercollege, 46 Makedonitissas Ave., P.O.Box 24005, 1700 Nicosia, Cyprus*
[4]*Chairman COST 219ter project, Ecole Polytechnique Fédérale de Lausanne, ELB 034 (Bâtiment ELB) Station 11, 1015 Lausanne, Vaud, Switzerland.*

COST 219ter is a European action group within the COST European Framework and is made up of scientific and technical researchers from over 30 countries across Europe and the rest of the world. Its main objective is to increase the accessibility of next generation telecommunication network services and equipment. It is aiming to achieve this by design or, alternatively, by adaptation when required. RNIB has been involved in this action for several years and has been particularly important due to knowledge of user trials and performing evaluations on products but specifically on mobile devices. Over the last 9 months RNIB together with Intercollege, Cyprus and Universidade Técnica de Lisboa have been performing work to test the evaluation process that it uses, this will include end user evaluations with mobile devices. This paper provides a brief outline of evaluation techniques to show how an evaluation process for evaluating mobile phones has been produced.

Introduction

The origins of COST 219ter date back to 1986 when the first COST 219 action was started. The aim of this action was to look at *"Future Telecommunications and teleinformatics facilities for disabled people and elderly."* This action ran for ten years and was hugely successful in this arena. COST 219 was immediately followed by a new COST 219bis Action which ran for five years ending in 2001 with the title of *"Telecommunications: Access for disabled and elderly people."* The current COST219ter Action started in 2003 and is set to run until 2007. The focus of this action is *"Accessibility for All to Services and Terminals for Next Generation Networks."* COST 219ter has four working groups; Information collation, Accessibility of Emerging Information Society Technologies, Testing for Accessibility/Assessing Special Solutions and Smart Houses & Home Networking. Each group has their own set of objectives and tasks to complete. Working Group 3 (Testing for Accessibility) has been tasked with reviewing current methods for evaluating mobile phones and (where possible) the production of a process to do this for the next generation devices. This group identified three areas, which should be tested, to determine whether a next generation terminal or service is accessible or not. These areas were; hardware, software and services.

Why is testing for accessibility important?

Although there are hundreds of different types of handset available to the end user, generally speaking, the designers of these handsets have designed them for aesthetic values rather than their practicality. Devices such as the mobile phone have been made possible due to the miniaturisation of technology, Zwick et al, 2005. The miniaturisation trend has seen the mobile phone become smaller in size, the buttons closer together and the buttons to sit flush to the surface of the phone. This trend has meant that disabled peoples' needs in a mobile phone have been designed out in favour of consumer preferences. However it could be argued that disabled people have a stronger need for mobile communication due to their specific circumstances (e.g. they may not be able to walk to a pay phone or may not be able to remember numbers easily or may not be able to use voice telephony). So that disabled people are not excluded from accessing services and terminals for next generation networks it was felt that a process should be devised to combat current trends.

What is an evaluation?

Evaluation is a tool used to determine whether a system or product is fit for the purpose that it was initially designed to do. An evaluation can take several forms, can be performed during the design process of a product or system or after a system or product has been released on to market, can be performed by the designers or a specialised evaluator. An evaluation may involve actual users or not. There are many variations to evaluation and it is not possible to discuss and critique them in this paper. There are also a number of factors that can influence an evaluation such as the timescales for the evaluation, the types of users, the number of users as well as the type of evaluation route. Therefore this paper will highlight a background to evaluation and what general routes are available to show how an evaluation for mobile phones was arrived at. It is intended that this evaluation method will be used to evaluate present and the next generation mobile devices to include disabled end users. The main emphasis of this evaluation methodology is accessibility however usability issues will be included as it is felt the usability of the device is just as important as the accessibility of the device. For further discussion of user trials, the reader is referred to McClelland, 1999 and Christie et al, 1999.

Accessibility and usability

Accessibility and usability are tightly connected concepts and can influence each other. If the mobile devices were inaccessible to a large section of the population then these users would not be able to use them effectively. However, if usability for disabled users has not been considered in addition to accessibility then this may make the mobile phone difficult to navigate and difficult to use.

Accessibility

Accessibility is most commonly discussed in relation to people with disabilities, because this group are most likely to be disadvantaged if the principles of user centred design are not implemented. Failure to follow these principles can make it difficult or impossible for people with disabilities to access content whether that is in print or on a screen of a mobile phone. Creating accessible content should be an integral part of the design philosophy, and accessibility features should be incorporated into all aspects of the design process. Testing for accessibility should also be incorporated into any and all user testing regimes, and should never be seen as an isolated event that can occur after other user testing has taken place. Designing for accessibility is thus as much a strategic issue as a purely technical one. Therefore, in context of communication, accessibility can be defined as:

> "Successful access to information by people who have disabilities."

Usability

Usability determines the quality of the user experience with the product and/or system. In this traditional context, usability is perceived in terms of the design of the user interface, which facilitates efficient and effective completion of the user tasks. If you can use a product easily and correctly straight away, then the product has good usability. The product should also be simple and intuitive to use. Therefore, usability can be defined as:

> "... The extent to which a product can be used by specified users to achieve specified goals with effectiveness, efficiency and satisfaction in a specified context of use" (ISO 9241-11, 1998).

Categories of evaluation

Evaluation methods can be categorised into two types: Non-end User Evaluations (also known as Expert Evaluations) and End User Evaluations. Within these categories are a number of evaluation methods available to the evaluator. Some of these evaluation methods can only be used within a specific domain (e.g. website evaluation) whereas others are more generic and can be used across domains.

Whereas end user evaluations require the product or system to be near completion or to be completed, an expert evaluation can be performed very early in the design phase. This is advantageous to the design team as any major flaws that have been incorporated into the initial prototypes can be identified and rectified before they become costly to put right. Another advantage of the expert evaluation is that it can inform the team whether the product or system is suitable and ready for testing with end users.

Types of expert evaluation

There are many types of expert evaluation that can be applied to examine a product or system. The majority of these techniques are used within the field of Human Computer Interaction as a method of testing the usability of a system. Some of these methods can be transported and adapted for a different domain whereas some are domain specific. One such method is Jakob Neilson's ten heuristics for testing the usability of websites. Although evaluators of different domains have transported it, its primary usage is to test the usability of websites.

Using checkpoints to determine the state of the system is well known and can be transported with greater ease. Heuristics have been used in a wide arena from websites to naval vessels. However, the more complex the item being evaluated the more complex the heuristics need to be in order to fully examine the item. In order to do this, a specific set of principles or heuristics has to be drawn up in order to test the system or product in question. This may take considerable time and effort and any such heuristics will need to be scrutinised for validity and reliability before they can be used and trusted. Nevertheless once these heuristics have been scrutinised they will allow an evaluator to use them to quickly and efficiently test the system or product.

Types of End User Evaluations

Generally speaking the methods that are available to evaluate with end users are not so varied and vast as the methods of expert evaluations. McClelland, 1999 cites that the three methods that can be used are; Interview and questionnaire, direct observation and the collection of objective data. McClelland, 1999 cites that a traditional interview is "one on one", one end-user and one evaluator however he recognises that group interviews "are being used more frequently in product evaluation work." A group interview can also be called a "focus group" Christie et al, 1999. It is felt that the first method, the Interview (individual and focus groups) and questionnaire method is the most important as it can influence the other two methods. For instance direct observation cannot be used when a questionnaire methodology is used and the evaluator is not with the end user when they are completing the questionnaire. McClelland supports this as he adds, "The extent to which a user trial makes use of these data collection methods will vary although in practice a blend of these methods is often used." Evaluation teams that are part of COST 219ter, such as the RNIB and FMH University teams uses a blend of these techniques in that observation techniques are employed and objective data is collected when performing group and individual interviews, therefore using a blend of the methods highlighted by McClelland.

For an end user to complete a user trial, McClelland identified that they would need to "complete some tasks." McClelland added, "Tasks must involve all areas that are of interest, be relevant to the user and must be measurable." Task based scenarios is an method that is employed with considerable success within the RNIB Evaluation Team in order to add an element of reality to the evaluation and to gain valuable data on parts of the product or system which need specific attention.

Why are both types of evaluation needed?

The expert evaluation and end user evaluations can be used to identify different areas however the expert evaluation can be used to lead into the end user trials to shape the way end user trials are carried out. By this it is meant that the expert evaluation could highlight an area of concern for the evaluators and they can make specific questions or scenarios that will bring this area into the evaluation.

Proposed process for evaluating mobiles

The proposed process for evaluating mobiles can be broken down into the two stages previously highlighted; an expert phase and an end user stage. Unfortunately this paper cannot go into the exact detail of each stage but can give a summary of it and how it was arrived at.

Expert phase

Using the three areas put forward from COST 219ter (Hardware, software and services), three sets of Heuristics were written to determine whether a mobile phone was accessible or not. It was decided that rather than just writing top level heuristics, which could be vague, each heuristic should have a set of checkpoints which need to be satisfied. Once the first set was complete it was clear that a supporting document was needed in order to complete the heuristics. Therefore each set of heuristics has a completion form, which allows the evaluator to catalogue their comments about each checkpoint. This has lead to a complete Heuristic Evaluation Stage, with each part (hardware, software and services) having fourteen top-level heuristics, with each heuristic being broken down into individual checkpoints. A set of instructions has also been written which describes the process that the evaluator must go through and what the evaluator needs in order to complete the evaluation successfully.

End user phase

Once an expert evaluation phase has been completed it should be followed by an end user phase. The RNIB evaluation team always recommends an end user stage in any evaluation of a product or system as it recognises the importance of consulting and including end users. This can be supported by McClelland, 1999 who said that "User trials can provide information which is at least useful and, at times, essential if products are successfully to accommodate user requirements."

The information gained from the expert phase should be used to shape the user trials in terms of the evaluation route (Interview, questionnaire and focus group) and in terms of the types of tasks the user is asked to perform and questions asked. By considering the data gained from the expert phase the evaluator can target any areas of the product or system that failed during the expert phase and evaluate these areas further with end users.

Summary

This paper has given a brief background to evaluation to show how it arrived at the proposed evaluation process for the mobile phones. Currently this work is on going and will be finalised in March 2006. The heuristics have been written and are currently being tested to determine whether they are accurate or not. This is being done by a series of user trials using focus groups, interviews and questionnaires across several European countries. It is hoped that the outcome will be a preferred evaluation process to test mobile phones that can be adopted by evaluation teams and handset manufacturers alike.

References

Christie, B., Scane, R. and Collyer, J. (1999). Evaluation of Human-Computer Interaction at the user interface to advanced systems. In *Evaluation of Human Work: A Practical Ergonomics Methodology*, edited by J.R Wilson and E. N. Corlett. 2nd Edition (London: Taylor and Francis)

McClelland, I. (1999). Product Assessment and User Trials. In *Evaluation of Human Work: A Practical Ergonomics Methodology*, edited by J.R Wilson and E. N. Corlett. 2nd Edition (London: Taylor and Francis)

Zwick, C., Schmitz, B. and Kühl, K. (2005) *Designing for Small Screens: Mobile Phones, Smart Phones, PDA's, Pocket PCs, Navigation Systems, MP3 Players, Game Consoles.* Switzerland: AVA Publishing SA

DESIGNING FROM REQUIREMENTS: A CASE STUDY OF PROJECT SPECTRUM

Andree Woodcock, Darryl Georgiou, Jacqueline Jackson & Alex Woolner

Coventry School of Art and Design, Coventry University, Priory Street, Coventry CV1 5FB, UK

Children with Autistic Spectrum Disorders suffer from varying degrees of qualitative impairments in social interaction, communication and restricted patterns of behaviour. This is accompanied by hyper- and hyposensitivities in each of the senses. Given that each child seems to have a unique profile, there is a clear need to develop systems that may not only be of benefit and pleasure to them, but that are also tailorable to their individual characteristics. This paper outlines the research undertaken in understanding the characteristics of children with ASD and how such an understanding has led to the development of a low cost, multimedia environment for mainstream schools.

Introduction

In Georgiou et al (2003) we outlined our approach to the design of polysensory environments for children with autistic spectrum disorders (ASD). We hope to develop an environment, incorporating interactive media, that can be tailored to meet the needs of individual children and facilitate their engagement with their surroundings and other people. Central to this, is that we should not let technology lead the research and development, but should focus on the needs of the users – the children, their parents and carers. This paper commences by briefly summarising the nature of autism, moving onto requirements elicitation and presenting specimen results, before concluding with an illustration of how these have informed system design.

Introduction to autistic spectrum disorders

The American Psychiatric Association (DSM – IV, 1994) characterise Autistic Spectrum Disorders by qualitative impairment in:

- social interaction *e.g.* use of nonverbal behaviour, failure to develop peer relationships, lack of willingness to share experiences and lack of social or emotional reciprocity
- communication *e.g.* delay in speech onset, inability to engage in conversation, stereotyped, repetitive use of language, inability to engage in make believe or social imitative play and
- restricted, repetitive and stereotypical patterns of behaviour, interests, and activities *e.g.* preoccupation with one or more stereotypes, restrictive patterns of interest, inflexible adherence to routines and rituals, preoccupation with parts of objects.

With an onset before three years of age, the most effective time to mediate a break through in these patterns is in early childhood. However, given that symptoms may vary in both their pattern and extremity from one individual to another, and that children with ASD may not be able to articulate their needs, or even have their needs correctly identified, it is

Table 1. Specimen results from the quantitative data.

	Lower functioning children	Higher functioning/Aspergers
Preferences	Red Round shapes Nursery rhymes, meditation music Smooth, soft and downy textures Mirrors Soft play areas Sound/light equipment	Blue Circular shapes Rock/pop music Smooth soft and downy textures Projected light effects Soft play areas Sound/light equipment
Dislikes	Sticky, slimy or prickly textures Loud noises and specific noises Sensitivities to smell Interaction, engagement with others	Sticky, prickly, slimy, rough textures Loud noises and specific noises Smells, certain lighting Interaction with others

very hard to adopt a user centred approach to system design, although one is clearly needed.

Requirements elicitation

User centred design can be undertaken using three different approaches (Eason, 1992); namely design for users, by users or with users. Given the nature of the user group we relied primarily on the first of these, namely design for users, based on an understanding of the end user population through personal experience, observation, semi structured interviews (with parents and, where possible children) and questionnaires. Where possible iterative design may be undertaken in conjunction with users.

A web-based questionnaire was used to ascertain the profile of children with ASD, their sensory preferences and previous experience of multi-sensory rooms. From the 500 responses we established a profile of the intended user group and the levels of tailorability needed to accommodate most of the children (see Table 1). These findings were corroborated through observation of eight children from different parts of the spectrum, playing in traditional environments. To add depth to the data, 25 semi structured interviews were conducted; 10 with teenagers with Asperger's Syndrome or High Functioning Autism and 15 with parents of children on various places of the autistic spectrum. Also, in order to build a rich picture of the life of a child with ASD for the designer, and contextualise the system, detailed descriptions of a "day in the life" of 5 children were created to show how ASD affects each child and how the use of an interactive environment could be of benefit.

As well as Jackson collecting these materials and presenting them in summary tables, Woolner (the designer) felt the need to immerse himself at a deeper level. This was in order to develop a working relationship with the user group, derive his own research material and develop his own knowledge of the community. He felt it was impossible, even when working in collaboration with an expert, to act solely from the information and direction received from others. Woolner therefore worked as artist-in-residence at a special needs school, provided technical support on similar projects and forged links with local schools.

Overview of requirements

From Table 1 it can be concluded that children with an ASD have sensory issues in terms of olfactory, tactile, vestibular (movement), auditory and visual input. If the final system is to facilitate sensory integration then each of these areas has to be addressed and opportunity provided to gradually introduce some dislikes in order to decrease sensitivities.

Observations made in traditional, multi sensory environments showed that some children derived benefit from these, displaying both enjoyment and relaxation. However there were noticeable differential effects caused for example by lighting, on those with Asperger's Syndrome and those with "classic" autism. Some parents reported that although their children enjoyed the experience they became overstimulated, hyperactive and aggressive for the rest of the day. However all children became visibly calm and more relaxed from tactile input such as immersion in the ball pool, being squashed under soft bean bags or spun around in an encasing hammock. We may conclude from this that although tailorable digital media may be useful, there is also a need for concrete, tangible objects to be used, perhaps at the start of the sessions for relaxation. This might allow the children to be more focused and able to work and interact with the visual and auditory stimuli offered to them.

From the interviews with parents the following themes emerged:

- The association of colour with mood and behaviour.
- Widespread spinning behaviour through all the group – of either self or objects.
- Differences in movement and co-ordination. A high proportion of the higher functioning children had coordination problems, whereas children on the lower end of the spectrum were seen as agile and active, but with their own distinct pattern of movement and needed to repeat certain movements in each environment.
- The need for an environment over which the children could exhibit some control.
- Predictability made the children feel secure and reduced anxiety. An environment in which the child knows what is to happen next and possibly author such changes themselves, can empower the child and give them a feeling of security.
- Interaction with others was a widespread problem.

The interpretation of these is dependent on the child; *e.g.* preference for "spinning" can have a different meaning to each child. Some children may like to spin small wheels on a car whilst other may like to spin themselves. Additionally, from an ethical perspective should we be reinforcing a behaviour that may be viewed as unwanted or abnormal in certain circumstances? The results confirmed the need for an environment that is sufficiently tailorable and adaptable to accommodate and benefit children at all places on the spectrum.

Communicating requirements to the designer

The requirements were presented in a number of ways to the research team – reports, summary tables, case studies and discussions. This approach was adopted over formal methods because of the complexity and level of detail that needed to be conveyed to the designer, before he could understand the complexity of the subject area. Additionally there was little enthusiasm for producing or receiving formal specifications once material been presented in other ways.

Developing the modules

The discussion about requirements and the need to feed these quickly into system design modules led initially to a series of poorly integrated early prototypes, which were technology based, stand alones. For example, the discovery that a lot of children liked spinning, red, circular shapes, and had poor eye hand co-ordination led to the production of a simple module in which a series of virtual cogs could be interlinked and spun in different directions. Although this, and similar ideas enabled Woolner to produce initial prototypes, this bottom up approach failed to create the immersive environment we required

The "breakthrough" for the project came after 18 months when during a face-to-face, brainstorming session we stepped back from the immediate user requirements to just consider how we imagined children would use the space. We all agreed that Project Spectrum at its most basic was an empty room in, for example, a school, that could provide refuge and tailorable experiences away from mainstream activity. Into this we could add material that would help children to become more engaged with the world. Positioning our space in this context generated further requirements related to the number of users, timetabling, affordability and usage.

It may be argued that this approach is not novel – for example there are, Snoezlen environments and soft and multimedia play areas. However, in some cases use of these is restricted to parents who can bring their child to the installation; the installation is expensive, large and requires skilled technicians to operate it; there has been little evaluation of the benefits the children derive from being in the environments; and technology led projects may not be at all intuitive in terms of their cause and effects (so confusing the children), and many cannot be tailored to benefit children on different places on the spectrum. If our project is to make a contribution it will be in identifying these areas as ones that can be addressed through the creation of tailorable, affordable rooms located in mainstream schools, that are accessible to all children.

Building the environment

With the above aims in mind, we located a primary school in Birmingham, which had a room we could "make over". Obviously we would have liked to be able to design and build a room to our own specifications, however, in terms of ecological validity being provided with a typical classroom, and overcoming its limitations showed that it should be possible to do this in any school.

The room used we took over is approximately 6 m square, has three large windows that open onto a playground which is noisy during break times. It has a high ceiling, lit by strip fluorescent lighting. The floor was covered with an ageing nylon carpet, the walls painted beige and covered in posters, pin up boards, black boards and an old interactive whiteboard.

This was converted by Woolner over the summer into a low stimulation sensory room (see Figure 1). The walls were stripped and repainted white, the floor replaced with natural marmoleum. Blinds were made from white blackout material to block out light and noise from outside. A custom projection screen was built and installed along with a data projector, positioned to allow for interaction with digital content. Two cameras, speakers and a computer system were installed to deliver the digital content. The strip lighting was removed and replaced with daylight bulbs and an LED lighting system also installed to allow for control of the ambient light colour. Furniture was minimal and standard, and organised in such a way as to allow individual and paired working, both in the context of the classroom and when participating in the interactive modules.

All the digital modules (such as the one shown in Figure 2) have been developed according to the researched requirements of the children. They are designed to add to the palette of activities a teacher or carer may use to engage with the child. Based around the senses, the modules engage the children through vision, sound, movement and touch. The digital system allows the child to receive immediate feedback from their actions, creating a cycle of interaction that empowers the individual through an immersive control system. The software that controls the system has been designed to

Figure 1. Interacting in the room.

Figure 2. Movement and colour.

be simple and intuitive with a small learning curve, so that teachers can start using it immediately, without technical support.

Future work

We are currently iteratively developing our environment – bringing in different lighting solutions and soft furnishings. The room is used as a base room for one child and his support worker and we are introducing him to the research team, experimental protocols (cameras etc.) and the modules. In 2006 we will invite more children into the space and introduce video conferencing to show what is happening in lessons. Our evaluation strategy is likewise evolving. We still hope to show engagement and pleasure, but have quickly realised that the everyday problems faced by these children will overwhelm our results.

References

American Psychiatric Association, 1994, *Diagnostic and Statistical Manual of Mental Disorders – Fourth Edition (DSM – I, 1994.)* Accessed on 28th Nov '05, from url http://www.asatonline.org/about_autism/autism_info01.html.

Eason, K.D., 1992, The Development of a User Centred Design Process: A Case Study in Multi-Disciplinary Research. *Inaugural Lecture.* Loughborough University. 14th October.

Georgiou, D., Jackson, J., Woodcock, A. and Woolner, A., 2003, The design of polysensory environments for children with autistic spectrum disorders, in McCabe, Contemporary Ergonomics; Springer-Verlag.

INCLUSIVE DESIGN IN TRANSPORT

MY CAMERA NEVER LIES!

David Hitchcock[1], John Walsh[2] & Victoria Haines[3]

[1]*CEDS, The Acorn Centre, 51 High Street, Grimethorpe S72 7BB*
[2]*Photarc Surveys, Beech House, Beech Avenue, Harrogate HG2 8DS*
[3]*ESRI, Holywell Building, Holywell Way, Loughborough LE11 3UZ*

The Centre for Employment and Disadvantaged Studies was commissioned by the Mobility and Inclusion Unit of the Department for Transport to conduct a study to measure occupied wheelchairs and scooters to determine the characteristics of their users and to assess the changes that are taking place in the design of the devices. The project team used the technique of photogrammetry to collect the data for over 1,000 devices. In addition to explaining the technique, this paper gives a recount of the project and explores the potential application of photogrammetry in ergonomics projects.

Introduction

The Centre for Employment and Disadvantage Studies (CEDS) is the research division of yes2work, a social firm working with those who have disabilities or are otherwise disadvantaged. CEDS was commissioned by the Mobility and Inclusion Unit of the Department for Transport (DfT) to conduct the 2005 survey of occupied wheelchairs and scooters to determine their overall masses and dimensions.

This study was the third in a series (previous studies were conducted in 1991 and 1999) to identify and report on trends in wheelchair and scooter designs. The two previous studies concentrated on adult devices and collected data principally at the Mobility Roadshow a regular and well attended event which provides the opportunity to try out mobility products, drive adapted vehicles and find out the latest information from a wide variety of charities, interest and research groups. CEDS was given a broader remit to include children's wheelchairs and to collect data at the Mobility Roadshow and through a number of site visits around the UK.

Photographic data capture

All three studies required efficient data collection in order to capture a high number of representative devices and their occupants with the minimum of inconvenience. In all three cases photographs were taken of each participant for subsequent analysis.

The 1991 and 1999 studies used single-image photogrammetry. In the 1991 study each wheelchair user was required to manoeuvre into a right angle formed by two checkerboards in order for a side profile and a front profile photograph to be taken. Because some users found it difficult to make the necessary manoeuvre this technique was slightly revised for the 1999 study so that the wheelchair only needed to be positioned alongside a single white checkerboard, while the same profiles were photographed.

Each of the photographs taken were used to manually calculate the various dimensions – height, length, width, ankle height, shoulder width, knee height and axle spacing. The previous researchers reported that the two photographs of each subject measurements were taken directly from the prints and the dimensions were calculated by scaling from checkerboards and using trigonometry (Stait et al, 2000). To validate this method, photographs were taken of a person in a wheelchair and then actual measurements were taken in situ.

For the most recent CEDS study, it was decided that an alternative use of photography was desirable which enabled greater portability for the site visits and was not reliant on accurately positioned checkerboard or devices and provided opportunity to extract a wider range of measurements. The most appropriate technique selected was that of multi-image photogrammetry, which is more accurate and flexible than the single-image method used before.

History and development of photogrammetry

The principle of drawing via projection has been around since ancient Greece, but the first developments in photogrammetry occurred shortly after Daguerre created photographic prints on glass plates. The man regarded as the father of photogrammetry is Colonel Laussedat who between 1849–1859 experimented with the use of photography for topographic mapping. Since then, the techniques have become extremely advanced and the instrumentation increasingly miniaturised.

Photogrammetry has been used in many fields including, but not limited to; aerial mapping, building survey, forensic science, traffic accident investigation, forestry, environmental monitoring, automobile, aircraft and submarine manufacture, and shipbuilding. In short, anything that can be photographed can be measured. Photogrammetry may be crudely split into stereoscopic and multi-image processes.

Most mapping has been carried out by stereo-photogrammetry, where a series of overlapping images are taken from an aircraft with a large-format survey camera. The maps are plotted from these images with stereo-plotting instruments. Stereo techniques may also be employed with ground-based (terrestrial) photography. Drawing building elevations is one such application.

Stereo plotting instruments used to be quite large, took a great deal of skill to operate and were very costly. The cameras used for stereo photogrammetry were also large and expensive. Thus photogrammetry has traditionally been the domain of specialised personnel.

The recent advent of multi-image photogrammetric software that runs on a PC has changed this situation. Rather than strips or pairs of overlapping photographs, multi-image photogrammetry involves taking images of an object from several different angles. It is an easier technique for the non-specialist.

These packages are usually cheap compared to more traditional equipment and are generally easier to learn. Coupled with the digital camera revolution, these packages have brought photogrammetry within the reach of more people. The new software suits those who want to extract measurements from photographs without spending a long time becoming experts.

How multi-image photogrammetry works

The software used for this project was PhotoModeler Pro5, written by EOS systems in Vancouver. It is a multi-image photogrammetry package that has been written to be easy to learn and use. The basics of using PhotoModeler are:

- A digital camera is calibrated using PhotoModeler's in-built calibration module.
- Several images are taken of the desired object from different viewpoints with the camera. At least one scaling distance is measured on the object. It is essential that

each image is taken at a good angular separation from the other images. There should not be less than 30 degrees between images in order to obtain accurate results.
- The images are imported into a PhotoModeler project and several common points are marked in the images. Each photo must share at least six common points with adjacent photos and they must all hang together as an inter-linked image bundle. The points will include the end-points of the scaling distance.
- Processing then takes place and if the points are well marked, an accurate 3D model will be formed. Further points and other features such as lines and curves can then be added to the model, and further processing carried out.
- After the scaling distance is applied, the model will have true real-world scale and accurate measurement can then be taken from it. The model can also be exported to several CAD formats for further manipulation.

Data ranging from simple dimensions all the way up to rendered 3D models can be extracted from photos. An object can be measured with little or no physical contact as long as it has enough identifiable points to mark. The photography can be taken in a fraction of the time needed for more direct physical measurement, and the data capture takes place in the comfort of the office.

The practical use of photogrammetry

Having elected to use photogrammetry, CEDS adopted the following method to the data collection exercise.

1. A process for taking the photographs was developed to ensure that sufficient images for data extraction could be taken in the minimum of time – throughput rate of participants needed to be one per minute to meet target numbers. The minimum number of images was found to be seven, using the camera arrangement shown in Figure 1. In order to ensure that both the markings of the "wheelchair zone" (needed for scaling) and all of the pertinent wheelchairs features were captured in each image, the minimum distance between the zone and the camera needed to be at least 1.6 m. All photographs were taken from a standing position except in positions 1 and 7 which required the photographer to squat in order to achieve sufficient angular separation.
2. Those responsible for the data collection (the actual taking of photographs) attended a half-day training session to provide awareness of the technique and the specific

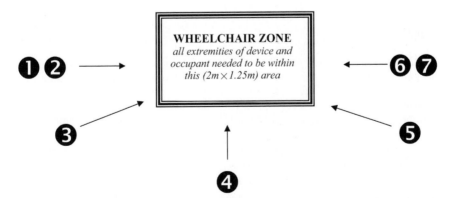

Figure 1. The seven camera positions to capture images of rear, front and side profiles.

needs of those who would subsequently use the PhotoModeler software to extract the dimensions, and to practice using the cameras.
3. Those who would use the PhotoModeler software each received an initial two-day training course and a one-day follow-up after initial practice.

The pros and cons of the photogrammetry technique

In considering the usefulness of photogrammetry within the context of ergonomics projects it is perhaps helpful to review the lessons learned through this particular project.

It may be argued that taking seven photographs was more time consuming than taking the two (side and front profile) in the previous studies. Even so, the photographs for each wheelchair could all be taken in less than one minute. This was a key selling point in attracting over 1,300 participants. Another appealing aspect, particularly for child participants, was that at no stage during the process, was it necessary to touch the occupant or wheelchair.

In order to both ease the efficient positioning of devices and achieve sufficient angular separation between the photographs it was necessary to create a working area of at least $5.5\,\text{m} \times 2.8\,\text{m}$. This is actually quite a large area, which for example may be beyond the size of stands that may be available at pertinent exhibitions.

Pilot trials were conducted to practice the technique. Some of the images taken during these were either over or under-exposed. Also, some images were out of focus, which indicated that too wide an aperture had been used. This reveals the need to train the photographers in the basics of photography exposure. With hindsight, if a tripod and cable shutter release had been used to take the shots, a narrow aperture in combination with a slow shutter speed could have been used and thus avoid the need for harsh lighting or potentially upsetting flash.

Without doubt the key benefits of photogrammetry prevailed themselves at the data extraction stage. Processing the photographs using PhotoModeler took approximately 15 minutes per device. However, obtaining the dimensions from the processed photographs was typically both quick and straightforward. Also although final analysis was not available at the time of preparing this paper, indications suggested that the process was very accurate. Photogrammetry made it possible to adopt an iterative approach to data extraction; selecting new or additional dimensions to be attained as the project progressed. Future needs from manufacturers or other researchers may also be satisfied as all the photographs are catalogued and processed ready for further dimensions to be acquired as necessary.

Conclusion

The time and resources required for acquisition of equipment, familiarisation, training and piloting to overcome teething troubles have all been outweighed by the creation of a large data set of processed photographs which can be used repeatedly for the extraction of accurate dimensions. The authors are convinced that the technique may be used in other projects where there is a need for the efficient and precise recording of dimensions.

References

Stait, R.E., Stone, J. and Savill, T.A., 2001, *A survey of occupied wheelchairs to determine Their overall dimensions and weight: 1999 Survey*, TRL Report 470

Wolf, P.R., 1987, *Elements of Photogrammetry, with air photo interpretation and remote sensing*, Second Edition (McGraw-Hill)

COLLECTION OF TRANSPORT-RELATED DATA TO PROMOTE INCLUSIVE DESIGN DOOR-TO-DOOR

Ruth Sims[1], J. Mark Porter[1], Steve Summerskill[1], Russ Marshall[1], Diane E. Gyi[2] & Keith Case[3]

[1] *Department of Design & Technology,* [2] *Department of Human Sciences,* [3] *Wolfson School of Manufacturing & Mechanical Engineering, Loughborough University, Leicestershire LE11 3TU*

A computer-based inclusive design tool (HADRIAN), developed under the EPSRC "EQUAL" initiative, is being expanded through the EPSRC Sustainable Urban Environments programme. This development will result in the tool including data on transport usage and related issues, providing a database of physical, emotional and cognitive information for 100 individuals, including those who are older and/or physically disabled. The collection of anthropometry by use of body scanning technology, as well as issues concerning the collection of physical capability data, whether by field observation, questionnaire response, or laboratory trials, are discussed. The work detailed is ongoing, and presented here are the methodological and ethical issues arising from consideration of the needs of those wishing to make journeys, and the collection of data to facilitate better design and policy to ease that process. This paper should be read in conjunction with Porter et al (2006) also presented at the conference.

Introduction and aims

The work being conducted will involve data collection from 100 individuals, including those who are older and/or physically disabled. The data collected will consist of anthropometry, joint constraints and reach range volumes, which are used within HADRIAN to construct individual virtual human models of all the participants, and physical task capability data (including postures and behaviours) in kitchen-related bend-reach-lift tasks and transport-related stepping up/across and reaching tasks. These behavioural elements allow the tool to predict success or failure of virtual tasks for each of the participants. In addition to these data, a questionnaire will also be administered in order to discover some of the cognitive and emotional barriers to travel, as well as coping strategies and further details of physical capability (walking, carrying luggage, climbing stairs, and so on). These data will be entered into the existing HADRIAN software, expanding on data collected previously whenever possible, and additional individuals where this is not possible.

The aim of this paper is to present and discuss some of the methodological issues arising from the aims of the project as a whole, ethical considerations and concerns with using older and/or physically disabled participants, and in obtaining information about the extent of their physical capabilities without putting them at risk of exceeding their capabilities. The future scope of the project will also be presented.

Methodology

Participant selection

There were a number of issues that arose when the issue of participant selection was considered. It was felt that the complexity of carer interactions would be too difficult at this stage to quantify and model within the software tool. As such the decision was made to exclude those who are not physically able to, alone, get themselves out of their house and onto the pavement to join the transport system. Clearly, such people still have needs and aspirations for transport usage, and a future study could exclusively explore the considerations involved with individuals and carers. It was hoped that it would be possible to revisit as many of the 100 people who had previously participated and see if they would be willing to participate again. This would enable some study of the longitudinal aspects of age and disability, due to the number of years since the first study data collection took place. However, some participants might have sadly died in the interim, some might no longer be physically able to get outside without assistance, and some might have changed contact details.

All participants were required to complete a medical screening questionnaire before the trials. However, medical screening questionnaires typically preclude the involvement of people with current medical conditions, yet due to the older and/or disabled nature of many of the participants it was expected that some would have multiple medical complaints, and it is of interest to this work to investigate the problems that these might cause the person when travelling. However, it was required that the experimenters be aware of any conditions that would cause a problem during the course of the trial, so that steps could be taken to reduce the risk and adapt the trials procedure accordingly. As an example, those participants with vertigo/dizziness were not asked to bend low in case this exacerbated the condition.

Questionnaire

A detailed questionnaire was developed to get rich, detailed information regarding a participant's physical abilities, and also to tap into their cognitive and emotional issues surrounding transport usage. Participants were asked a number of questions concerning their physical abilities, based on the Office of Population Censuses and Surveys scale (1988), transport usage (and reasons for not using, if appropriate) for trains, buses, trams, London-style taxi cabs and minicab taxis, walking distances, as well as issues surrounding taking luggage on the different transport modes (and different amounts of luggage, including suitcases and pushchairs), the types and frequency of journeys made, stairs, lifts, escalators, and timetable usage. The questionnaire also included a request for information about problems experienced in the local area. Any local areas that participants identify as causing problems when travelling will be visited by the experimenters, measurements taken and so on, to verify the reports from the participants. In short, the questionnaire aims to provide information concerning issues that may arise at any point during the whole journey process, from leaving home to wait at a bus stop, walking between transport modes at an interchange, making that change, and arriving at the destination.

Vehicle rig design

When making a journey using public transport a person might expect to be met with a variety of step heights and handle locations during ingress and egress. A rig was designed to assess participants' ability in these situations, and decisions concerning which heights and handle positions to study were made after referencing the relevant public transport regulations and field observations within the Midlands area. Train carriages and trams are covered by the Rail Vehicle Accessibility (Amendment) Regulations, 2000. This states the maximum step height should be 200 mm, with handrails placed internally on either side of the external doorways, vertically between 700 mm and 1200 mm above the floor. From observations it was found that step heights into trains varied between 180 mm and 280 mm, with the one example of trams having no step at all. London Underground state that the maximum step height on their lines is 240 mm.

Figure 1. Photograph of the vehicle rig.

Buses and coaches (carrying more than 22 people for public usage) are covered by the Public Service Vehicle Accessibility Regulations, 2000. This states that the maximum step height from pavement to bus should be 250 mm, with the first handrail inside the bus being within 100 mm of the entrance and between 800 mm and 1100 mm above ground level. Observed step heights for buses were found to vary between 170 mm and 300 mm, and for coaches between 270 mm and 370 mm.

A rig consisting of an entrance and exit was designed to reflect these measurements and assess what participants are capable of. This allowed for different "door" widths on each side: one side narrower to simulate an older-style bus or train, or coach entrance, and the other side wider to simulate the wider access of newer buses and trains. The handles on each side could be placed in a choice of two positions, on the narrow side they could be set at 100 mm or 200 mm from the entrance to the "vehicle", on the wider side they could be set at 300 mm or 400 mm from the entrance. The step heights were given as 150 mm, 250 mm, or 350 mm to reflect the worst-case scenario. There was also a 100 mm gap horizontally between the ground and the "vehicle" on both sides, to reflect the horizontal gap between pavement/platform and the body of the vehicle.

The design of the rig raised a number of ethical considerations. In order to get a true idea of participants' abilities they would be required to step onto and across the rig. In order to keep the trials as safe as possible a strict protocol was developed and trialled (using able-bodied participants). Participants would complete the questionnaire before attempting the rig, giving information about what the person would attempt and what would cause them problems. The rig was then set up according to these responses: able-bodied participants had both step heights set at the maximum 350 mm, with handles set at 200 mm and 400 mm respectively. Less able participants had lower step heights and handle heights adjusted to their ability. Once the rig was set correctly, participants were first asked to observe an experimenter demonstrating the task. An experimenter would be standing on each side to offer assistance if required, and it was reinforced that participants should only attempt if they were happy to do so and they should take their time. A rest could be taken if required, and when it came to stepping down participants were asked to first look at the required step and state whether they were happy to continue, before doing so in a controlled, safe manner. Care was taken to reinforce this as participants proceeded with the task. Anyone who felt unsure about the task was obviously free to stop, and steps could be removed if required during the trial.

Body scanner

The body scanner ($[TC]^2$ NX_{12} Body Measurement System) has the potential to be used to quickly and accurately collect body dimensions, of use in constructing anthropometrically correct virtual human models of individual participants. Participants are required to undress in an enclosed private cubicle into brief clothing that is neutral to their skin tone, as high contrast with skin tone causes problems in attaining a complete scan. To achieve accurate measurements the clothing should fit closely to the body shape. Once inside the scanner itself participants stand and then sit in given postures (demonstrated to them beforehand by the experimenter). Each scan itself takes a matter of seconds, and then the person can exit and dress again. The scan is light-based and therefore

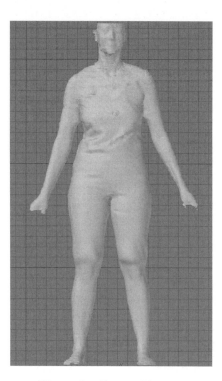

Figure 2. Scanned image.

harmless, however it was decided to ask participants prior to the trials if they are epileptic, and any who are will not be required to be scanned.

Traditional external anthropometry was collected to enable comparison with measurements extracted from the scanner to see if the use of the technology made the process quicker and more accurate for able-bodied participants. However, when considering less able participants, they may require assistance to undress and get dressed again (an assistant must be provided by themselves for ethical reasons), and participants may not be willing or able to get undressed/redressed within the scope of the trial. The temperature of the room and how long the process would take them are other issues for concern. Less able participants might also not be able to get into the required positions, and there are issues concerning those in wheelchairs, as the scanner dislikes reflection or high contrast, so a "stealth" wheelchair would be required for those needing to remain in the chair for the duration.

Future work

Having completed the lengthy piloting stage of the data collection, and addressed the issues discussed in this paper, it is now expected that the data collection trials will progress in the near future. The data collected will then be entered into the HADRIAN software tool. Future work includes comparison of anthropometry collected using traditional methods versus that obtained from the scanner, to ascertain the accuracy and speed of data collection for different abilities of participants. A separate study of scanning a group of less able participants is also proposed, in order to investigate issues such as the time required and any problems with, for example, parts of a wheelchair obscuring the occupant and the accuracy of the scan, and so on. This in turn will inform future work.

References

Marshall, R., Porter, J.M., Sims, R.E., Gyi, D.E. and Case, K., 2005, "HADRIAN meets AUNT-SUE", *Proceedings of INCLUDE 2005*, INCLUDE 2005, Royal College of Art, London, UK, April 2005, pp 1–7, ISBN 1-905000-10-3, [Also on CD-ROM].
Martin, J., Meltzer, H. and Elliot, D., 1988, *OPCS surveys of disability in Great Britain: The prevalence of disability among adults*, (Office of Population Censuses and Surveys, Social Survey Division, HMSO).
Porter, J.M., Marshall, R., Sims, R.E., Gyi, D.E. and Case, K., 2003, "HADRIAN: a human modelling CAD tool to promote 'design for all' ", *Proceedings of INCLUDE 2003: inclusive design for society and business*, 6, Royal College of Art, London, 2003, pp 222–228, ISBN 1 874175 94 2, [CD-ROM].
Porter, J.M., Case, K., Marshall, R., Gyi, D.E. and Sims, R.E., 2004, "Beyond Jack and Jill": designing for individuals using HADRIAN, *International Journal of Industrial Ergonomics*, 33, pp 249–264.
Porter, J.M., Case, K., Marshall, R., Gyi, D.E. and Sims, R.E., 2006, Developing the HADRIAN inclusive design tool to provide a personal journey planner, *Contemporary Ergonomics*, Proceedings of the Ergonomics Society Annual conference, 4–6th April 2006, Cambridge UK.
Public Service Vehicles Accessibility Regulations, 2000, Statutory Instrument No.1970 (Department for Transport).
Rail Vehicle Accessibility (Amendment) Regulations, 2000, Statutory Instrument No.3215, (Department for Transport).

DEVELOPING THE HADRIAN INCLUSIVE DESIGN TOOL TO PROVIDE A PERSONAL JOURNEY PLANNER

J. Mark Porter[1], Keith Case[2], Russ Marshall[1], Diane E. Gyi[3] & Ruth Sims[1]

[1]Department of Design & Technology, [2]Wolfson School of Manufacturing & Mechanical Engineering, [3]Department of Human Sciences, Loughborough University, Leicestershire LE11 3TU

HADRIAN is a computer-based inclusive design tool that was originally developed through the EPSRC "EQUAL" initiative, and is being enhanced with funding from the EPSRC Sustainable Urban Environments programme. The tool is being expanded to include data on an individual's ability to undertake a variety of transport-related tasks, such as vehicle ingress/egress, coping with uneven surfaces, steps, escalators, lifts, street furniture and complex pedestrian environments. A specific feature of the enhanced HADRIAN tool will be a journey planner that compares an individual's physical, cognitive and emotional abilities with the demands that will placed upon that individual depending on the mode(s) of transport available and the route options.

Introduction

HADRIAN (Human Anthropometric Data Requirements Investigation & ANalysis) is a computer-based inclusive design tool that was created as the main deliverable of an EPSRC grant (1999–2002) under the Extending Quality of Life (EQUAL) initiative. The tool has been described in detail in previous publications (e.g. Porter et al, 2003; Porter et al, 2004; Marshall et al, 2005). The purpose of this paper is to describe how the tool will be further developed with current funding from the EPSRC Sustainable Urban Environment Programme.

AUNT-SUE consortium

Accessibility and User Needs in Transport for Sustainable Urban Environments (AUNT-SUE) is a multi-disciplinary consortium of researchers from London Metropolitan University, Loughborough University, University College London, RNIB, London Borough of Camden and Hertfordshire County Council. The range of experience includes ergonomics/human factors, accessibility, transport planning, transport policy, computer aided engineering, civil and environmental engineering, tourism and cultural diversity. The aim of the consortium is to improve our understanding of the needs, abilities and preferences of people who experience transport-related exclusion in towns and cities. Better empathy with disadvantaged users and would-be users will be encouraged through an AUNT-SUE "toolkit" to support planners, designers, operators, user groups and others working to make urban transport and street design more inclusive.

The research team will develop these decision-support tools using a combination of human modelling CAD, Geographic Information Systems (GIS), laboratory testing of prototype designs, and observation and analysis in real-life "testbeds" within Camden and Hertfordshire. To help direct the work of the consortium, an AUNT-SUE practitioner network has been set up that comprises people involved in policy and planning, design and operation and those representing user groups. This network facilitates exchange on best practice and members critically review the emerging toolkit and resources through our annual symposia.

"Design for all" or inclusive design

From a philosophy to design practice

"Design for all" or inclusive design needs to move from being a philosophical viewpoint to being a central feature of design practice. Key to this is establishing empathy between designers (who usually start their careers whilst young, healthy and able) and the people who would primarily benefit who are often older, in poor health and unable to achieve all the tasks they would like to with ease and confidence (if at all).

Current design practice frequently involves using anthropometric and biomechanical databases that present percentile values for body size and strength, joint ranges for mobility and so on. These numbers do not motivate the designer to vigorously explore design solutions that are more inclusive. In fact, the commonly accepted view for mainstream design is to cater for only the 5th to 95th percentile users of a product or service. This is designing for numbers, not people. Today, we believe that it is no longer acceptable to continue this approach of deliberate "designing out" of people who are in the top or bottom 5% of size or ability.

Furthermore, people can be excluded from using a product or service because of a wide range of factors including their personal cognitive and emotional dimensions. For example, a person may be able to reach a control but not able to manipulate it as required; able to lift an item but not have the balance to carry it safely; able to see a timetable but not able to plan a journey route; able to walk 100 metres but not confident enough to cross a busy road; able to climb stairs but not willing to walk past a group of teenagers in an environment dominated by graffiti; and so on.

Continuing development of the HADRIAN inclusive design tool

The AUNT-SUE project has enabled us to continue developing HADRIAN. The current prototype tool includes the following:

a) detailed information on the size, shape, abilities and preferences of a wide range of people, each presented as individual datasets (see Figure 1). This information comes as a set of screen displays for each person, including video clips showing them undertaking a variety of tasks such as lifting a baking tray with oven gloves on, reaching to food stuffs on a low or high shelf etc. These data are both informative and foster empathy between the designer and these future users/consumers. Basic cognitive and emotional data related to "activities of daily living", such as shopping, cooking and making a journey, will also be included.
b) a simple task analysis framework whereby the designer can specify a series of task elements (such as look at, reach to, lift to, walk to, climb up) with reference to physical items in a CAD model of an existing product or a early prototype of a new design.
c) an automated analysis whereby each individual in the database is assessed in terms of their ability to complete successfully each task element. This procedure deals with the multivariate nature of interactions with products, integrating the relevant physical, cognitive and emotional issues.

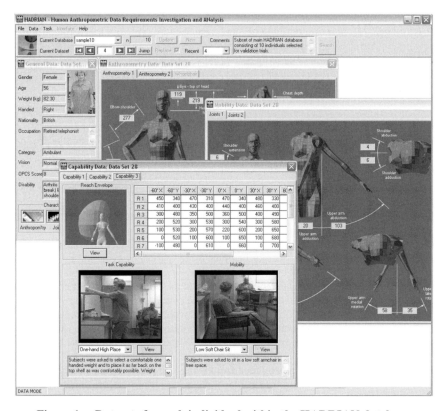

Figure 1. Datasets for each individual within the HADRIAN database.

d) those people who cannot achieve a critical task element, and are effectively "designed out", are brought to the attention of the designer – this includes a virtual simulation of the problem(s) encountered by each person. This visualisation of the person and the problem encourages the designer to explore potential solutions by modifying the CAD model and repeating the analysis.

We are currently expanding HADRIAN to include data on an individual's ability to undertake a variety of transport-related tasks, such as vehicle ingress/egress, coping with uneven surfaces, steps, escalators, lifts, street furniture and complex pedestrian environments. The tool will provide a database of physical, emotional and cognitive information for 100 individuals, carefully selected to cover a very wide range of abilities. This will allow those who are planning, designing and operating transport vehicles and systems to maximise social inclusion and access through consideration of the issues being faced by the users of public transport as they move to and fro between home, work, education, medical care, leisure and shopping. We plan to undertake some case studies within the testbed areas, or elsewhere if more suitable, where HADRIAN is used by designers and planners to "try out" different options for a specific design problem, be it a ticket barrier, a train station or access to the Olympics 2012. The tool should be able to identify the superior design option in terms of inclusivity, and to give direction as to how to improve it further, if appropriate.

Developing the HADRIAN inclusive design tool 473

Figure 2. Examples of environments and tasks that have physical, cognitive and emotional dimensions when travelling.

The development of a personalised "journey planner"

A specific feature of the enhanced HADRIAN tool will be a journey planner that compares an individual's physical, cognitive and emotional abilities with the demands that will placed upon that individual depending on the mode(s) of transport available and the route options. For example, some people do not use trains or buses because of difficulties experienced getting on and off or in getting to their seat before the vehicle moves off; some prefer not to walk through busy public areas or places with graffiti where they feel vulnerable; some experience problems finding their way on unfamiliar routes; some are unable to walk far or to climb steps with confidence; some are reluctant to cross busy road junctions; and so on (see Figure 2).

If a particular desired journey is unachievable or very difficult, either unaided or with support from others, then that person is likely to feel socially excluded. The proto type journey planner should allow people to predict problems that they may experience before deciding to make the journey. Hopefully, a suitable alternative route and choice of transport mode(s) can be identified using the planner such that the task demands fall within the person's abilities.

Whilst the database will only comprise the 100 individuals forming the HADRIAN database, it is envisaged that a web-based planner could be made available. Members of the public would need to complete an on-line questionnaire to provide relevant personal data on their body size (i.e. clothing sizes), general health, abilities and transport preferences. A major issue in such a planner would be compiling a database of the specific demands that would be placed on the traveller as a function of the exact geographic locations and the transport modes available for a particular journey. It is hoped that a pilot trial can be run in the testbed areas, with transport nodes, shopping areas, museums, theatres, cinemas and restaurants providing data for their surrounding areas (i.e. from the nearest bus stops, train station, taxi rank etc.). The data would include distances by foot, details of steps/lifts/escalators, perceived safety, performance of street lighting and signposting, quality of the pavement/street surface, etc.

Progress to date

Progress to date on the HADRIAN tool includes: the design and construction of rigs for assessing ease of ingress/egress to a variety of vehicles; setting up our newly acquired body scanner to be able to create 3D virtual human models of the individuals who will form the HADRIAN database; designing and piloting our "journey planner" questionnaire that identifies physical, cognitive and emotional issues that are experienced when travelling using a variety of modes of transport; developing a detailed experimental protocol for the data collection phase and gaining approval from our ethical committee. We have learnt a lot from the above activities and some important issues are discussed in detail in Sims et al (2006).

References

Marshall, R., Porter, J.M., Sims, R.E., Gyi, D.E. and Case, K., 2005, HADRIAN meets AUNT-SUE, *Proceedings of INCLUDE 2005*, Royal College of Art, London, UK, April 2005, pp 1–7, ISBN 1-905000-10-3 [Also on CD-ROM].

Porter, J.M., Case, K., Marshall, R., Gyi, D.E. and Sims, R.E., 2004, "Beyond Jack and Jill": designing for individuals using HADRIAN, *International Journal of Industrial Ergonomics*, 33, pp 249–264.

Porter, J.M., Marshall, R., Sims, R.E., Gyi, D.E. and Case, K., 2003, HADRIAN: a human modelling CAD tool to promote "design for all", *Proceedings of INCLUDE 2003: inclusive design for society and business*, Royal College of Art, London, 2003, pp 222–228, ISBN 1 874175 94 2 [CD-ROM].

Sims, R., Porter, J.M., Summerskill, S., Marshall, R., Gyi, D. and Case, K., 2006, Collection of transport-related data to promote inclusive design door-to-door, *Contemporary Ergonomics*, Proceedings of the Ergonomics Society Annual conference, 4–6th April 2006, Cambridge UK.

CAN A SMALL TAXI BE ACCESSIBLE? NOTES ON THE DEVELOPMENT OF MICROCAB

Peter Atkinson

School of Art & Design, Coventry University, CV1 5FB

Microcab has been developed with the objective of safely and comfortably carrying a wheelchair user, in addition to meeting the needs of passengers with a variety of abilities, within a vehicle for a total of three passengers. Evaluation of the design by people with disabilities at an early stage was influential in establishing the format, layout and detailing of the vehicle. The empirical testing of full-size mock-ups by a number of individuals from regional user group Coventry and Warwickshire Accessible Transport Committee was recorded and analysed using photos, video, questionnaires and observation. The results revealed that the small space could be formatted to allow a wheelchair user to manoeuvre into a comfortable and safe travelling position, and that a clear low flat floor, large entry/exit doorway, integrated ramp and well positioned hand-holds are key to accessibility.

Introduction

Inspired by the rickshaw culture of the East, Microcab was originally conceived as a very lightweight part of the public transport infrastructure, particularly for cities such as London with their attendant problems of congestion and pollution. Hydrogen fuel cells would provide back-up energy to silently propel the electrically assisted 3-wheeler and two passengers seated behind its pedalling "rider". From 1996 John Jostins prototyped and developed the Microcab. With increasing confidence in the design, the project was brought into Coventry University's ADAM Lab in 2003 to enable a wider team to contribute to its development.

At that time the decision to make Microcab fully electrically powered was taken with the initial focus on a low-weight solution as offered by a 3-wheeled design, however a variety of concepts were under consideration, including a 4-wheeled layout. The 4-wheeled variant was recognised as having the potential for wheelchair access capability and a paper based space planning exercise was conducted to package the vehicle, with a volume representing the wheelchair user based on the Transport Research Laboratory's (TRL) work on the safe carriage of wheelchairs in vehicles (Le Claire et al, 2003). Initially entry and egress of the wheelchair user through a rear tailgate was also considered, however this was disliked owing to the potentially vulnerable location of the wheelchair user in the road and the difficulty of picking passengers up directly from the pavement. This option was later precluded with the decision to locate the hybrid drive-train motor, batteries and hydrogen (fuel) tank to the rear of the emerging 3-wheeled concept. This paper considers the research undertaken since then to establish the requirements of passengers using a wheelchair and how they could be accommodated in the Microcab.

A collaborative approach

In the development of the Microcab we have received close co-operation from members of the Coventry and Warwickshire Accessible Transport Committee (CWAT), who campaign and liaise both locally and nationally with government, transport and infrastructure operators. Individuals from this group have participated in a series of four user-trials with the vehicle mock-ups to date, which have aimed to derive qualitative information and highlight areas for improvement in the proposed design.

First user trial

A wooden mock-up was constructed over a period of 8 weeks from June 2004, representing the interior space of the 3-wheeled micro-taxi proposal. This design was influenced by a number of visual design exercises conducted by design professionals, academics and students incorporating visual elements from the 2003 prototype H3 vehicle (Figure 1a). Originally, the main focus was on issues for walking passengers, including aspects of entry and exit of the vehicle relating to the floor height and the detailing and comfort of passenger seating. Access for a wheelchair user was explored speculatively in recognition of the 1995 Disability Discrimination Act's (DDA) potential application to the vehicle's use as a taxi.

This led to the first user trial in September 2004 with four participants who had restricted mobility. Participant 1 (P1) had birth palsey, paralysis of the right arm, could

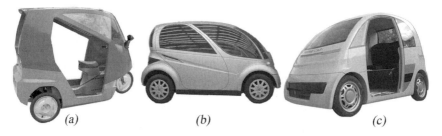

Figure 1. Microcab design development: (a) H3 prototype 3-wheeled vehicle from 2003; (b) full-size 4-wheeled clay styling model, 2004; (c) the H4 prototype in 2005.

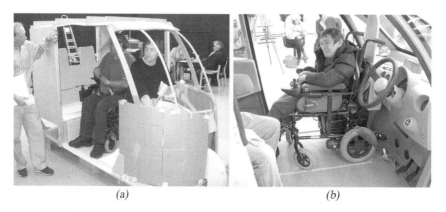

Figure 2. Wheelchair users evaluating the design: (a) first trial with a wheelchair in September 2004; (b) inside the H4 prototype in 2005.

only grip with the left hand and walk short distances with a stick; P2 no sight, with guide dog Sheba, P3 was partially sighted; P4 wheelchair user from spinal injury in 1967, stroke affected legs, and left arm.

Procedure

Background information on each of the participants was collected regarding the nature of their disability. In the trial they were asked to enter the vehicle, seat themselves in the vehicle as if for travel, comment on comfort related issues, then exit the vehicle. A team of 4 academic staff, 2 project designers and 4 students acted as observers, interviewers, and camera operators and ensured the safety of participants during interactions with the mock-up. Throughout the process observations were made of problems in the user's interaction with the mock-up. This led to a number of immediate minor modifications, including adjustment of the passenger seat height and the provision of an impromptu step.

Other factors were identified such as: P1 and P3 (both ambulatory) commented on the relatively high step of 250 mm into the vehicle; this subsequently led to consideration of a fold-out step integral with the floor. The 25 mm high lip at the edge of the sill was identified as a potential trip hazard and removed from subsequent mock-ups. It was noted that the door and A-pillar formed potential handholds; these and other potential handhold positions inside the vehicle were marked with coloured stickers. These were required more on exiting than on entry.

Wheelchair interactions

Using a short ramp, the wheelchair user entered, like the ambulant participants through the side passenger doorway. He commented on the usefulness of handholds in stabilising himself during a journey and enquired about the possibility of their integration into the ceiling. The possible requirement for wheelchair restraints was not explored at this stage.

The 1.3 m external width of the vehicle was slightly too narrow to accommodate the wheelchair user comfortably in the sideways facing orientation attempted. The elliptical door aperture was inclined too far from vertical and reduced the potential door aperture usable by a wheelchair, which could only enter once part of the B-pillar had been removed. The drivers seat base and backrest rear facing surfaces were identified as needing to be scuff-resistant with large-radiused corners which would not damage or injure wheelchairs, passengers, or guide-dogs.

User Trials 2, 3 and 4

A further three trials have been undertaken – two with wooden mock-ups, and more recently a static test with a prototype vehicle. Information from the trials has been fed back to the designers and incorporated in the vehicle's subsequent iterative development. The issues that we are currently working on, are summarised below.

Wheelchair types

Although Stait et al (2000), in a survey of visitors in wheelchairs to Mobility Roadshows, noted that the percentage of electric wheelchairs had remained fairly stable over a period of 8 years, this trend may not be reflected amongst public transport users, where the increase in availability of electric models is likely to reflect an increasing usage trend over the past few years. The many different designs of wheelchairs (both manual and electric) need to be tested. The way in which the chair can manoeuvre is not easily assessed simply through modelling and the user's experience is important.

Passenger orientation

It is clear from the previous research into a wheelchair accessible taxi design (Hall et al, 1986) that most users would prefer to travel facing forwards, this is feasible from a safety perspective as full Wheelchair Tie-down and Occupant Restraint (WTORS) systems have been found to be effective in crash testing (Le Claire, 2003). It has been demonstrated that this orientation can be accommodated within a floor area 1250 mm wide by 1420 long, provided that the driver's seat can be folded or stowed so that most of the floor space is available for the wheelchair user to turn. We consider that provision should be made for rearward facing orientation of the wheelchair, with the inclusion of a backrest as recommended; this may be a more practical solution should the need arise for an internal partition separating the driver as this manoeuvre consumes less of the total floor area (with those wheelchairs tested to date).

Manoeuvring the wheelchair

We have observed that the wheelchair users make a number of rapid, minor adjustments to arrive at the correct travel position. This needs to be measured through video analysis to compare the time taken by the wheelchair user's manoeuvre with the overall journey time, a factor which may affect the driver's attitude. We will investigate further whether the interior space needs to be increased to enable simpler manoeuvring. To date we have worked with a very small number of wheelchair users, whose data has not as yet been compared with that compiled in the TRL surveys (Stait et al, 2000), however our aim is to ensure that we can accommodate wheelchairs up to the dimensional limits specified within ISO 7193, as well as most of the varied wheelchair designs now available.

Integrated ramp

From observation of the wheel-chaired participants entering the taxi, they started to turn the chair before it fully entered the cab. This has led to a redesign of the ramp which now has two differently sloping surfaces, at its upper end the surface is near horizontal to allow pre-entry manoeuvring, while its lower surface slopes at an incline of 1:5.3 for a length of 1 m, the maximum incline considered acceptable in previous user-trials (Hall, 1986). Manoeuvrability has improved, but the steep approach angle remains a concern. Attempts will be made to reduce the incline to around 1:8 in the design for a future prototype ramp, to enable the vast majority of self-propelling and attendant pushed wheelchair users to ascend and descend un-assisted following guidance from other research into ramps (Sweeny et al, 1989). Hall et al (1986) observed that a ramp was useful for those with limited walking ability. This could remove the need for a separate step.

Interior detailing

The first two users trials provided most information about packaging, the third and fourth trials highlighted a number of details which need to be addressed to accommodate better the footplates of wheelchairs when turning; a number of collision points have been identified, including the driver's brake pedal and lower internal surfaces of the A and B-pillars (door pillars). Following the third user-trial the dashboard mounted drive switch (analogous to an automatic car's gear lever) was repositioned away from the left hand side of the driver when its proximity to the wheelchair user's elbow was identified, the parking brake will be similarly re-positioned.

Conclusions

From its conception as a human powered vehicle, ergonomic mock-ups have been a part of the design process, but their usefulness has been highlighted in gathering information on the needs of disabled people through our user trial participants whose response to the experience has been particularly constructive. Consideration for, and collaboration with disabled people has led to a radically different approach to the design of a small road vehicle; putting aesthetic styling second behind the need for good levels of accessibility may yet lead to an iconic design!

Acknowledgements

My thanks go to Matt Smith for the photographs of mock-ups, Andree Woodcock and Karen Bull for advice on production of this paper, and John Jostins, Elaine Mackie, Sarah Davies, and students of Coventry University for help with the user trials.

References

Disability Discrimination Act 1995, (c. 50) Part V – Public Transport – Taxis, Elizabeth II, 1995

Hall, M.S., Silcock, D.T. and James, J.G. 1986, *Operational Trials with a Wheelchair Accessible Taxi*, (TRRL, Crowthorne)

ISO 7193-1985 (E), *Wheelchairs – Maximum Overall Dimensions*, International Organisation for Standardization

Le Claire, M., Visvikis, C., Oakley, C., Savill, T., Edwards, M. and Cakebread, R. 2003, *The Safety of Wheelchair Occupants In Road Passenger Vehicles*, (TRL Limited, Crowthorne)

Stait, R.E., Stone, J. and Savill, T.A. 2000, *A Survey of occupied wheelchairs to determine their overall dimensions and weight: 1999 Survey*, (TRL Limited, Crowthorne)

Sweeney, G.M., Clarke, A.K, Harrison, R.A. and Bulstrode, S.J., 1989, An Evaluation of Portable Ramps, *British Journal Of Occupational Therapy*, **52**(12), 473–475

METHODS AND TOOLS

KNOWLEDGE REPRESENTATION FOR BUILDING MULTI-DIMENSIONAL ADVANCED DIGITAL HUMAN MODELS

Niels C.C.M. Moes

Delft University of Technology, Fac. Industrial Design Engineering, Section CADE, Landbergstraat 15, 2628 CE Delft, The Netherlands

To develop an Advanced Digital Human Model (ADHM) for application in the design of consumer products much specific knowledge is needed. This knowledge much be processed and related for increased abstraction level. Managing the complexity requires the creation of a core model and attached submodels, each covering a specific aspect of an ADHM. A procedure to represent the different pieces of knowledge, including data, mathematics and algorithms is presented.

Introduction

Digital human models are often used in a design process, in medical examination, training and demonstration, and in animation. Our field of application is designing consumer goods such as supports, tools and protective means. ADHM should allow (i) acceleration of the design process, (ii) assessing mechanical and physiological loads inside the body and in the contact area between body and an artefact, (iii) optimisation of product properties. Existing DHM-s do not allow (i) to simultaneously consider the stresses and deformations of tissues, and the physiological effects of such tissue loads, and (ii) to apply the model for optimisation of product properties. We concluded that adequate models must be more knowledge intensive to cope with such requirements. It is our goal to stepwise develop an ADHM, based on natural data such as shape and material properties, that allows (i) the evaluation of internal stresses and deformations, tissue relocations, muscle activation and the effects on the physiological tissue functioning under external loads, (ii) the application in an optimisation process. The knowledge, that is needed in a specific ADHM, depends on the application at hand. The basic problem was formulated as follows: *What knowledge is needed to build a quasi-organic model of the human body and how must this knowledge be managed?* This paper presents (i) the requirements for such model, (ii) the knowledge, that is needed to build such model, (iii) the procedures to reduce and synthesise the knowledge, (iv) summary of the first experimental results.

Requirements for ADHM

An ADHM is based on algorithms that process and relate the knowledge from several disciplines. Since human properties are uncertain, and their measurements incomplete and biased, the model must be defined in *vague* terms (e.g. fuzzy or statistical). Since the model must allow the representation of different persons or groups of people it must

---

C.C.M.Moes@IO.TUDelft.nl

consider variable properties such as the shape of the body and the internal tissues, the material properties of the different tissues and the physiological criteria for tissue functioning. Depending on the application at hand the model must be multi-functional by allowing the inclusion of specific submodels. In other words, the model must be *adaptable* to variability and application.

Knowledge needed for building ADHM

The model can only consider available knowledge. Knowledge that is not available can possibly be provided in future. In order that it can then be included, the model must be *extendable*, which arises requirements for the core of the model and the knowledge representation of the submodels.

In figure 1 the basic setup of an ADHM is shown. The central control is called the core. To the core several pieces of knowledge can be attached. This core contains the algorithms to (i) process the knowledge that is delivered by the submodels, (ii) communicate between the submodels, and (iii) to take specific decisions. It does not necessarily contain specific data, that must be delivered by attached submodels.

In an ADHM several types of knowledge, provided by the submodels, can be attached, see figure 1. *Anatomical* knowledge considers the location and the function of the tissues, internal structure, contained active and passive elements, and functional relationship with other tissues. Knowledge on *morphology* considers the shape of the body and the internal tissues, and their connections and contact properties (geometric relationships). *Physiological* knowledge describes the functioning of tissues such as fluids

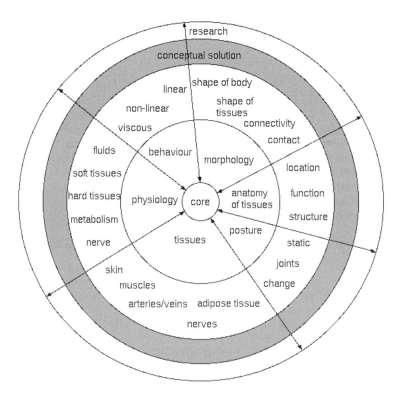

Figure 1. The ADHM consists of a core and number of submodels.

(blood, lymph, interstitial fluid), soft tissues (muscles and adipose tissue), hard tissue (bones), metabolic processes and the nerve system. The mechanical *behaviour* is described by (i) the material properties, which are usually elastic, non-linear and/or viscous, (ii) the activation of muscular structures. The *posture* submodel describes changes of the positions of the joints. A change of the posture modifies the shape of body and tissues, relocates tissues, and deviates the transmission of forces through the body. As a matter of fact figure 1 does not show a complete model. For specific applications new submodels can be attached and others removed.

Formal theories and procedures

Figure 2 shows our basic solution for building an ADHM. It consists of the reduction, formalisation and utilisation of knowledge and a pilot implementation. An action that converts knowledge to a higher level of abstraction, is called a Knowledge Engineering Action (KEA). The diagram consists of two parts. The left part represents the conceptual solution (from aggregated knowledge towards algorithms) for building an ADHM, and the right part the testing and implementation of the developed algorithms.

KEA 1: Reduction and structuring of the knowledge

Aggregated knowledge e.g., raw measurement data, are usually not very descriptive for the underlying phenomena. They have to be converted to a higher level of abstraction using for instance statistical techniques, so that they can be used in higher abstraction levels of knowledge representation. For example, shape data can represent the shape domain of a group of people using, for instance, vague discrete interval modelling (VDIM) (Rusák, 2003), which enables describing shape in terms of the location of a set of spatial surface points as a function of, for instance, body characteristics (Moes, 2004).

In order to manage the complexity, the total model can be divided in a set of submodels, that are able to handle coherent parts of the total knowledge, see figure fig:Extendable. For our ADHM this has been realised using a morphological model, a behavioural model and a product shape model. The morphological model describes shape of and connectivity between the tissues, and the contact conditions. The behavioural model describes the

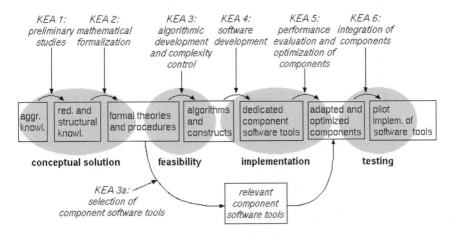

Figure 2. The procedural model for the conceptual solution and tool development. The level of abstraction increases from left to right.

effects of external stimuli on the body. If the model is used in a product design environment, the artefact in question must be modelled in conjunction with the ADHM as well as their physical interaction, while the product shape model is used to derive the shape of the contact area and the mechanical properties of the product surface from the loaded ADHM. This requires the evaluation of the interaction between the ADHM and an accordingly modelled product.

KEA 2: Formal representation of the knowledge

The basic ideas of a conceptual solution are usually not very difficult to understand. We know intuitively reasonably well what data should be collected, how they should be related, etc. However, the actual implementation needs mathematical expressions that describe the concrete processing of the knowledge: every detail of the processes inside an ADHM must be expressed in mathematical terms. It requires a profound knowledge of the human anatomy, physiology, biomechanics, etc.

KEA 3: Algorithmic representation

The formalised knowledge can be operationalised by the conversion of the mathematical expressions of the modelled processes into practical algorithms. Consider for example the algorithmic representation of the mentioned submodels of the ADHM. The morphological model is based on measured shape data (Moes et al, 2001). The point clouds of the individual subjects must be aligned according to a set of measured bony landmarks by rotation and translation of the point clouds, which can be computed using matrix operations. Then the resulting total point cloud is analysed for inner and outer hulls, and converted to a shape model of distribution trajectories and statistically defined location indices (Moes, 2004). The resulting model can be used to (i) describe the shape of the body of a (group) of subjects, and (ii) to generate new shapes. In order to support the computation the mathematical expressions are converted to algorithms, and suitable software is used for the actual implementation.

KEA 4–5: Software

Commercially available software must often be adapted and optimised for the application at hand. In our application commercial Finite Elements Modelling (FEM) software (Marc, 2001) and statistical software were used, but the software for the geometric alignment (Moes, 2004) and the VDIM (Rusák, 2003) were developed at location.

KEA 6: Pilot implementation

A pilot implementation is needed to test the model for finding uncertainties, errors, erroneous assumptions, etc. For instance, the constitutive models for the mechanical behaviour of human tissues are quite complex. In the past they have been developed for different applications (for instance prostheses), obtained under varying measurement setup, for different subjects, and for specific interpretation of the measured stress-strain relationships. Therefore we tested many constitutive equations for our FEM (Moes 2004).

Application of ADHM in ergonomics design

The ADHM must allow optimisation of product properties for ergonomics criteria. This requires the definition of an ergonomics optimisation criterion, or Objective

Optimisation Functional (OOF). This functional, that was earlier called the *Ergonomics Goodness Index* (EGI) (Moes & Horváth, 2002), contains the relevant quantities that contribute to the ergonomics quality of a product. By changing product properties (design parameters) an optimised EGI can be computed for a specific user product interaction. For instance, the shape of a seat can be modified so that mechanical stresses inside a body will reduce the risk of collapsing blood vessels and the arising of decubitus.

What has been realised so far?

We have developed a first ADHM for the upper leg and buttock regions of the human body. VDIM was used to create a generic shape model (morphological model), based on scanned shape data of the skin of living subjects (Moes et al, 2001) and scanned data of bones using the images of the Visible Human Project (VHP 1997). This ADHM allows predicting the shape of the body surface and the shape of the bones in terms of distributed spatial points (Moes, 2004). The adaptability of the bone model was based on fitting special bony landmarks. A generated shape was used as geometric input for creating a solid FEM. The validation of the constitutive equations was carried out by a comparison of the measured pressure distribution data (Moes, 2006) of sitting subjects with the computed data (Moes and Horváth 2002). The relationships between the shape of a seat and the stresses and strains inside the body were investigated, compared with real medical knowledge, and the results were transferred to a virtual chair (Moes, 2004).

Conclusions

Although a theoretical and experimental study have proven that the proposed approach for an ADHM is feasible, we are far from a final realisation. Much knowledge is still missing, additional mathematics and algorithms have to be developed and elaborated, and only limited software is available. Nevertheless, significant results have been obtained for a specific application (seat). Further research is needed to (i) elaborate the core model, (ii) improve the existing submodels and develop further submodels, (iii) testing and optimisation of the model, and (iv) application of the model in design practice.

References

MARC (2001). *MARC Volume A: Theory and Users Guide.* MARC Analysis Research Corporation, Palo Alto, CA 94306 USA, version 2001 edition.

Moes CCM (2004). *Advanced human body modelling to support designing products for physical interaction*, Ph.D. Dissertation, Delft University of Technology, ISBN: 90-9018829-0.

Moes CCM (2006). Modelling the Sitting Pressure Distribution and the Location of the Points of Maximum Pressure for Body Characteristics and Rotation of the Pelvis. *Ergonomics*, ?(?):? Status: accepted for revision.

Moes CCM and Horváth I (2002). Estimation of the non-linear material properties for a finite elements model of the human body parts involved in sitting. In Lee DE, editor, *ASME/DETC/CIE 2002 proceedings*, pages (CDROM:DETC2002/CIE-34484), Montreal, Canada. ASME 2002.

Moes CCM and Horváth I (2002). Optimizing the Product Shape for Ergonomics Goodness Index. Part I: Conceptual Solution. In McCabe Paul T., editor, *Contemporary Ergonomics 2002*, pages 314–318. The Ergonomics society, Taylor & Francis.

Moes CCM, Rusák Z, and Horváth I (2001). Application of vague geometric representation for shape instance generation of the human body. In Mook DT and Balachandran B, editors, *Proceedings of DETC'01, Computers and Information in Engineering Conference*, pages (CDROM:DETC2001/CIE-21298), Pittsburgh, Pennsylvania. ASME 2001.

Rusák Z (2003). *Vague Discrete Interval Modelling for Product Conceptualization in Collaborative Virtual Design Environments*. PhD thesis, Delft University of Technology, Fac. Industrial Design Engineering.

VHP (1997). The Visible Human Project. URL: http://www.nlm.nih.gov/research/visible/visible_human.html.

CONDUCTING RESEARCH WITH THE DISABLED AND DISADVANTAGED

David Hitchcock[1], Victoria Haines[2] & Susan Swain[1]

[1]*CEDS, The Acorn Centre, 51 High Street, Grimethorpe S72 7BB*
[2]*ESRI, Holywell Building, Holywell Way, Loughborough LE11 3UZ*

Based on their own project work, this paper presents a pragmatic and experience-based review of the advantages and difficulties of conducting research with and on behalf of those with disabilities or who are otherwise disadvantaged. It considers the financial aspects, health and wellbeing issues and the factors which should be considered in the preparation of proposals and the management of projects.

Introduction

The Centre for Employment and Disadvantage Studies (CEDS) is the research division of yes2work, a social firm working with those who have disabilities or are otherwise disadvantaged. Many of CEDS projects stem from the need to identify practical solutions to barriers faced by those seeking to attain and maintain employment. Other research commissions derive from a wider remit to increase the understanding of the factors which need to be considered by those responsible for the design of accessible activities, equipment and facilities. In both cases, CEDS research is founded on ergonomics principles for robust information collection and analysis.

The ergonomics approach demands an accurate analysis of the activities of concern, including the environment in which they are performed, and a thorough understanding of the humans involved. This presents a fundamental difficulty for CEDS and others similarly involved in conducting research with the disabled and disadvantaged – just what are their pertinent characteristics? In mainstream ergonomics projects, despite an increasing wealth of literature there often appears to be a lack of directly relevant data. Projects involving the disabled and disadvantaged typically have even less to draw upon; with a distinct lack of specific anthropometry and capabilities data.

This article does not attempt to be definitive, and the authors recognise the excellent work of others in this area, but it seeks to draws on CEDS' experience, to promote discussion, particularly among the ergonomics community, as to how best serve and benefit from the non-mainstream population.

A matter of understanding

In common with all projects – irrespective of disability – the starting point needs to be to ensure that at the project negotiation stage, the client, the researcher and the beneficiaries (the users) are all clear and in agreement as to the aims, objectives, means and outcomes of the work.

The client

In CEDS' experience most clients appear unsure as to precisely what they mean by the terms "disability" and "disadvantaged". For example, it is not uncommon for clients to think of disability issues to be restricted only to the difficulties and needs of those in wheelchairs. In some instances, the client's concept of disability may also extend to the needs of the blind or even the hard of hearing. This is not intended to sound harsh or critical; it is simply a reflection on the realities of commission with a client having a specific objective. Without up-front direction from the researcher it seems that projects can set off "on the wrong foot", ultimately leading to inaccurate outcomes. A project can begin with the best of intentions, aware of the parameters that have been set (e.g. evaluating a design only on the aspects of wheelchair access) but several months down the line when the report is published it is all too easy for the findings to be presented in such a way as to give the impression that all aspects of accessibility have been considered. The cynic might argue that this is the inevitable outcome of a client paying lip service to the needs of those with disabilities in order sell more product, or to prove a particular point. However, it is probably more likely to be a combination of misunderstanding of definitions, the passing of time, and the twists and turns which both pure and applied research projects can take.

Although clearly requiring redress, this limited view of disability is perhaps less troublesome than one which recognises a need to address all aspects of all disabilities and disadvantage. Where a client appreciates that disability and disadvantage address a wide array of issues and user needs, the project objectives may become fuzzy and the overall approach somewhat iterative. With so many factors to consider the project can be slow to define while all stakeholders debate and agree exactly what impairments should be given priority. In an ideal world it would be good to consider all disability types to be of equal importance. This, however, is rarely practical or necessary. Where inclusive design is the goal then the wider remit is essential, but for more targeted projects certain impairments are more relevant than others. Clarity is the key.

Before going further, it is worthwhile attempting to define the terms. The Disability Rights Commission (www.drc-gb.org) suggest that 20% of people of working age in the UK are considered by the government to be "disabled" in that they have a disability or a long-term health condition that has an impact on their day to day lives.

This includes people with cancer, diabetes, multiple sclerosis and heart conditions; people who have a hearing or sight impairment or a significant mobility difficulty, or who have mental health conditions or learning difficulties. It is this vast range of conditions which can often extend beyond the mindset of the client or researcher and, therefore, the project remit.

CEDS adopts a simple and common definition for "disadvantaged", as being deprived of basic social rights and security through poverty, discrimination, or other unfavourable circumstances. While this embraces an even more diverse range of conditions, the client tends to focus on specific aspects such as alcoholism, drugs use, long-term unemployment, minority groups and the so on. As a result, there appears to be less confusion regarding projects aims and objectives. Of course, from the ergonomics perspective, the range of influential factors does not decrease, but in general they are less overt at the onset of the project. For simplicity, therefore, this article will focus on the disability matters, although the points raised may translate well to the issues of disadvantage.

The researcher

The researcher can mirror the client's perspective of disability, and as such can present similarly undesirable constraints or otherwise to the research if they are not clear as to the project inclusions. Beyond this, in the authors' experience it is the expectations placed on the researcher which can cause difficulties. The ergonomics researcher is often viewed upon as an expert in disability rather than ergonomics. Their knowledge is unrealistically expected to extend to cover all kinds of disability by client and subject alike. This can lead to frustration from the client and all manner of communication obstacles with subjects. Under these circumstances, the researcher – and the project – may be made unnecessarily vulnerable and may be compromised.

The subjects

Subjects should be made just as clear about the project purpose and objectives. Even though they may also be the beneficiaries this is not always straightforward. People are affected by disability or health conditions in different ways and the onset of disability varies. It could be from birth, due to an accident or a sudden incident such as a stroke or a gradual process. In other words, disability adds even more differences to what the ergonomist recognises as human variability.

Furthermore, and most importantly, the person is not defined by their disability or the researcher's view of them (Michailakis, 2003). Psychosocial factors are given increasing consideration in ergonomics studies so it should be of little surprise to find that while it might be convenient for the researcher to categorise each subject by their primary disability, it may be secondary disabilities or other issues which are more important to the subject. In conducting research into the usability issues of an inclusive pub, the authors discovered that disabled customers were as concerned about the "quality of the beer" and the "provision of condom machines" as they were accessible bars, tables and toilets. In another study looking at the successes and otherwise of people with various disabilities attaining and maintaining fulfilling employment; issues such as finance, organisational culture and inconvenient working hours were identified as influential factors alongside those specific to their disabilities.

A matter of planning

Representation

User consultation is central to the ergonomics approach. Ensuring the representation of those participating in the research is fundamentally important. Convenience can be the compromise here. Using a simple example; perhaps the subjects need to be wheelchair users. It can be easiest to gather sufficient subjects from, for example, the local wheelchair basketball team who may be relatively young and fit and therefore not necessarily representative of the target user group.

Several agencies, including CEDS, maintain subject databases but the researcher must be aware that recruiting representative participants is complex and time-consuming for a variety of reasons; geographical spread, travel difficulties, lack of independence, communication constraints, fear of losing benefits through paid participation and ethical approval. This means that simultaneous multi-user activities such as focus groups are difficult to arrange. If project resources allow then the benefits of individual consultation can be achieved, particularly if the researcher can travel to the subject rather than the other way around. Familiarity of surroundings for the subject can create comfort and enable greater accuracy of response if, for example, the questions poised by the researcher are in context with the environment and activities.

Working "For" or Working "With"

It is well established that there are benefits from involving people with disabilities in research about them – motivation to participate (Joseph et al, 2005), improved quality of data (Osher et al, 2001) and greater understanding of actual user needs (Hitchcock and Taylor, 2003). Emancipatory research – where those studied make decisions about the research design and data analysis – offers an approach for true involvement (Good, 2001). CEDS endeavours to engage researchers with appropriate control whenever possible and have found that clients are particularly keen to see such involvement, but the practical problem for the ergonomist is the limited availability of disabled people within the discipline.

Flexibility

The adverse effects of medication, "bad days" and external circumstances such as hospital appointments require a flexible approach to project planning. Participants may cancel at short notice and if an individual feels overloaded they may even need to pull out of the project altogether. Over reliance on any one individual in a relatively tight timeframe can be a mistake. However, careful and flexible planning can enable projects to run smoothly, particularly if milestones do not have to necessarily be sequential and different strands of the project can be run in parallel.

Unfortunately, if an inflexible approach is adopted and problems do arise, the inevitability may occur when researchers are left with little choice other than to make assumptions, rely on literature alone or try simulation in some way. At best these approaches can raise awareness and promote further questioning. They are not a substitute for user involvement.

A matter of perspective

Probably one of the biggest challenges facing the ergonomics researcher is that of not making assumptions about their subjects. This is not unique to the disability arena, but is perhaps emphasised, not least because, as already mentioned, it may be unfairly assumed by the client or subject that the researcher is an expert in disability rather than ergonomics. It is possible that having been involved in the project or area for a while that even the researcher begins to assume this too! It should be remembered that there may be considerable perceived and actual differences in understanding and power between researchers and subjects (Bollard, 2003) which can make qualitative research problematic (Llewellyn, 1995), particularly for those with intellectual disability. Steps should be taken to ensure that all those involved in a project are working from a level playing field and that additional efforts and resources are provided where necessary; disability awareness training for researchers, providing assistants for those with memory loss, developing communication skills for those with speech or hearing impairments, using alternative formats for those with learning difficulties and so on.

Concluding thought

Conducting research with the disabled and disadvantaged is far from straightforward but with clear objectives set by all those involved, methods devised with participants and a fluid timetable it is possible to generate realistic, accurate and beneficial outcomes.

References

Bollard, M., 2003, *Going to the Doctor's: The Findings from a Focus Group with People with Learning Disabilities,* Journal of Learning Disabilities; 7(2) 2003, 156–164
Disability Rights Commission, www.drc-gb.org
Good, G.A., 2001, *Ethics in research with older, disabled individuals,* International Journal of Rehabilitation Research; 24(3) 2001, 165–170

Hitchcock, D.R., Taylor A.J., (2003). *Simulation for Inclusion – true user centred design?* Include 2003 Conference Proceedings (The Helen Hamlyn Research Centre)

Joseph, D., Wailoo, M.P., Jackson, A., Petersen, S.A., and Anderson, E.S., *Participation of disadvantaged parents in child care research,* 2005, Child Care, Health and Development; 31(5) 2005, 581–587

Llewellyn, G., 1995, *Qualitative research with people with intellectual disability,* Occupation Therapy International; 2(2) 1995, 108–127

Osher, T.W., van-Kammen, W., and Zaro, S.M., 2001, *Family participation in evaluating systems of care: Family, research, and service system perspectives,* Journal of Emotional and Behavioral Disorders; 9(1) 2001, 63–70

AGILE USER-CENTRED DESIGN

Marc McNeill

Thoughtworks, 9th Floor Berkshire House, 168–173 High Holborn, London WC1V 7AA

Agile methods are becoming increasingly common in application design, with their collaborative customer focus and iterative, test driven approach. Whilst they share many common principles, it is rare for Agile methods to incorporate user-centred design and human factors approaches. Similarly, there are many agile techniques that are well suited to user-centred practices. This paper discusses how the two approaches can be incorporated. It introduces practical techniques such as the use of stories to capture information needs; collaborative planning; visual modelling; rapid, time-boxed iterations; stand-ups and retrospectives. It advocates how using such techniques, useful and usable applications can be developed at greater speed with less business risk.

Introduction

Martin (2002) identifies common fears that are present on many projects; the project will produce the wrong product, the product will be of inferior quality, the project will be late, the team will work excessive hours, commitments will be broken and ultimately the project will be painful for all involved. Processes, constraints and deliverables are added to projects to help mitigate these fears; however they often become an end to themselves, making projects even more cumbersome and likely to fail. In an effort to overcome this project overload, a group of industry experts came together as the Agile Alliance and drafted a manifesto for a new way of developing software (Beck et al 2001). Key to the manifesto were;

- **Individuals and interactions** over processes and tools
- **Working software** over comprehensive documentation
- **Customer collaboration** over contract negotiation
- **Responding to change** over following a plan

The intention was not to deny value in the things on the right hand side but to place greater value on the things on the left hand side.

Agile methods are lightweight software development processes that employ short iterative cycles, involve users to establish, prioritise and verify requirements and rely on knowledge within a team rather than documentation (Boeham and Turner 2004). They have developed in reaction to traditional software engineering that is seen as overly bureaucratic and slow. Rather than investing a large amount of time in up-front design, rigidly capturing and documenting requirements, Agile methods are adaptive and people orientated. Whilst change is an inevitable and often painful aspect of IT projects (e.g. scope "creep/reduction) Agile welcomes it, allowing the project to adapt to changes as and when they happen.

Agile and User-centred design

User-centred design (UCD) shares many of the characteristics of Agile (Table 1), however they are rarely combined and there can be conflict between the two practices

Table 1. Similarities between features of Agile methods and User-centred design.

Principle	Agile	User-centred design
Customer Focus	All activities are focused on providing tangible business value. The customer is typically defined as a representative from the business	All activities are focused on providing (business) value through ensuring a useful, usable and engaging product. The customer is not defined as just the project stakeholders, but the end users as well.
Iterative Development	Early and frequent delivery of working software (often weekly) contributes to project visibility, reduces project risk via regular feedback, fosters continuous improvement and enables early realisation of business benefits.	Develop, test and refine the user interface via regular feedback to the end users. The focus is upon the business risk as well as the technical risk. When lo-fi prototyping with storyboards iterations are typically one to two day cycles.
Test-Driven Development	Testing plays an integral role in every phase of the project life cycle.	User testing plays an integral role in the development of the interaction design.
Collaboration	Collaboration between customers, product managers, analysts, developers, and QA maximises team efficiency.	Even more collaborative with the sharing of ideas and models in addition to stories and code.
Visibility	All stakeholders are provided with maximum visibility into project progress via regular showcases and retrospectives as the project progresses.	More rapid visibility; interaction design is the premier communication tool, defining the outward appearance of what the product will do.

(e.g. Nelson 2002). The greatest source of contention is whether UCD processes constitute "big up-front design", an anathema to Agile. Agile practitioners argue that traditionally an inordinate effort goes into the design which will undoubtedly change as the project develops. From a UCD point of view there is an inherent risk in this approach. Focussing upon building discrete functional components to be stitched together as they evolve (rather than considering the application a holistic user experience from the outset) risks delivering a product that is inconsistent and confusing. This almost inevitably results in an inefficient, error prone and ultimately unfulfilling user experience.

Given the similarities between Agile principles and UCD principles, it is argued that UCD need not be seen as big up-front design, rather a "quick start" to galvanising project success. The overall Agile process, that of rapid iterations delivering value to customers is in fact very compatible with the UCD approach. This paper introduces how the approaches can be combined, and how Agile techniques can be used to increase UCD input into projects.

Agile user-centred design processes

Communication

At the centre of Agile user-centred design is facilitated communication. Rather than producing long, wordy documents which are so often produced at the early stages of a project this process instead uses visual techniques that are engaging and allow all stakeholders to give rapid feedback.

"I'm glad we're all agreed then."

Starting with findings from Contextual Inquiry techniques (Beyer & Holtzblatt 1998), identifying "roles and goals," (how different persona may use the system), process modelling and simple tools such as whiteboards and PowerPoint, an understanding of the issues are elicited and shared with all stakeholders.

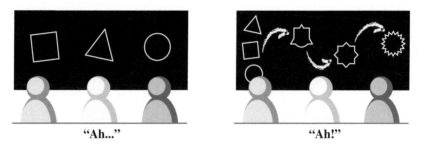

"Ah..."   **"Ah!"**

Once the issues are clearly articulated, facilitated workshops are run to create solutions. The output of this process may initially be process mapping, but rapidly develop into storyboards; low fidelity prototypes/visual representations of how the GUI may appear.

After several iterations, often shared with a wider user community, a genuinely shared understanding of the problem, solution and approach are gained.

"I'm glad we're all agreed then."

For example, in developing a new account opening application for a bank, storyboards helped refine the proposition. The requirements capture process was significantly shortened; rather than eliciting requirements in a void, a tangible model enabled all stakeholders to see what they might get (and change it accordingly). Perhaps the most powerful result of the storyboards was the ability to place them in front of end users and ask them to complete simple processes in a linear fashion. This user testing rapidly demonstrated that the new process would significantly reduce the time to complete the process by more than 50% providing greater validity to the business case.

Stories

Stories support the storyboards. They are small pieces of discrete functionality appearing in the storyboards, relating them to business value with testable criteria. Thus rather than a requirement for "the application to be easy to use", it must be expressed in criteria that can be tested, such as the user can complete a goal within n seconds.

In its most elementary form a story identifies who wants the story, what it needs to do and why it is valuable to have: As a [type of user] I want [some particular feature] so that [some benefit is received]. For example: "As a bank customer I want to view my current account balance so that I know if a recent cheque has cleared."

Collaborative planning

It is a painful reality that not every requirement will make it to the final application. Functionality is stripped out as the project progresses and time frames and budgets get stretched. GUI requirements stated up front may not reflect downstream changes in the project.

In Agile the stories are written on index cards and are physically shuffled according to their priority and business value. This business value is usually driven by the "so that" statement of the story. This prioritisation exercise helps inform the sequence that stories are developed. This is not to say that some low value stories may not be played. Placing the user at the centre of the design may require lower priority stories to be included to enable a coherent user journey. Thinking in terms of feature usage and criticality helps inform this process (Patton 2005). The development team, by estimating the effort required to complete each story, set a cut-off point for the number of stories that can be addressed in that release/iteration. A release strategy is crafted and stories are "played" in short weekly iterations during the release. The storyboards inform the usability of iteration output, enabling the user interface to be continually evolving, but always focussed upon the end user.

Rapid, time-boxed iterations

Applications are developed in Agile through small, regular incremental iterations, continually testing both the form (i.e. is it delivering on business requirements,) and function (i.e. does the code work). Storyboarding allows the application form to be tested quickly and cheaply, ensuring that the development iteration focuses upon delivering quality code with minimal need for re-work because it does not meet the customer or client expectations.

Showcases

At the end of each iteration there is a showcase where all stakeholders are invited to trial the stories that have been developed. This often includes end users who can validate the usability as it is being developed. If usability is not inherent, stories may not be signed off and require further iterations to get right.

Stand-ups

Every morning the team has a stand-up meeting. These are focussed meetings that communicate daily status, progress, and plans to the team and any observers; identify obstacles more quickly so that the team can take steps to remove them; set focus for the rest of the day and increase team building and socialization (Yipp 2006). Often during stand-ups interface issues will be identified. Rather than leaving the developers to interpret ambiguities in style guides or usability guidelines, the stand-up offers the UCD team member the chance to help work through GUI issues as they come up rather than at the end of the iteration.

Retrospectives

Retrospectives are held anywhere from weekly to monthly to assess how well the team is working with regards to its process. It is an opportunity to take the time to discuss "what has gone well", "what we should do differently" and "what puzzles us" in a structured manner. This is extremely helpful for the team to adjust its process (McKinnon 2006) and provides a voice for the UCD team member that is often lost in projects. Key to Retrospective success is the "safety valve"; attendees anonymously identifying how honest and comfortable they feel with their feedback. For example a scale from "No Problem, I'll talk about anything" to "I'll smile, claim everything is great and agree with managers".

Conclusions

Agile methods are gaining acceptance in IT organisations as an efficient and effective means to developing applications that deliver on the business's requirements. Agile is usually a development-centric philosophy, espousing engagement with the business and using stories and code as the model for communication. User-centred design extends the approach; rather than using code as the model it uses visualisation to articulate the solution. Through collaborative workshops, creating stories and translating them into storyboards and low-fidelity prototypes enables iterations to be showcased on a daily rather than fortnightly basis. Engaging all stakeholders in the process ensures that when the developers start cutting code the focus will be on ensuring code quality, mitigating the risk of business driven changes that could not be articulated without having something tangible to evaluate.

References

Beck, K., Beedle, M., van Bennekum, A., Cockburn, A., Cunningham, W., Fowler, M., Grenning, G., Highsmith, J., Hunt, A., Jeffries, R., Kern, J., Marick, B., Martin, R., Mellor, S., Schwaber, K., Sutherland, J. & Thomas, D., 2001, *Manifesto for Agile Software Development.* http://Agilemanifesto.org/.
Beyer, H. & Holtzblatt, K. (1998) *Contextual Design – Defining Customer Orientated Systems.* (Morgan Kaufman, CA.)
Boeham, B. & Turner, R. 2004, *Balancing Agility and Discipline.* (Addison-Wesley).
Martin, R.C., 2002, *Agile Software Development, Principles, Patterns, and Practices.* (Prentice Hall).
McKinnon, T. (2006) *Retrospective agility*, Objective View, **8**, 10–17, http://www.ratio.co.uk/objectiveview.html.
Nelson, N. (2002) Extreme Programming vs. Interaction Design, http://www.fawcette.com/interviews/beck_cooper/default.asp.
Patton, J. (2005) *It's All in How You Slice It*, Better software, January 2005, 16–40.
Yip, J. (2004) *Patterns For Daily Stand-Up Meetings*, http://www.thoughtworks.co.uk/PatternsDailyStandupJason%20Yip.pdf.

MANIKIN CHARACTERS: USER CHARACTERS IN HUMAN COMPUTER MODELLING

Dan Högberg[1] & Keith Case[2]

[1] School of Technology and Society, University of Skövde, SE-541 28, Skövde, Sweden
[2] Mechanical and Manufacturing Engineering, Loughborough University, Loughborough, Leicestershire LE11 3TU, UK

Design methods that support the development of a good understanding of user requirements are essential for product development. Traditional methods involve direct interaction with users but a product development team can still gain great benefit from applying what might otherwise be considered "second best" methods. Among the range of methods and tools available are *User Characters* and *Human Computer Modelling*. Why not combine them? This paper reviews the two approaches in the context of research conducted in user diversity considerations, and then discusses opportunities for integration of the two methods into a new approach that might be called *Manikin Characters*.

Introduction

To design products that meet or surpass the range of future product user needs, designers can gain from utilising methods that support the development of a good understanding of these requirements. Direct interaction between product developers and users is an acknowledged approach for supporting this understanding. This is complementary to appropriate information about user aspects that is communicated to designers via ergonomists and market research, since these are also important sources of information to enhance understanding and support the design task. Even though direct interaction between designers and users is acknowledged as the preferred situation, there might be situations where it is hard to achieve. One argument is that today's designers often work at a distance from widely diverse populations of users. Under such circumstances, and indeed as a complement to direct interaction, a product development team can gain from applying what might be considered as "second best" methods. Among the range of methods and tools available to support designers are two approaches that have generated increased industrial and academic interest in recent years: *User Characters* and *Human Computer Modelling*. Both are intended as means for integrating ergonomics into the product development process, and are particularly aimed at supporting the design task at virtual stages of the development process. However, the methods are very different, for example in the type of human characteristics they try to represent and in their presentation to designers.

User characters

The *User Characters* (also known as *Personas*) method (Buur and Nielsen, 1995; Nielsen, 2002) is aimed at evoking user consideration among designers, ultimately to the benefit of both the company and future product users. The method is based on the

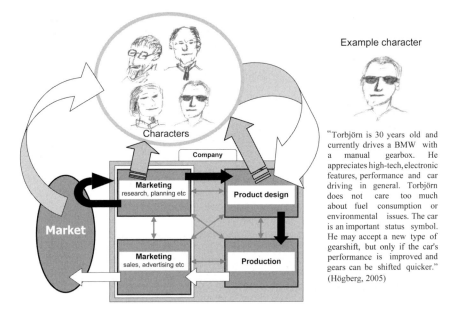

Figure 1. **User characters as a transitional system in product development.**

development of hypothetical characters that describe targeted users. The design team uses these characters throughout the design process to animate the users so as to perceive, consider and communicate user aspects and user diversity in a richer way. The user characters are comprised of firm knowledge of real users, and described in a way that generates a feeling of knowing the person. Between three and seven characters are generally proposed, as is the use of pictures to illustrate the characters (Buur and Nielsen, 1995; Pruitt and Grudin, 2003).

User characters can be seen as parts of a transitional system where the characters support discussions within a product development team about user aspects. Even in cases where there has been direct interaction between project members and users, the characters can be a way to gain from, and store, real experiences and to support understanding and communication, e.g. with other project members who have not met users, and to evoke empathy for the users throughout the project. Figure 1 tries to describe the transitional system, portraying the fictional direct contact between designers and users. Small grey arrows within the company represent common information flows, making it possible to understand that contact of designers with product users commonly goes through the marketing function, indicating the possible cause for misconception when market researchers communicate user related information to designers, e.g. through product design specifications.

Human computer modelling

A common approach in contemporary industry is to move product development work into virtual environments. To reduce the risk of time consuming and expensive iterations or products that do not fully meet the ergonomics specification the use of *Human Computer Modelling* (also known as *Human Simulation Tools*) has become essential (Porter et al, 1995; Chaffin, 2001). The system used in this study is RAMSIS

Figure 2. Human computer modelling for vehicle accommodation.

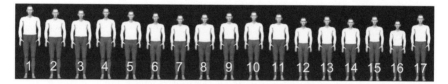

Figure 3. Example of a male manikin family.

(Seidl, 1997), which is mainly intended for evaluating human accommodation in car interior design (Figure 2). The tools include a *computer manikin*; an advanced computer model of the human body, typically with modifiable size, shape and posture. Common features of the functionality are clearance and reach verification, posture prediction, vision analysis, comfort assessment and biomechanical analysis.

A difficulty in using human computer modelling tools is the determination of manikin dimensions required to correctly represent the anthropometric diversity of the targeted users (Högberg, 2005). This applies particularly to non-expert tool users, and especially in multivariate design problems, such as automobile cockpits. One approach is to perform simulations including all members of a predetermined manikin family, defined to represent the targeted users. This would be similar to, and indeed complementary to, having a group of real test persons recruited to assess products being developed. One essential difference is that the virtual test persons will always be available, even concurrently in different places.

The A-CADRE manikin family is based on statistical treatment of anthropometric data, resulting in descriptions for 17 manikins which represent anthropometrical diversity in a proficient way (Bittner, 2000). By using these manikins as user representatives in design, a high level of accommodation can be achieved. A-CADRE is used here to exemplify the approach of using manikin families (Figure 3). Figure 4 illustrates an application of manikin families and how the approach can support the consideration of anthropometric diversity when defining a seat adjustment area required to accommodate targeted car drivers (Högberg, 2005).

Discussion of concepts for expanded assessment

Human computer modelling tools represent anthropometric and biomechanical characteristics, and connected aspects such as posture and comfort prediction. Ideally

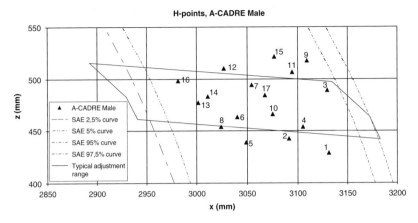

Figure 4. Hip-point locations of male manikin family.

though, a fuller representation of the human is needed through the future combination of physical and cognitive models (Plott et al, 2003; Bubb, 2004), into something that might be called *Virtual Humans*. This is an exciting prospect, but is difficult, since humans are very complicated in perception, thinking and operation of the body.

A more readily achievable approach would be to apply the user characters method to computer manikins, developing manikins into characters. This might be called *Manikin Characters*. In essence, such manikin characters could be the means to inform and inspire designers of many user related aspects. Unlike *Virtual Humans*, *Manikin Characters* would leave much of the design task with the designer, as the method supports the designer with appropriate information and stimulates creativity, rather than offering direct solutions. This may not be bad as a characteristic of designing is the creation of solutions to complex and ill-defined problems. All people, even designers, meet and interact with people, have opinions about people, recognise "types" of people and are able to imagine how people might respond to certain situations and inputs. However, a fallacy among designers is that they consider themselves representative of the user and as being typical average users (Pheasant and Haslegrave, 2006). Eason and Harker (1991) argue that, when designing products aimed at a wide market there may be no specific user population to participate in the design process and a danger that the ergonomics requirements are based on an ill-founded "model" of the user. The manikin characters approach has the potential to reduce the effect of these tendencies.

Porter et al (1995) argue that the best means of supplying the ergonomics input in a complementary fashion to the engineering input is to develop CAD systems with facilities to model both products and people. Manikin characters support this by "getting the product users into the CAD systems"; enabling not only the products to be considered within the tool but also a range of human product interaction aspects.

The emotional responses of users to products is becoming a more important aspect of customer satisfaction (Jordan, 2000), so designers will need to become involved with such issues, in conjunction with functionality and usability requirements. The manikin characters method is believed to assist in considering possible user responses and diversity, e.g. emotional responses, when generating and evaluating design proposals.

The validity of the ideas behind manikin characters is a challenging research question, as is how manikin characters might best be created, communicated to, and used by designers. One area to investigate would be how user characters should be linked to computer manikins, i.e. should the anthropometric models and character descriptions be separate features or associated, e.g. by mapping certain characters to certain manikins. A related issue would be manikin character family structure. The number of members in a manikin family to represent both anthropometric and personal diversity is unlikely to be the same. It would be convenient if it were, but there could be risks of sub-optimising if it was an objective to make the numbers match.

Inherent in the approach presented in this paper is the risk that the designers will bias the results. However, the approach should be seen as a way to assist reasoning about user responses and diversity, perhaps when time and money for more thorough studies are lacking, and to support communication within a design project team.

References

Bittner, A.C. (2000). *A-CADRE: Advanced family of manikins for workstation design.* XIVth congress of IEA and 44th meeting of HFES, San Diego. 774–777.

Bubb, H. (2004). Challenges in the application of anthropometric measurements. *Theoretical Issues in Ergonomics Science* 5(2): 154–168.

Buur, J. and Nielsen, P. (1995). *Design for usability – adopting human computer interaction methods for the design of mechanical products.* 10th International Conference on Engineering Design, Praha. 952–957.

Chaffin, D.B. (2001). Introduction. *Digital human modeling for vehicle and workplace design.* D.B. Chaffin, Ed. Warrendale, Society of Automotive Engineers. 1–16.

Eason, K.D. and Harker, S.D.P. (1991). Human factors contributions to the design process. *Human Factors for Informatics Usability.* S. Richardson, Ed. Cambridge, Cambridge University Press. 73–96.

Högberg, D. (2005). *Ergonomics integration and user diversity in product design.* Department of Mechanical and Manufacturing Engineering, Loughborough University. Doctoral thesis.

Jordan, P.W. (2000). *Designing pleasurable products: an introduction to the new human factors.* London, Taylor and Francis.

Nielsen, L. (2002). *From user to character – an investigation into user-descriptions in scenarios.* DIS'02, Designing Interactive Systems, London. 99–104.

Pheasant, S. and Haslegrave, C.M. (2006). *Bodyspace: anthropometry, ergonomics and the design of work. 3rd ed.* Boca Raton, Taylor & Francis.

Plott, B., Hamilton, A. and Laughery, R. (2003). *Linking human performance and anthropometric models through an open architecture.* Warrendale, Society of Automotive Engineers. SAE Technical Paper 2003-01-2203.

Porter, J.M., Freer, M.T., Case, K. and Bonney, M.C. (1995). Computer aided ergonomics and workspace design. *Evaluation of Human Work: A Practical Ergonomics Methodology.* E.N. Corlett, Ed. London, Taylor & Francis. 574–620.

Pruitt, J. and Grudin, J. (2003). *Personas: practice and theory.* Conference on designing for user experiences, San Francisco, ACM Press New York. 1–15.

Seidl, A. (1997). *RAMSIS – A new CAD-tool for ergonomic analysis of vehicles developed for the German automotive industry.* Warrendale, Society of Automotive Engineers. SAE Technical Paper 970088.

EFFECTS OF VIEWING ANGLE ON THE ESTIMATION OF JOINT ANGLES IN THE SAGITTAL PLANE

Inseok Lee

*Department of Safety Engineering, Hankyong National University,
Anseong, Republic of Korea
Visiting Faculty in Department of Human Sciences, Loughborough University,
Loughborough, Leicestershire LE11 3TU*

In assessing risks related to working posture, pictures of postures are taken from various directions, which can be a source of observation error. Joint postures of the neck, low back, knee, shoulder, and elbow were taken from 7 different viewing angles and 19 observers estimated joint angles after observing the pictures in 2-dimensional display. The joint angles were also measured using an optoelectronic motion measurement system. The estimation error increased as the viewing angles increased, but the patterns differ according to which joint angles were being observed. In general, it is strongly recommended to maintain the viewing angle within 40 degrees from the sagittal plane, while taking pictures of postures from the behind the individual is least recommended.

Introduction

The association of poor body postures with work-related musculoskeletal disorders (WMSDs) has been reported in a number of studies (Bernard, 1997). In order to identify and assess risk factors of WMSDs, observational methods have been widely used in industry. They neither interfere with job process nor require expensive equipment, while being relatively reliable (Genaidy et al, 1994; Kilbom, 1994). They depend on the analyst's ability to estimate the angular deviation of a body part from the neutral position and there can be some errors in estimation. There are several factors affecting the reliability of observation, such as the analyst's ability, the method of observation, the number of items to be observed, characteristics of the postural classification scheme, the analyst's viewing angle, and so on (Kilbom, 1994).

Currently, video recording is more widespread than direct observations. However, video recording can be less accurate than direct observation when estimating joint angles, since human vision is three-dimensional and video recording reduces the image to two dimensions (Kilbom, 1994). To secure sufficient accuracy in two-dimensional assessment of postures from photographs, some proper guidelines for the reduction of perspective errors should be employed (Paul and Douwes 1991). One of the guidelines is to record the posture while keeping the viewing angle perpendicular to the plane composed by the body parts of the joint of interest. The deviation of joint angles between two-dimensional and three-dimensional measurements for the same joint was shown to increase as the viewing angle deviated from the perpendicular (Paul and Douwes, 1991). This finding is based on a calculated simulation study, without human observers being involved in the estimation of joint angles. In reality, it's very difficult to maintain one viewing angle because the observed individual moves during work and a space constraint may exist in work sites. A more detailed guideline for the recording of working posture based on observational experiments would be helpful for more reliable assessment.

In this study, a laboratory experiment was conducted to analyse the effect of viewing angle on the estimation of joint angles when the body postures are photographed and displayed in a two-dimensional display. The aim is to give some practical guidelines for the users of observational methods for the assessment of risk of WMSDs. Ideally this should reduce the error of observation and increase the reliability of assessing the risk factors.

Methods

Observers

Nineteen undergraduate students (2 females and 17 males) observed photographs of postures and estimated joint angles in the experiment. Their mean ($\pm$ SD) age was 24.6 ($\pm$3.0) years, ranging from 22 to 36 years. The observers were informed of the object and contents of the experiment and gave documented consent.

Experimental factors

Three factors were varied in the experiment: viewing angle, joint posture, and range of joint angle. Viewing angle is the angle between focus line of the camera and the frontal plane of the human body (Figure 1). The postures were photographed in 7 different viewing directions: $-60, -40, -20, 0, 20, 40$, and 60 degrees.

The joint postures for angle estimation were; flexion of the neck, low back, knee, shoulder and elbow, and the extension of the neck, low back and shoulder. They are the main motions at each joint in the sagittal plane. The flexion and extension of the neck, low back and shoulder were differently treated because they are different in the direction and range of motion, even though they are in the same plane.

The ranges of motion for the joint postures were measured and divided into 4 ranges: $0\sim25\%, 25\sim50\%, 50\sim75\%$ and $75\sim100\%$ of full range of motion. For a joint posture, a joint angle was selected randomly from each range of joint angle, at each viewing angle. Therefore, 4 pictures of a joint posture with different joint angles were selected for a given viewing angle, resulting in a total of 28 different pictures (4 joint angles $\times$ 7 viewing angles). The range of joint angle was adopted as an experimental factor to investigate whether the magnitude of angular deviation from the neutral position affects the estimation of the joint angle.

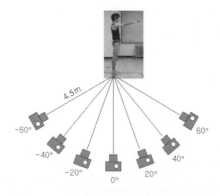

Figure 1. Viewing angles in the experiment.

Measurement of joint angles and photographing

An optoelectronic motion measurement system with 7 infra-red cameras (ProReflex, Qualisys, Sweden) was used to measure the position of joints. Reflective markers were put on 14 landmark positions: right and left sides of the head, the 7th cervical vertebra, sternum, right and left acromions, right elbow, right wrist, right and left great trochanters, right and left hip joints, right knee, and right ankle joint.

In addition, seven digital camcorders were used to take photographs of joint postures at different viewing angles. They were positioned 4.5 m apart from the centre point between both heels of the subject, with the centre of lens being 1.2 m above the floor.

A 25-year old male subject was asked to move the body parts comfortably to the maximum position. The motions were recorded by the motion measurement system and digital cameras simultaneously. The digital camera in the viewing angle of 0 degree was connected to the motion measurement system so that they were synchronized automatically.

The joint angles were calculated using the position data measured by the motion measurement system. The frame for the selected joint angle was automatically calculated and the picture was captured from the video recordings.

Experimental procedure

Before the main experiment, line and angle estimations were carried out by the observers as a pre-test in order to test their ability of ratio-scale estimation and make them accustomed to angle estimation. They were asked to estimate the length of nine lines of different lengths, in ratio scale, and the angles of nine figures with different angles.

The definition of joint angles at each joint was explained to the observers with figures and descriptions. They were given sufficient time to completely understand and remember the definition of joint angles.

In the main experiment, the observers were presented with pictures on a 17-inch CRT colour monitor, from which they estimated joint angles. For each joint posture, 28 pictures were presented consecutively with the maximum observation time of 15 sec for a picture. The pictures were presented in a random order and the sequence of joint postures was also randomised. The observers were allowed to have a rest between two consecutive joint postures and it took about 1 hour for an observer to complete the estimation.

Data analysis

An absolute error of observation was defined as the absolute value of the difference between the estimated joint angle by the observer and the measured angle using the motion measurement system. With the absolute error values as the measure for the error of observation, an analysis of variance (ANOVA) was conducted based on a model with three main effects of viewing angle, joint posture, and range of joint angle, and their interaction effects.

Results and discussion

The ANOVA showed that all the main effects and their interaction effects are statistically significant ($p < 0.0001$). For the significant main effects, Duncan's multiple range test were conducted at the significant level of 0.05. Figures 2 show the mean values of absolute error for each viewing angle and joint posture.

The mean absolute error of the viewing angle of –60 degree was significantly higher than the other viewing angles, while viewing angles of 0 and –20 degree showing significantly the lowest values of error (Figure 2(a)). The absolute error showed an increasing trend as the viewing angle increases in both directions, except the viewing of 60 degree of which absolute error was lower than those of viewing angles of 40 and −40 degrees, with no statistical significance.

The viewing angle over 40 degree in any direction is presumed to decrease the performance of estimation of joint angle. However, the effects of viewing angle differ according to the related joint posture. Figures 3 show the effects of viewing angle on the absolute error of estimation in joint postures of neck and shoulder. The absolute errors in −40 and −60 degrees increased for the neck flexion, and in −60, 20, 40, and 60 degrees for neck extension. For the shoulder flexion, it increased in −60, 20, 40 and 60, while it increased in −60, −40 and 60 degrees for the shoulder extension. Although the pattern of error increasing is not as clear as calculated by Paul and Douwes (1991), it can be found to be as the viewing angle in the opposite direction of the joint posture increases.

The absolute error of estimation for knee flexion (KF) showed the highest value, while shoulder extension (SE) showed the lowest mean value of absolute error (Figure 2 (b)). The estimation of joint angle for the other postures did show much difference among them. The presumed main reason for the highest absolute error for knee flexion is that the posture involves the simultaneous motions of thigh and shank links, while other postures involve motion of one link. Another possible reason may be related to the definition of the joint angle. The joint angles were defined as the angular deviation of the body part from the neutral position. For example, the knee flexion angle was defined as the angle between the thigh link and the extended shank link. The definition seems to be recognized as natural in the neck, low back, and shoulder, while not in the

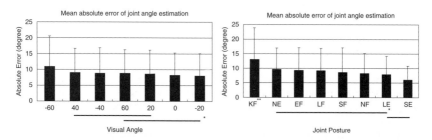

Figures 2. Mean absolute errors stratified by (a) viewing angles and (b) joint postures (*no significant difference at alpha = 0.05; **NF and NE for the neck flexion and extension, LF and LE for the low back, KF for the knee, SF and SE for the shoulder and EF for the elbow).

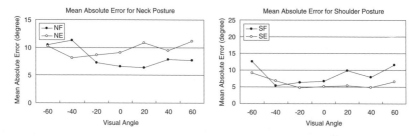

Figures 3. Mean absolute errors stratified by viewing angles for (a) the neck and (b) shoulder postures.

elbow and knee flexions. The absolute error for elbow flexion (EF) was also relatively higher than other postures. Between the flexion and extension in the same joint, the estimation of joint angle for the extension was better than for the flexion in the low back and shoulder, whilst the opposite is true for the neck. It is presumed that the smaller range of motion makes the estimation easier. For the neck extension, the range of motion for the flexion and extension is not much different. It is presumed that people find estimating neck extension angle difficult, since there is no certain landmark point for the neck joint and the head-neck link.

Conclusion

The effects of the viewing angle on the estimation of joint angle are significant, though it is not as clearly demonstrated as in the simulated study (Paul and Douws, 1991). The reason for the differences might be related to the innate human error of observation. The effects patterns differ according to the related joint postures. In general, it is strongly recommended to maintain the viewing angle within 40 degree from the sagittal plane, while taking pictures of postures from behind the individual is the least recommended.

Acknowledgment

This work was supported by the Korea Research Foundation Grant funded by the Korean Government (MOEHRD)(D00775).

References

Bernard, B. (ed.) 1997, *Musculoskeletal disorders and workplace factors: A critical review of epidemiologic evidence for work-related musculoskeletal disorders of the neck, upper extremity, low backs*, DHHS (NIOSH) Publication No. 97–141, US Department of Health and Human Services.

Genaidy, A.M., Al-shedi, A.A. and Karwowski, W. 1994, Postural stress analysis in industry, *Applied Ergonomics*, **25**, 77–87.

Hagberg, M., Siverstein, B., Wells, R., Smith, M.J., Hendrick, H.W., Carayon, P., Perusse, M., Kuorinka, I. and Forcier, L. (eds) 1995, *Work-related musculoskeletal disorders (WMSDs): A reference book for prevention*, (Taylor and Francis, London).

Kilbom, A. 1994, Assessment of physical exposure in relation to work-related musculoskeletal disorders – what information can be obtained from systematic observations?, *Scandinavian Journal of Work Environment and Health*, **20**, 30–45.

Paul, J.A. and Douwes, M. 1993, Two-dimensional photographic posture recordings and description: a validity study, *Applied Ergonomics*, **24**, 83–90.

EVIDENCE-BASED ERGONOMICS

Ash Genaidy[1] & Judy Jarrell[2]

[1] *University of Cincinnati, College of Engineering, Cincinnati, Ohio, USA*
[2] *University of Cincinnati, College of Medicine, Cincinnati, Ohio, USA*

> In recent years, ergonomics practices have increasingly relied upon the knowledge derived from field studies. Thus, the objectives of this research are to develop and test a general purpose "Epidemiological Appraisal Instrument (EAI)" for evaluating existing studies or as a tool to design new observational studies using a critical appraisal system rooted in epidemiological principles. The EAI consisted of 43 questions grouped into six scales. Results indicated that an assessor with basic background in epidemiology and biostatistics would be able to correctly respond on 4 out of 5 questions. The EAI proved to be a valid and reliable instrument that may be used in various applications.

Introduction

Historically, the knowledge generated in the field of ergonomics has evolved from its early inception from strict reliance on experimental studies to the use of information from field or epidemiological studies (also known as observational) studies. Today, ergonomics practices are derived from the knowledge gained from both types of studies. In this regard, ergonomic practices may greatly benefit from the advances made in "Evidence-Based Medicine" defined as "the conscientious, explicit and judicious use of current best evidence in making decisions about the care of individual patients". To the clinician, the practice of evidence-based medicine integrates individual clinical expertise with the best available external clinical evidence from systematic research. To the ergonomic practitioner, the practice of "evidence-based ergonomics" should integrate his/her individual experience as well as the best external available evidence from systematic research.

Similar to evidence-based medicine, evidence-based ergonomics may consist of four steps: (1) formulate a clear question from a user standpoint; (2) search the literature for relevant ergonomic articles; (3) evaluate (i.e., critically appraise) the evidence for its validity and usefulness; and, (4) implement useful findings in ergonomics practice. With respect to evidence-based evidence medicine, medical practice and health policy have been based upon data from individual and meta-analyses of randomized clinical trials (RCTs) as these are the gold standard. Therefore, checklists have been developed to evaluate the methodological qualities of individual RCTs for use in meta-analyses, and studies have investigated the impact of different methodological qualities on outcome measures. Recently, it has been argued that observational studies should be seen as complementary where it is not feasible to conduct RCTs. Currently, there is limited research devoted to the exclusive evaluation of the methodological qualities of observational studies. Based on a comprehensive search of MEDLINE from 1966 to March 2004, we located only one study by Downs and Black (Downs et al 1998) who developed an instrument to simultaneously evaluate RCTs and observational studies.

As pointed out earlier, ergonomics practices have started to rely more and more on the evidence derived from observational studies. Contrary to experimental studies, there are several challenges to the conduct of epidemiological or field studies because of the potential biases which may threaten both the internal and external validity if not accounted for at the design and analysis stages of these types of investigations. Therefore, there is an urgent need for instruments for the evaluation of observational studies.

The aim of this study was to develop and test a general purpose "Epidemiological Appraisal Instrument (EAI)" for evaluating existing studies or as a tool for designing new observational studies using a critical appraisal system rooted in epidemiological principles. The types of observational studies that may be evaluated by the EAI include: cohort (prospective and retrospective), intervention (randomized and non-randomized), case-control, cross-sectional, and hybrid (e.g., nested case-control).

METHODS

Figure 1 presents a road map for the development and testing of the EAI pilot and revised versions. The nine steps of the road map are detailed in the following sections.

The EAI is comprised of 43 questions grouped into five scales:

1) Reporting (17 items) – the study is clearly described allowing the reader to make an accurate evaluation of the study including study hypothesis/aim/objective, study design, exposure/intervention, outcome, covariates and confounders, statistical tests, and main findings.
2) Subject/record selection (7 items) – subject selection and recruitment have been conducted to minimize bias.
3) Measurement quality (10 items) – methods have been uniformly followed for measuring and recording of data and applied with the same precision to all groups.
4) Data Analysis (7 items) – the study has analyzed the data using appropriate statistics and accounted for confounders and covariates.
5) Generalization of results (2 items) – results are applicable to the eligible population and can be extended to other groups.

To respond to each item, the assessor is provided with two or more of the following levels to choose from:

1) "Yes" – designates the information is complete.
2) "Partial" – designates the information is partially complete.
3) "No" – designates the information is not described but should have been provided.
4) "Unable to determine" – is applicable whenever the information provided is unclear or insufficient to answer the question.
5) "Not Applicable" – a means for skipping an item. An example is an item that targets a study design different from the one evaluated.

The EAI provides detailed explanations for each choice of answer. These explanations were drawn from epidemiological sources and based on the experience of our research team consisting of epidemiologists, physicians, and biostatisticians (an electronic file is available upon request from the authors) (Genaidy 2004). For example, if demographically similar, internal controls are considered the top choice for a comparison group, and are assigned "Yes" versus a "No" assigned to a very dissimilar population. Although these explanations for choices of answers were a great challenge in the EAI development, these are essential to provide objective appraisal criteria.

Results

The overall baseline degree of agreement between the raters and the first author (AG) was 0.79. Four of the five scales had values between 0.80 and 0.97, with the generalization of

Evidence-based ergonomics 511

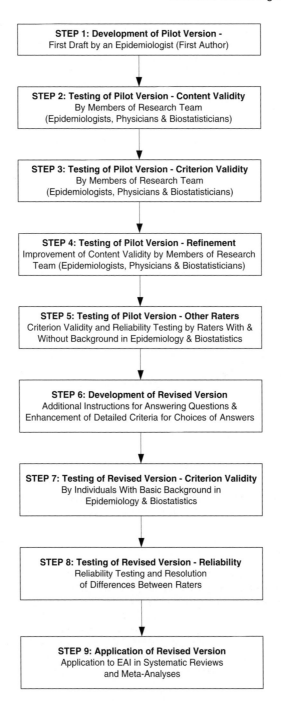

Figure 1. Road map for development and testing of the EAI pilot and revised versions.

results scale scoring the lowest value (0.66). The overall degree of agreement was very similar for the good and average articles (0.82), and lower for the poor article (0.74). In addition, the Spearman correlations ranged between 0.73 and 0.94.

The baseline internal consistency for the revised version was comparable to that for the pilot version for the reporting, measurement quality, and overall scales, higher for the data analysis and generalization of results scales, and lower for the subject selection scale. The overall quality scores for the articles were: (a) good article –1.40 ($\pm$0.30); (b) average article –1.33 ($\pm$0.23); and, (c) poor article –0.90 ($\pm$0.18). The differences were statistically non-significant at the 5% level for the "good" and "average" articles, but both were statistically different from that for the "poor" article.

Additional testing indicated that the inter-rater degree of agreement was between 80% and 100% for all responses, except "partial" which achieved a score of 54% with the remainder misclassified as "Yes" (32.86%) or "No" (12.86%). The overall inter-rater reliability was 0.90 (95% confidence interval or CI 0.87–0.92). The weighted Kappa for the five scales were in the excellent range (0.80–1.00): (a) reporting –0.80 (95% CI 0.75–0.86); (b) subject selection –0.95 (95% CI 0.91–1.00); (c) measurement quality –0.88 (95% CI 0.81–0.95); (d) data analysis –0.87 (95% CI 0.79–0.95); and (e) generalization of results –1.00 (95% CI 1.00–1.00). The weighted Kappa for study execution, consisting of the latter four scales, was 0.92 (95% CI 0.89–0.95).

Concluding remarks

The present study developed a comprehensive appraisal instrument that may be used in various applications, such as systematic reviews and meta-analyses. The following conclusions could be drawn:

1. An assessor with basic background in epidemiology and biostatistics would be able to correctly respond on 4 out of 5 questions.
2. On average, the assessment may improve with the quality of the article.
3. The preliminary application of EAI to a meta-analysis suggests that the inclusion of the overall quality into the calculations may reduce heterogeneity and improve the meta-risk estimate.

Further research is encouraged to test its reliability and validity on a larger scale and on several dimensions.

References

Downs S.H. and Black N. 1998, The feasibility of creating a checklist for the assessment of the methodological quality both of randomised and non-randomised studies of health care interventions, *Journal of Epidemiology and Community Health,* **52**, 377–384.

Genaidy A. 2004, *A Guide to the Development and Use of Epidemiological Appraisal Instrument (EAI)©* Version 3.0 Thesis/Dissertation, University of Cincinnati, Division of Epidemiology and Biostatistics.

VALIDITY OF DUAL-ENERGY X-RAY ABSORPTIOMETRY FOR BODY COMPOSITION ANALYSIS

J.A. Wallace, K. George & T. Reilly

Research Institute for Sport and exercise Sciences, Liverpool John Moores University, Henry Cotton Campus, 15–21 Webster Street, Liverpool L3 2ET

The aims of this study were to validate the use of dual-energy X-ray absorptiometry (DEXA) in the measurement of percent body fat (%BF) and to establish whether DEXA total %BF or DEXA subtotal %BF (without the head) has a better agreement and less error compared to hydrodensitometry. Percent body fat of 30 females was measured using DEXA and hydrodensitometry (using both predicted and measured residual volume [RV]) methods. Results displayed a close agreement and strong correlation between the methods, the DEXA subtotal %BF demonstrated the least measurement error and systematic bias in relation to hydrodensitometry. Results also showed that RV should be measured rather than estimated where possible when using the hydrodensitometry method.

Introduction

Body composition, especially the body fat component is important in the assessment of nutritional status, disease risk, health, athlete assessment and the effectiveness of interventions in physical fitness programmes. Many institutions conducting assessment of athletes are using more in-depth methods that offer measures of lean mass and bone density in addition to body fat, such as dual-energy X-ray absorptiometry (DEXA). This approach is growing in its application and is fast becoming a preferred method of body composition analysis. Hydrodensitometry is considered as the gold standard, criterion method for determining percent body fat. The use of DEXA to determine adiposity has gained popularity, yet the validity and reliability of this method have yet to be determined. The purpose of this study was to validate whole-body measures of % body fat using fan-beam DEXA scans by comparing them with values derived from hydrodensitometry using both measured and predicted residual volume (RV) in women.

Methods

Lung residual volume and underwater weight were measured and a DEXA scan obtained (total and subtotal data [without the head] recorded) during a single test session in 30 female subjects (age: 20.8 ± 1.6 years; mass 71.0 ± 11.4 kg; height 1.68 ± 0.06 m). Body volume, body density (Durnin and Womersley, 1974) and percent body fat (Siri, 1956) were calculated from the subjects' underwater weight using both measured RV (Helium dilution) and predicted RV (Boren et al, 1966). Data were analysed for differences in percent body fat using one-way ANOVAs. Correlations and relationships were established using Pearson's product moment correlation and bivariate linear

Table 1. Comparison of the DEXA method with hydrodensitometry.

Method Comparisons	MD	SDD	SEE	TE	CV (%)	TEM
DEXA total	−0.73	−0.23	1.42	4.02	3.33	1.12
DEXA subtotal	−0.12	0.03	1.40	0.66	3.28	0.99

regression. Mean difference (MD), standard deviation of the difference (SDD), standard error of the estimate (SEE), total error (TE), coefficient of variance (CV %) and technical error of measurement (TEM) were calculated for each method. The procedures received approval from the University's Human Ethics Committee. Participants' written informed consent was also obtained.

Results

Results showed that there were no significant differences in % body fat between hydrodensitometry (30.55 ± 6.19%) and either DEXA total (29.82 ± 5.94%) or DEXA subtotal (30.43 ± 6.22%). The MD and SDD from hydrodensitometry for % body fat values of DEXA total were greater than from DEXA subtotal values. The negative MD demonstrates a small underestimation in %BF by the DEXA method. The SEE of % body fat between hydrodensitometry and both DEXA total and DEXA subtotal were comparable, the CV% of % body fat between methods was also minimal and comparable (Table 1). Measures of the DEXA method for subtotal % body fat values, showed a smaller TE and TEM than when using the DEXA total values, compared with the criterion (hydrodensitometry) (Table 1). Results demonstrated strong correlations between the methods. The strongest correlation occurred between hydrodensitometry and DEXA subtotal ($r = 0.974$, $P < 0.01$), although the correlation between hydrodensitometry and DEXA total was only marginally smaller ($r = 0.973$, $P < 0.01$). Bivariate regression analyses yields slope values for hydrodensitometry with DEXA total and DEXA subtotal as not significantly different from 1.0. Bland-Altman analysis indicated that the systematic difference between hydrodensitometry and both DEXA total and DEXA subtotal values of % body fat were unaffected by the absolute mean % body fat of the two methods. A significant difference occurred between the subjects' predicted residual volume (1.33 ± 0.15 l) and their measured volumes (1.00 ± 0.21 l). Hydrostatic % body fat was significantly higher ($P < 0.05$) when using the measured values of residual volume (30.6 ± 6.2%) rather than the estimated values of residual volume (28.3 ± 6.5%).

Discussion

The major finding of this study was that percent body fat as estimated by DEXA agreed well with percent body fat as estimated by hydrodensitometry in young women. No significant difference in % body fat was found between hydrodensitometry and either DEXA total or DEXA subtotal. The low MD between hydrodensitometry and DEXA was considered clinically small and within the limits of practical application. On the basis of Lohman's (1996) criteria, body composition estimates from DEXA total and DEXA subtotal showed good accuracy based on the low SEE (<3%). Lohman (1996) has suggested that regression equations with TE greater than 3.3% are inaccurate. The high TE for DEXA total % body fat was greater than the SEE, indicating systematic error in the prediction and a limited accuracy, when including the head in the results. In contrast the very low TE for DEXA subtotal indicates a low systematic error in the prediction and good accuracy. The high correlation between DEXA and hydrodensitometry, combined with the low SEE, suggests that the DEXA method is viable for predicting percent body fat. The low CVs and low TEM between DEXA and hydrodensitometry mean that the estimations of percent body fat in this study were within the limits of acceptable variation between two methods. Again it should be noted that DEXA subtotal values demonstrated a marginally greater validity than DEXA total values.

Goss et al (2004) compared measured and predicted RV in determining body composition and found that only one out of the three available RV prediction equations did not differ from the value obtained using measured RV. The different values of percent body fat derived from predicting from actually measuring and measured RV underlines the importance of always measuring true RV.

Conclusions

The close agreement and strong correlation between hydrodensitometry and DEXA techniques mean that DEXA can be used to determine % body fat accurately and precisely in females with a range of body compositions. Subtotal % fat should be favoured over total % fat to reduce the error associated with including assumptions about the head when using the DEXA method furthermore, RV should always be measured rather than predicted when estimating % body fat by the hydrostatic method.

References

Boren, H. G. and Kory, R.C. (1966). The veteran's administration army cooperative study of pulmonary function tests: II. The lung volume and its subdivision in normal men. *American Journal of Medicine*, **41**: 96–114.
Durnin, J. and Womersley, J. (1974). Body fat assessed from total body density and its estimation from skinfold thickness: measurements on 481 men and women aged from 16 to 72 years. *British Journal of Nutrition*, **32**: 77–97.
Goss, F., Robertson, R., Swan, P., Harris, G., Trone, G. and Utter, A. (2004) Comparison of measured and predicted residual lung volume in determining body composition of collegiate wrestlers. *Journal of Strength and Conditioning Research*, **18**: 281–285.
Lohman, T.G. (1996). Dual energy X-ray absorptiometry. In A.F. Roche, S.B. Heymsfield and T.G. Lohman (eds.) *Human Body Composition*, (Human Kinetics, Champaign; IL) 63–78.
Siri, W. (1956). The gross composition of the body. *Advanced Biology Medicine and Physiology*, **4**: 239–80.

BODY COMPOSITION IN COMPETITIVE MALE SPORTS GROUPS

J.A. Wallace, E. Egan, K. George & T. Reilly

Research Institute for Sport and Exercise Sciences, Liverpool John Moores University, Henry Cotton Campus, 15–21 Webster Street, Liverpool L3 2ET

The aim of this study was to examine the effects of sport participation on body composition and bone density in professional male athletes. Body composition of 85 soccer and Rugby Union players and 43 control subjects was determined by dual-energy X-ray absorptiometry (DEXA). The soccer and Rugby players had significantly greater lean mass and lower percent body fat than controls and they also had a significantly greater bone mineral density (BMD). Among the Rugby players, forwards were taller, heavier and had a greater lean mass and BMD than the backs. The observations imply that sport has beneficial effects on the development and maintenance of peak bone mass and body composition. They suggest that athletes who experience a greater frequency and intensity of impact forces through their sport participation express superior bone densities.

Introduction

Sport participation can have a beneficial effect on the development and maintenance of peak bone mass and causes sport-specific physiological adaptations (Andreoli et al, 2001). There is no consensus regarding which sports are more beneficial for increasing peak bone mass and appendicular muscle mass.

Rugby and soccer are two sports that differ in their loading characteristics and desirable physique. Soccer is characterised by impact loading forces and also involves large amounts of torsional and compression forces being placed on the lower limbs. Rugby is characterised by large amounts of compressive forces on both the upper and lower limbs, as well as incorporating a ground reaction force on the lower limbs from running. The aim of this study was to measure body composition and BMD in male professional soccer and Rugby Union players compared with sedentary controls, and to determine the difference in body composition between forwards and backs in Rugby.

Methods

One hundred and twenty two male subjects (24.7 ± 4.3 years; 1.81 ± 0.08 m; 82.19 ± 9.09 kg) were measured, consisting of premiership Rugby Union players (n = 47), Premier League soccer players (n = 38) and sedentary controls (n = 43). Participants wore lightweight shorts without zips, buttons, or any other metal and removed all jewellery for the scanning. The procedures received approval from the University's Human Ethics Committee. Participants' written informed consent was also obtained.

Subjects arrived at the laboratory in the morning, after fasting for 3 hours. Subjects' age, height (m), and weight (kg) on scales were measured. Body composition (total

Table 1. Body composition variables (mean ± SD).

	Controls	Soccer	Rugby	Forwards	Backs
Mass (kg)	84.42 ± 10.71	84.12 ± 7.38	101.07 ± 2.09	109.2 ± 9.5	92.9 ± 8.2
Lean mass (kg)	65.14 ± 7.14	69.06 ± 6.34	81.93 ± 8.79	87.9 ± 7.1	76.3 ± 6.3
Fat mass (kg)	16.24 ± 5.39	13.89 ± 12.65	15.36 ± 4.60	17.2 ± 0.1	13.3 ± 2.9
Fat (%)	18.93 ± 4.55	13.45 ± 2.73	15.04 ± 3.26	15.6 ± 3.7	14.2 ± 2.5
BMC (g)	3040 ± 409	3538 ± 454	3896 ± 476	4184 ± 421	3607 ± 392
BMD (g.cm^2)	1.31 ± 0.09	1.45 ± 0.10	1.50 ± 0.11	1.55 ± 0.11	1.45 ± 0.11

mass, fat mass, lean mass, percent body fat [%BF], bone mineral content [BMC] and BMD) was then measured using DEXA (Hologic QDR series, Delphi A, Bedford, Massachusetts). All scans were analysed by the same investigator.

Results

Body mass was significantly higher ($P < 0.001$) in the Rugby players, but not in the soccer players ($P > 0.05$) compared to sedentary controls and lean mass was significantly greater in both Rugby ($P < 0.001$) and soccer players ($P < 0.05$) compared to the controls. There was no significant difference ($P > 0.05$) between the controls and either Rugby or soccer players for fat mass; however, % body fat was significantly lower than control values (18.9%) for both sport groups (Rugby: 15.03%; soccer: 13.45%). Bone mineral content was significantly higher ($P < 0.001$) in Rugby and soccer players compared with the sedentary controls, with Rugby players displaying the greatest BMC. Bone density was also significantly higher ($P < 0.001$) in the sports groups (Rugby: 1.49 g.cm^2; soccer: 1.45 g.cm^2) compared with controls (1.31 g.cm^2).

Irrespective of playing position the Rugby players had significantly greater values for total mass, lean mass, % body fat, bone area, BMC and BMD compared with the soccer players; no difference in fat mass was observed between sports. There was a marked contrast in body composition between forwards and backs in Rugby Union (Table 1); forwards had a significantly greater total mass, lean mass, fat mass, BMC and BMD, but no significant difference in % body fat, compared to the backs.

Discussion

The main finding of this present study was that participation in sport at an elite level is associated with a greater lean body mass, a reduced body fat percentage and increased BMC and BMD compared with normal values. This observation supports the belief that participation in sport causes beneficial physiological adaptations (Andreoli et al, 2001). In this study lean mass was increased by 6% and 25% and percent body fat was reduced by 5.5% and 4% in soccer and Rugby Union players respectively compared to control subjects. These trends are in agreement with Calbet et al (2001) and Elloumi et al (2005) who established similar differences for their soccer and Rugby players, compared to controls. However, Babic et al (2001) reported sizeably higher % body fat values in Croatian Rugby players. The differentiating levels of sport participation and professionalism between the players in these studies is the likely the cause of these disparities.

The current study also demonstrates that participation in professional sport is associated with a high whole-body BMC and BMD, regardless of the sport and loading characteristics. Calbet et al (2001) found 13% and 10% greater BMC and BMD values in soccer players compared to controls, which supports our findings of 16.4% and 10.7% higher values respectively. Elloumi (2005) also found significantly higher BMC (40%) and BMD (16%) in the legs of Rugby players compared to controls; these values are slightly higher than found in this current study for the whole-body (BMC: 28% and BMD: 14.5%).

Within the Rugby group, forwards were taller, heavier, had a greater % -body fat and a greater total lean mass and fat mass than backs. Likewise BMD and BMC for the whole-body were greater in the forwards, supporting the findings of Elloumi et al (2005) and Babic et al (2001). These differences demonstrate that greater BMC and BMD values occur when athletes experience a high frequency and intensity of compressive forces due to physical impact, along with ground reaction forces, rather than the ground reaction forces alone, as experienced by soccer players.

Conclusion

In this study, participation in professional soccer and Rugby Union was associated with an increased bone density, lean mass and a reduced percent body fat compared with sedentary controls. These differences imply the beneficial effects of sport participation for optimising peak bone mass. Notwithstanding the likelihood of body composition types being attracted to specific sports, Rugby Union (irrespective of playing position) may be more effective in maximising lean mass and bone density compared with soccer.

References

Andreoli, A., Monteleone, M., VanLoan, M., Promenzio, L., Tarantino, U. and DeLorenzo, A. (2001). Effect of different sports on bone density and muscle mass in highly trained athletes. *Medicine and Science in Sports and Exercise*, **33**, 507–511

Babic, Z., Misigoj-Durakovic, M., Matasic, H. and Jancic, J. (2001). Croatian rugby project-Part 1. Anthropometric characteristics, body composition and constitution. *Journal of Medicine and Physical Fitness*, **41**, 250–255

Calbet, J.A., Dorado, C., Diaz-Herrera, P. and Rodriguez-Rodreguez, L.P. (2001). High femoral bone mineral content and density in male football (soccer) players. *Medicine and Science in Sports and Exercise*, **33**, 1682–1687

Elloumi, M., Courteix, D., Sellami, S., Tabka, Z. and Lac, G. (2005). Bone mineral content and density of Tunisian male rugby players: Differences between forwards and backs. *International Journal of Sports Medicine*, **In publication**

OCCUPATIONAL HEALTH AND SAFETY

"LET'S BE CAREFUL OUT THERE" THE HONG KONG POLICE OSH SYSTEM

Michael Dowie[1], Paul Haley[2], Nigel Heaton[3], Rod Mason[4] & Duncan Spencer[5]

[1]Senior Assistant Commissioner of Police, Hong Kong Police,
[2]Senior Inspector of Police, Hong Kong Police,
[3]Director, Human Applications, [4]Superintendent of Police,
Hong Kong Police, [5]Senior Manager, Human Applications

In 1997 the Hong Kong (HK) Government passed legislation binding on HK organisations requiring them to adopt a pro-active approach to risk management. The HK Police, (HKP) in common with other Government departments, was included within the scope of the legislation. The Force HKP determined to ensure that they were an exemplary organisation with respect to complying with the requirements of the legislation. Initial benchmarking highlighted a number of areas where the HKP could improve its Occupational Safety and Health (OSH) management. In 2002 the HKP decided to adopt a comprehensive Top-Down, Bottom-Up approach to risk management. The HKP engaged in a five year programme to embed OSH within the organisation. The programme includes training, auditing, an Intranet based OSH system and external validation of their OSH standards.

Introduction

Hong Kong (HK) is a Special Administrative Region of the People's Republic of China (SAR). In 1997 the HKSAR Government passed legislation to address health and safety concerns in the HKSAR. This took the form of an additional chapter in legislation known as OSH Ordinance Cap. 509. The scope of Cap. 509 was broad. It required organizations to adopt a pro-active approach to risk management and to implement a suitable system for managing risk. In addition, subsidiary regulations were made; Cap. 509 A set out detailed health and safety regulations for the workplace and manual handling; and Cap. 509 B set out detailed health and safety regulations for the use of display screen equipment. Whilst Cap. 509 allowed for the HKSAR government departments to be immune to criminal prosecutions (analogous to the UK's Crown Immunity) it explicitly required departments to comply with the legislation and its statutory instruments (improvement and suspension notices), which made government departments and individuals potentially civilly liable for any health and safety failures. The HKP, in common with other government departments, was required to respond to Cap. 509 and to ensure that the HKP was doing enough to protect the health, safety and welfare of its own staff and of other people affected by its operations.

The initial audit

The Hong Kong Police undertook at large-scale audit in 2001–2002 identifying gaps in its management of occupational health and safety and producing a plan to fill those gaps in. The work was undertaken by the HKP's Support Wing. The audit look at what the Force HKP was doing to manage health and safety, what accidents and incidents were occurring and what structures would need to be in place to allow the HKP to discharge its duties and to protect better its staff. The review highlighted sufficient deficiencies, particularly in the way that decisions were documented. It also highlighted weaknesses in learning from mistakes once made. Finally it recognized that there was an opportunity to significantly improve health and safety performance within the HKP by changing behaviours and attitudes. Part of the audit looked at the work of other similar Police Forces throughout the world. One of the Police Forces that was visited was the Metropolitan Police, who at the time had a Commissioner and an ex-Commissioner being prosecuted by the HSE. One of the lessons that the Metropolitan Police were keen to emphasize was the importance of a "joined up system".

The main recommendations to emerge from the audit were:

- to undertake a 5-year programme to deliver a comprehensive occupational safety and health management system
- to brief the most senior officers in the HKP on OSH issues, and to do this regularly
- to put in place structures that would allow the HKP to develop internal expertise on OSH issues
- to work with external experts to drive through an organisational-wide change in OSH within the HKP.

Phase one

The first phase of the project was concerned with setting the agenda and winning the hearts and minds of the senior officers. It was recognised from a very early stage that the programme had to be a long term one. The HKP employs over 34,000 people and the OSH programme was deployed to change attitudes to OSH and to alter behaviour. From the start, it was recognised that a management of change programme would not deliver results in the short term.

The most important stakeholders were the senior officers who run the HKP. Phase one of the project kicked off with a high level presentation to the Commissioner and the main management board. The aim of the presentation was to reach a consensus on what a good OSH system might look like and what it would achieve. The aim was to ensure that the senior decision makers bought into a vision of OSH that would be cascaded down through their Departments.

One of the challenges from the management board was to explain why an OSH system was required and what it would achieve. Preparation work undertaken by Support Wing was able to demonstrate how many officers were injured and what the current bill for OSH related claims was.

There was agreement that a programme was essential but that it had to be simple, driven by policing needs, must complement operational policing and be embedded into existing management structures and systems.

All senior police officers were required to attend a one and a half day course in risk management within the HKP. This was a course designed to meet the Institution of Occupational Health and Safety's (IOSH's) attainment standard for risk management. Some 400 senior officers attended the initial training. The message from the course was that OSH was an important part of any operational policing operation, senior officers were required to manage OSH issues and there were simple tools and techniques that were being implemented throughout the HKP to support effective OSH management.

Phase two

The second phase of the project was about delivering resource and support to senior officers. To achieve these goals a number of separate strands were developed.

Risk assessor training

Nearly 400 officers were given basic risk assessment training. Again, the course was tailored to the HKP but designed to meet the IOSH attainment standard for general risk assessment. The course was run over three and a half days and included input from experienced police officers to help set the context and to provide illustrations from a HK policing perspective.

Developing HKP policy

The HKP needed explicit guidance on how to implement OSH and the normal channels for this type of guidance is the headquarter order. Three headquarter orders were produced.

- Headquarter Order No. 6 (HQO 6) – the occupational safety and health policy
- HQO 7 – occupational safety and health risk management
- HQO 8 – display screen equipment risk management.

The headquarter orders explained how OSH would be managed throughout the HKP. They contain the basic concepts and explain the mechanics of the system to those involved at all levels.

To support the training and HQOs, the HKP began to develop a central list of tasks that were considered to contain significant risk. This task list would be shared throughout the HKP on the local Intranet and officers such as Formation Commanders would be able to upload local risk assessments and download sanitised template assessments for adaptation to their respective local circumstances to reflect operational and situational risks in any one particular area.

Phase three

Phase three of the project was aimed to provide the HKP with a cadre of internally trained officers who were able to deliver key training in three critical areas – general risk assessment; manual handling risk assessment and DSE risk assessment. The cadre would be supported by external experts who would quality assure the delivery of the training and the output of the programme. The benefits of this approach were viewed in terms of officers training officers had more creditability than training run by external contractors. Firstly, the cost of the programme was significantly reduced. Secondly, it became viable to train an additional 2000 officers.

The training of trainers required the HKP to commit to the deployment of a cadre of specialist OSH trainers who would be able to meet the rigorous training standard and assimilate sufficient OSH expertise to be able to train other officers. A selection process was followed which identified some 20 potential officers, of whom some 15 met the final standard across all areas.

These officers were supported in the development of their learning materials, training notes and delivery methods. Each member of the cadre was required to undergo written and practical testing and those who were not able to meet the standards were provided with additional support or were given the opportunity to be re-deployed.

The final cadre undertook a number of practice sessions prior to being formally appraised and assessed by an external verifier (see below).

Phase four

The cadre's training was reviewed by an external verifier. All course materials, training notes and project deliverables were included in the verification. This helped ensure that a consistent message was received by delegates and that the standard of training was to an acceptable level.

The HKP viewed external verification as an essential part of its OSH strategy. Meeting externally recognised standards would enhance the image of the OSH process and would also ensure that officers were not attending a training course that had been invented solely for the HKP.

The external verifier identified a number of areas that could be improved but was satisfied with the level of knowledge and techniques of the presenters.

Discussion

The challenges to the HKP were many. It was clear that to be successful, a large-scale management of change programme, engaging officers at all levels, was essential. There must be commitment from the top, both in terms of being seen to support the process as well as providing the resource to staff the process. There also needed to be a change of attitude in the middle management structure. Whilst the HKP has to tolerate some risk, it could not afford to be too risk accepting. Persuading middle ranking officers to put health and safety on the planning agenda and encouraging them to change both their own behaviour and that of their team was viewed as a major challenge. From the start of the project it was recognised that a project like this would only be successful if officers felt the need to do something. Training, in itself, might not provide sufficient impetus to effect change. A twin strategy was agreed. Firstly, the process was to be made as easy as possible with expert support available centrally. A new OSH team was created to take the process through the early stages. This was located in the HKP's Support Wing. The function of the OSH team was to staff training, provide expert opinion and local support.

Secondly, it was agreed to incorporate OSH into the local and strategic audits that the HKP routinely conducts. By integrating OSH within an existing process, officers would be encouraged to meet OSH goals and would receive feedback on how well they were doing. As the cycle times for audit are relatively long (typically 18 months), officers would be sufficiently prepared to meet the audit targets and OSH would be seen as just another part of operational policing.

It was clear that whilst in the UK most organisations have a 30 year history of complying with OSH regulations and implementing OSH controls, no such culture existed in HK. However, the HKP were in a position similar to that found by many UK Police Services in the late 90's when they were suddenly required to meet health and safety regulations for all their officers. The HKP was able to learn from the very recent experience of other UK Police Services and benefited from seeing what worked and what didn't in the UK.

Significant achievements

Since the project began, there have been a number of significant achievements. In 2005, the Interpol Heads of Training Symposium was hosted by the HKP. Attached to the symposium was an exhibition at which the achievements of the OSH project were heavily advertised. To compliment this, the Director of Management Services made a keynote address to the symposium detailing the achievements of the OSH project and presenting the vision for the future.

Also in 2005, the HKP entered the project into the HK occupational safety and health council annual OSH awards. The HKP was awarded the silver medal out of some 147 entrants for its work to develop an OSH system.

In December 2005, Hong Kong hosted the World Trade Organisation (WTO) 6th Ministerial Conference with the inherent potential for violent anti-WTO protest and clashes with the HKP. For the first time OSH risk assessment and risk management played a significant role in the planning and preparation of this event involving over 10,000 police officers in a public order role. The strategic impact upon operational deployments was taken into consideration within the structure of the new Force Safety Management System.

In 2006, the UK's Metropolitan Police Service is undertaking a fact finding tour to share information on best practice and to benchmark against a similar organisation.

Conclusion

The process is now beginning to become embedded within the Hong Kong Police. There is a still a programme of work until 2008 designed to support OSH activities. It is clear that the HKP can still improve, even though it has come a long way.

The HKP have demonstrated that it is possible to design, develop and implement a sophisticated OSH risk management system from nothing. In some respects, not having to deal with a legacy of different systems and different legislation was of benefit to the HKP. However, the early recognition that the project was not simply about rolling out training but was about changing the culture, had the greatest impact. The HKP is now able to adopt a single, uniform methodology for identifying and controlling OSH risk without impacting its operational effectiveness. OSH issues are not delegated to the health and safety manager, they form a core part of the agenda of all operational officers. In future, OSH training will be routinely delivered to all levels within the HKP and the OSH system will continue to become more sophisticated in content and simpler to use.

A METHOD FOR TESTING PRESSURES ON FINGERS IN ENTRAPMENT HAZARDS

Edmund Milnes

Health & Safety Laboratory (HSL), Harpur Hill, Buxton, Derbyshire SK17 9JN

Trip-nip bars are used to prevent fingers/hands being drawn into in-running nips on machines such as printing presses. However, if trip-nip bars are improperly designed, installed or maintained, they can represent a significant entrapment hazard if the fingers/hand are drawn against them and excessive force is required before the safety mechanism trips, or even after tripping has occurred. Current standards and industry guidance do not advise a suitable method to test whether the crushing pressures on fingers at trip activation are acceptable. This study determined the pressure-pain tolerance limits from a range of subjects. A set of test fingers was designed and calibrated using pressure sensitive film. Field tests of an entrapment hazard using the test fingers were successful and the test method is designed to be repeatable by industry during maintenance checks.

Introduction

Trip-nip bars are a common guarding device on machines where there is an in-running nip hazard (i.e. two adjacent surfaces which narrow towards each other and create a net in-pulling effect into a crushing zone). Trip-nip bars are designed and installed such that if a persons hand starts to be drawn in to the in-running nip, the movement of the hand into that area moves the trip-nip bar which activates an automatic stopping mechanism, halting further movement of the counter-rotating surfaces. Although these devices are commonly found in the printing industry on lithographic presses, there are numerous pressing machines which rely on this safety device. This research resulted from Health & Safety Executive (HSE) concerns about a trip-nip bar encountered on a printing press during a routine inspection visit. An assessment by an HSE Mechanical Engineering Specialist Inspector identified that an entrapment hazard existed between the edge of an in-cut in a blanket roller and the trip-nip bar. The key problem was that the stopping mechanism only activated in the final 2–3 mm of trip-nip bar travel, however the roller required approximately 8–9 mm to stop fully. The additional 6 mm of roller travel once the trip-nip bar is fully deflected results in crushing forces on the fingers/hands. The most likely time for this to occur is during cleaning of the rollers when operators clean the surfaces with a cloth. At this time the rollers are allowed to move in slow-crawl mode (17 mms^{-1} maximum surface speed, HSE Printing Information Sheet 1). However, this speed is still sufficient to create a risk of entrapment between the trip-nip bar and the edge of the in-cut if an operators hand is resting in the in-cut and some distraction occurs or simply if the operator miss-judges the time available to withdraw his hand. HSL Ergonomics section was asked to advise HSE on acceptable trip-nip entrapment pressures on fingers and how these could be tested.

Pre-study status of pressure tolerance literature and test methods

A review of literature on crushing forces and pressure tolerance revealed two main types of data; Pressure-Pain Thresholds (PPT) and Bone Surface Crushing pressures

(BSC). PPT levels are the point at which pressure begins to be perceived as painful. BSC pressures are those which could strain the surface cortical bone beyond its elastic limit and cause permanent deformation.

Three previous studies have recorded similar values for PPT. Fischer (1986) applied pressures to the fingers using a 10 mm^2 probe and recorded average PPT's of 148–175PSI in male subjects and 112–145PSI in female subjects. Bremmun (1989) using a 0.28 cm^2 probe recorded average PPT's of 92PSI (males) and 63PSI (females), and Frannson-Hall (1993) using a 1 cm^2 probe recorded average PPTs of 119PSI (males) and 80PSI (females). However, there is strong evidence that the application area and profile of application significantly effect the perception of pain and pressure tolerance. Woodworth (1938) indicates that much higher localised pressures can be sustained without excessive pain or injury (e.g. 425PSI generated at the tip of a needle). This is also intuitive based on pressure-pain assisting judgement of applied force as well as pressure; the same pressure over greater areas can only result from greater forces and thus greater energy input to the body, with greater potential for injury as the energy is dissipated. As one might expect, BSC Pressures are much higher than PPT (e.g. approximately 30.9kPSI) (Currey, 2002).

The current relevant British Standards and Printing Industry Guidance do not describe an effective test of potential entrapment pressures. BS EN 1010-1:2004 indicates that to test the activation of the trip mechanism a low-compressible 15 mm diameter "test-finger" should be used. This is only a functional test and it is unlikely to provide any effective indication of the pressures that could be generated on the fingers.

Test development method

Pressure-Pain Tolerance Limit (PPTL), the point at which surface pressure pain becomes intolerable, was identified as a more appropriate acceptable safety threshold than PPT or BSC pressures. The relatively low PPT levels, as a threshold, could lead to excessive accidental tripping. However, if the BSC pressures were used as a threshold, there is a strong possibility that soft tissue injuries could occur before the required tripping force is exerted. It was also decided that because of the potential significance of entrapment profile and area, information on PPTL needed to be gathered with forces applied by a similar entrapment profile to the hazard itself.

Overview of approach taken

A rig was constructed which replicated the trip-nip entrapment profile, in order to record PPTL data. A range of subjects' PPTL data was recorded (variables recorded were total crushing force (N) at PPTL, uncompressed finger depth (mm) and approximate pressure levels (PSI)). The relationship between finger dimension (depth) and crushing force was calculated. Test fingers were designed and constructed; anthropometric data was used to identify a range of appropriate diameters for a range of test fingers and 3 sets of designed with different skin properties. Test fingers were calibrated by applying a crushing force to each test finger using the test rig. The force applied was derived from the regression statistics (finger depth vs. crushing force at PPTL). This was considered to generate the average PPTL and approximate pressure levels were recorded using pressure sensitive film, i.e. a calibrated pressure indication sheet was made for each test finger. Test fingers with pressure sensitive film on the surface were inserted into entrapment hazard and compared the film results with the calibrated sheets to identify whether entrapment pressures were higher than average PPTL figures.

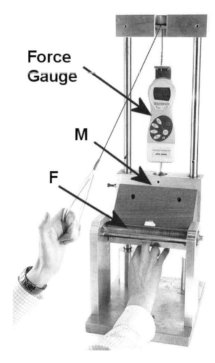

Figure 1. Test rig for measuring peak crushing force at PPTL.

Measurement of PPTL

Figure 1 shows the rig which was used to test subjects PPTL. Subjects were asked to insert their fingers between the moving block (M) and the fixed "trip-nip" bar (F). They were then asked pull on the pulley which raised the block and pressed their finger between the block and the trip-nip bar. They were asked to gradually increase the tension in the pulley (i.e. the crushing force) until they found the pressure on their finger intolerable. The peak tension in the pulley was recorded using a force gauge.

Prior to testing, subjects uncompressed finger depths at the points where the pressure was exerted were measured with vernier callipers. Subjects were asked to position their fingers so that pressure was exerted on the distal and proximal joints and medial phalange of the middle finger and little finger. In total 5 subjects volunteered to participate in this study; 2 females (aged 25 and 27) and 3 males aged between 25 and 50. Tests were alternated between fingers to try and minimise any learned or increased tolerances or increased sensitivity to localised pressure following a test. For each test a strip of Pressurex® (pressure sensitive film) was placed on the dorsal surface of the finger and the trip-nip bar (where the highest pain levels were perceived). The Pressurex® indicates pressure using a chemical reaction which generates a red colour on the film. Higher pressure causes more chemical to be released and a darker shade of red is produced.

Figure 2. shows a scatterplot of the finger joint depth vs. compressive force at PPTL.

Regression analysis indicated a significant linear relationship ($p < 0.05$) between the two factors represented in figure 2 with 17.1% of the variance in peak compression

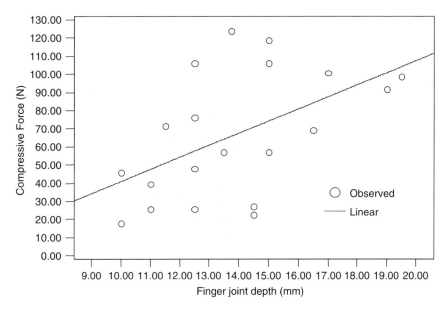

Figure 2. Results of finger PPTL compression tests.

Table 1. Finger skin surface pressure (PSI) at PPTL.

Finger	Mean PSI (&SD)	Range (PSI)
Little (both joints)	356 (*102*)	149 to 428
Middle (both joints)	346 (*136*)	70 to 569
Little Distal	321 (*136*)	149 to 428
Little Proximal	391 (*38*)	320 to 428
Middle Distal	300 (*86*)	150 to 402
Middle Proximal	392 (*167*)	70 to 569

force at PPTL predicted by finger depth. The linear relationship shown in Figure 2 is described by the following equation:

- Peak Compressive Force at PPTL = −25.562 + (6.652* Finger Joint Depth).

Surface pressure on the fingers was estimated based on the colour changes of the Pressurex® films placed between finger and trip-nip bar during tests of PPTL. Table 1 contains the key pressure estimates.

Test finger design & calibration

3 sets of test fingers were designed. Each set consisted of a small, medium and large cylinder of stiff urethane with a specific artificial skin on the surface; Cold Shrink Rubber Tubing, Polychloroprene Sealing Strip and a Thermoplastic Elastomer with a Shore-A hardness of 60, similar to human skin. The skins were selected on a trial and error basis to fulfil a need for the stiffness of the skin to compensate as much as possible for the circular cross-section of the urethane tube (which tends, unlike real fingers to concentrate pressure in a ridge following the tubes' long axes).

The test fingers ranged between 13 and 20 mm in diameter (including skin thicknesses between 1.5 and 3 mm). The test fingers were calibrated by using the rig to apply a compression force calculated based on their diameter/depth. Pressurex® films were recorded from the surface of the test fingers during these compressions. These calibration films provided the benchmarks which were compared with Pressurex® films taken from tests in the real trip-nip entrapment hazard.

Conclusions

Tests on the printers' trip-nip bar indicated that the trip-nip bar/roller in-cut entrapment is likely to be significantly higher than PPTL for the majority of people. Pressures of 1400PSI and above were estimated based on Pressurex® films.

Although the subjects in this study had a wide range of finger sizes based on the working population, the total number of subjects is relatively low and it would be useful to gather further data. The estimates of pressures at PPTL (Table 1) are based on matching the darkest areas of colour on Pressurex® films with colours on a chart provided by the film manufacturer. Although this is useful for comparing pressures in two different exposure conditions, further tests in this area may be able to use electronic pressure sensing equipment, which could give a more accurate real-time measure of surface pressures. A real challenge with any of this equipment is over-protection of the skin by the measuring surface itself, which may affect peoples' sensitivity to an applied force.

It is important to note that the PPTL results in table 1 could be quite specific to the profile of the surface of the trip-nip bar (curve radius 15 mm). However, they nevertheless give an indication of the levels of pressure that become intolerable over areas of approximately 4 to 5 mm^2 (the average approximate area which sustained the test pressure).

The general method used in this study should be useful and replicable for entrapment of any body part, providing that suitable test media can be identified and calibrated (i.e. items with suitable/representative overall stiffness and surface compressibility).

References

Brennum, J., Kjeldsen, M., Jensen, K., and Jensen, T. S. 1989 Measurement of human pressure-pain thresholds on fingers and toes. *Pain*. Volume 38. pp 211–217.
BS EN 1010-1:2004. Safety of Machinery – Safety requirements for the design and construction of printing and paper converting machines – Part 1: Common requirements. British Standards Institute.
Currey, J. D. 2002 *Bones; Structure and Mechanics*. Princeton University Press.
Fischer, A. A. 1986 Pressure tolerance over muscles and bones in normal subjects. *Archives of Physical Medicine and Rehabilitation*. Vol 67. pp 406–409.
Fransson-Hall, and C., Kilbom, A. 1993 Sensitivity of the hand to surface pressure. *Applied Ergonomics*. 24(3) pp 181–189.
HSE Printing Information Sheet No.1. Safe systems of work for cleaning sheet-fed offset lithographic printing presses.
Woodworth, R. S. 1938 Quoted in *Handbook of Human Engineering Data 2nd Edition*, Tufts College, Technical Report SDC-199-1-2 1951.
© Crown copyright (2006)

SLIPS AND TRIPS IN THE PRISON SERVICE

Anita Scott & Kevin Hallas

Health and Safety Laboratory, Harpur Hill, Buxton, Derbyshire SK17 9JN

Slips and trips account for approximately 20% of injuries to staff in the Prison Service. The aims of this study were to establish underlying causation for falls to prison staff and to provide recommendations to reduce their occurrence. Analysis of Slip, Trip and Fall (STF) accident data was conducted at a national level (RIDDOR – Reporting of Injuries, Disease and Dangerous Occurrences Regulations) and also at a local level (accident records from seven prisons). Prison visits to seven locations were also conducted which provided opportunities to talk with local health and safety representatives. Discussion points to emerge from the accident records analysis and the visits included: cleaning regimes, maintenance work, footwear policy, stair safety and compensation claims. Recommendations designed to reduce STFs have been made.

Introduction

Slips and trips are a major cause of injury for staff in the prison service. RIDDOR data shows that in 2004/05 slips and trips accounted for 19% of major injuries and 20% of over 3 day injuries, with 2003/04 data also showing similar percentages (20% for majors and 21% for over 3 day injuries). The aim of this piece of work was to establish underlying causation for falls to prison staff and to make recommendations to help reduce them.

Methodology

An analysis of accident records was conducted on a national level and on a local level. 452 slip and trip RIDDOR accident reports were analysed. The RIDDOR data was provided by the Health and Safety Executive (HSE) for the Standard Industrial Code (SIC) Justice and Judicial activities (SIC75230) for the year 2003/04. Accident types included over 3 day (365) accidents and major accidents (87). Local accident records were reviewed during visits to seven UK prisons. One year of data was considered from each prison. The prisons visited included five adult male prisons, a young offender and juvenile prison and a women's prison. The visits provided the opportunity to talk to health and safety staff, prison officers and ancillary workers (eg, kitchen staff and works department staff). Slipperiness measurements of flooring surfaces were also taken during the visits using the Stanley Pendulum (measures the coefficient of friction of the surface) and a micro-roughness meter (measures the roughness of the surface).

Accident records analysis

RIDDOR data

Most falls were initiated by a slip (51%), with 25% initiated by a trip. Common slip hazards included water (or wetness), food and ice. Common trip hazards were temporary

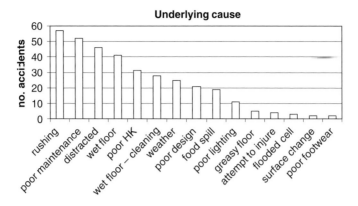

Figure 1. Underlying causation for falls to prison staff.

obstacles (eg, food trays, entrance matting), uneven ground and permanent obstacles (eg, gate fixings). Activities when falls frequently occurred were: walking on stairs or steps, walking on the level, responding to an alarm and running (Alarm calls may be raised if, for example, there is a disturbance between prisoners, or if a staff member is being threatened. Alarm calls usually result in prison officers rushing to the scene.). Incidents frequently occurred outdoors and on steps and stairs. Underlying reasons for falls were identified as: rushing (eg, in response to an alarm), poor maintenance (eg, of pathways), distractions (eg, whilst restraining a prisoner), wet floors, poor housekeeping (HK, eg, obstacles or slip hazards not cleared away) and poor weather (Fig 1).

Local data

Analysis of local accident records supported the information gleaned at a national level. The physical characteristics of each prison site reflected in the type of incidents seen in the accident books. Sites that were spread widely in geographical terms, for example, sites built on old RAF bases, had higher relative proportions of trip accidents on uneven outdoor surfaces, whereas sites which were contained in a smaller area had higher relative numbers of slip accidents.

Prison visits

Seven prison visits were conducted, with one day spent on each site. The visit involved discussion with prison health and safety representatives covering topics such as cleaning regimes, footwear policy, maintenance work and compensation claims. A tour of the site followed by slipperiness measurements of various floor surfaces were undertaken. Local accident records were also examined. A number of common themes emerged from the prison visits and these are described below.

Flooring

Common flooring types used on prison wings were vinyls, which generally provided poor slip resistance in water-wet conditions. In some cases, safety vinyls were used which offered improved slip resistance. There did not seem to be a common flooring specification policy in place which gave consideration to obtaining anti-slip surfaces for prison buildings.

Footwear

The footwear policy of the Prison Service is to provide officers with £50 per year to contribute towards footwear. The advice for the selection of footwear is that it should be "black and sensible". A vast array of footwear was seen, the majority was of a safety type (ie, with steel toe caps), but minimal consideration seemed to be given to slip resistance as a criterion in shoe selection. One of the prisons visited had conducted a small-scale trial of slip resistant shoes which included 16 staff members. The reaction to the shoes was very positive, and subjectively the shoes were rated to perform well in slippery conditions, for example, in the wet.

Cleaning

Prisoners normally perform cleaning duties. Mop and bucket is the most common technique used, with scrubber dryer machines being used occasionally. Scrubber driers however are prone to breaking or vandalism. A large proportion of falls are a result of officers slipping and falling on floors which remain wet after cleaning. Effective use of barriers and signs is therefore important to ensure that pedestrians cannot walk over wet areas. More consideration should be given to the drying of floors subsequent to wet cleaning which might be achieved by the use of scrubber driers. In addition, there seemed to be a lack of policy in relation to how spills (such as food and beverage spilled at meal times) are dealt with.

The quality of cleaning depends, to a degree, on the type of prisoner that carries out the cleaning work. For example, areas of prisons with Lifers tend to be kept in better condition than those prisons with Juveniles, or "holding" prisons where there is a high turnover of prisoners. Lifers tend to take more care over their environment, because they have to live in it for a longer amount of time. There were also reports of competition between prison wings to achieve the "shiniest" floors via cleaning and polishing. This practice of obtaining a high "sheen" has obvious implications for slip accidents.

Since prisoners fill the role of cleaning staff, training in appropriate cleaning techniques is an important consideration. This training need has started to be addressed by training prisoners on the BICS (British Institute of Cleaning Science) course. Supervision of cleaning duties is also an important element in ensuring quality of cleaning tasks. Supervision is conducted by prison officers, so it is important that officers are familiar with effective cleaning techniques. It is intended that prison staff will also be allowed to complete the BICS training course.

Stairs

Accident records analysis and discussions with health and safety personnel highlighted that stairs are common sites for falls. The stairs observed in the prisons varied considerably. Stair modifications to avoid falls can include the application of slip-resistant and colour-contrasting, square nosings, and also to ensure adequate lighting. A frequent occurrence is for stairs on the wings to become contaminated with food. Food and beverage is spilt on the stairs when prisoners carry their meals back to their cell. This contamination makes the stair tread slippery and may initiate a slip and fall. A simple solution to prevent the spillage of food and beverage could be to provide lids for cups and plates.

Grounds maintenance

The maintenance of floors, stairs and lighting to be kept in a good condition is important in the prevention of falls. Grounds maintenance work is usually conducted by the Works Department on site. Small maintenance jobs can be completed in relatively short time scales (days). Larger pieces of work, for example, resurfacing of car parks, require a bid to be put forward. Frustrations were aired among health and safety representatives at the delays in completing these larger jobs.

Compensation

Compensation claims for injuries sustained in accidents are becoming increasingly common among both prison officers and prisoners. Settlements can be in the order of thousands of pounds, with equivalent amounts spent on the associated legal costs. The increasing occurrence of claims can be used as an incentive to target accidents such as slips and trips, which can be avoided by applying simple, cost-effective measures.

Recommendations

A series of recommendations were designed for reducing STFs to prison service staff, these included:

- When cleaning is in progress use barriers and signs to indicate wet areas. Use cleaning methods which dry the floor where possible (eg, scrubber driers).
- Continue to train prisoners and prison officers in the BICS (British Institute of Cleaning Science) cleaning course.
- Introduce a spills policy such that at meal times, food and beverage spills are monitored and dealt with promptly.
- The use of entrance matting at doorways will help to remove moisture from pedestrians' shoes. The matting should be large enough such that there are no wet footprints left on the floor after a person has crossed the entrance mat.
- Trial a larger sample of slip-resistant footwear. Recommend that prison officers select anti-slip work shoes.
- In icy weather, grit outdoor areas in advance. Include car parks and footpaths. Keep up to date with weather conditions using local weather reports.
- Where flooring is specified for new builds or for replacement, anti-slip properties should be a major consideration, especially in foreseeably wet areas such as in kitchens and shower rooms. Education of staff to understand slip potential and slipperiness test methods will be useful for those involved in flooring procurement.
- Where possible, stairs should have colour-contrasting, anti-slip and square nosings. Consistency in stair risers and goings is also important.
- When responding to alarm calls, systems should be implemented to ensure that an appropriate number of officers attend the scene and in an organised fashion.
- Where there are building workshops in prisons, consider training prisoners in pathway laying and repairs.

Discussion

There is a hierarchy of control that should be considered when reducing slip and trip risks (HSE, 1996). The first stage should be to prevent contamination getting onto the floor in the first place. Common routes for contamination in prisons include wetness from rain and ice, wet cleaning activities and spillages of food and beverage. To prevent this contamination, practical interventions include: entrance matting to prevent ingress of moisture from inclement weather; cleaning systems to ensure that pedestrians cannot cross wet areas by use of signs and barriers, or the use of dry cleaning methods. Systems for the monitoring and clearing of spot spillages will also be useful (for example, dedicated prisoners to monitor spillages at meal times). It is also important to maintain the inherent slip resistance of the floor by using the cleaning methods recommended by the supplier, this will require a certain amount of training for cleaning personnel, ie, prisoners. If the slip resistance of the floor is too low, it may be necessary to replace the floor altogether. At this point, it may be useful for staff who are involved in procurement to have received some training which covers slip potential and the measurement of slipperiness. In relation to the wider environment, it is important that lighting systems ensures good visibility of the flooring surface. The selection of anti-slip footwear could also be used to reduce the likelihood of slips. Finally, training and information should be supplied to employees to raise awareness of slips and trips. It should be made apparent that behaviours such as running to alarm calls often result in falls. Practical measures to avoid trips include: eliminating holes, slopes or uneven ground; remove obstacles that pose trip hazards and provide suitable lighting so that obstacles can be seen.

Reference

HS(G)156. (1996) Slips and trips. Guidance for the food processing industry. HSE books.

FALLS FROM VEHICLES

Kevin Hallas, Anita Scott & Mary Miller

Health and Safety Laboratory, Harpur Hill, Buxton SK17 9JN

Over the last five years, accident reports sent to HSE and local authorities show that nearly 60 employees were killed and 5000 seriously injured in the haulage and distribution sector – simply doing their job. The greatest number of vehicle fall accidents occur when the worker is accessing or on a trailer during loading/unloading/sheeting operations The Health and Safety Laboratory (HSL) have undertaken a study of the underlying causes of falls from vehicles, encompassing access and egress routes, slip resistance of surfaces and driver behaviour. This paper focuses on the physical design aspects that affect the drivers' risk of falling, allowing recommendations for improvements to existing and new vehicles.

Introduction

Over the past five years, accidents reported to HSE and local authorities show that nearly 60 employees were killed and 5000 seriously injured in the haulage and distribution industries (HSE, 2003). A further 23,000 workers suffered injuries serious enough to keep them away from work for over three days (HSE, 2003). These figures represent a higher rate of accidents to employees in either the construction or agricultural industries, which are both regarded as hazardous industries (HSE, 2003).

Falls from vehicles represent around one third of workplace transport accidents and are spread across a wide range of industries. An analysis of HSE accident data for "goods type vehicles" (Walker, 2004) suggests that at least one third of these accidents are caused by an initial slip or trip. In Walker's analysis 448 accident reports were examined. At least 124 accidents involved a slip before the fall and 16 involved a trip before the fall. Walker also highlighted that 9% of incidents occurred when drivers were using steps.

Methodology

Surface slipperiness measures, vehicle access systems and trip hazards were recorded from 64 Large Goods Vehicles (LGVs) including 14 flatbeds, using the following methodologies.

1. Slipperiness data from the load area, tail lifts and cab steps were taken where possible. Slipperiness was indicated by measures taken with the Pendulum, a device that measures the surface coefficient of friction (CoF).
2. Surface microroughness measures were also taken where possible. Microroughness gives an indication of surface slipperiness in contaminated conditions. In general terms, the greater the microroughness value, measured in μm, the lower the potential for slip. Where it is not possible to measure using the Pendulum, Rz roughness gives a good indication of slipperiness.

3. Step dimensions (heights and depths) to the vehicle cab and the 5th wheel area were measured along with the provision and placement of handholds. Access equipment to the load area was also recorded, taking note of the materials used, their state of repair and evidence of contamination.
4. Where tripping hazards were present, details were recorded and photographed.

Findings

Cab access

The largest cab step height was always between the ground and the first step, ranging from 340 mm to 530 mm. Only four of the cabs measured had consistent step heights. The majority of step heights (46 out of 48) were less than 400 mm, with most in the range 250–400 mm. The cab step depth ranged from 130 mm to 270 mm. In general terms, the larger the step depth the better the likelihood that a user can gain a good footing on the surface. 14 of the 48 cab steps measured had step depths less than 200 mm. Most cab steps were made from metal with a profiled surface (others were plastic, plastic coated metal and rubber). A limited number of steps were seen with modified surfaces, such as abrasive tapes and high roughness fibreglass sheets.

For metal steps with highly worn areas, surface roughness measures of below 10 μm (Rz parameter) were commonly recorded. This indicates that the potential for slipping is high in wet conditions (UKSRG, 2005). Metal steps with very low levels of wear were often lightly corroded, leading to roughness values of 20 μm or more. Whilst this higher level of roughness would suggest adequate slip resistance in wet conditions, it should be considered that areas in this condition are not routinely contacted by the drivers' shoes. The most heavily worn area tended to be on the outer edge of the step where the boot rotates over the step edge. As a result, this area tended to be the smoothest, just where the friction demand is greatest. Profiled steps give the possibility of a physical interlock with the cleats of the drivers' shoe. However, this is difficult to predict because the design of the cleats on shoe soles vary in pattern, dimensions and depth.

The majority of cabs had handholds and in most cases, handholds were situated on both sides of the cab. Handhold design varied in whether they were a continuous length just inside the cab door or a more discrete, shorter length. The height of handholds varied from 1180 mm to 1760 mm from the ground. On the whole, vehicle cabs appeared to have satisfactory handhold provision.

5th wheel area

The 5th wheel is the device that couples the trailer to the tractor unit of a large goods vehicle. Drivers (and occasionally maintenance personnel) are required to climb onto the 5th wheel area, in order to link up the airlines and electrical connections from the cab to the trailer unit. There is a great deal of variation in the layout and the surfaces used in the 5th wheel area. Normally there are steps to the 5th wheel area, and often there will be a handhold.

Step height data shows the first step height was always the greatest. The lowest step height measured was 120 mm, but this varied considerably. The step depths measured in the 5th wheel area tended to be relatively small, with 100 mm being a common measure.

Due to its position next to the diesel tank on the vehicle, the surfaces in the 5th wheel area tend to become contaminated with diesel. Where steps are inset into the fuel tank, it is difficult, to both position the foot firmly in the constrained space of the foothold, and to avoid slipping due to contamination. The surfaces used in covering the 5th wheel area (including battery unit) tend to be profiled metal or plastic surfaces. Often, a "patchwork" of surface materials is used and in this situation a particular shoe may interact well with one surface, and poorly with another, giving the operator an inconsistent level of grip. Often, underfoot accidents occur when there is a significant change in available friction, such as when walking from a good, slip resistant surface to a slippery surface. Consistency in adjacent surfaces would help reduce this risk.

Consideration should also be given to the manual handling tasks undertaken in this area. Attaching the airlines for the trailer braking system requires reasonable physical force from the driver, which will increase the friction demanded of the surfaces. Sometimes, the 5th wheel region has areas that are not covered, for example between the chassis members, which leaves the driver to walk on a very uneven surface, made up of chassis rails, compressed air tanks, and the fuel tank. The surfaces of these items are generally smooth (slip hazard), and uneven (trip hazard).

Tail lifts

Mechanical tail lifts are fitted to the rear of an LGV to assist with loading operations. Observation revealed two main designs, either those folding underneath the rear of the vehicle, or those that fold up against the doors. The platforms of the tail lifts tend to be made from aluminium, with a ribbed or profiled surface.

Pendulum and roughness data suggests that tail lifts offer a low slip potential in clean dry conditions. In wet conditions, tail lifts tend to present a high slip potential. The profile of the lifts tends to be directional, so that they provide greater slip resistance walking directly to and from the vehicle, but very little slip resistance when walking across from one side of the vehicle. Commonly they are of insufficient size and the operator has to manoeuvre the load in all directions. For roadside deliveries, the load may be taken off the side of the lift onto the pavement, again necessitating the use of the surface in all directions. The tail lifts which did offer satisfactory slip resistance in wet conditions were those which were older and had suffered more wear and scratching, increasing the surface microroughness.

Two vehicles were seen with handrails guarding the sides of the tail lift, to prevent drivers from falling. In one case, the handrails were hinged from the platform, and the driver only lifted one side into place, risking tripping over the other whilst working on one half of the platform.

Load area surfaces

The materials used for the surface of the vehicle load beds varied. Many load beds were made from layered plywood, finished with a resin coating. Often the resin surface has a light profile. Surfaces used on operating goods vehicles should be capable of providing adequate friction to prevent loads from moving and to protect operators from slipping, particularly under contaminated conditions. The surfaces generally used in refrigerated load areas provided good slip resistance. Wooden planks, epoxy coatings and metal profiles are also common materials used for load beds. All the materials used present a low slip potential in clean dry conditions. When wet contamination was introduced, the slip potential was increased. On some surfaces, the slip potential remained low (Pendulum SRV > 36), but on others the slip potential was increased to a high level (Pendulum SRV < 25), and as such pedestrians would be expected to have difficulty walking on these surfaces. Most materials were measured in various states of wear, and the slip resistance varied accordingly. The materials have been ranked, from best to worst, in terms of typical slip resistance values with water contamination:

1. Resin & Aggregate
2. Square Profile Resin Plywood
3. Durbar Profile Resin Plywood

4. Hexagonal Profile Resin Plywood
5. Plywood
6. Painted Plywood
7. Wooden Planks
8. Wide Aluminium Strip.

Surface microroughness measurements show that slip potential is significantly increased around the edges of most load areas, where the surface material changes to the smooth steel structure of the vehicle. This smooth threshold is the critical area because the operator would fall from the vehicle if a slip occurred. The shape of the threshold is important: square edges are better than rounded edges because they increase the effective tread depth for pedestrians. This reduces the likely overhang of the foot or shoe, thus keeping the pedestrian's centre of gravity within their base of support. The slip resistance of these areas is improved when the surface microroughness is higher, which often occurs due to scratching of the surfaces during loading and unloading operations or light surface corrosion. Aesthetics are a consideration for manufacturers and operators, with some surfaces finished with high gloss paints, or polished metal. Vehicles were observed with modified surfaces, designed to tackle the issue of pedestrian slips. Anti-slip paints and abrasive tapes have been used to give a high roughness finish to critical areas of vehicles, such as the threshold around the load area.

The condition of the load area surfaces varied between vehicles. The surface often becomes damaged with wear and tear from the load. Forklift trucks may also dent the side impact barriers and the wooden surface of the load area. Some of the wooden planks were in a very poor state of repair with broken planks and protruding nails presenting foreseeable trip hazards. Drivers commented that the load areas sometimes become covered in a film of algae during wet weather conditions if the trailers are left outdoors.

Load area access

The only access routes to the load bed of the flatbed vehicles examined, were the 5th wheel access and the rear or side impact barriers. The side impact barriers are not considered a safe means of access since they are often positioned slightly under the load bed, so that the driver has to negotiate the lip of the load bed. Both the side steps and rear bumper on flatbeds are positioned away from any handholds, further increasing the difficulty of accessing the load. With a load in place, it is likely that the operator would use the load itself, or strapping as a handhold, which may not be strong enough to support the driver. The height of the load bed above ground is typically 1–1.3 m, thus the consequences of a fall are likely to be serious.

Observation of vehicles with contained load areas revealed that occasionally, steps or ladders are fitted to assist with access onto the load area. The robustness of these ladders or steps varies. In some cases, fold-out steps are provided, which give substantial footings up to the vehicle load. In other cases, the ladders may be relatively flimsy, and prone to damage. The condition of the steps or ladders varies, with some having been damaged over time, and others never used. It was observed that handholds rarely accompany the steps or ladders fitted on to vehicles. If the steps or ladders are positioned to one side of the rear of the vehicle, the driver may be able to use the side of the vehicle as a handhold to assist entry. Where designated steps or ladders are not present, drivers may use the rear bumper as a step. Although not designed as a step there is evidence of anti slip paints being added to the bumper.

Side impact barriers are usually made from steel or aluminium, sometimes finished with paint. Microroughness values as low as 1.4 μm were recorded for a freshly painted side impact barrier. Depending on the degree of paint wear or metal wear, the upper level of surface roughness for the barriers was 29.7 μm. The minimum surface microroughness required for slip resistance in wet conditions is 20 μm, suggesting that very few of the side impact bars observed would provide satisfactory slip resistance. The profiled surfaces tend to be very subtle, compared with cab steps for instance, so this is unlikely to improve the slip resistance through interlock with shoe cleats. The use of side impact barriers as a means of access to the vehicle bed, can be seen from evidence of wear on the barriers themselves. Barriers are often fitted with anti-slip finished suggesting that manufacturers are aware of their use as a step.

Conclusion

The risk of the driver falling whilst operating a large good vehicle could be reduced significantly by considering the drivers' interaction with the vehicle at the design stage. Designers need to acknowledge the need for drivers to access load areas, and improve access and egress provision accordingly. Vehicle operators can also reduce the risks associated with existing vehicles by making simple modifications, improving maintenance and housekeeping of vehicles.

References

HSE. *Health and Safety in Road Haulage.* INDG379, 2003 (HSE).
Walker, D. *Major accidents involving falls from goods vehicles.* HSE 2004.
HSE. *The assessment of pedestrian slip risk. The HSE Approach.* Slips1, 2004 (HSE).
UKSRG, *The Measurement of Floor Slip Resistance – Guidelines Recommended by the United Kingdom Slip Resistance Group*, Issue 3, 2005.

HUMAN VARIABILITY AND THE USE OF CONSTRUCTION EQUIPMENT

P.D. Bust[1], A.G.F. Gibb[1], C.L. Pasquire[1] & D.E. Gyi[2]

[1]*Department Of Civil And Building Engineering,
Loughborough University, Leicestershire LE11 3TU*
[2]*Health And Safety Ergonomics Unit, Department Of Human
Sciences, Loughborough University, Leicestershire LE11 3TU*

> The use of ergonomics in the construction industry is at a lower level than in office work or manufacturing. One area in which workers can benefit from is good ergonomic design of construction equipment. Previous work by the authors has revealed that the equipment designers are innovative and ready to produce equipment to assist workers by, amongst other things, removing manual handling tasks but are unable to do this within a conservative construction industry As part of a larger project looking at global construction health and safety a number of focus groups were held. Inviting leading construction equipment designers to discuss what allowances for human variability were being made in the design process. This included how the equipment is used in the UK, if it is being developed differently for developing countries and what has to be considered when foreign workers use the equipment on UK sites.

Introduction

The UK Government's Health and Safety Executive (HSE) reported that 3900 out of 100,000 people whose current or most recent job was in construction in the last eight years suffered from an illness which they believe was caused or made worse by this job (HSE 2005). The use of equipment in construction is a major contributor to these figures.

Nearly 5 million people are exposed to hand-transmitted vibration in a one-week period in the UK. Many of these work in the construction industry where hand-held power tools are widely used. These are a major source of vibration exposure, which can lead to hand-arm vibration (HAV) injury (HSE 1999).

Being struck by a moving vehicle accounted for 15% of all construction workplace fatalities in the year 2002/03 which was second only to fall from heights (46%) (HSC 2003). Fatalities have also occurred from maintenance of the hydraulic parts of heavy earth moving equipment and accidents frequently occur during loading and unloading of construction equipment from support vehicles.

Failure to use personal protective equipment (PPE), its misuse or use of the wrong type also has a an impact on the number of injuries (eye damage, cuts and lacerations) and potential health problems (respiratory, dermatitis, hearing loss).

Any current attention to ergonomics in the development of construction equipment is restricted. Consideration in the development process of safety, health, physical workload and productivity is of a lower standard in the construction industry and there are opportunities for improvement (Vedder and Carey 2005).

We therefore need to investigate how the design of construction equipment is currently being carried out before we go on to advise how to apply ergonomic principles to best effect. The design of construction equipment used to replace manual handling

Table 1. Focus Group Exercise Results

Do the following worker/operator characteristics have an impact on the overall design of construction equipment? – 100% = greatest impact

Characteristic	Score %	Give example/s of equipment affected
Hand size	92	Power tools – trigger. Vehicles – joystick, switches and levers; respirators; ear plugs
Height	75	Vehicles – pedal reach, operator access, view Powered respirator; Floor breaker
Weight	75	Vehicles – suspension seating, access/egress Any belt mounted PPE Clothing suits
Memory	75	Contiguous design consistency; intuitive control; Some specialist power tools; most PPE
Gender	58	Reach and strength. Most things designed for men PPE
Age	67	Optimum at 40 Affects competence and fitness
Handedness	67	Hand tools Dominant population customer base
Risk awareness	92	All equipment for full lifecycle (packaging through to use)

operations was considered in previous research (Bust et al 2004) and the use of equipment was identified as a significant factor in the management of health and safety at the beginning of research into construction health and safety in developing countries. An investigation of its effect was then included in focus group work and as part of a questionnaire survey.

Methods

Focus Groups

Three focus group meetings, with a number of industry professionals (total n = 18), were held to discuss topics associated with work in developing countries. The areas covered included culture, language, use of construction equipment, skills and training.

An exercise was used to provide a break around half way through the focus groups. Members of the group were asked about the relevance of worker/operator characteristics to the design of construction equipment and to give examples of what equipment would be affected. See table 1.

Meetings were recorded and later transcribed. The transcripts were used to identify themes and salient points which are covered later in this paper.

Questionnaire

A questionnaire was distributed to industry professionals with experience of working in developing countries. The main section of the questionnaire consisted of 20 questions which were derived at a workshop with steering group members from the European Construction Institute Safety Health and Environment Task Force.

Four questions each were constructed, on a five point Likert scale, with aspects relating to individuals, tasks carried out, equipment used, organisational issues and the environment worked in see table 2. Results were coded so that overall positive and negative responses to the questions were shown and a Pearson's product-moment correlations comparison carried out to identify any similarities between the responses to the questions.

Table 2. Equipment Questions in Questionnaire.

No.	Question
3	Construction vehicles were used in a safe manner
4	Workers were able to provide the standard of electrical work required
11	Equipment normally used in developed countries was not available
17	The workers use of power tools was satisfactory

Findings

Focus Groups

A summary of points raised in the focus groups is as follows:

- Behaviour – it was said that equipment designers assume that the their equipment will be abused on construction sites and try to build this in as far as possible to their design. Whilst they make the machines as safe as they can it has been noted that the users tend to take more risks with the equipment as it becomes safer to use.
- Variability – designers have to strike a balance between designing something that is safe to use by a population and training the population to use the equipment safely. It was also commented that equipment is still predominantly designed for use by men.
- Organisational – Simple systems (No Hat – No Work) of safe use are required to aid enforcement/supervision of the use of equipment because of widespread violations by the workforce. Designers are unable to counter all possible misuse of equipment so a multifaceted (management system, culture, education/training and support of good behaviour) approach is required.
- Global – While UK companies are asking for power tools design to consider vibration levels, their mainland European counterparts are mainly interested in the levels of power available. Developing countries are starting to get to grips with safety aspects of equipment but health issues are not even "on the radar". Developing countries are starting to adopt European systems for safety so equipment manufacturers are selling the European regulations along with their goods.
- Ergonomics – Some of the more successful equipment manufacturers in the industry have adopted the use of ergonomics principles –
  – Mechanical handling – iterative approach
  – Power tools –
  – MEWPs – varied operation of equipment for different competence level
  – Excavators – user groups and questionnaires
  – PPE – colour coding of equipment for easy identification of correct usage

The results from the focus group exercise can be seen in table 1. The highest scores for impact on design of worker characteristics were for hand size and risk awareness while the lowest score was for gender.

Questionnaire

There were 87 responses to the questionnaire from construction professionals with experience of working in 36 different developing countries. Their overall positive and negative responses were more pronounced for questions relating to equipment (safety leaning) than to the individual (health leaning). See Figure 1.

Out of the 20 questions from the main section of the questionnaire there was significant correlation between the equipment questions, between the environment questions and between the equipment and environment questions.

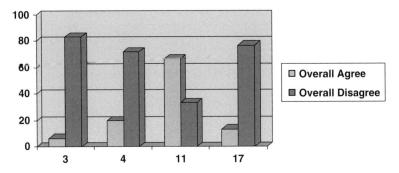

Figure 1. Responses to Equipment Questions.

Discussion

A UK based study of 100 accidents in construction states that shortcomings with equipment, including personal protective equipment were identified in over half (56%) of the incidents (Haslam et al 2005). The results of this research referring to the expected abuse of equipment by workers and unsafe use of equipment in by workers in developing countries would seem agree with this.

Many work tasks and associated equipment and tools in the construction industry are not designed with ergonomic principles in mind (Vedder and Carey 2005). However, the designers in the focus groups identified a number of issues relating to worker variability and behaviour and recommended organisational requirements to the management and supervision of construction equipment and some of the more successful equipment manufacturers in the industry have adopted the use of ergonomics principles.

It is interesting to note that UK equipment design requirements are being influenced by litigation (Vibration white finger claims) and that European equipment is being sold as a package with European safety legislation to developing countries.

The focus group exercises indicated that physical characteristics (hand size, height and weight) as well as behavioural /cognitive characteristics scored higher than gender which reinforces focus group comments that construction equipment is mainly designed for men.

The questionnaire results explain that use of equipment in developing countries has a greater response, coupled with the environment it is used in and the responses significantly related. The questionnaire results showed that health and safety issues in the developing countries more readily identify safety issues relating to equipment use before health issues relating to individuals.

Conclusions

Construction equipment as designed by some of the leading manufacturers makes allowances for variation in workers' physical size and behaviour but still largely for use by a predominately male worker population.

Simple solutions for health and safety management are required to allow for abuse of equipment by construction workers and the design for safe equipment needs to be supported by a multifaceted approach.

References

Bust, P.D., Gibbb A.G.F., Haslam, R.A. 2005. Manual handling of highway kerbs – focus group findings. *Applied Ergonomics,* 36, 417–425.

Haslam, R.A., Hide, S.A., Gibbb A.G.F., Gyi, D.E., Pavitt, T., Atkinson, S., Duff, A.R. 2005. Contributing factors in construction accidents. *Applied Ergonomics,* 36, 401–415.

HSE 1999, *CRR232/1999 Hand-transmitted vibration: occupational exposures and their health effects in Great Britain* Her Majesty's Stationery Office.

HSE 2005, *Health and Safety Statistics 2004/05* Her Majesty's Stationery Office.

HSE 1999, *The High 5 – Five Ways to Reduce Risk on Site* Her Majesty's Stationery Office.

Vedder, J., Carey, E. 2005. A multi-level systems approach for the development of tools, equipment and work processes for the construction industry. *Applied Ergonomics,* 36, 471–480.

THE RELATIONSHIP BETWEEN RECORDED HOURS OF WORK AND FATIGUE IN SEAFARERS

Paul Allen, Emma Wadsworth & Andrew Smith

Centre for Occupational and Health Psychology, Cardiff University, 63 Park Place, Cardiff CF10 3AS

A range of seafarers working on different vessels were asked as part of a questionnaire survey whether they had ever deliberately under-recorded their working hours to comply with regulations. Compared on six measures of health and fatigue the 40% of respondents who admitted at least occasionally under-recording their working hours were found to be significantly more fatigued/less healthy than the 37.3% who reported never under-recording. Whilst the percentage of a predominantly British sample who admitted under-recording working hours is extremely worrying, the fact that those who reported under-recording were also found to be on average more fatigued and less healthy raises serious questions concerning the credibility of the current system.

Introduction

When approaching the issue of seafarers' fatigue working hours have often been a central area of focus as reflected in current legislation. Following implementation of the EC (European Community) working time directive in June 2002, a record of hours of work/rest now needs to be kept for all seafarers working on EC flagged vessels. As part of the third phase of the Seafarers' Fatigue Project a question was included in the survey questionnaire to assess the extent to which seafarers might be working in excess of working hours regulations whilst doctoring records to hide infringements. In addition to assessing the extent of under-recorded working hours, analysis was conducted to consider how under-recording seafarers might fair in terms of fatigue and health outcomes, key indicators of success when evaluating welfare targeting regulation.

Method

In phase three of the Seafarers' Fatigue Project survey data were collected from deep sea (n = 331), short sea (n = 129) and offshore support (n = 97) volunteers. The questionnaire employed was designed to address all key areas of seafaring life as well as incorporating a number of standardised health and fatigue scales and a question directly addressing the issue of under-recorded working hours, as follows:

> *To your knowledge, have your working hours ever been deliberately under-recorded in order to comply with international regulations on hours of work and rest? Don't know/Never/Occasionally/Frequently/Always*

Prevalence of under-recorded working hours

From the cross sector sample 37.3% reported never under-recording working hours, 28.3% reported under-recording occasionally, 11.7% reported frequently or always under-recording and 22.8% answered "don't know". Whilst the 40% of seafarers who reported at least occasionally under-recording working hours might appear high, it must be considered that in reality the figure is almost certainly higher with many seafarers reluctant to admit regulation infringement even in an anonymous questionnaire.

One of the most alarming facts about the prevalence of under-recorded working hours in the current survey is that the sample in question represents what could arguably be described as the better end of the industry. From the sample of 558 seafarers 75.2% reported working on British flagged ships, 94.0% were British/Irish, 94.3% were officers and 70.2% earned more than £30,000 a year. With 40% of such a sample of highly paid, well trained and highly ranked seafarers admitting to under-recording working hours it is not difficult to imagine the situation being considerably worse elsewhere.

Working hours recording and fatigue

For the purposes of analysis the sample was divided into those who had at least occasionally under-reported working hours (n = 223) and those who never under-reported working hours (n = 208) with the comparison based on a range of fatigue and health scores. The groups were compared in terms of three fatigue scales derived from survey questions (fatigue at work, fatigue after work and fatigue symptoms), the profile of fatigue related symptoms fatigue scale (PFRS-F), the cognitive failures questionnaire (CFQ) and the general health questionnaire (GHQ). On all six comparisons the group who reported under-recording working hours were shown to be significantly more fatigued/less healthy than the non under-recording group, as shown in table 1 below.

In terms of accounting for the result shown in table 1 it might be suggested that under-recording is associated with a particular sub-group of seafarers however analyses were conducted which challenge this proposition. The under-recording and non under-recording groups were compared in terms of a number of key factors and it was shown that in terms of nationality, flag of vessel, job type and tour classification the two groups showed no significant differences. Out of all the key comparisons only one was found to be significant with those in the under-recording group earning less on average than those from the non under-recording group (chi square = 10.19 (2), $p < 0.01$).

Table 1. Fatigue and health scores for mis-recording and non mis-recording groups.

Scale	Non under-recording Mean (SE)	Under-recording Mean (SE)	F (1 df)	P (ANOVA)
Fatigue at work	3.44 (.06)	3.64 (.05)	7.44	<.01
Fatigue after work	2.33 (.03)	2.58 (.03)	26.16	<.001
Fatigue symptoms	2.57 (.05)	3.09 (.05)	54.62	<.001
PFRS-F	24.67 (.86)	27.29 (.80)	4.95	<.05
CFQ	33.90 (.88)	36.93 (.78)	6.74	<.05
GHQ	1.15 (.16)	1.80 (.17)	7.18	<.01

(Note: for all scales a higher score = higher fatigue or poorer health status).

Discussion

It is clear that the current system for recording seafarers' working hours is fundamentally flawed with company intermediation preventing honest disclosure. Seafarers are currently finding themselves in an impossible situation where, through unspoken sanction of employment prospects, they are forced to under-record how long they work to keep peace with legislators. The problem is, however, that without any honest disclosure of working hours there is no warning light to come on for enforcement authorities to spot, leaving the industry to deteriorate behind a façade of compliance. No seafarer is willing to make a stand against the current status-quo because to do so would only single themselves out as a "non-coping" individual by comparison to other employees. With many seafarers required, when necessary, to "flog" working hours sheets, a warped picture emerges concerning the state of the industry with the definition of "good practice" skewed by misrepresentative paperwork. The current research shows that not only are a large proportion of seafarers under-recording working hours, but that seafarers who under-record are actually more fatigued and less healthy than their non-under-recording counterparts. If the recording of working hours was brought in as a proxy means of assessing the health and welfare of seafarers then it appears the procedure is failing. A study by the Marine Accident Investigation Branch (MAIB) on bridge watchkeeping came to the conclusion that "…the records of hours of rest on board many vessels, which almost invariably show compliance with the regulations, are not completed accurately" (p.13) and the present results confirm this. The requirement for employees to work compulsory over-time is undesirable but necessary on occasion, however when the same employees are obliged to present records with fictitiously reduced schedules of work the situation might be classed as exploitative. Ironically, the very completion of working hours sheets appears to achieve little more at present than increase the work load for those whom the system was designed to monitor and potentially help.

Acknowledgments

This research is supported by the Maritime and Coastguard Agency, the Health and Safety Executive and NUMAST. We would like to acknowledge the contribution made by those who participated in the research. The views expressed here reflect those of the authors and should not be taken to represent the views or policy of the sponsors.

Reference

Marine Accident Investigation Branch (MAIB). 2004, *Bridge watchkeeping safety study.* (Department for transport, London)

OIL, GAS AND CHEMICAL INDUSTRIES SYMPOSIUM

THE IMPACT OF PSYCHOLOGICAL ILL-HEALTH ON SAFETY

Chiara Amati & Richard Scaife

The Keil Centre Limited, 5 South Lauder Road, Edinburgh EH9 2LJ

This paper describes a project which aims to understand the impact of psychological ill-health on safety performance, and what can be done to mitigate these effects. The project focuses on anxiety and depression – often reported as "stress" – which are the main causes of psychological ill-health in the UK working population. The effects of anxiety and depression can be similar (e.g. negative impact on memory) or opposite (e.g. anxiety can lead to quicker reactions, or over-reactions, whereas depression can lead to slower responses and lack of sufficient reactions). Both anxiety and depression affect cognitive processes and therefore have negative impacts on decision-making. The work on this project is on-going and due for completion in summer 2006.

Introduction

In the UK, there has been a recent increase in attention to psychological ill-health at work, and work-related stress in particular. The UK Health and Safety Executive (HSE) has funded extensive research and published guidance for employers in this area (see Smith et al, 2000). The research has resulted in a good level of understanding on the main "risk factors" that can cause stress at work.

The last few years have also seen a growing consideration of the impact of human performance on accidents and safety-related behaviours, focussing on the impact of factors external to the individual on safety performance (e.g. safety culture, quality of procedures, equipment, etc), and on the way personality or attitudes may impact on safety. Overall, the first field of enquiry has arguably proved more fruitful.

What is missing is a comprehensive understanding of the links between psychological wellbeing, human performance and safety. This should address two separate issues, whether psychological ill-health at work can impact on safety behaviour, and how the presence of sources of work-related stress may affect safety performance. This article presents the research done to date on these two issues; the work is on-going and due for completion in summer 2006.

Prevalence of psychological ill-health

The estimates of the extent of psychological ill-health in the UK population vary, but a report from the Mental Health Foundation (Bird, 1999) concluded that:

- One in four people will experience some kind of mental health problem in the course of a year;
- Between 10 and 25% of the general population present with mental health problems every year;
- One in six people suffer from depression and 1 in 10 people suffer from anxiety in the UK.

15% of the general population with mental health problems will be working and are more likely to suffer from common neuroses – depressive and anxiety disorders. Evidence for the prevalence of psychological ill-health at work indicates that:

- Approximately 26% of the UK working population exhibit symptoms of minor psychiatric disorders;
- Approximately 20% consider their work to be very or extremely stressful resulting in a negative impact on their health;
- Approximately 16% of managers report having taken time off work in the last year due to stress;
- An average of 0.5 day's absence a year is caused by stress;
- The most common forms of psychological ill-health are depression and anxiety.

These figures indicate that psychological ill-health is a significant problem in the UK working population and that it is likely that the a significant proportion of people experiencing symptoms are still at work.

Psychological ill-health and performance

Since the most common forms of psychological ill-health in the UK working population are depression and anxiety, this research therefore focussed on these conditions. The following is a summary of the main findings to date.

Anxiety

The human body is designed to react when we perceive a real or potential threat, to help us either fight the threat or flee from it (the "fight or flight" response). When we react to a valid stimulus, this is healthy. When we react in the absence of a valid stimulus, or when the individual experiences difficulty returning to a "normal" state, this can be unhelpful. The experience of high arousal combined with feelings of worry (e.g. fear of failure) tends to be classed as anxiety.

Typically, anxiety can affect performance in a number of ways (Eysenck, 1983, Eysenck, 2000, Hanin, 1980):

- Change in focus of attention (Increase in focus on the perceived source of threat, decreased ability to divide attention, decreased ability to integrate information from more than one source, leading to poor judgements);
- Decrease in attentional capacity (e.g. Sephton et al, 2003). (Decrease in the amount of information that can be held in short-term memory, reduction in the ability to recall information from long-term memory);
- Increased perception of events as threats (increased reactivity to real threats but more "false alarms", increased conservative judgements, greater expressions of aggression, increased the number of errors, increased number of irrelevant thoughts);
- Decreased efficiency in task performance, increasing feelings of fatigue.

There is also some evidence to suggest that anxiety can increase the likelihood of certain types of musculoskeletal disorders (MSDs) (Devereux et al, 2004).

The effects of anxiety and high arousal vary from task to task, dependent on the specific skills required for accurate performance, and complexity of the task. For example, a highly complex task requiring integration of information from multiple sources would be more severely affected than a simple task.

Depression

When diagnosing an individual with depression, clinicians would usually be looking out for the experience of the majority of the following symptoms on a regular basis:

- Fatigue or loss of energy;
- Depressed mood most of the day;
- Significant weight loss or gain with no particular attempt to do so;
- Psychomotor agitation or retardation;
- Recurrent thoughts of death;
- Feelings of worthlessness or excessive guilt;
- Diminished interest or pleasure;
- Insomnia or, to a lesser extent, hypersomnia;
- Diminished ability to think or concentrate or indecisiveness.

Sufferers of depression often shy away from social contact so common implications are: difficulty communicating with others, avoiding non essential communication and interaction, difficulty remaining calm under pressure, difficulty getting on with others.

The effects of depression on performance tend to include:

- Negative cognitive focus (tendency for biased recall of information, greater recall of negative (mood-congruent) information);
- Decreased general cognitive resource (difficulty sustaining attention, frequent attention shifts, greater number of errors, decreased ability to recall information);
- Decrease in energy/general resources (significant decrease in motivation to perform tasks well, significant increase in fatigue).

Evidence suggests that performance on both simple tasks (for example needing fast reaction time) and complex tasks (requiring co-ordination of effort or decision making) will be negatively affected in the depressed worker (Martin et al, 1996; Wang et al, 2004).

Psychological ill-health and accidents

Humans are strongly influenced by the conditions in which they operate, and certain conditions are known to have an affect on performance. It is theoretically possible that the likelihood of these conditions negatively influencing performance could be exacerbated by symptoms of psychological ill-health, hence the likelihood of error could be increased.

Evidence gathered to date from the USA National Transportation Safety Board database of airline accidents suggests psychological ill-health is often implied in the investigation of accidents. It contains 44 accident reports where anxiety, depression or taking associated medication were cited. In around 50% of these, psychological illness[1] was either a causal or contributory factor. Of these, 30% were suspected suicides. Excluding the suicides, 64% of the accidents involving psychological ill-health resulted in fatalities. In 17% of the analysed investigations, anxiety was a significant factor; of these, nearly 60% referred to an occurrence of panic in either the pilot or the passengers. These occurrences tended to be those with less severe consequences.

Overall, the review of this database suggested that accident reporting for psychological ill-health does not fit with the accident triangle – all accidents seem to be very serious, suggesting under-reporting of minor accidents.

---

[1] Defined for the purpose of these analyses as either: knowledge that the individuals involved suffered from depression or anxiety; or presence of related medication in their body.

Discussion

The results of the research to date suggest that psychological ill-health definitely has an influence performance, and that this influence can be sufficiently powerful to result in accidents. This has a number of implications for safety managers, line managers and individuals working in safety-related roles.

Psychological ill-health can influence the occurrence of accidents by affecting the ability of the individual to perform effectively and timeously and their ability to think and interpret information accurately. Psychological ill-health may also interact with other task factors to increase the likelihood that the person could make an error. The effects of psychological ill-health can also affect non-clinical populations (i.e. people from the general population who have not been clinically diagnosed with a form of psychological ill-health).

The organisation needs to be aware when personnel are experiencing symptoms of psychological ill-health, in order to be able to monitor the work of the individual. In addition, the individual should be assisted in developing greater understanding of common psychological states that are more likely to lead to errors or accidents.

The characteristics of tasks more likely to be negatively affected by psychological ill-health are:

- Having to complete multiple or complex tasks or divide attention;
- Having to concentrate for long periods of time;
- Working under high levels of noise or distraction;
- Having to use complex procedures;
- Having to co-ordinate work with other team members;
- Not having had recent training for critical aspects of the task;
- Tasks for which the individual is inexperienced;
- Tasks where environmental factors are already known to influence performance.

The next phase of the project will involve integrating the literature reviewed to date with findings from an industry partner accident database. Specifically, the project will focus on actual accidents that have been caused by psychological ill-health to identify common situations, psychological states or similar that are commonly contributing to incidents.

References

Bird, L (1999) *The Fundamental Facts...all the latest facts and figures on mental illness*. The Mental Health Foundation. ISBN 0 901 944 637

Devereux, J et al (2004); "*The role of work stress and psychological factors in the development of musculoskeletal disorders – The stress and MSD study*"; HSE Research Report 273

Eysenck, MW (1983) *Anxiety and Individual Differences* in Hockey, GRJ (Ed) *Stress and Fatigue in Human Performance* John Wiley & Sons Ltd: London

Eysenck, M W (2000) *Anxiety and cognition : a unified theory* Hove : Psychology Press

Hanin, YL (1980) *A study of anxiety in sport* in WF Straub (Ed) *Sport Psychology: An Analysis of Athletic Behaviour* Movement Publications: NY

Martin, JK; Blum, TC; Beach, SRH & Roman, PM (1996) *Subclinical depression and performance at work* Social Psychiatry & Psychiatric Epidemiology 31: 3–9

Sephton, SE; Studts, JL; Hoover, K; Weissbecker, I; Lynch, G; Ho, I; McGuffin, S & Slamon, P (2003) *Biological and psychological factors associated with memory function in fibromyalgia syndrome*. Health Psychology, 6, 592–597

Wang, PS; Beck, AL; Berglund, P; McKenas, DK; Pronk, NP; Simon, GE & Kessler, RC (2004) *Effects of Major Depression on Moment-in-Time Work Performance* American Journal of Psychiatry, 161: 10

CHECKING FAILURES IN THE CHEMICAL INDUSTRY: HOW RELIABLE ARE PEOPLE IN CHECKING CRITICAL STEPS?

Jamie Henderson & Steve Cross

Human Reliability Associates, 1 School House, Higher Lane, Wigan WN8 7RP

The last line of defence against adverse incidents in many industries is a human check. People are used at critical points in processes to verify that inputs, processes and outputs match planned conditions, and also to capture previous failures before they result in costly incidents. This paper describes, by means of a case study, one way in which checks can fail. Possible reasons for the failure are given and new design principles for reducing the likelihood of checking failures are presented.

Introduction

The last line of defence against adverse incidents in many industries is a human check. People are used at critical points in processes to verify that inputs, processes and outputs match planned conditions, and also to capture previous failures before they result in costly safety, quality or environmental incidents.

This issue is of particular importance for plants, such as those in the chemical and pharmaceutical industries, which operate batch processing systems. This paper describes a charging incident with potentially significant consequences that occurred at such a plant, although this was a serious issue, the particular incident did not have any direct consequence for quality or safety. The case study illustrates how the value of a check can be systematically eroded by a combination of human factors issues. The final part of the paper provides guidelines for managers designed to reduce the likelihood of checking failures.

Description of incident

In this incident the wrong chemicals were charged to a process vessel. Space constraints in the storage area meant that mixed pallets of material had to be stored together, on one occasion, a storeperson inadvertently delivered a mixed pallet to the charging area. Formal checks had been designed to capture this type of error. However, the checks failed and the charge took place. Although the error was recognised, significant costs were incurred in saving the batch.

Following this incident Human Reliability was invited to conduct a human factors review of the charging process. The review included a task analysis, error analysis and review of performance influencing factors. Task observations were undertaken and interviews conducted with operators, team leaders and managers. The review yielded some valuable insights into the factors affecting a successful checking system.

Planned defences against mischarging

Prerequisites of a successful charge are as follows (EPSC, 2003):

- The right substance needs to be present.
- The right amount of substance needs to be present.
- The substance needs to be added to the correct location.
- The substance needs to be added at the correct time.
- The process conditions need to be correct.

For this process, the main focus was on ensuring that the right substance was present. The defences took the form of independent, repeated checks made by operators at several points during the charge. In addition, some checks took place after the charge to mitigate the consequences of a failure. Table 1 focuses on the pre-charge identity checks.

The procedure specified that each of the four checks should be repeated by a second operator, prior to the charge, and that each operator should sign their name to confirm the check. This practice is known as the *second signatory* or *four-eyes-principle* (EPSC, 2003). It aims to ensure an independent confirmation of the substance identity.

Actual defences against mischarging

The interviews and observations of the charging process made it clear that there were substantial deviations from the checking procedures. The main deviations were:

- The four checks were rarely repeated by a second operator.
- The barcode identity check was either not carried out at all, or not repeated by a second operator.

These deviations had significant implications for the reliability of the process. However, from the operators' perspective, there were clear, sensible reasons for their actions.

Possible reasons for deviation from specified practice

The four checks were rarely repeated by a second operator

This was a result of operators working in pairs and doing checks as a team (i.e. rather than check the identity *independently* they were *both* present when the check was carried out). There were three main factors that contributed to this way of working.

Table 1. Pre-charge checks specified in charging procedure.

Check	Description
1. Visual identity check	Visual comparison of the substance details on the drum label with the substance details on a process sheet.
2. Quantity check	Visual check that the substance amount present tallies with the amount listed on the process sheet.
3. Barcode identity check	Scan of barcode labels on each drum using a hand held scanning gun. Identity of material appears on the gun's screen and is compared with the identity described on the process sheet.
4. Barcode quality check	With the same scanning gun used for Check 3, the operator checks that the substance is within its expiry date and that it has been approved for use.

Firstly, the procedure required two people for the job but did not explicitly state that they should work independently. In a busy working environment, the most efficient way of organising the work was to find a second person and take them to the charge preparation, rather than start the preparation and then find a second person.

Secondly, the value of independent checks for system reliability was neither dealt with explicitly in the procedure nor communicated to the operators via training.

Finally, when operators signed the process sheet, their perception was that they were signing to take *responsibility* for the material to be charged, rather than confirming that *two independent checks* had been made. This perception was reinforced by the use of the term "countersign" on the process sheet. This term favoured an unquestioning approach by the second operator and reduced the likelihood of the repeat check taking place.

The barcode identity check was not carried out at all, or repeated by a second operator

There were three main reasons for this omission. Firstly, there were a number of known issues surrounding the credibility of the barcode scanning system. In particular, it was frequently offline due to routine system maintenance, often during the middle of a shift when charging typically takes place. Despite this unavailability, there was no expectation that the charge should stop until the system was available again. Therefore, operators were told that the barcode system was an important identity check, but expected to carry on when it was not working. Additionally, the barcode scanning guns were notorious amongst operators for usability problems.

Secondly, following the mischarging incident, the superiority of visual identity checks over barcode identity checks had been strongly emphasised to the operators. One hypothesis was that the barcode scanning system, only recently introduced, had been seen by the operators as a replacement for the visual identity checks. In fact the principal reason for its introduction was for stock control purposes and its use as a tool for checking identity was a way of maximising the benefit from a new tool.

Consequences for reliability of charging process

These differences had significant implications for the reliability of the charging process. As a result of the lack of independence in the checking process and the absence of the barcode identity check, the likelihood of undetected error was substantially higher than expected. This is illustrated in Figure 1.

The expected probability for the failure of the identity check is a product of the four individual failure probabilities, assuming independence between the checks. If the procedure is followed, the expected overall failure probability is 0.0001, or one in ten thousand. However, as a number of the checks are likely to be omitted, the actual failure probability is significantly higher, as shown in Figure 1, at 0.1, or one in ten.

It should be stressed that the numerical values, whilst derived from the most relevant sections of a standard human reliability text, are designed to be illustrative. Whatever the actual failure rate for this specific task, however, it is clear that the actual probability may be up to three orders of magnitude worse than the expected probability.

Guidelines for preventing human checking failures

Independence

A major problem with the specified procedure for charging in this case study was a failure to take account of the potential influence of *human error dependency*. Human

Illustrative expected error probabilities of the identity check

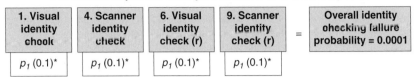

Illustrative actual error probabilities of the identity check

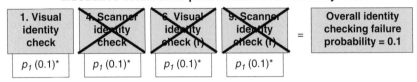

Figure 1. Illustrative error probabilities for the identity check (p_1 = expected probability of failure of check).[1]

error dependency may be understood as the degradation of the independence of checking as a result of over-reliance on either other people or on equipment (HSE, 2001). A number of factors are known to increase human error dependency, these include: checks that are done jointly, deference to colleagues, trust, peer pressure, check fatigue, erosion of manual checks following the introduction of technology and lack of reinforcement.

In designing a checking procedure, these factors must be taken into account. There are also a number of strategies that can be applied to increase independence:

- Require checks to be carried out at different times.
- Require checks to be carried out by unrelated people (for example, in this context, team leaders, storepeople or a designated checker).
- Where checks are carried out together, introduce a communication protocol and associated training to increase the consistency of the checking process.
- Provide training (for example, training in human error dependency).

Active linking

Where several items must be checked, a major failure mode is an incomplete check. For the charging process, this is the failure to check every drum. Actively linking the check action to the checked object should significantly reduce this likelihood. As an example, the procedure could require operators to interact with each object to be checked such that they have to record a unique identifier or add/remove something from the checked item. However, this approach should be followed with caution (see discussion on *Simplicity*).

Human-centred design

The design task design should try to take into account the operators' preferred working method. In this case, it was observed that the charging of a vessel would typically be carried out by two operators. This seemed to be the preferred and most efficient method.

---

[1] From chapter 20 of NUREG 1278, the most appropriate table of error probabilities is 20–22, Estimated Probabilities That A Checker Will Fail To Detect Errors Made By Others. This gives the probability of failure in checking routine tasks with written materials as 0.1 (p_1).

Clearly, however, conforming to worker-preferred systems may be in conflict with other necessary design requirements – in this case the requirement for independence.

Simplicity

Complex systems are far more likely to fall into disuse or be abused than simple systems. This is particularly true where the complexity does not add appreciable benefit for the users. In addition, complexity can also introduce additional systems that require additional resource. For these reasons, proposed changes to the system must be assessed to see whether the resource burden it introduces outweighs the benefits it brings.

Emphasis on critical steps

Proper emphasis must be given to critical steps in a task. This can be achieved by singling out the most critical checks for special treatment (for example, by using double checks for only these steps). It may also be possible to increase emphasis on critical steps by removing checks that are designed to prevent less significant failure modes.

References

European Process Safety Centre (2003) *Misidentification of chemicals: causes and precautions for storage, transport and production plant.* International Section of the ISSA on the Prevention of Occupational Risks and Diseases in the Chemical Industry: Heidelberg, Germany.

HSE (2001) *Preventing the propagation of error and misplaced reliance on faulty systems: A guide to human error dependency.* Offshore Technology Report 2001/053.

NUREG/CR-1278 (1983) *Handbook of Human Reliability Analysis with Emphasis on Nuclear Power Plant Applications*, Nuclear Regulatory Commission: Washington D.C.

EARLY HUMAN FACTORS INTERVENTIONS IN THE DEVELOPMENT OF AN FPSO

Jonathon Fisher & W. Ian Hamilton

*Human Engineering Ltd, Shore House, 68 Westbury Hill,
Westbury-On-Trym, Bristol BS9 3AA*

A New Floating, Production, Storage and Offloading (FPSO) vessel is due to be commissioned off the coast of West Africa in 2007. Members of the project's operations development team raised concerns over the proposed manning levels of the central control room (CCR). A predictive workload study was conducted in order to identify the optimum complement of technicians for the CCR. A predictive human error analysis was also conducted on four critical activities undertaken by CCR personnel. The analysis highlighted several significant risks to both production and the environment that could cost the oil field operating company substantial amounts of money. The study predicted that the proposed manning level would be insufficient for efficient management of the new offshore facility. That a change in manning levels has been identified two years prior to the commissioning of the new vessel proves the benefit of early human factors interventions.

Introduction

Where the ocean is too deep for a conventional oil rig a Floating, Production, Storage and Offloading (FPSO) vessel is often used. An FPSO is a large ship with oil extraction and process facilities built onto the deck. The processed oil is stored in large tanks in the hull and offloaded every three to four days by visiting oil tankers.

Commissioning and the first 12 to 18 months of operation is the busiest period during an FPSOs lifecycle. During this time the frequency of maintenance, equipment testing, equipment failure, new well commissioning and general "teething" problems are at their highest. In addition operators will be learning how the new wells and oil reservoirs behave in terms of oil flow. The CCR technicians are responsible for coordinating all maintenance and process related activities across the vessel. It is therefore of paramount importance that the manning level of the CCR is correct when the FPSO enters commissioning. A failure to establish a suitable manning level for the CCR when the FPSO goes into operation could result in unnecessary shutdowns and place unacceptable workload on CCR personnel. Human Engineering Limited was commissioned to provide a quantitative prediction of the workload within the CCR, to indicate any necessary changes to CCR DCS[1] equipment, and to conduct human error analyses on a series of critical scenarios identified by the project team.

---

[1] Distributed Control System.

Method

In order to quantify the anticipated workload of the CCR a timeline analysis method called the Time Pressure Assurance Method (TPAM) was used.

TPAM

TPAM evaluates workload as a percentage of time that an operator is required to perform tasks (total time occupied divided by total time available). The technique uses a conceptual model of a control room technician's (CRT) job role. It identifies two key elements to the CRT job role; the physical tasks and the cognitive tasks.

The physical tasks are the elements of the CRT job role that can be observed. These are event driven tasks performed by the CRT such as reading pressures and temperatures, checking the status of fire and gas detectors, valve manipulations, issuing/checking maintenance permits, communications and alarms handling.

The cognitive tasks are the elements of the CRT's job role that cannot be observed. These are the on-going monitoring and supervisory activities such as planning, anticipating and decision making, which constitute the majority of the tasks performed by the CRT.

TPAM uses objectively measurable data for the estimation of an operator's "busyness". The "busyness" score is calculated as the total percentage of shift time that is taken up with performing event driven duties.

Task timing estimates for event driven tasks are developed from a database of known values or are calculated from engineering psychology equations (Card, Moran and Newell, 1983). Total time occupancy within the CCR is then calculated by multiplying task frequencies by task durations.

Alarms and communications handling form an important and large part of the CRT job role. However, they are very difficult to predict until the FPSO is being commissioned and a history of the behaviour of both equipment and oil reservoirs can be established.

The majority of alarms require the CRT to carry out an action such as communicating with an outside operator, checking pressures/temperatures/levels/flows or changing a control set point. EEMUA 191 (1999) states that (based on user experience) the minimum average time performing actions relating to an alarm is one minute (during normal operations).

Table 1 shows benchmark levels for assessing average alarm rates (taken from EEMUA 191).

At the time of the project the FPSO operations team was undertaking an alarms rationalization process to create an alarms system that will be optimised for the CRT. The aim was to produce an alarms system which meets the EEMUA 191 guideline of no more than 6 alarms an hour (during normal operations). An early review of the alarms rationalisation provided a good indication that the operations team was close to meeting project targets for an efficient alarms system.

Table 1. Acceptability of various rates of alarms (taken from EEMUA 191).

Long term average alarm rate in steady operation	Acceptability
More than 1 per minute	Very likely to be unacceptable
One per 2 minutes	Likely to be over-demanding (industry average in HSE survey)
One per 5 minutes	Manageable
Less than one per 10 minutes	Very likely to be acceptable

As a result of positive indications from the alarms rationalisation study, the EEMUA 191 guidelines were used (in conjunction with an approximation of 1 minute of activity for every alarm) to build up workload predictions for alarms and communications within the CCR. These alarms and communications predictions are used in conjunction with the objectively measured event driven tasks to develop a time occupancy percentage for the CRT.

Where the percentage figure is below 50% this would be regarded as acceptable. The 50% value is chosen because the time occupancy calculation is limited to observable tasks and does not include monitoring and planning activities for which sufficient time must be allowed.

Where the time occupancy value falls between 50% and 80% this is regarded as marginally too high. In these cases further assessment using task analysis of critical performance scenarios will be required to demonstrate the level of acceptability of the demands on the operator.

A time occupancy value of above 80% indicates that the workload is unacceptably high, and redesign, operational change or manning changes will be required to address this problem.

The workload boundaries utilised by the TPAM approach have been drawn from the research into pilot workload carried out by Parkes and Boucek (1989).

The limitation of the TPAM approach is that it does not take into consideration the possible conflict between tasks that must be performed concurrently. Concurrent task demands can cause unacceptable conflicts of attention and performance. Obviously, the greater the level of time occupancy within the CCR then the greater the likelihood of concurrent task conflicts occurring. Given that the FPSO had yet to be built and the control system finalised, a workload modelling approach that addressed task conflicts and scenario analysis was considered to be unfeasible.

It is acknowledged that the level of workload experienced over the course of a shift is not static. Workload is fluid with peaks and troughs over the course of the 12 hours. TPAM indicates the average level of workload experienced by a CRT over the course of a 12 hour shift.

Number of DCS screens required

From task analyses and subject matter expert interviews the number of DCS monitors required for major CCR activities was identified. Following on from this a conflict matrix was developed which cross referenced all major CCR activities with one another in order to identify potential simultaneous activities and thus the maximum number of DCS screens in use at any one time.

The number of screens assessment was used in conjunction with the results from the workload study in order to give guidance on optimising the DCS for the required number of CRTs.

Human error analysis

Human Engineering performed a screening exercise with operations personnel to identify critical activities that can occur simultaneously (thus compounding the CRT's workload) and have potentially serious consequences in terms of the environment, production or safety.

Four activities were identified by operators as involving high workload due to the number of tasks involved and the time pressure associated with them.

For each critical activity identified, a detailed task analysis was created. A human error analysis was performed on each task analysis to identify any significant human errors that could impact on the activities.

Table 2. Predicted workload within the CCR.

	Low-level of permit issuing	Typical level of permit issuing	High-level of permit issuing
Baseline level of workload without alarms and communication	32.71%	41.24%	49.77%
6 alarms an hour	42.71%	51.24%	59.77%
12 alarms an hour	52.71%	61.24%	69.77%
30 alarms an hour (industry average in HSE survey)	82.71%	91.24%	99.77%
60+ alarms an hour	132.71%	141.24%	149.77%

Results

Workload analysis

A major task identified for control room technicians was the issuing of permits for maintenance. Because of the early stage of the project it was unknown how variable the level of maintenance would be. By examining historical records of existing similar installations and obtaining subject matter expert opinion low, medium and high estimates of permit issuing were developed. Table 2 shows the predicted time occupancies taking into account differing levels of permit issue and alarms.

Human error analysis

The human error analysis highlighted 22 human errors with the potential for significant safety, production or environmental consequences. Recommendations on how to eliminate, control or mitigate each error were provided in a human error issues register.

Number of DCS screens assessment

An assessment of the number of screens required for efficient process control identified that the current proposed number was insufficient to handle upset plant conditions without adversely affecting both the CRTs performance and situational awareness.

Following the results of the workload study a recommendation was made to increase the number of DCS screens in order to optimise the control equipment to the new manning requirement.

Conclusions

The numbers of alarms and permits issued are the crucial elements of the workload for the CRT. Only if the number of alarms per hour met the highest standard of EEMUA 191 and there was a low level of permit issue would the CRT's baseline level of workload fall below the 50% time occupancy boundary. Therefore it was predicted that the CRT's baseline level workload will at least be in the "yellow zone" of time occupancy.

If alarm rates exceed 30 an hour (the industry average – EEMUA 191) then it is predicted that the redline boundary of workload will be exceeded. It is predicted that the CRT will not be able to perform all tasks without a significant amount of concurrent working, therefore increasing the likelihood of human error. The first 12 to 18 months of operations will be the busiest period for the CRT. It is anticipated that there will be a large volume of maintenance ongoing as equipment is started-up, shut down and safety systems are tested. It is therefore not unreasonable to assume a high level of permit issue.

If a high level of permits is assumed then the key factor in determining the appropriate manning level for the FPSO CCR is the alarms rationalisation process. It is impossible to predict the response of the oil reservoirs and new equipment when the FPSO first starts-up. However, an alarms system that has been optimised through a logical alarms rationalisation process should enable the CRT to avoid alarm flooding and react accordingly.

During the first 12 to 18 months of operations an increase in manning levels was recommended within the CCR. After this period it is thought that the workload level will drop. However, whether the drop in workload will be significant enough to warrant a reduction in manning is in doubt. Therefore it was recommended that the manning level of the CCR is reviewed again at that stage.

The study shows that despite the project being at a relatively early stage of the design process it is still feasible for human factors to offer an objective contribution that can provide benefit to a project. The human factors interventions applied within this project provide assurance to both the project safety case and the maintenance of production targets. Commissioning a petrochemical asset with an inadequate manning concept can potentially result in vast sums of money being lost by the operating company.

The TPAM approach is objective and simple to use. Its output is framed within an engineering context that systems engineers and other non-human factors specialists can appreciate.

It is acknowledged that there are weaknesses to the method (such as the inability to consider concurrent task working) but at such an early stage in the design process such an approach to workload modelling is often impossible. When used in conjunction with a range of other human factors techniques, such as SME workshops and human error analysis, TPAM can be used to develop an early indication of workload demands within a control room.

References

Card, S.K., Moran, T.P., and Newell, A. (1983). The *Psychology of Human-Computer Interaction*, Lawrence Earlbaum Associates, Hilldale, New Jersey.

The Engineering Equipment and Materials Users Association Publication 191, 1999. *Alarms Systems: A guide to design, management and procurement* (EEMUA).

Parks, D.L., and Boucek, G.P. (1989). *Workload Prediction, Diagnosis, and Continuing Challenges*. In McMillan. G.R., Beevis. D., Salas. E., Strub. M.H., Sutton. R., and van Breda. L. *Applications of Human Performance Models to System Design*. Defence Research Series Volume 2. (Plenum Press) 47–63.

SHIFTWORK ON OIL INSTALLATIONS

Andy Smith

Centre for Occupational and Health Psychology, School of Psychology, Cardiff University, 63 Park Place, Cardiff CF10 3AS

The present study investigated psychological response to different work schedules on offshore oil and gas installations. Performance and mood were evaluated before and after each shift using portable computerised tests. The 14-day protocol was undertaken by workers on the following shift schedules: 14 days (0600–1800, N = 15), 14 nights (1800–0600, N = 13), night–day swing shift of 7 nights and 7 days (N = 12) and day–night swing shift of 7 days and 7 nights (1200–2400, 0000–1200, N = 5). Those doing 14 days showed a slowing of reaction times over the course of the tour. Those doing 14 nights showed very low alertness at the end of the first night but appeared to be adapted by the end of the first week. Those doing swing shifts showed impairments when the shift changed.

Introduction

The effects of shift-work on performance, safety and well-being have been widely studied onshore but there is little information on this topic in the offshore oil industry. Recent research has, however, shown that shift patterns, job types and installation characteristics are significant predictors of safety, work and health measures (Parkes et al, 1997). Assessments of sleep, alertness, mood, workload and psychomotor and cognitive performance showed that adaptation to night work was evident in the 14N group during the second week offshore. However, rollover patterns gave rise to significant impairments compared with the fixed shift schedules. Although the sleep, performance and mood data showed a significant advantage of the 14D/14N schedule other factors were relevant. A survey showed a strong preference for the 7N + 7D pattern as this allowed adjustment to a normal routine before starting leave. Operational constraints, such as crew change arrangements or helicopter schedules, further complicate decisions about shift schedules. It is now essential to confirm and extend these findings in further studies that also include physiological measures of adaptation.

The health and safety problems of shift workers on offshore oil installations are unique to this population, since complete biological adaptation to night shift is seen with some schedules (Bjorvatn et al, 1998). The advice of the European Work Directive is not appropriate in this case as it is based on the premise that night shift workers do not adapt (as is usually the case onshore). Previous research at the University of Surrey indicates that adaptation to shift-work offshore is possible, but rate and extent vary between individuals and shift schedules. The present project continued to investigate the psychological response to different work schedules on offshore oil and gas installations. Performance and mood offshore were evaluated by use of portable computerised tests (Smith et al, 2001). A compromise between field studies of actual work performance and artificial laboratory task performance can be achieved by asking shift-workers to perform artificial laboratory-type tasks immediately before and after coming on shift. Using this technique, Wojtczak-Jaroszowa et al (1978) showed performance on the nightshift to be worse than morning or afternoon shifts. The present study extended this approach to offshore workers. In these same individuals, biological adaptation, risk factors for heart disease (including diet and temporal patterns of nutrition) and metabolic responses were assessed. The multidisciplinary and integrated approach permitted the simultaneous evaluation of psychological and physiological variables in the same individual while offshore. A pilot study had shown that the proposed research was feasible and that the procedures were applicable for use on the rigs.

The final objectives were to advise the Industry and the Regulator as to a) work schedules which maximise performance and b) strategies to minimise risk to health and safety.

Method

A fourteen-day study protocol was undertaken by male offshore oil and gas installation workers on one of four 12-hour, 14 day (or 14 days of a 21 day offshore tour) shift schedules: night-shift 1800 h–0600 h (14N), day-shift 0600 h–1800h (14D), night–day swing-shift of 7-nights 1800 h–0600 h and 7-days 0600 h–1800 h (7N7D), day–night swing-shift of 7 days 1200 h–2400 h and 7 nights 0000 h–1200 h(7D7N). Details of assessment of biological adaptation are given in Arendt et al (2004). Mood and performance were assessed before starting work and after finishing work. Mood was assessed using bi-polar visual analogue scales presented on a computer (after Herbert et al, 1976). Measures of alertness, hedonic tone and anxiety were obtained. Three performance tasks were carried out. These measured simple reaction time, choice reaction time, momentary lapses (errors and occasional long responses), focused attention and visual search, speed of encoding of new responses and organisation of responses (after Broadbent et al, 1986).

Sample

The final sample consisting of the following groups:
14 days: N = 15
14 nights: N = 13
7days/7 nights: N = 5 (4 working 12.00–24.00, 1 working 06.00–18.00)
7 nights/7 days: N = 12.

Results

14 days

Those doing 14 days showed slower simple reaction times as the tour progressed (see Table 1). This was observed for both the pre-and post-work measures. The slowing of reaction time was not accompanied by increased errors, lapses of attention or changes in selective attention. However, there was slower encoding of new information as the tour progressed (as indicated by the difference between alternating and repeated stimuli) which could account for the slower reaction times later in the tour.

14 nights

Those starting work on a night shift soon after arriving on the installation were very fatigued at the end of the first shift. This probably reflects both the effects of working at night and fatigue due to travelling to the installation (and possibly the partial sleep deprivation due to an early start from home). At the start of the tour alertness was higher at the start of the shift and declined after it. This pattern was present until day 9

Table 1. Simple RT (msecs) for those working day shifts (scores are the means, s.e.s in parentheses).

Day 2		Day 7		Day 14	
Before work	After work	Before work	After work	Before work	After work
342 (13)	321 (13)	364 (28)	343 (14)	415 (29)	367 (21)

Table 2. Alertness ratings for those working nights (scores are the mean alertness ratings, s.e.s in parentheses High scores = greater alertness)

Day 2		Day 5		Day 9		Day 13	
Before work	After work	Before work	After work	Before work	After work	Before work	After work
253 (17)	189 (18)	254 (15)	216 (13)	234 (16)	230 (13)	211 (15)	204 (16)

when there appeared to be no difference between pre and post shift ratings (see Table 2). A similar pattern was observed for other aspects of mood. Overall, this suggests adaptation to night work. Towards the end of the tour alertness declined supporting the view that cumulative fatigue occurs on the installations.

The simple reaction time data showed that reaction times were slower after work than before it. This effect was no longer significant at day 5, confirming the view that there was adaptation to night work. The choice reaction time tasks also showed that there was adaptation to night work by the end of the first week. This was found for both average reaction time and long responses. Evidence of cumulative fatigue over the course of the tour was also found for this shift, with new information being encoded more slowly as the tour progressed.

7 nights/7 days

The alertness data suggests that this group did not adapt by day 7. Following this there was a general drop in alertness with the switch to days. Alertness then remained low over the course of the day shifts. The simple reaction time data also suggested that there was no adaptation to night work. With the switch to days there was a slowing of simple reaction time and evidence of a cumulative fatigue as the tour progressed. The number of long responses declined over the course of the 7 nights suggesting that some adaptation was occurring. However, the number increased with the switch to days and continued at a high level until day 13. Speed of encoding of new information decreased over the tour suggesting that this shift pattern also leads to cumulative fatigue.

7 days/7 nights

It should be noted that there were much smaller numbers in this sample than those doing the other shifts. In addition, most of the sample worked 00.00 to 12.00 rather than 06.00–18.00. As this sample started their day shift at 12.00 noon they were less alert post-shift than pre-shift. Day 8, the first day after the switch to nights, was associated with low alertness which was followed by some adaptation to night work. The simple reaction time data also showed impaired performance on the first night shift, as well as evidence of cumulative fatigue. The small sample size meant that none of the other analyses of the performance tasks revealed significant effects.

Conclusions

Overall, these results confirm earlier findings and suggest that there is some adaptation to night work offshore. The results also show that there is a large initial problem with night work and this may reflect a combination of factors (travel to the installation, no sleep prior to the shift). Similarly, the cumulative fatigue observed even in those doing day work shows that shift scheduling is not the only factor influencing performance offshore. The present methodology can now be used to assess possible changes to shifts, such as a 0300–1500 shift, which is recommended on the basis of physiological adaptation.

The present study confirmed effects reported in earlier studies. Adaptation to night work does occur in those working a 14 night schedule and this confers a major benefit compared to schedules which involve changes from nights to days (or vice-versa) during the tour. Indeed, the results also confirm previous findings showing impairments at the start of the swing shift. With regard to start times, the results support the notion that a later start time may be beneficial but they do not allow evaluation of the 0300–1500 shift proposed to maximise adaptation.

One of the major findings of the study has been the cumulative fatigue over the course of the tour. This has been observed in most shifts and suggests that even if one solves the issue of shift scheduling one is still left with the problem of long working hours. Furthermore, the first night shift was found to be particularly fatiguing and this probably reflects a combination of factors: having to work at night; travel from home; travel to the rig; problems associated with a rapid handover etc.

The present project also examined physiological adaptation to shift work offshore and the methodology can now be used to assess alternative shifts and methods of maximising adaptation. The results also show that fatigue offshore is likely to reflect many other factors than just the shift schedule. Indeed, it is likely that a combination of factors will induce fatigue offshore and possible countermeasures aimed at preventing or reducing fatigue (e.g. napping) need to be evaluated. Training in awareness and prevention of fatigue is also important and the topic needs the same serious consideration as is given to other health and safety issues.

Acknowledgements

The research described in this paper was funded by the Health and Safety Executive. Its contents, including any opinions and/or conclusions expressed, are those of the author alone and do not necessarily reflect HSE policy.

References

Bjorvatn, B., Kecklund, G. and Akerstedt, T. 1998, Rapid adaptation to night work at an oil platform, but slow readaptation after returning home. *Journal of Occupational and environmental Medicine*, **40**, 601–608

Broadbent, D.E., Broadbent, M.H.P. and Jones, J.L. 1986, Performance correlates of self-reported cognitive failure and obsessionality, *British Journal of Clinical Psychology*, **25**, 285–299

Herbert, M., Johns, M.W. and Dore, C. 1976, Factor analysis of analogue scales measuring subjective feelings before sleep and after sleep. *British Journal of Medical Psychology*, **49**, 373–379

Parkes, K.R., Clark, M.J. and Payne-Cook, E. 1997, Psychosocial aspects of work and health in the North Sea oil and gas industry: Part III – Sleep, mood and performance in relation to offshore shift rotation schedules. HSE Offshore Technology Report (OTH96 530)

Smith, A., Lane,T. and Bloor, M. 2001, Fatigue Offshore: A comparison of Offshore Oil Support Shipping and the Offshore Oil Industry. Cardiff: Seafarers International Research Centre Report

Wojtczak-Jaroszowa, J., Makowsa, Z., Rzepecki, H., Banaszkiewicz, A. and Romejko, A. 1978, Changes in psychomotor and mental task performance following physical work in standard conditions and in shift-working conditions. *Ergonomics*, **21**, 801–810

ASSESSING THE IMPACT OF ORGANISATIONAL FACTORS ON SAFETY IN A HIGH HAZARD INDUSTRY

Johanna Beswick[1], Shuna Powell[1], Martin Anderson[2] & Alan Jackson[2]

[1]*Health and Safety Laboratory, Harpur Hill, Buxton, Derbyshire SK17 9JN*
[2]*Health and Safety Executive, Stanley Precinct, Bootle, Merseyside, L20 3RA, Edgar Allen House, 241 Glossop Road, Sheffield S10 2GW*

> The Health and Safety Executive (HSE) and the Health and Safety Laboratory (HSL) conducted a human factors intervention at a high-hazard industry site. A team approach was adopted. Two data collection techniques were utilised; focus groups and semi-structured interviews. Interview question sets and format were informed by a review of site accident data, existing knowledge of site inspectors, relevant literature and the experience of the specialist inspectors. The complementary and novel use of these two techniques using an independent facilitator in this applied setting allowed a larger sample of the organisation to be accessed than would otherwise have been possible and allowed both individual and group opinion to be assessed.

Background

This paper describes the research methods employed during an intervention to assess the impact of organisational and human factors on safety undertaken at a high-hazard site. The team comprised Health and Safety Executive (HSE) inspectors, HSE Human Factors (HF) specialist inspectors and Health and Safety Laboratory (HSL) HF scientists.

For the 12 months prior to the intervention, there had been an increase in incidents at the site, e.g., loss of containment. In response to HSE inspectors' increasing concern over the safety performance and general attitude towards safety of the high-hazard site, HSE felt that a major structured intervention focusing on human factors and organisational issues was required. The results of this intervention would contribute to the evidence required for HSE to effectively address the concerns they had over safety at the site.

Objectives of intervention

- To determine the organisational and human factors that had contributed to the poor safety performance of the site in recent times.
- To inform HSE inspectors' decisions on how to interact with site personnel in the future.
- To pilot the effectiveness of team working between HSE and HSL in this type of intervention.

Project process

HSE inspectors, HSE HF specialist inspectors and HSL HF scientists met to share their experience and knowledge of the site and discuss methods of collecting data the intervention required. It was decided that the following methods would be used:

- Review of site accident data
- Semi-structured interviews
- Focus groups
- Feedback to site

The project team decided on a sample of site staff whom they would like to interview or participate in the focus groups. The groups of staff covered the seven main categories of staff. This ensured the sample used in this intervention was representative of personnel on site. The site then arranged for the appropriate sample of staff to be interviewed.

The project team were on site for a week. The semi-structured interviews were carried out by an HSE specialist inspector and a site inspector. HSL staff observed some interviews. Detailed notes were taken during interviews and focus groups.

Project team organisation

Historically, for an intervention of this kind, HSE and HSL colleagues had not worked together as part of one project team. However, it was felt that the nature, scale and timings of the required intervention, along with the complementary skills of the HSL and HSE technical HF specialists, meant that a team approach involving site and specialist inspectors from HSE and psychologists/ergonomists from HSL was highly appropriate. All project team members brought important knowledge to the project:

- HSE specialist inspectors have expertise in Human and Organisational Factors that is specifically applied and relevant to high-hazard industries. These specialist skills are used to support site inspectors when required. They took the lead on the project, drawing on their experience in the field of similar interventions, and their close links with HSE site inspectors and HSL scientists.
- HSE site inspectors have worked with the high-hazard site for a number of years. They are responsible for the regulation of health and safety at major hazard sites. They have detailed knowledge of the history and operation of the site and the personnel involved.
- HSL scientists brought experience of researching human and organisational factors in a range of industries using a variety of methodologies. They are also viewed as being unbiased as they are perceived as independent of HSE, acting purely as researchers and have no legal enforcement role.

HSE specialist inspectors drew on the knowledge and experience of all team members to inform methods of data collection to be used in the intervention. An overriding aim was for the project team to engage with and gather data from as many individuals, from different sectors and levels on the high-hazard site as possible, to increase the reliability and validity of the conclusions drawn.

Data collection techniques

Semi-structured interviews

HSE have successfully used semi-structured interviews in similar interventions, and felt they would be an appropriate technique to use to collect the majority of information for this intervention. The project team were interested in assessing the impact of organisational and human factors on safety, and therefore felt that techniques to collect qualitative data on these factors were most appropriate, given that they had quantitative data on safety at the site. Semi-structured interviews have pre-determined questions, but the order can be modified, question wording changed and questions can be omitted or included based on who is being interviewed (Robson, 2002). They are ideal to use where investigators are interested in the meaning of particular phenomena to the participants; where individual perceptions of processes within a team, department or organisation are studied using several interviews; and where individual historical accounts are required of how a particular phenomenon developed (King, 1994).

The project team developed a set of questions based on previous questions sets used in similar interventions and building on the categories highlighted by Jacobs and Haber (1994). These included topics such as resources, organisational culture and organisational style. In this way, a degree of consistency was achieved in how HSE HF specialists interact with personnel at high-hazard sites. Other topics were included based on discussions with the site inspectors, such as the site personnel's perception of HSE. In addition, HSL reviewed a series of accident reports from the company and pinpointed where they saw recurring human factors issues that had contributed to the accidents. The HSL team members analysed each report independently, bearing in mind factors that can impact on safety such as those highlighted by relevant industry reports including over-reliance on safety procedures and managing contractors. They then discussed their findings as a team. They collated the findings, which HSE used to further inform the question sets for the semi-structured interviews.

From the large pool of questions, specific items were selected for each participant depending on their job in the organisation. Time was also allowed for participants to raise any particular concerns.

Focus groups

Focus groups are essentially group interviews, where a moderator or facilitator guides the discussion based on the topics that the facilitator raises. They generate a comprehensive understanding of participants' experience and beliefs (Morgan and Kruger, 1998). Whilst they can be used as the primary data collection method in a study, they are often used alongside other methods, as in the present intervention.

The team felt that focus groups would provide useful supplementary data on group and individual opinions to those gathered in the semi-structured interviews. The interaction of participants produces rich data, which can provide a different and complementary perspective on phenomena than that revealed by one-to-one interviews. Another important advantage of utilising focus groups was that they enable a large amount of data to be collected from a number of participants in an efficient manner.

The team decided that the focus groups should follow a less structured format as this would allow the participants to raise concerns and organisational factors that were important to them, and not "pre-judged" by the intervention team to be important (as described by Morgan and Kruger, 1998). Following this less structured format allowed for the generation of data that could be compared to those produced by the interviews, thus allowing for triangulation of data during analysis. The question set used in the focus group was informed by project team discussions during the formulation of the semi-structured interview question set, along with HSL observation of a sample of the interviews.

The moderator or facilitator plays an important role in a focus group. They have a number of functions such as introducing topics for discussion; controlling flow of discussion and summarising and reflecting to check understanding.

It was important that HSE gathered honest opinions on the organisational and human factors that were impacting on safety at the site, and for this reason it was important for the focus group facilitator to be perceived as independent. Using HSL facilitators in the focus group helped to provide a forum for site staff to express views that they may not otherwise have felt comfortable in expressing in front of the HSE team.

Themes from the focus groups were compared with the data from the semi-structured interviews to check for consistency in emerging themes between the data generated by two methodologies used by HSE and HSL.

Experienced HSL staff were used to facilitate the focus group; two facilitators were used – one to primarily guide and manage the discussion, with the co-facilitator primarily recording the discussion and interjecting when necessary. Each focus group comprised personnel of similar grades to encourage participants to feel comfortable discussing relevant issues.

Analysis

There are a number of techniques available to analyse the qualitative data generated by this project. The data in this project were analysed using a template approach, whereby key themes or codes were determined on a priori basis (derived from theory and experience). The themes then served as a template on which the data analysis was based, which can be modified as analysis progresses (Drisko, 2000).

Interview data

Each inspector compiled their notes taken during the interviews under key theme headings. These headings were then rationalised into a common set of emergent themes and each interviewers' notes were collated in a spreadsheet indicating which themes emerged from which category of staff. The spreadsheet also indicated how many of the interviewers highlighted a particular theme. The list of emergent themes was then panel ranked by the team members based on perceived importance expressed by the interviewees and frequency with which it was raised. These were the key issues and themes as perceived by the site. The ranked emergent themes were then compared with the results of the focus groups for consistency. Finally, a second ranking was carried out by the HSE inspectors to prioritise the emergent themes in line with the requirements of the regulatory regime in which major hazards sites operate. This final list of key themes and issues was used to target HSE inspection and enforcement action to aid the site in securing compliance with health and safety law and improve their safety performance in the control of major accident hazards.

Focus group data

Immediately after each of the three focus groups the two facilitators discussed and listed the main themes emerging from that group. Based on the notes taken during the focus group and the summary, each facilitator independently did a thematic analysis of each focus group discussion. The results were then triangulated and common emerging themes were noted. The themes were further separated into themes common across all three groups, across two groups and in themes specific to one particular group.

Qualitative data analysis can have some disadvantages, primarily due to the use of the individual as the primarily analyst, rather than a statistical procedure. However, utilising several researchers to analyse the data, and triangulation through multiple methods of data collection can help to verify conclusions.

Findings and conclusion

A number of organisational factors were found to impact negatively on the safety performance of the site. These included:

- the site being reactive to safety concerns and incidents, rather than proactively identifying safety concerns and improvements;

- some of the existing procedures used by personnel were felt to be time consuming and not workable, and were therefore not always followed;
- an inadequate competence assurance system;
- insufficient resources; and,
- a focus on personal safety, rather than major accident hazard potential.

The complementary use of focus groups and semi-structured interviews in this applied setting allowed a larger sample of the organisation to be accessed than would otherwise have been possible and allowed both individual and group opinion to be assessed. This facilitated triangulation of data analysis and increased the validity of the data collected. The project team approach enabled the intervention to comprehensively and efficiently meet its objectives by drawing on the different and complementary skills, experience and knowledge of all team members.

References

Drisko, J. W. 2000, *Qualitative data analysis: It's not just anything goes!* In Robson, C. *Real World Research*. Second Edition, 2002. Blackwell Publishing Ltd.

Jacobs, R. and Haber, S. 1994, Organizational processes and nuclear power plant safety. *Reliability Engineering and System Safety*, 45, 1–2, pp75–83.

King, N (1994) The qualitative research interview. pp 16–17.In C. Cassel and G. Symon, eds, *Qualitative Methods in Organizational Research*. London: Sage. In Robson, C. *Real World Research*. Second Edition, 2002. Blackwell Publishing Ltd.

Morgan, D. L. and Kruger, R. A. (1998) *The focus group kit,* London: Sage.

Robson, C. (2002) Real World Research. Second Edition. Blackwell Publishing Ltd.

INTEGRATING HUMAN FACTORS IN AN OIL PLATFORM CONTROL ROOM DURING ORGANISATIONAL CHANGE

Zoë Mack & Liz Cullen

Atkins, WS Atkins House, Birchwood Blvd, Birchwood, Warrington WA3 7WA

As part of an ongoing study a survey to identify Human Factors (HF) problems in the central control room of a North Sea Oil Platform was undertaken. The focus was on issues that not only pose health and safety risks to staff working in the control room but which, by extension, could affect process safety by making human error more likely. HF issues identified were in relation to the following: design and layout of the room and its equipment, interface design, the working environment, training needs, workload, stress, fatigue and morale. In order to reduce the risk of human error which could lead to unsafe conditions on the platform, recommendations were made to address the HF issues. This work was carried out during a time of organisational change which had implications both on the findings of the HF survey and on how the recommendations were addressed.

Introduction

This paper relates to work carried out in relation to the Central Control Room (CCR) of a North Sea oil production platform. This CCR provides a centralised monitoring and control facility for the platform processes of original design circa 1970's. In order to minimise the risk of process shutdowns and major accidents, it is essential that the control room is well designed and supports the operators working there in performing their jobs safely. In this case a number of changes related to process control, shutdown and metering systems had been implemented, which had a significant impact upon the CCR. Specifically, new Visual Display Unit (VDU) based systems were provided for the control and monitoring of the new processes. These were installed alongside the existing VDU systems and hard-wired panels. Additionally, new tasks were imposed upon the operators, and structural changes to the control room significantly impacted on the work environment. Despite the significant impact the project was having on the CCR and its operations, HF was not actively or formally integrated into the project and it appeared that there had been insufficient consultation with operations personnel. However, since commissioning, the operator has recognised the need (as required by The Offshore Installations (Safety Case) Regulations, Health & Safety Executive, 2005) to carry out remedial work and hence contracted Atkins to carry out the HF CCR survey.

The aim of the survey was to identify HF problems and derive recommendations to address these. The focus was on issues that that not only pose health and safety risks to staff working in the control room, but which, by extension, could affect process safety by making human error more likely. Although the risk of a major accident is managed by the provision of automatic safety systems (e.g. emergency shutdown), it is unwise to rely too heavily on these and there are financial implications of their use (in terms of production downtime). Additionally, operator performance is particularly critical in cases where automatic systems are bypassed or fail to function. It is necessary, therefore, that the control room conforms to HF best practice in terms of its design, layout, and general environment such that the risk of human error is adequately controlled.

Methodology

The initial stage of the work involved data collection during a visit to the platform. This included observation of control room activities, discussions with control room staff and HF assessment of features within the control room against relevant standards and HF best practice. The analysis (including Task Analysis), findings and associated recommendations were detailed in a report. However, as these were relatively high level, further work was carried out to address the main risks identified during the initial assessment. This involved a second visit to the platform and more detailed assessment of the human-machine interfaces and other aspects of the CCR.

Findings

A review of some of the main deficiencies found is detailed below with a summary of the recommendations made to mitigate these problems.

Environment

Noise was a significant problem in the control room. This was associated with a number of sources including:

- HVAC (Heating, Ventilation and Air Conditioning) plant located in a room contained in the CCR.
- Generators located adjacent to the CCR.
- Communications (e.g. deck crew communications which were audible on the VHF radio and face to face conversations in the control room).
- Staff passing through the control room to access the Maintenance Supervisor's office or use the control room facilities.

The CCR noise levels resulted in difficulties during face to face verbal communications as well as vibration of the desk and other control room items. Under incident conditions these issues would be exacerbated by the General Platform Alarm which continues to sound throughout the incident's duration. It was recommended that a full noise survey be carried out to objectively assess the noise levels against Health and Safety Executive (HSE) guidelines and steps taken to remove sources of noise. Incorporating the suggested functional layout recommended during the study (see below) would reduce at least some of the noise produced by staff entering the CCR.

Control Room and Equipment Layout

There was open access from the control room to a stairway leading up to the maintenance supervisor's office and (via a door) to the rear of the panels where the Instrument Technicians work. This unrestricted access to the control desk makes the Control Room Operator (CRO) too accessible to other personnel, and increases distraction as other staff pass through the control room. This is aggravated by staff entering the main control area to use equipment there (e.g. photocopier, PC for email etc). The operations supervisor office was also located in the corner of the CCR with no wall or door and only a temporary partition. This meant staff meeting with the operations supervisor were directly audible to the operators in the CCR causing more distraction while working. To combat this problem it was recommended that the operations supervisor should have a fully enclosed office and that a partitioning wall be installed between the control area and the door. This would allow staff to become more segregated according to their function and reduce the unnecessary interactions with and possible distractions to operators.

There were a number of issues relating to equipment layout on or near the desk, including the following:

- The equipment on the control desk was not positioned according to use. For example, equipment often used together was not co-located on the console. With the existing desk layout, this required the operator to move frequently between different PC based systems to collect data from one (either recording it on paper of memorising it) for subsequent entry into another system. Alternatively another member of staff was used to read out the data to the operator entering it into the system.
- The desks were cluttered with non-critical equipment that did not need to be on the desk e.g. satellite phone (this is provided as back up only) and two printers.

The study recommended an alternative layout for the equipment on the main console, grouping equipment according to its importance, frequency of use, functionality and sequence of use and relocating non-critical equipment to an alternative desk nearby. It was also recommended that the gas compression console be moved closer to the CRO to minimise workload and human error risks.

Alarms & Interface design

A combination of hardwired panels and VDUs were provided in the control room. These were mainly used throughout the shifts to obtain information on plant status and on alarms which are acknowledged and reset using buttons on the panel, a desktop box or a function on the VDUs. The problems identified with the design of the panels and VDU screens, particularly in relation to alarms, are only detailed at a high level here:

- There was a lack of consistency in the Human Machine Interfaces (HMIs) installed within the CCR and many cases where the individual HMIs were not designed in line with HF best practice. Although all operators were aware of this discrepancy, there is an obvious risk of confusion particularly in times of stress and high workload. It is also a concern that new staff may not be sufficiently aware of this difference.
- There was a significant amount of redundant equipment on the panels. Whilst most of this was covered or labelled as redundant it is a source of distraction to CROs, and could impede visual search. There was also evidence of poor functional grouping where equipment appeared to be positioned according to space availability rather than any logic.
- There was a significant amount of duplication with panel alarms which were required to be acknowledged and reset twice (i.e. using the alarm control on the desk, and also a VDU alarm acknowledge and reset buttons). This increases operator workload and frustration.
- It was not possible to differentiate between alarms in terms of criticality and there was no prioritisation as all alarms in the list on screen were the same colour making it difficult to detect the significant critical alarms.

Detailed recommendations were made to improve the interfaces and alarms following HF best practice and the relevant standards. This enabled the operator to work with the vendors of these systems to improve the VDUs and panel design.

The influence of organisational change

During this HF survey, the platform was mid-way through a period of organisational change which involved restructuring of roles and responsibilities, making some roles redundant. This appeared to have an adverse effect on some staff due to concerns over job security, and the proposed organisational changes. There is a high risk of concentration being affected, potentially resulting in operator error, and as such, organisational change is recognised by the HSE as a key area of importance in relation to HF (Health & Safety Executive, 2003). Specific issues for the platform are discussed below.

A common feature of organisational change is an increase in staff turnover due to perceptions of job insecurity. There was some evidence of this effect on this platform. Uncertainty about how individuals will be affected by organisational change can act as a stressor on staff, particularly in cases where job security is a real or perceived issue. In addition to this, the physical distance between platform staff and onshore personnel, and a lack of face-to-face contact because of this, can contribute to a feeling of isolation and uncertainty especially if staff are not kept informed at every stage of the change. Because of these issues, there was identified to be a risk of increased control room workload due to temporary staff shortages and a competency gap, particularly in relation to platform-specific knowledge. Consequently training has become a critical issue for the platform in order to ensure sufficient competent personnel area available across the platform and particularly in the control room to manage the required range of task demands. This turnover has more recently reached a plateau and the duty holder has addressed this issue with various initiatives so the number of staff working per shift has increased.

Similarly, the definition of roles and responsibilities across the platform is also a critical area following the reorganisation. The restructuring resulted in changes being made to assigned roles and responsibilities. Where this occurs there is often the risk of discrepancy between what the new role entails, what is detailed in written job descriptions, and the role as perceived by the appointed staff member. As well as causing confusion and making workload difficult to asses, this may also have a poor effect on morale as staff may feel their roles are not recognised by fellow colleagues.

There was also some evidence that the reorganisation adversely affected morale on the platform and by consequence increased the risk of human performance deficiencies related to motivational issues and deteriorated working relationships. In particular, inaccuracies in job descriptions specifying roles and responsibilities may have resulted in onshore staff not fully understanding the roles of offshore staff and therefore having less appreciation of the problems faced in their jobs. Improving the empathy between the two parties should be a priority in order to build more co-operative and effective working relationships, reducing workload for both and improving communications.

Finally, as a result of the organisational change delays were incurred on the action taken in response to the HF survey. This was due to practicalities of obtaining authorisation from the correct party to commission work. This meant that staff perceived a lack of action to improve the CCR following initial work undertaken by Atkins and other contractors, becoming frustrated after repeatedly discussing the problems that exist in the CCR. Ultimately this is likely to have caused a reduction in motivation and morale leading to an increased risk of performance deficits in undertaking control room operations.

Conclusions

The survey carried out on this platform highlighted HF issues including design and layout of the room and its equipment, interface design, the working environment, training needs, workload, stress, fatigue and morale. Collectively these deficiencies presented a substantial risk of human error that could cause or contribute to unsafe conditions, in certain circumstances, contributing to a major incident event. This suggested that that the major accident hazard risk for the platform was not being controlled or managed to a level that was as low as reasonably practicable (ALARP). Recommendations were made to address the deficiencies identified.

A number of these recommendations have now been taken forward and further measures are underway. However, given the practical constraints on the changes that could be made within the control room, in some cases it was not possible to justify the disruption and expenditure that the proposed modifications would involve. In other cases, the modifications themselves imposed an increase level of risk on the platform.

Fortunately the company responsible for operating the platform recognised the problems and were proactive in taking the HF work forwards. Going beyond the recommendations made in the HF work programmes, a decision has been taken to commission a new CCR on the platform whilst commissioning interim projects to manage and improve the conditions within the existing control room. This provides an opportunity to address the identified HF deficiencies through early HF integration into the project. Specifically a project plan has been established to integrate HF and end user representation throughout the project lifecycle from an early stage to avoid the need for later remedial work and its associated implications in terms of costs and major accident hazard risk.

References

Health & Safety Executive, 2003, *Organisational Change and Major Hazards*, CHIS7, Series 7, (HSE Books)

Health & Safety Executive, 2005, The Offshore Installations (Safety Case) Regulations 2005, ISBN 0110736109. (The Stationery Office Limited)

WHY DO PEOPLE DO WHAT THEY DO?

Ronny Lardner[1] & Graham Reeves[2]

[1]The Keil Centre, 5 South Lauder Road, Edinburgh EH9 2LJ
[2]BP plc, Chertsey Road, Sunbury on Thames TW16 7LN

The "WHY DO PEOPLE DO WHAT THEY DO" Workshop will introduce a Human Factors Analysis Toolkit that is used by an international energy company's incident investigators to analyse behaviour, and to subsequently put in place more effective recommendations to prevent a similar accident happening again. The "Toolkit" comprises an ABC analysis for intentional behaviours and a human error analysis for unintentional behaviours. Case studies are presented to illustrate the benefits obtained, including a violation which led to an environmental release, and a human error near-miss concerning spillage of hydrocarbons. It is anticipated that the methods described will be of interest to other process industry organisations, who wish to deepen their understanding of how to apply human factors methods to improve performance.

Introduction

The Human Factors Analysis Toolkit was developed to address shortcomings with the existing root-cause analysis methodology, which was recognised as being very effective at identifying what happened and addressing technical failures, but not so successful in consistently identifying why people behaved as they did. The existing process involved structured evidence gathering, interviewing by trained staff, development of an incident time-line, identification of critical factors, and the application of a root cause analysis model to guide recommendations. The root cause analysis method included consideration of violations, but did not address human error.

The following design principles were applied to the design of the analysis toolkit:

- Tools to be based on sound analytical methods, supported by existing research.
- Methods designed to help the investigator reach their conclusions on the basis of evidence gathered.
- Methods to be suitable for use by trained investigators, who are not human factors specialists.
- Toolkit capable of being imparted via a 2-day training course, delivered by internal company personnel.
- Toolkit to permit analysis of intentional and unintentional unsafe behaviour.
- Provides written support, guidance and examples for investigators.

A four-step process was developed, supported by structured worksheets, which allowed incident investigators to:

1. Accurately define and describe the behaviour(s) they wished to analyse.
2. Determine, on the basis of the evidence available, whether it appeared the behaviour(s) were intentional or unintentional.
3. For intentional behaviour (often termed a violation), apply ABC analysis.
4. For unintentional behaviour (often termed an error), apply human error analysis.

ABC analysis

ABC analysis is so called because of the three elements involved in understanding why people intentionally behaved as they did (Komaki et al, 2000; Health and Safety Executive, 2002). **A** refers to **A**ntecedents, which come before the behaviour and prompt or trigger behaviour. **B** refers to the specific **B**ehaviour we are interested in. **C** refers to the **C**onsequences of that behaviour for the person involved. The ABC model assumes the following 3 propositions are true:

- Behaviour is largely a function of its consequences.
- People do what they do because of what happens to them when they do it.
- What people do (or do not do) during the working day is what is being reinforced.

Most unsafe behaviours do not involve people deliberately intending to harm themselves or others. From their point of view, their behaviour usually makes perfect sense. ABC analysis helps the investigator understand, from the other person's point of view, the antecedents (which triggered the unsafe behaviour), and consequences (which reinforced the unsafe behaviour). Once this is understood, antecedents and consequences can be rearranged (and written into recommendations) in such a way that will make it more likely that the person involved, and others in a similar situation, will behave safely in the future.

Antecedents trigger the behaviour or enable the behaviour to occur once. Consequences encourage the behaviour to occur regularly. Arguably much traditional health and safety management activity is devoted to providing antecedents for desired behaviours (e.g. training, suitable equipment, signs, procedures), and less attention is given to how consequences reinforce safe and unsafe behaviour. For this reason, ABC helps the safety professional gain additional insight into what influences safe and unsafe behaviour.

An ABC analysis begins by defining the antecedents of the behaviour. Antecedents can be the presence or absence of factors such as suitable tools and equipment, other peoples' example and procedures. After the antecedents have been defined, the consequences of the behaviour are described from the perspective of the person who was involved. Examples of consequences include getting injured or harmed, saving time and getting approval from a supervisor or manager.

Each consequence is then assessed for the following, from the perspective of the person who performed the behaviour:

Positive/Negative	– from their perspective, if this consequence occurred, would it be positive or negative? Note that getting injured or harmed will usually be assessed as negative.
Immediate/Future	– from their perspective, does this consequence occur immediately after the behaviour (now or soon) or in the future? Note that getting injured or harmed will usually be assessed as something that will happen in the future, not now.
Certain/Uncertain	– from their perspective, is it relatively certain that this consequence will occur, or somewhat uncertain? Note that getting injured or harmed will usually be assessed as something which is uncertain (i.e. it has not happened to me yet, so it won't happen today).

Positive, Immediate and Certain consequences influence behaviour much more strongly than Negative, Future and Uncertain consequences do.

Having fully described the problematic behaviour, the next step in the process is to define a safe alternative to this behaviour, which antecedents will help to ensure that this behaviour is triggered, and the type of consequences that will help to reinforce the behaviour. The results of the analysis can then be turned into practical recommendations to reduce unsafe behaviours and introduce new, safe alternatives to replace them.

Human error analysis

The human error analysis (HEA) technique adopted is loosely based on TRACEr-lite (Shorrock and Kirwan, 2002). The human information-processing model proposed by Wickens, (1992) describes four stages of human information-processing and performance, namely perception, memory, decision-making and action. When performing any task, people perceive information about the outside world using all of the senses, and may use this information along with information retrieved from memory to arrive at decisions that are used to determine and execute action. A human error can occur as a result of a failure in any of these four stages, as the following process industry examples illustrate:

- Perception error – misperceive a reading on a display screen.
- Memory error – forget to implement a step in a procedure.
- Decision error – fail to integrate various pieces of data and information, resulting in misdiagnosis of a process upset.
- Action error – inadvertently activating the wrong switch.

To find out why these four types of error happen, it is necessary to establish what caused the failure in that part of the human information-processing system, i.e. what were the underlying psychological factors? As well as explaining why an error has occurred, the underlying psychological factors also give us strong indications as to what we can do to prevent such errors, or reduce their impact.

It is also necessary to be mindful of the fact that human performance in general is very heavily influenced by the conditions under which people perform. Such conditions are known as performance-shaping factors, and can help to further clarify why an error occurred, and also provide a great deal of extra information to help specify a practical solution. Examples of performance-shaping factors, which may increase the likelihood of error, include very high workload, poor ergonomic design of equipment and displays, and inadequate training.

The human error analysis technique described, supported by a worksheet, allows the investigator, on the basis of the evidence gathered, to:

- Classify which type of error was involved (Perception, Memory Decision, Action).
- Identify any performance-shaping factors.
- Understand the underlying psychological error cause, each of which is linked to a set of example solutions, which can be further developed for specific circumstances.

To help the investigator, each error type is accompanied by industry and everyday examples, and a comprehensive checklist of performance-shaping factors.

Analysing a violation – an example

Process control operators applied a series of safety over-rides to maintain production, without first conducting a risk assessment and involving their supervisor, as specified in plant procedures. As a result, product vented from a knock-out drum, resulting in an environmental release. The initial investigation focused on the intentional violation by the operator, and recommended discipline, briefings, re-writing the procedure, and re-training.

Further analysis by an investigator trained in the Human Factors Analysis Tool revealed that plant management had tacitly encouraged the application of over-rides to maintain production, and had inadvertently reinforced this practice. Also, it was established that the over-ride key was kept in a readily-accessible location, which allowed over-rides to be used without supervisory involvement. Additional recommendations, which flowed from the human factors analysis, ensured management's role in ensuring production vs. safety conflicts was strengthened, and required removal of the key to the supervisor's custody. Without these additional recommendations, the initial recommendations would have limited effect.

Analysing an error – an example

An incident occurred where the lid of a rail-tanker filled with hazardous liquids was not closed prior to the train's departure, resulting in a potential for spillage. Initial investigations had focused on the "carelessness" of the loading operator. A human error analysis was requested.

It was established that this type of incident had occurred on a number of occasions, and had involved several different operators, all of whom had hitherto been considered careful and competent employees. It was established that the rail-car filling operation was a complex task, with many procedural steps. The radio communications system was not working, and the back-up communication system required the operator to interrupt the loading task to access a phone. The human error analysis categorised the problem as a memory error, influenced by the performance-shaping factor of interruptions from the phone.

A simple solution was proposed, which involved issuing a plastic seal for each rail tanker lid, to be fitted after loading of each tanker was complete. If the loading operator was left with any plastic seals, this indicated a lid remained open. The radio communication system was also fixed. Interestingly, management's reaction to these findings and recommendations were that they had expected something more complex and expensive to implement.

Conclusion

Current analysis of human behaviour in incident investigation is often relatively superficial, the causes are seldom identified, thus missing opportunities to improve human performance and prevent incidents recurring. A specific weakness is the understanding of human error, which is much better understood and managed in other domains. The Human Factors Analysis Toolkit represents the start towards the development of a computerised toolkit that will enable incident investigators to analyse both intentional (often termed violations) and unintentional (often termed errors) behaviours, and to identify appropriate interventions. However, before this can happen, the existing toolkit will need to be streamlined and a balance truck between simplicity and ease-of-use, and maintaining sufficient rigour. It is anticipated that the "toolkit" will also be supportive of a "Just Culture" and help managers and supervisors to decide what consequences can be applied to an individuals behaviour.

References

Health and Safety Executive, 1999, *Reducing error and influencing behaviour HS(G)48*, (HSE Books, Norwich)
Health and Safety Executive, 2002, *Strategies to promote safe behaviour as part of a health and safety management system, CRR 430/2002*, (HSE Books: Norwich)

Komaki, J. et al 2000, A rich and rigorous examination of applied behaviour analysis research in the world of work, *International Review of Industrial and Organisational Psychology*, **15**, 265–367

Lardner, R., Fleming, M. and Joyner, P. 2001, Towards a mature safety culture. In *Hazards XVI, Analysing the past, planning the future*, (Institute of Chemical Engineers, Rugby)

Shorrock, S.T. and Kirwan, B. 2002, Development and application of a human error identification tool for air-traffic control, *Applied Ergonomics*, **33**, 319–336

Wickens, C.D. 1992, *Engineering Psychology and Human Performance*. 2nd Edition, (Harper Collins, New York)

PHYSICAL ERGONOMICS

PHYSICAL ERGONOMIC DESIGN ASPECTS OF COMPUTER WORKSTATIONS: A LIECHTENSTEIN PERSPECTIVE

Els Kessler[1], Stella Mills[2] & Siegfried Weinmann[1]

[1]Liechtenstein University of Applied Sciences, Fürst-Franz-Josef-Strasse,
FL-9490 Vaduz, Principality of Liechtenstein
[2]Staffordshire University, Beaconside, Stafford ST180DF, United Kingdom

> This poster uses ergonomic requirements from health and safety legislation in Liechtenstein as well as criteria from ergonomic theory to analyse physical ergonomic aspects of computer workstations in the office and to evaluate the risks upon users.

Introduction

There has been little research of the ergonomic design of computer workstations with regard to Liechtenstein, a tiny country situated on the north side of the alpine chain and a member country of the European Economic Area. However, the Council Directive of the European Communities of 1990 on the minimum health and safety requirements for work with display screen equipment (hereafter Directive 90/270) was transferred into law in 1998. Office work can not be regarded as completely safe, since a static posture is usually adopted for long hours at a computer workstation (Kroemer and Grandjean, 1997). Poorly designed computer workstations often lead to discomfort such as musculoskeletal pains and eye fatigue and can be a contributory risk factor to health problems of employees (e.g. Smith et al, 2003).

Method

The research was designed to evaluate the ergonomic design of computer workstations, the physical environment in which the office work was carried out and the interaction of the users with their workstation equipment and furniture. Individuals employed by the Liechtenstein University of Applied Sciences (LUAS) freely agreed to allow their workstations to be evaluated. Seven computer workstations (n = 7) out of 70 were assessed in December 2004. The seven workstations were located in different multiple office rooms in the university building. Because of the narrow sample there is no guarantee that the findings below would generalise to all office workers in Liechtenstein or even all employees of the LUAS. A combination of data collection techniques was employed. A high degree of structure was built into the assessments of the workstations by using a questionnaire comprising 97 questions. Data regarding noise and satisfaction with thermal and lighting conditions were gathered from the subjective responses of the users expressing their feelings about these issues during a structured interview. The study also used workstation measurements, observational techniques and digital photography.

Findings and recommendations

The five main problems identified during the field study are summarized below.

None of the users had the keyboard and the mouse at elbow height with their shoulders relaxed and upper arms hanging relaxed to the sides as recommended (e.g. Sanders and McCormick, 1992). In fact, the elbows were constantly resting on the work surface during observation or the arms were resting on the edge of the work surface. These postures can cause the ulnar nerve to be compressed at the point of contact leading to pain at the inner side of the forearm down to the little finger (e.g. Pheasant, 1991). Moreover, when the keyboard is above elbow height, unnecessary strain is exerted in the arm, shoulder and neck muscles, because the shoulders are lifted up or the elbows are raised and spread (Kroemer and Kroemer, 2001). Because all users had fixed desk heights, they were recommended to raise the level of their seats and to place keyboard and mouse within a convenient arm's reach at elbow height (whilst shoulders and upper arms are relaxed). The four users that were putting documents in front of the keyboard were advised to make use of a document holder (Directive 90/270). In addition, shorter users were recommended to use a footrest (Directive 90/270), if the feet could no longer rest firmly on the floor, since pressure to the underside of the thighs may restrict blood circulation in the lower legs (Bridger, 2003). Shorter users may also sit only on the edge of their seats being unable to lean backward.

The majority of the users (six out of seven) did not make proper use of their backrest or had their backrest even locked, although ergonomically designed chairs were provided. Except in one case, all seat backs were tiltable (Directive 90/270). Three users demonstrated an upright seating posture during the observation, but it is known that most people are unable to sit erect in the 90 degree posture for longer periods (e.g. Smith et al, 2003) and they soon adopt a slumped posture characterized by a lumbar kyphosis (e.g. Oborne, 1995), forward head posture and rounded shoulders (Szeto et al, 2002) like most of the other users. This posture exerts considerable strain on the upper back and shoulder muscles and results in increased pressure on the intervertebral discs which is not desirable (e.g. Oborne, 1995). The users were informed about the benefits of the adjustments of their working chair. They were recommended to assume a variety of postures and to frequently rest their back against an inclined backrest, because this sitting posture reduces both disc pressure and static strain of the back and shoulder muscles (e.g. Kroemer and Grandjean, 1997). In order to prevent aches and pains from prolonged seating the sitting posture could be alternated with the standing posture (ISO 9241-5, 1998).

The display screens of the workstations fared little better. At five out of seven workstations the viewing distance was not optimal. Preferably, if the task requires a significant amount of reading, the workstation design should permit the display to be used at a distance where the characters height subtends approximately 20 to 22 minutes of arc (ISO 13406-2, 2001). At five out of seven workstations the line of sight of the user was rectangular (and not parallel as recommended) to the window panes or lighting tubes. Two users had the top line on the display above eye level i.e. too high and at the three laptop workstations separate height-adjustable screens were missing.

Five users showed extended or ulnar deviated wrists when using the keyboard. An unnatural position of the wrists during highly repetitive finger movements is a contributory factor to RSI (Kroemer and Kroemer, 2001). In order to keep the wrists in a straight neutral posture the users were advised to take the use of wrist supports and split keyboards into consideration. If a laptop is used for an hour or longer, a separate keyboard should be provided (Directive 90/270).

In addition, six out of seven users experienced environmental discomfort including problems due to low humidity and noise.

Conclusions

Many of the ergonomic requirements of the European Directive 90/270 regarding equipment, furniture and environment were met at the majority of the analysed workstations at the LUAS. The users were provided with ergonomically designed office chairs and flat panels with good image quality. However, workstation elements should not to be considered in isolation from the total working system. Although each workstation has unique features, there were several ergonomic shortcomings that appeared to occur frequently. The results tentatively suggest ergonomic issues that might be addressed in ergonomic training programmes and redesign programmes at the LUAS.

References

Bridger, R. 2003, *Introduction to ergonomics*, Second Edition (Taylor and Francis, London)

Council of the European Communities 1990, Council Directive 90/270/EEC of 29 May 1990 on the minimum safety and health requirements for work with display screen equipment, *Official Journal of the European Communities*, L 156/14

ISO 9241-5 1998, *Ergonomic requirements for office work with VDTs. Workstation layout and postural requirements* (International Standard Organisation, Geneva)

ISO 13406-2 2001, *Ergonomic requirements for work with visual displays based on flat panels* (International Standard Organisation, Geneva)

Kroemer, K.H.E. and Grandjean, E. 1997, *Fitting the task to the Human. A Textbook of Occupational Ergonomics*, Fifth Edition (Taylor and Francis, London)

Kroemer, K.H.E. and Kroemer, A.D. 2001, *Office ergonomics*, First Edition (Taylor and Francis, London)

Oborne, D. 1995, *Ergonomics at Work: Human Factors in Design and Development*, Third Edition (Wiley, Chichester)

Pheasant, S. 1991, *Ergonomics, Work and Health*, First Edition (Macmillan Academic and Professional Ltd, Hampshire)

Sanders, M. and McCormick, E. 1992, *Human Factors in Engineering and Design*, Seventh Edition (McGraw-Hill, New York)

Smith, M., Carayon, P. and Cohen, W. 2003, Design of Computer workstations. In J. Jacko and A. Sears (eds.) *The human-computer interaction handbook* (Lawrence Erlbaum Associates Publishers, New Yersey), 386–395

Szeto, G., Straker, L. and Raine, S. 2002, A field comparison of neck and shoulder postures in symptomatic and asymptomatic office workers, *Applied Ergonomics*, **33**, 75–84

EXPOSURE ASSESSMENT TO MUSCULOSKELETAL LOAD OF THE UPPER EXTREMITY IN REPETITIVE WORK TASKS

Ulrike M. Hoehne-Hückstädt, Rolf P. Ellegast & Dirk M. Ditchen

*Berufsgenossenschaftliches Institut für Arbeitsschutz (BGIA),
BG-Institute for occupational Health an Safety, Alte Heerstr. 111,
53754 Sankt Augustin, Germany*

The purpose of the presented study is to perform a comparative exposure assessment to musculoskeletal load of the upper extremity by applying selected evaluation schemes. Several workplaces that involved repetitive tasks had been examined using the CUELA measuring system. The exposure assessment for each workplace was repeatedly conducted using the same data but other evaluation schemes. The results that remarkably differ in some points detect the varying significance of the evaluation schemes for dissimilar repetitive work tasks and in mutual comparison. They may provide criteria for compiling a catalogue that contains rapid, inexpensive methodology as well as more sophisticated one for exposure assessment to musculoskeletal load of the upper extremity in order to follow scientific and preventive goals.

Introduction

Physical work load may provoke musculoskeletal complaints, disorders or even diseases. According to national labour statistics, health-related absenteeism caused by musculoskeletal disorders have accounted for a nearly constant 25- to 30-percent rate in Germany over the last several years and still lead the field. The reported musculoskeletal complaints are localised almost half-and-half in the upper extremity including the neck and upper back as well as in the lower back and extremity (OSHA, 2003). Aside from the physical work load caused by manual materials handling such as lifting and carrying loads, the impacts of repetitive tasks are increasingly acknowledged as risk factors for work-related musculoskeletal disorders of the upper limb. For this reason, the German statutory accident insurers initiated the investigation of workplaces in the field of the textile and clothing industry which involved repetitive tasks.

Methods

A literature review revealed the following risk factors: the repetitiveness, awkward and/or static posture and movements, forceful actions/exertions, lack of recovery and work organisation as well as several additional factors, e.g. vibration, use of (unsuitable) gloves, and so on. Registering these risk factors is complicated by the number of parameters which need consideration and the different recommendations to measure the risk factors; e.g. the repetitiveness could be expressed in terms of cycle times, frequency of joint movements or mean power frequency. Therefore, the investigations of the different workplaces were conducted by using the CUELA measuring system

(Computer supported registration and long-term analysis of musculoskeletal loads) (Ellegast, 1998; Herda, 2002). This system consists of sensors and a miniature computer that can be attached to the test person's working clothes. Long-term measurements of the joint movements are made with a sampling rate of 50 Hz and the data is stored on a memory card. The measurements are additionally documented on video which will be synchronised with the other measurements. So the CUELA system and the linked software allow for simultaneously registering the numerous factors and for analysing the data with respect to the diverse parameters required when applying several evaluation schemes. Such evaluation schemes that integrate these parameters into an overall assessment were selected from the literature, i.e. RULA (McAtamney and Corlett, 1993), HAL TLV's (© American Conference of Governmental Industrial Hygienists, 2001) and OCRA Index (Colombini et al, 2002).

Results

In the following table 1, the results of exposure assessment to musculoskeletal load for two workplaces (cutting of filter materials and sewing in the footwear industry) are depicted. The sewing workplace had been investigated before and after ergonomic intervention.

The results concerning the same workplace, i.e. cutting of filter materials, differs remarkably because RULA emphasises the posture – here it is the high percentage of working time spent in an extreme trunk bending – as the main risk factor, which is not regarded at all when using HAL TLV's or OCRA INDEX for exposure assessment. Although these evaluations schemes are restricted by focussing on the postures and movements of the arms respectively the distal upper extremity, they integrated more risk factors in a sophisticated way. For this reason, they could show a slight improvement after ergonomic intervention in the sewing workplace.

These findings may be useful to develop a new concept for a catalogue of evaluation schemes containing rapid, inexpensive methods for first investigations as well as specified methods for in-depth examinations.

Table 1. Results.

Evaluation scheme	Right arm Cutting of filter materials	Left arm Sewing footwear before intervention	Sewing footwear industry after intervention
RULA	Final score = 6 Investigate further and change soon	Final score = 6 Investigate further and change soon	Final score = 5 Investigate further and change soon
HAL TLV's	Hand activity level = 5 Normalised peak force = 2 Intersection is below the action limit	Hand activity level = 5 Normalised peak force = 3 Intersection is above the action limit but below the threshold limit	Hand activity level = 5 Normalised peak force = 2,5 Intersection is on the action limit
OCRA INDEX	OCRA Index = 1,8 No risk area	OCRA Index = 2,4 Low risk area	OCRA Index = 2,1 Transition zone between low and no risk

Acknowledgement

The authors wish to thank the Institute of Ergonomics of Munich University of applied sciences (Prof. Lesser), the engineering studio Schwan Frankfurt (Mr. W. Schwan) and the institution for statutory accident insurance and prevention in the textile clothing and leather industry for their cooperation and support.

References

American Conference of Governmental Industrial Hygienists (ACGIH) 2001, *Hand Activity Level TLV's*. In *Threshold Limit Values for chemical substances and physical agents & Biological Exposure*. (© ACGIH, Cincinnati, OH, USA)

Colombini, D. et al 2002, *Risk Assessment and Management of Repetitive Movements and Exertions of the Upper Limb*. (Elsevier, Amsterdam)

Ellegast, R.P. 1998, *Messsystem zur automatisierten Erfassung von Wirbelsäulenbelastungen bei beruflichen Tätigkeiten. BIA-Report 5/98*. Hauptverband der gewerblichen Berufsgenossenschaften (ed.) (Sankt Augustin)

Herda, C.A. 2002, Entwicklung eines personengebundenen Systems zur Erfassung komplexer Haltungen und Bewegungen der Schulter-Arm-Region bei beruflichen Tätigkeiten. Inaugural dissertation (Johannes Gutenberg-Universität, Mainz Fachbereich Medizin)

McAtamney, L. and Corlett, E.N. 1993, RULA: a survey method for the investigations of work-related upper limb disorders. *Applied Ergonomics* **24**, 91–99

OSHA 2003, *Aktueller Bericht der Bundesregierung über den Stand von Sicherheit und Gesundheit bei der Arbeit und über das Unfall- und Berufskrankheitengeschehen in der Bundesrepublik Deutschland im Jahre 2003* http://de.osha.eu.int/statistics/ statistiken/bericht_zum_stand_von_sicherheit_und_gesundheit_bei_der_arbeit/

DATA MANAGEMENT SYSTEM FOR ANALYSIS OF OCCUPATIONAL PHYSICAL WORKLOAD

Ingo Hermanns & Rolf P. Ellegast

BG-Institute for Occupational Safety and Health (BGIA), Alte Heerstrasse 111, 53754 Sankt Augustin, Germany

For some years the German BG Institute for Occupational Safety and Health (BGIA) and several technical consultancy services of the German institutions for statutory accident insurance (Berufsgenossenschaften, BGs) apply the CUELA measurement system for long-term recording and analysis of physical workload in various occupational sectors.

The software of the CUELA system can be used to analyse one or several data sets. The data management system presented in this paper enables the investigator to analyse and evaluate a much larger set of long-term measurement data. For this purpose a specific software tool was developed recently, which conforms to these requirements and, in addition, makes the intersectional evaluation of physical workload data possible.

Introduction

In 1993 the German BGs began to compensate employees for work-related degenerative diseases of the lumbar spine caused by heavy-load handling or working in extreme trunk-flexed postures. For the evaluation of these occupational diseases as well as for preventive purposes the knowledge of work-related mechanical exposure is needed in order to estimate the health risk for the musculoskeletal system during occupational tasks. Therefore the BGIA developed a measurement system named CUELA (Computer-assisted registration and long-term analysis of musculoskeletal workload), which was successfully used for whole-shift recording and analysing of mechanical loads in various occupational work sectors.

Lately the software was improved by integrating a new data management system, which makes comparative assessments of long-term exposure data sets possible.

Methods

CUELA system

CUELA is a mobile long-term field measurement system for the recording of working postures, joint angles and load weights (Ellegast and Kupfer, 2000). To meet different requirements in the field the CUELA system can be used in different stages of expansion. There are different versions of CUELA systems to detect:

- spine and lower limb movements and handling of load weights (see Figure 1)
- spine, upper/lower limb and head movements and handling of load weights.

The registration of heart rate, EMG signals and hand force data can be realized by means of add-on modules.

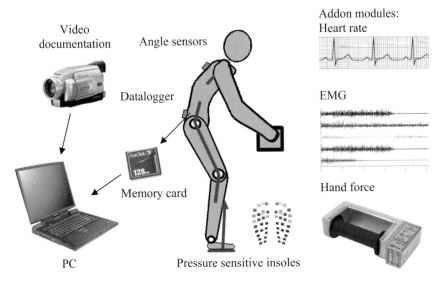

Figure 1. CUELA system.

In order to relate the measurement data to the corresponding occupational situation, the measurement sessions are recorded on video. After finishing the measurement the video can be displayed synchronously to the measurement data.

CUELA software

The CUELA user interface consists of three main windows (see Figure 2). One window shows time graphs of the measured body and joint angles or a posture code – classified by the OWAS method (Karhu et al, 1977) – or further calculated results, e.g. lumbar compression forces and handled load weights. The second window contains a three-dimensional figure, which is animated by the measurement data. In the third window a synchronized video of the workplace situation is displayed.

Based on the large number of different input parameters the CUELA software generates various statistical results, which are automatically presented in charts. These include body and joint angles, load weights, lumbar compression force, OWAS, heart rate and EMG with chart types histograms, boxplots, duration of static levels or postures and statistic values.

In general, all statistical outcomes are classified with reference to the literature and relevant standards (e.g. ISO 11226, EN 1005-1, EN 1005-4, Drury 1987).

The charts are processed into an HTML or a Microsoft Word document. Such a document may contain more than 50 single charts. Although the output is customisable, the number of charts is often too large to get a good overview over the work-related exposure. For prevention tasks this output is very detailed but no automated comparison between several measurements had been possible.

The data management system

To achieve this comparison among a set of measurements a data management system with an integrated scoring system was developed.

For risk assessment each chart is classified into a score system that ranges from 1 (low exposure) to 9 (high exposure). It should be pointed out that the score has got a relative ranking. The measurement of a set or project with the lowest exposure indicator

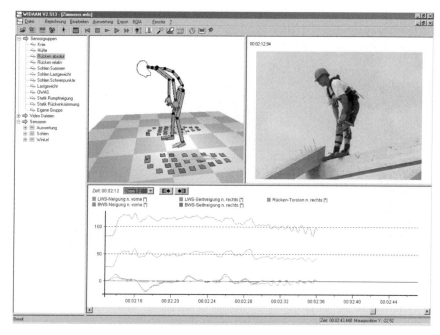

Figure 2. CUELA software: measuring data, animated figure, synchronised video.

Table 1. Examples of score calculations.

Score type	Formula	Remarks
Cumulative load weights	$\sqrt{\dfrac{\sum m_i^2 \cdot \Delta t}{T}}$	m_i: load weight (kg), Δt: duration of handling, T: measurement duration
OWAS score	$\dfrac{\sum ActionCode_i \cdot \Delta t}{T}$	According to OWAS, action codes: AC2 = 2, AC3 = 3, AC4 = 4
Lumbar compression force	$\sqrt{\dfrac{\sum F_i^2 \cdot \Delta t}{T}}$	F_i: compression force (kN), Δt: sample rate, T: measurement duration
Trunk flexion angle	$\dfrac{\sum AngleZone_i \cdot \Delta t}{T}$	According to EN 1005-4, Angle zones: 2 (<0°), 1 (0°–20°), 2 (20°–60°), 4 (60°–90°)

has got a score of 1 and the measurement with the highest exposure indicator has got a score of 9.

The calculation of the score values should be understood as an ongoing process which depends on relevant standards and results of research processes. The current version of the CUELA software uses the following formulas to generate the score values (see Table 1). The maximum values of each score are used to scale all other results to the range from 1 to 9. If there is a new measurement with a score which exceeds the previous maximum value of 9 all score values must be recalculated.

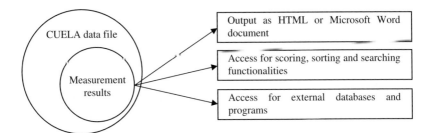

Figure 3. Extract of the measurements results embedded in the CUELA data file.

Table 2. Draft ranking list based on OWAS score.

Job	Activity/work area	OWAS ranking
Airplane Mechanician	Wiring	9.0
Carpenter	Assembling rafters	8.7
Concrete Worker	Drawing of pattern	7.6
Nurse	Internal medicine	5.3
Concrete Worker	Durn construction	5.3
Potter	Making pottery	4.6
Warehouseman	Commissioning	3.6
Electrician	Cable ducts	2.5
Carpenter	Commissioning	2.2

Because of the complex structure of the CUELA measurement file and the memory usage, no common database is applied. The searching and scoring features for the measurement comparison and risk assessment are implemented as follows.

The statistical results, which are generated automatically by the CUELA software, are not only exported as HTML or Microsoft Word document but, also stored as an embedded extract in the CUELA data file (see Figure 3). This accelerates the access to the measurements statistics, because no recalculation of the statistical results is needed.

Results

With this data structure all requirements for the data management system are fulfilled. It can now provide functions like searching and sorting for one score value or any combination of scores. This method can be applied to a huge number of measurements with regard to the various exposure indicators.

As a result of this processing it is possible to conduct a comparative analysis of physical workload data. The user obtains a ranking list of occupational tasks based on single (see Table 2) or multiple exposure indicators or on the risk scores. The underlying evaluation score system is a tentative draft. Our future aim is to conduct a review of the correlation between the ranking lists and epidemiological outcomes in order to progressively refine the scoring system.

References

DIN EN 1005-1 2005, *Sicherheit von Maschinen – Menschliche koerperliche Leistung – Teil 1: Begriffe*. (Beuth, Berlin)

DIN EN 1005-4 2002, Menschliche koerperliche Leistung – Teil 4: *Bewertung von Koerperhaltungen und Bewegungen bei der Arbeit an Maschinen*. (Beuth, Berlin)

Drury, C. G. 1987, *A Biomechanical Evaluation of the Repetitive Motion Injury Potential of Industrial Jobs.* Seminars in Occupational Medicine, **2**; 41–49.

Ellegast, R. and Kupfer, J. 2000, Portable Posture And Motion Measuring System For Use In Ergonomic Field Analysis. In: *Landau Ergonomic Software Tools in Product and Workplace Design.* (Ergon, Stuttgart)

ISO 11226 2000, *Ergonomics – Evaluation Of Static Working Postures*, International Organization of Standardization.

Karhu, O., Kansi, P. and Kuorinka, I. 1977, *Correcting working postures in industry: A practical method for analysis*, Applied Ergonomics, **8**; 199–201.

EFFECTS OF NON-NEUTRAL POSTURE ON HUMAN RESPONSE TO WHOLE-BODY VERTICAL VIBRATION

Gerry Newell[1], Setsuo Maeda[2] & Neil Mansfield[1]

[1]*Environmental Ergonomics Research Centre, Loughborough University, Loughborough, UK*
[2]*Department of Human Engineering, National Institute of Industrial Health, Kawasaki, Japan*

Off-road machinery drivers are exposed to a number of occupational hazards. Whole-body vibration and awkward postures are two of the most prevalent in this population; this paper addresses potential risks of combined postural constraints with whole-body vibration exposure. Common off-road working conditions were simulated on a multi-axis shaker system to determine the influence of non-neutral twisted postures and whole-body vertical vibration on the transmissibility of vibration from the seat to the head of 14 subjects. Results for vertical transmissibility indicated a higher degree of inter-subject variability for the normal posture compared to the twisted posture. Both conditions showed comparable resonance frequency (~5–6 Hz), however, above the frequency of the peak the similarity between conditions was lost. Results suggest clear differences between relative head motions for the two posture conditions tested. In order to reduce the combined risk of whole-body vibration and constrained posture the precise nature of the head motion needs to be fully characterised.

Introduction

Operators of earthmoving machinery are regularly exposed to a range of occupational hazards. The nature of their working task can expose them to unsafe magnitudes of whole-body vibration and shock, in conjunction with a variety of postural constraints. This has been highlighted previously by Newell et al (2005), who, through field research identified concomitant risk factors for the operators: driving in postures with elements of twisting in the back and neck whilst simultaneously exposed to whole-body vibration. Drivers who frequently adopt driving postures in which the trunk is considerably twisted or bent forward have been found to have a greater risk of low back pain (Hoy et al, 2005). Vibration exposure has also been associated with back pain, yet previous research has failed to address these two risk factors in combination.

One way to develop the understanding of the combined effects of whole-body vibration and posture on the human body is to consider vibration transmission through the body. The transmissibility of the human body can indicate the biomechanical response to whole-body vibration (Paddan and Griffin, 1998). Seat-to-head transmissibility and investigation of the rotational movement of the head can also provide an indication of the level of disturbance an operator may experience while operating earthmoving machinery. Head motions of most interest may be in other axes particularly pitch (flexion/extension of neck), especially as this can have a greater impact on vision (Griffin, 1990).

Many studies have been performed to investigate the transmissibility of vibration from the seat to the head for subjects exposed to translational whole-body vibration (WBV) (e.g. Paddan and Griffin, 1993; Matsumoto and Griffin, 1998). One consistent finding across studies is that seated subjects' fundamental resonance frequency exists in the region around 4–5 Hz for vertical vibration exposure. Researchers are interested in the resonance frequency of the human body as the vibration at this frequency will be amplified by a build-up of stored energy in the repeated stretching and compression of tissue (Mansfield, 2005). Posture has been mentioned in a number of the studies as a factor that could influence the differences found between subjects (e.g. Paddan and Griffin, 1988). However, none of the studies have attempted to reflect the typical postures adopted by off-road machinery drivers.

Zimmerman and Cook (1997) investigated the effects of vibration frequency and postural changes on subjects' response to seated WBV exposure. They found significant interactive effects between pelvic orientation, vibration frequency and seated human's response to WBV. The authors concluded from the results that proposed standards should consider occupant posture during vibration exposure if their intent is to decrease the incidence of WBV associated low back pain. Kitazaki and Griffin (1998) reinforces this notion by suggesting that any forces causing injury from WBV will not be well predicted by biomechanical models incapable of representing the appropriate body motions and the effects of body posture. Therefore it is important to understand the interactions between vibration exposure and non-neutral postures to improve understanding of the biomechanical responses of the human body. This paper reports the results from a study specifically designed to investigate these interactions.

Methods

Thirteen male subjects and one female subject participated in the experiment. Subjects had a mean age of 23.4 (s.d. 1.2) yrs, a mean stature of 170 (s.d. 10) cm and a mean weight of 66.4 (s.d. 11.4) kg. The experiment was approved by the Research Ethic Committee of the National Institute of Industrial Health, Japan.

Subjects were exposed to 60 seconds of vertical random vibration within the frequency range 1–20 Hz. The magnitude of the vibration was $1.0 \, \text{m/s}^2$ r.m.s (root mean square, unweighted). This vibration magnitude is representative of typical vibration magnitudes found in a variety of off-road machines. Subjects sat in two different postures: "normal" forward facing posture and "twisted" non-neutral posture. In both conditions the hands rested on the lap and the back was off the backrest. In the "normal" condition, subjects were instructed to sit in a relaxed upright posture facing straight ahead. In the "twist" condition, subjects were instructed to look over their right shoulder in the coronal plane (Figure 1). The trials were repeated 3 times for each posture condition and the order of test conditions was randomised using a Latin-square design.

Forward Posture   Reversing Posture

Figure 1. Two postures used in the experiment. Forward facing posture (normal) and reversing non-neutral posture (twist).

Subjects held a bite-bar tightly in their mouth comprising accelerometers mounted on the left and right side and at the back of the head. The accelerometers measured the vibration at the head in the fore-and-aft, lateral and vertical directions. Seat-to-head transmissibilities were calculated in order to determine the ratio of the input acceleration at the seat (vertical vibration) and the output acceleration at the head (fore-and-aft, lateral and vertical vibration). The ratio gives a measurement of the extent to which the vibration has been attenuated or amplified by the spinal system. If the ratio is greater than 1 then the vibration has been amplified. The transfer function was calculated using the cross-spectral density method. Calculations of roll (lateral bending of head) and pitch (flexion/extension of head/neck complex) motion at the head were also completed.

Results and discussion

Intra-subject variability was very small between the three repeat trials for both posture conditions. The inter-subject variability was greater between subjects during the normal posture condition compared with twisted posture (Figure 2).

The normal posture condition indicates a peak in the transmissibility of vertical vibration around 5–6 Hz in all subjects with the exception of subject 1. This subject had a peak in transmissibility at 12.5 Hz; it is difficult to determine what would cause this shift in peak transmission as posture was controlled during all trials. This anomaly is removed during the twisted posture condition where all subjects had a peak in transmissibility between 5–6 Hz. This is comparable to the normal posture condition.

For a rigid system, seat vertical to head vertical transmissibility would equal 1.0 at all frequencies. Seat vertical to head horizontal transmissibility would equal 0.0 at all frequencies. Observations of Figure 3 show that both the fore-and-aft (x) and lateral (y) axes have transmission values greater than zero. This is the result of cross coupling in the system as the human body is a flexible, non-rigid system. Transmissibility in the vertical (z) axis is greater than 1 at the peak frequency for both the normal and the twisted postures. This amplification is sustained in the normal posture even at higher frequencies; the twisted posture presents a distinctly different pattern where less amplification is experienced after the peak frequency of 5 Hz. This shows that during the twisted posture condition there is increased damping occurring within the system. This damping may result from vertebral rotational components while the spine is twisted.

Figure 3 shows differences between postures for the horizontal and rotational axes of vibration. Transmission to fore-and-aft and pitch motion at the head was slightly elevated in the normal posture compared with the twisted posture. This contrasts to a large

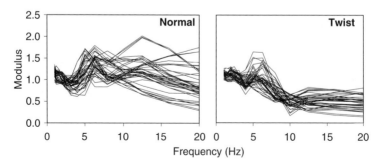

Figure 2. Transmission by modulus for vertical seat-to-head vibration of all subjects trials while seated in a normal posture and twisted posture.

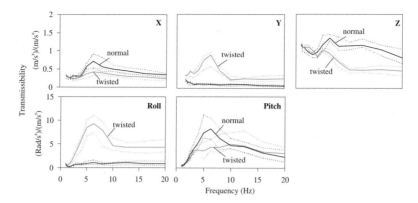

Figure 3. Transmissibility of vibration from the seat to the fore-and-aft (x), lateral (y) and vertical (z) axes at the head, with roll and pitch motion of the head. Values are presented as median (solid line) and upper and lower quartiles (dashed lines) of 14 subjects. Black lines denote normal posture and grey lines the twisted posture.

increase in lateral and roll motion at the head during the twisted condition. There was negligible transmissibility in the normal posture compared with the twisted posture; lateral transmission changed from ~0.0 to 1.0 and there was a substantial increase in roll motion at the head.

The increased forces and moments created during the twisted posture could place additional stress on the spinal column, as the stabilisation of the head-neck complex becomes more difficult. The cervical spine can withstand the highest axial compressive loads and sustain the highest load (and strain) magnitudes when it is in the straight neutral position (White and Panjabi, 1990). This is important to take into consideration when interpreting the attenuation of vertical vibration observed for the twisted posture. Sitting in a normal upright posture with the head in the neutral position causes a low load on the cervical spine; the load movement is balanced by muscle forces and tension of the passive structures. The more the head departs from neutral, the more the load increases (Thuresson, 2005). Therefore in the neutral position the head and neck will be more adapted to loading, this will be the case when exposed to vibration and shocks. This could mean that while the spine is rotated away from the neutral position there may be a greater impact on the structures even with a lower vibration transmission.

There is one important consideration of this study in the understanding of the relative head motions during exposure in the two different postures. During the twist condition the bite-bar held in the subjects' mouth would be rotated ~90° to the right and slightly below the horizontal compared to the normal position; this would cause the accelerometers to change orientation with respect to the pelvis. Considering the complex motion of the neck during the twist condition it is likely the x-axis values have transposed to the y-axis and in addition roll motion appears as pitch motion. This makes individual interpretation of either roll or pitch a challenge. To investigate this phenomenon, further research is required to determine the dynamics of the head using 3D motion analysis.

Conclusions

It was found that resonance frequency for the vertical transmission of vibration from a seat to the head was similar for a neutral and a twisted posture. There was lower vertical vibration transmission for the twisted posture compared with the normal posture; however, this was counteracted by the large increase in lateral transmission and roll motion at the head. The findings highlight concerns of the increased loading that can be placed on the spinal units during twisted postures and whole-body vibration exposure.

Acknowledgements

This work was supported by the Japan Society for the Promotion of Science and the National Institute of Industrial Health.

References

Griffin, M.J. 1990, *Handbook of Human Vibration*. First Edition, (Academic Press, London)

Hoy, J., Mubarak, N., Nelson, S., Sweerts de Landas, M., Magnusson, M., Okunribido, O., and Pope, M. 2005, Whole body vibration and posture as risk factors for low back pain among forklift truck drivers, *Journal of Sound and Vibration*, 284, 933–946

Kitazaki, S., and Griffin, M.J. 1998, Resonance behaviour of the seated human body and effects of posture, *Journal of Biomechanics*, 31, 143–149

Mansfield, N.J. 2005, *Human Response to Vibration*, First Edition, (CRC Press, Boca Raton)

Matsumoto, Y., and Griffin, M.J. 1998, Movement of the upper-body of seated subjects exposed to vertical whole-body vibration at the principal resonance frequency, *Journal of Sound and Vibration*, 215 (4), 743–762

Newell, G.S., Mansfield, N.J., and Notini, L. 2005. Inter-cycle variation in whole-body vibration exposures of drivers operating track-type machines, *Journal of Sound and Vibration* (in press)

Paddan, G.S., and Griffin, M.J. 1998, A review of the transmission of translational seat vibration to the head, *Journal of Sound and Vibration*, 215 (4), 863–882

Paddan, G.S., and Griffin, M.J. 1993, Transmission of vibration through the human body to the head: a summary of experimental data. *ISVR Technical Report 218*, University of Southampton

Thuresson, M. 2005, On neck load among helicopter pilots – Effects of head-worn equipment, whole-body vibration and neck position, *Thesis*, Neurotic Department, Division of Physiotherapy, Karolinska Institutet, Stockholm, Sweden

White, A.A., and Panjabi, M.M. 1990, *Clinical Biomechanics of the Spine*, Second Edition (Lippincott-Raven, Philadelphia)

Zimmerman, C.L., and Cook, T.M. 1997, Effects of vibration frequency and postural changes on human responses to seated whole-body vibration exposure. *International Archives of Occupational and Environmental Health*, 69, 165–179

A SYSTEMATIC REVIEW AND META-ANALYSIS OF LOWER BACK DISORDERS AMONG HEAVY EQUIPMENT OPERATORS

Mbulelo Makola & Ash Genaidy

University of Cincinnati College of Engineering, Cincinnati, Ohio, USA

Critical appraisal demonstrated that the quality of health studies on heavy equipment operators were marginal to average. There was an association between working as heavy equipment operator and lower back pain. Further epidemiological studies are warranted on heavy equipment operators. The findings should be used for the improvement of heavy equipment design from an operator standpoint.

Introduction

Heavy equipment is engineering vehicles designed to execute specialized, heavy-duty tasks such as engineering and construction activities. The operation of heavy equipment may expose its driver to multiple ergonomic risk factors, including static work postures (e.g., trunk and neck twisting, stooping, deep sideway trunk-bending), whole-body vibration, shock (also called jarring and jolting), physical work demands (e.g., walking, pulling, lifting), climatic conditions (e.g., heat, cold), and psychosocial factors (e.g., job satisfaction). These ergonomic risk factors are known to be associated with musculoskeletal disorders. This research deals with this issue among heavy equipment operators and seeks an answer to the following question: "Are whole-body vibration/shock and working postures associated with lower back and neck pain among heavy equipment operators (i.e., forklift, crane, tractor, and earth moving equipment) while accounting for individual (i.e., age, gender, prior history of back or neck disorders) and occupational confounders (i.e., material handling, climatic conditions, psychosocial conditions)?". Numerous reviews have been published on the subject. The great majority of these reviews was qualitative and did not include the ingredients of an evidence-based methodology. This paper deals with a critical appraisal and meta-analysis designed to provide an answer to the aforementioned question.

Methods

Publications were retrieved by a search of electronic databases including PubMed, NIOSHTIC, and NIOSHTIC-2 (up to May 2005) using the following keywords: (a) forklift/fork-lift and back/neck pain; (b) tractor and back/neck pain; (c) crane and back/neck pain; (d) earth movers; (e) operating engineers and back/neck pain. In addition to the electronic search, bibliographies in identified papers were reviewed for additional studies. The critical appraisal was conducted through the use of an "Epidemiological Appraisal Instrument". A meta-analysis was conducted using the fixed-effect model if there was no heterogeneity at the 10% level; else, the random-effects model was applied.

Results and discussion

Thirteen articles were found which matched the inclusion criteria clearly specified in the research question. The critical appraisal demonstrated that the overall quality ranged mostly between marginal to average. It is worth noting that the number of studies for forklift operators was twice the number of studies for crane, tractor or earth moving operators. Three studies measured whole-body vibration and postural loading as exposures. Lower back pain was investigated as an outcome in all studies. Neck pain was only measured in one study.

Heterogeneity was found for all groups except forklift operators. Therefore, the random-effects models were used for crane, tractor, and earth moving operators. The fixed-effect model was applied for forklift operators. The meta-relative risk for the different groups was as follows from highest to lowest:

(1) Tractor – 2.983 (95% CI 1.595–5.582)
(2) Forklift – 2.077 (95% CI 1.645–2.624)
(3) Crane – 1.854 (95% CI 0.893–3.848)
(4) Earth Moving – 1.839 (95% CI 0.897–3.773)

The overall estimate was only significant at the 5% level for the forklift and tractor operators.

Two major conclusions can be drawn from the above results: (1) the quality of epidemiological studies should be improved for ergonomic studies in the field; (2) attention should be paid to exposure and outcome measurements; (3) the findings should be used in the improvement of heavy equipment design for its operators.

TRANSPORT

ENCOURAGING CO-OPERATION IN ROAD CONSTRUCTION ZONES

Tay Wilson

Psychology Department, Laurentian University, Ramsey Lake Road, Sudbury, Ontario, Canada P3E 2C6

Co-operation (or not), with its immediate and following ramifications, at lane constriction merge sites during road construction is highly conspicuous to drivers. Observed co-operation at such lane constriction sites on the A1(M) motorway is reported. Comparison with some data in Canada is made. A second order linear control theory model of a gradually dampening delay of traffic resulting from non co-operation is developed and presented. Finally, a programme for encouraging more effective cooperation and subsequent traffic flow is developed and presented.

Introduction

A particularly noticeable and often frustrating instance of unsociable driving occurs at temporary lane closing sites when some drivers fail to merge appropriately into the single lane, electing rather to rush down the cleared space and force themselves into the line of traffic at the last moment. Let us call this manoeuvre "queue hopping" and the manoeuvre of actively assisting a driver to enter a queue before an impending lane closure as co-operation.

Wilson and Godin (1993) studied co-operation in lane merging situations within the city of Sudbury, Canada. Overall, sixty per cent of trapped merging drivers were assisted by timely speed adjustment or lane change. However, when opportunity existed for an assisting lane change by a potentially cooperating driver, assistance rate was only half that of when only speed change was available for co-operation. This result indicated the possibility of inertial driving in the former situation. In contrast, on a single trip in 2005 of several hundred kilometers from Gretna Green, Scotland to Cambridge, England along the A1M route, amid the usual variety of lane closures to which that route has been subjected over recent years, only one instance of co-operation was recorded – at the end of Stainmoor.

Condoning queue hopping appears to be particularly deleterious for the general encouragement of sociable, safe driving for at least two reasons. First, since large numbers of drivers see the offending overtaking vehicle gain position and time, the reward for the manoeuvre seems likely to encourage the less scrupulous to follow suit and thus to induce further unsociable behaviour down the road when lanes are returned to service. Second, the activity, accompanied by inevitable negative delaying social reaction in the queue, is likely to slow down overall traffic flow rates.

Three topics are covered. First, an introduction is given to some applicable mathematics relevant to assessing flow rates. Second, suggestions are given for measuring induced unsociable behaviour beyond lane restrictions. Third, measures to reduce queue hopping are proposed.

Tel (705) 675-1151, fax (705) 675-4889

Mathematics of linear control systems

The techniques of engineering control theory have been usefully applied, in ostensibly non engineering contexts. [For instance, Wilson (1969) successfully modelled optokinetic nystagmus as the output of a second order linear system.] Most simply control theory represents a time varying system by an input modified by feedback from earlier outputs, an output and a mathematical law or transfer function determining the relation between the input and output. The important states of any time dependent system are its transient and its steady states; correspondingly the two common techniques of linear control system analysis are concerned with transient responses of the system to step inputs and steady state responses to sinusoidal inputs.

In both cases, three types of systems are relatively easily studied: zero-order linear systems, containing no differentials and obeying equations of the type

$$ky = f(t), \tag{1}$$

first-order linear systems obeying the differential equation

$$T\, dy/dt + y = 1/k\, f(t), \tag{2}$$

and second-order linear systems which obey the equation

$$\frac{1}{\omega^2}\frac{d^2y}{dt^2} + \frac{2\eta}{\omega}\frac{dy}{dt} + y = \frac{1}{K} f(t) \tag{3}$$

where eta is the damping ratio and omega is the natural angular frequency. The general solution to a second order linear system is either the sum of a pair of exponentials each multiplied by a constant or an exponential multiplied by a harmonic (involving sines and cosines) of the form (Swokowski, 1983):

$$y = e^{SX}\, (C^1\, \text{Cos}\, tx + C^2\, \text{Sin}\, tx) \tag{4}$$

For a zero order system, the output y depends only on the value of input $f(t)$. The response of a first-order system involves, for a step input, a delay in the attainment of the steady state response depending upon the value of the time constant T. A second-order system has four types of response depending upon the value of eta (η): undamped (eta = 0); underdamped (0 < eta < 1); overdamped (eta > 1) and critically damped (eta = 1). An undamped system tends to oscillate harmonically about some response value, an underdamped system oscillates with ever decreasing amplitude about some final response value. An overdamped system approaches a final response value gradually and monotonically. A critically damped system approaches the final response value as fast as possible without resulting in oscillatory behaviour. Of most interest here is that when sinusoidal (periodic) input is applied a zero-order system has no delay between output and input, a first-order system may have a phase shift of up to 90 degrees and a second-order system can have a phase shift up to 180 degrees at which point the response vector is in the opposite direction of the input vector. Higher phase lags can be modelled in terms of combinations of first and second order systems. Wilson (1969) provides an example of applying control theory to optokinetic nystagmus, the autonomic slow following – quick return of eye tracking to a series of lines or fence posts moving past an observer. He found eye tracking lagging stimulus back and forth movement by 90–180 degrees indicating a second order brain control system which processed acceleration information.

Applying linear control theory to unsociable driving at lane constrictions

Assume that the drivers cutting into the queue at lane constrictions form a periodic disturbance (which is at least Fourier analysable into sinusoidal components) to the slowly moving queue causing a propagating series of brake lights and eventually alternate stopping and starting down the queue. Then as many readers have personally noticed, drivers relatively far down the queue in such situations will be delayed in their braking relative to drivers farther up the queue. Many will also have noticed that at times they will actually be accelerating forward when vehicles far up the queue are braking and they are braking when vehicles up the queue are accelerating forwards. In terms at least of acceleration this motion involves a lag of near 180 degrees and suggests the possible presence of a second order linear control system. Moreover, if as one travels back along the queue, the number of vehicles increases between 180 degrees out of phase braking instances the system should be underdamped (eta < 1 in the second order differential equation).

Now since second order linear control systems may function in such autonomic phenomena as optokinetic nystagmus (see Wilson, 1969), the experience of second order phenomena may affect behaviour in a driving context which of course has a strong visual component. These effects could spread beyond the obvious and frequently occurring collision with the braking car in front while watching moving vehicles farther up the queue to post queue driving behaviour.

Moreover, this phenomenon could be studied on the road and as simulated on a computer or driving simulator. It is requested that someone take up the challenge. For instance, length of queues and cutting in frequency required to induce "phase lags" in braking of 0–90 degrees and 90–180 degrees could be determined. Appropriate manipulation of cutting in frequency should enable disentanglement of simple time delay and stop and go phase lag magnitude. Now, having specified the delay input to drivers, their driving style could be studied as they sort themselves out after the lane restriction has ended. Furthermore collision rates and types on post-lane restriction roadways could be related to theoretical time delay/phase lag characteristics developed from prior knowledge of traffic flow and "queue hopping" frequency.

Reducing queue hopping

Because it is frustratingly visible to so many at a time and because it clearly assists the transgressor it appears desirable to try to reduce, this occurrence in pursuit of a general more sociable driving style in the locale. It is suggested that queue hopping sites are ideal locales to apply cameras when combined with signs interdicting merging past a certain point. A public information campaign could be developed instructing drivers as to appropriate behaviour and consequences (viz., merge before interdiction sign, allow merging drivers before that point to interleave as seamlessly as possible, camera in use, violators will be prosecuted). Success of the programme could be assessed in terms of change in time delay of traffic as well as by collision rate and speeding rate at various points downstream from the lane constriction.

References

Wilson, T. 1969, *A Control Theory Analysis of Optokinetic Nystagmus*. PhD. Thesis. York University, Toronto, Canada

Wilson, T. and Godin, M. 1993, A study of cooperation extended to trapped merging drivers. In E.J. Lovesey (ed.) *Contemporary Ergonomics 1993*, (Taylor and Francis, London), 111–116

Swokowski, E.W. 1983, *Calculus with Analytic Geometry*, (Prindle, Weber, and Schmidt, Boston, MA, P. 914).

DEVELOPMENT OF A METHOD FOR ERGONOMIC ASSESSMENT OF A CONTROL LAYOUT IN TRACTORS

D. Drakopoulos* & D.D. Mann

*Department of Biosystems Engineering, University of Manitoba,
Winnipeg MB R3T 5V6, Canada*

Efficient operation of a machine depends upon the design of its control layout. The literature documents several factors that are of importance in the design of a control panel. Despite knowledge about each individual factor, it is not known how these factors should be combined to yield an overall ergonomic assessment of a specific control layout. The objective of this paper is to describe the initial steps in a process to establish a numerical index capable of comparing, on a mathematical basis, different control arrangements in agricultural tractors. The model will be based on information that has been gleaned from the published literature, with input from both professional ergonomists and experienced agricultural tractor operators.

Introduction

In old tractors, controls exclusively consisted of hand and foot levers which, from an ergonomic perspective, were not very friendly to the driver due to their placement, length, and the travel distance required to carry out a specific operation (Ely et al, 1956). As a result, numerous complaints were reported from operators regarding the lack of overall ergonomic design inside the cab (Purcell, 1980). Banks and Boone (1981) developed a mathematical formula called "Index of Accessibility" to arrange control devices in a control layout based on two factors: the user's reach envelope and the frequency of use of each control. It was further assumed that frequently-used controls would be located closest to the operator. The goal of research related to controls is to maximize the efficiency of operation while minimizing driver effort (Hansson and Oberg, 1996; Deisinger et al, 2000).

A new method capable of comparing different control arrangements on a mathematical basis is of great interest because the meaning of "control access" cannot be determined theoretically. There are no existing standards with which one can make a comparison between the control's arrangement and the driver's convenience of operation. In this study, the initial steps in a process to establish a model to express this concept will be presented. The model will include information related to: (1) design standards for various types of controls, (2) the frequency with which categories of controls will be accessed during operation of a tractor, and (3) human factors considerations. The model, which is currently under development, will combine the above determinants in a numerical "index of functionality", which will assess the degree to which tractors' controls are compatible with the expectations of the operator for easy, efficient, and convenient operation. For each type of control, an equation will be developed which incorporates the design characteristics that affect its operation efficiency and its rank in terms of frequency of use during a typical day of tillage operation.

---

* Email: umdrakop@cc.umanitoba.ca

Methods and Results

Design standards for various types of controls

Various types of controls are found in tractors (Purcell, 1980). It takes more than variety (in terms of control types), however, to make a given layout of controls ergonomically acceptable. Controls must meet specific design standards (Clark and Corlett, 1995), be located within the functional reach of the operator (Purcell, 1980), and be properly labeled according to the latest standards (ASAE S304.7, 2000). In addition, basic controls must be placed on the appropriate side of the operator (Purcell, 1980) and be suitable for the specific task that will be carried out (Sanders and McCormick, 1993).

By example, the published literature cites that the most important design dimensions of push buttons are the diameter and separation distance. A detailed review of design guidelines was conducted to discover their recommended dimensions (Table 1). From an ergonomic perspective, these values represent an ideal push button located in a control panel. The most recent publications were consulted in determining the following recommendation: diameter = 12–25 mm; separation distance ⩾50 mm. The current design standards better depict the latest ergonomic knowledge related to control design. As a result, an effort was made to include the most recent recommendations giving special consideration to the optimum values proposed.

As done by Banks and Boone (1981), we used the proportional principle to establish a score for each type of control. We assumed that the ergonomic score is related to the proportion of controls that meet the recommended dimensions. For push buttons, the score is calculated as the average of the proportion of buttons having a diameter between 12 and 25 mm and the proportion of buttons with a minimum separation of 50 mm. Each proportion is given equal weight when determining the overall ergonomic score for push buttons (Eq.1).

$$Bp = (Bp1 + Bp2)/2 \qquad (1)$$

where, Bp = ergonomic score for push buttons; $Bp1$ = proportion of push buttons with a diameter between 12 and 25 mm; and $Bp2$ = proportion of push buttons having a minimum separation of 50 mm.

The summarized design dimensions for seven different types of controls that have been gleaned from the published literature are presented in Table 2.

Table 1. Recommended push button diameter and separation distance(optimum values shown in parentheses).

Author	Year of publication	Diameter (mm)	Separation (mm)
Corlett and Clark	1995	12–25	15–22
Van Cott and Kinkade	1972	⩾12	12–50
Grandjean	1988	12–15	(50)
Sanders and McCormick	1993	⩾12	(50)
NASA–STD–3000	1995	⩽40	N/A
MIL–STD–1472F	1999	10–25	(50)
Pheasant	1986	12–15	(50)
Weimer	1993	10–19	(50)
Konz	1990	N/A	(50)

Frequency of use for categories of controls

To determine frequency of use, controls were categorized in five groups according to their general function (i.e., steering, control of the implement, control of the motion of the tractor, control of the internal environment, and control of the external environment) based on the opinion of ten experienced farmers. A questionnaire was distributed, asking each farmer to rank the categories of controls according to their frequency of use during a tillage operation. Although surveys were completed individually, all ten farmers provided an identical ranking. The steering wheel was ranked as the most frequently used control. In decreasing frequency of use, farmers ranked the controls as follows: controls related to the functioning of the implement, controls related to the motion of the tractor, controls related to the internal environment, and controls related to the external environment. In the final model, the ergonomic score for each type of control will be multiplied by the relative frequency of use calculated for that type of control.

Human factors considerations

The following four ergonomic factors will be included in the final model: i) placement of controls, ii) functional reach, iii) suitability of controls, and iv) labeling of controls. Placement of controls refers to the proportion of controls on the right-hand side of the operator. Functional reach refers to the proportion of controls within a radius of 750 mm, covering a 180° envelope, in front of the operator. Suitability of controls refers to the proportion of controls where mode of action is appropriate to function. Finally, labeling of controls refers to the proportion of controls having an associated label, icon, or symbol. Five professional ergonomists participated in a survey in which they were asked to distribute 100 points among the four factors according to their relative importance (Table 3). In the final model, each ergonomic factor will be multiplied by a weight factor (i.e., WPc, WSc, WFr, and WLc) that incorporates the expert weight.

Table 2. **Summarized design standards for seven different types of controls.**

Dimension	Rotary Switch	Toggle Switch	Rocker Switch	Knob	Push Button	Hand Lever	Steering Wheel
Length (mm)	25–76	–	≥ 12	–	–	–	–
Width (mm)	≤ 25	–	≥ 6	–	–	–	–
Height (mm)	12–75	–	–	12–25	–	–	–
Separation (mm)	≥ 25	≥ 50	≥ 19	≥ 50	≥ 50	≥ 100	–
Arm length (mm)	–	12–50	–	–	–	–	–
Diameter (mm)	–	–	–	12–100	12–25	–	400–510
Displacement (mm)	–	–	–	–	–	≤ 355	–
Activation force (kg)	–	–	–	–	–	≤ 16	–
Rim thickness (mm)	–	–	–	–	–	–	19–32
Tilt angle (°)	–	–	–	–	–	–	40–60

Table 3. Weight assigned to each ergonomic factor by professional ergonomists.

Ergonomic experts	Placement of controls	Functional reach	Suitability of controls	Labeling of controls
Expert 1	30	10	20	40
Expert 2	20	40	30	10
Expert 3	30	20	40	10
Expert 4	30	15	15	40
Expert 5	40	40	15	5
Average	30	25	24	21
Expert weight	0.3	0.25	0.24	0.21

Rationale for the Index of Functionality model

The numerical index developed is called the "Index of Functionality" (IF) (Eq. 2). It can take any value between 0 and 1, with 1 being defined as the optimum value in terms of the functionality of an entire control panel.

$$IF = \frac{[(Fc_1 \cdot c_1) + \ldots + (Fc_i \cdot c_i)] + [WP_c \cdot P_c] + [WS_c \cdot S_c] + [WF_r \cdot F_r] + [WL_c \cdot L_c]}{\sum^n F + (WP_c + WS_c + WF_r + WL_c)} \quad (2)$$

where, $c_1, \ldots, c_i$ = different types of controls; $Fc_1, \ldots, Fc_i$ = relative frequency of use of different types of controls; n = total number of control types; P_c = placement of controls; S_c = suitability of controls; F_r = functional reach, and; L_c = labeling of controls.

Intuitively, we know that ergonomic improvements have been made by manufacturers over the past several decades. With data collected from a variety of tractors of varying age, we will be able to determine whether the proposed "Index of Functionality" is capable of recognizing these ergonomic improvements. If so, the Index may be useful for current designers.

Summary

1. The relevant literature was consulted to review the most important current design standards of various control types. The summarized values will be used to define recommended dimensions for each control type.
2. Experienced farmers participated in a survey to rank the five control categories in terms of their frequency of use during a tillage operation. In the final model, a direct correlation between the most important design characteristics of controls and their frequency of use will be implemented.
3. Professional ergonomists participated in a survey to determine the relative importance of four ergonomic factors. An expert weight factor will be used in the final model to recognize the relative significance of each of these four factors.
4. The numerical index may take any value between 0 and 1, with 1 being defined as the optimum value in terms of the functionality of an entire control panel.

References

ASAE Standards. 2000, Graphical Symbols for Operator Controls and Displays on Agricultural Equipment, *ASAE S304.7*, (ASAE, St. Joseph, MI)

Banks, W.W. and Boone, M.P. 1981, A method for quantifying control accessibility, *Human Factors,* **23**(3), 299–303

Clark, T.S. and Corlett, E.N. 1995, *The Ergonomics of Workspace and Machines: A Design Manual*, Second Edition, (Taylor and Francis, London)

Deisinger, I., Breining, R., Roßler, A., Hofle, I. and Ruckert, D. 2000, Immersive ergonomics analyses of console elements in a tractor cabin. In proceedings Fourth Immersive Projection Technologies Workshop, Iowa: Ames, June 19–20

Department of Defense. 1999, Human Engineering Design Criteria Standard, *MIL–STD–1472F*, (Department of Defense, Washington, DC)

Ely, J.H., Thomson, R.M. and Orlansky, J. 1956, Layout of workplaces: Human engineering guide to equipment design, (Wright–Patterson Air Force Base–Wright Air Development Center, Ohio), Report 56–171

Grandjean, E. 1988, *Fitting the Task to the Man*, Fourth Edition, (Taylor and Francis, New York)

Hansson, P.A. and Oberg, K.E.T. 1996, A method for computerized three dimensional analysis of biomechanical load on a seated tractor driver, *International Journal of Industrial Ergonomics,* **18**(4), 261

Konz, S. 1983, *Work Design: Industrial Ergonomics*, (Horizons Inc, Arizona)

National Aeronautics and Space Administration. 1995, Man–Systems Integration Standards, *NASA STD–3000*, Revision B, (Johnson Space Center, Houston)

Pheasant, S. 1986, *Bodyspace: Anthropometry, Ergonomics and Design*, (Taylor and Francis, London)

Purcell, W. F. H. 1980, *The Human Factor in Farm and Industrial Equipment Design*, No. 6, (ASAE, St. Joseph, MI)

Sanders, M.S. and McCormick, E.J. 1993, *Human Factors in Engineering and Design*, Seventh Edition, (McGraw–Hill Inc, New York)

Van Cott, H.P. and Kinkade, R.G. 1972, *Human Engineering Guide to Equipment Design*, (U.S. Government Printing Office, Washington D.C.)

Weiman, J. 1993, *Handbook of Ergonomics and Design Factors Table*, (Prentice Hall, New Jersey)

WORKLOAD ASSOCIATED WITH OPERATION OF AN AGRICULTURAL SPRAYER

Asit K. Dey & Danny D. Mann

Department of Biosystems Engineering, University of Manitoba, Winnipeg R3T5V6, Canada

A task analysis was carried out using structured survey questionnaires that were sent out to a random group of sprayer operators in Manitoba, Canada. Twenty sprayer operators (<25 yr(5%), 26–30 yr(15%), 31–40 yr(10%), and 41–60 yr(70%)) participated in this study; on average they had 21 yr of driving experience, but only an average of 2.7 yr of experience with GPS guidance systems. The important elements of the eye-glance behaviour and visual workload included looking at the horizon, booms, lightbar, mapping system, application display, tachometer, temperature gauge, and pressure gauge. The contributing factors for physical workload were: solenoid on/off, boom height control, sprayer pump on/off, hydrostatic and transmission control, and throttle. Factors influencing the mental workload included: operating the tractor/self-propelled unit, searching for guidance information, monitoring rate controllers and pressure gauges, controlling the sprayer pump, GPS system, and boom height controls, and talking on cell phone. Guiding the tractor involved the highest amount of mental workload. Fatigue indicators identified were sore eyes, poor concentration, restlessness, yawning, moodiness, boredom, and sleeping at the wheel.

Introduction

Spraying of herbicides and pesticides is an important part of agricultural production. Sprayer operators make an average lateral error (skip or double application) of 10% of their implement width (Palmer and Matheson, 1988). To reduce this error, the use of Global Positioning System (GPS) as a guidance tool has become a common practice in Canada. The GPS guidance tool provides lateral error information using a lightbar that has a horizontal array of light emitting diodes (LEDs) on either side of central LEDs, which glow whenever a corrective action is needed. Despite the presence of this accurate guidance system, Young (2003) reported that the operators of agricultural sprayers still tend to seek guidance information from external field cues for heading information. Agricultural sprayer operators often perform night spraying to cope with time pressure. But, night driving is more difficult than day driving because of reduced visibility (Macaulay, 1988). Dey and Mann (2004) reported that the lateral error under night driving conditions, and in the absence of external heading information, was 11.5% more than the lateral error during day driving.

It is hypothesized that the workload of an operator changes whenever a new information cue (i.e., a lightbar) is introduced into the operators' workstation. The spraying operation involves different in-vehicle cognitive (i.e., talking on cell phone), visual (i.e., watching lightbar), and manual control manipulation (i.e., transmission control) tasks. Vehicle control and event detection are shown to degrade most if the in-vehicle tasks require spatial cognitive resources and/or if the activity requires visual perception and/or manual control manipulation (Boer, 2001). For this reason, the objective of this study was to carry out a complete task analysis among a random group of sprayer operators to understand their perception of the dynamic workload associated with the spraying job.

Materials and methods

The data collection method for the task analysis included survey questionnaires that were sent out to a random group of sprayer operators in Manitoba. In this paper, only twenty survey questionnaires received from experienced GPS users were analyzed.

Generally, sprayer operators perform two major tasks: 1) they steer the vehicle by scanning the lightbar for lateral position and an aiming point for heading information (Young 2003) and 2) they check the booms. These tasks are completed during day, dusk, and night driving conditions. Therefore, to understand the dynamic change in the workload, a survey questionnaire was developed which contained multiple choice items, matrix questions, mutually exclusive questions, rating scales to obtain subjective information about the covert workload associated with different spraying tasks, and ranking techniques to order controls according to some specific criteria. The different sections of the questionnaire are mentioned below.

The personal information section gathered information on the age, gender, and handedness of the subjects; and whether they had any form of colour blindness. The farm information section contained questions on spraying experience with and without GPS, spraying time, the in-cab environment, total width of boom, type of sprayer, and experience with previous guidance systems. The visual workload section addressed questions on the type of visual displays typically used for the spraying operation and the time spent for each task in a 60 min driving period. It further sought subjective information on the importance of aiming cues and lightbars under day and night driving conditions. The physical workload section asked the participant to include the physical tasks completed during the spraying job and to rate them on a 1–7 scale (1 = most important and 7 = least important tasks). The mental workload section included questions on the percentage of time used for different cognitive activities performed during spraying; the subjects were asked to rate them on a 0–10 scale with 10 being the most demanding tasks. The fatigue section was designed to provide information on the fatigue indicators experienced during spraying and common methods used to alleviate fatigue.

Results

Personal information

Twenty male operators from a wide range of age groups (i.e., <25 (5%), 26–30 (15%), 31–40 (10%), and 41–60 (70%)) participated in this study; on average they had 21 yr of driving experience, but only an average of 2.7 yr of experience with GPS guidance systems. Only 35% (4 near-sighted and 3 far-sighted) of the subjects used corrective lenses, but none of them were colour blind. All but one subject were right-handed and only two subjects performed custom spraying.

Farm information

In 2004, the participants sprayed an average of 2183 ± 1362 ha of land. One subject had auto-steer guidance, but the rest of the subjects used lightbar guidance systems. The previous guidance tools used by the subjects included: Foam Markers (25%), Tramlines (15%), Disk Markers (5%), and Flags (5%) (note: 50% of the subjects had no prior experience with guidance tools). All but one commented that GPS guidance is better than the previous guidance system. Most of the subjects had air-conditioned sprayers (Self propelled: 75%, Pull type: 25%) with a boom width ranging from 24.4 to 36.6 m (median = 27.4 m) and the noise level inside the cab was classified as audible (35%), slightly audible (40%), or completely inaudible (25%). During spraying season, an operator sprays for an average of 14.6 h (i.e., day (12.5 h) + dusk (0.5 h) + night (1.6 h)) in a day. On average, spraying is continuous for 63 min to empty a tank followed by a 15 min stop for refilling. The driver typically makes straight parallel swaths (95% of the time); the remaining 5% of the time is spent in contour mode.

Visual workload

The survey analysis further revealed that the important elements of the eye-glance behaviour included looking at the horizon, booms, lightbar, mapping system, application display, tachometer, temperature gauge, and pressure gauge. Thirty-five percent of the subjects responded that they spent more than 75% of their driving time looking at the horizon for guidance information, but spent 51–75% of their driving time looking at the lightbar (Table 1). All the subjects spent equal amounts of time looking at the left and right booms. For the other visual displays, the subjects spent less than 25% of their driving time. Figure 1 shows that under both day and night driving conditions the lightbar is the most important source of guidance information followed by field cues (20%). Only one subject felt that the mapping device is important in spraying.

Physical workload

The analysis showed that the important contributing factors for physical workload, in decreasing order of importance, were: hydrostatic control and solenoid on/off, boom height control, transmission control, and throttle control.

Mental workload

The contributing factors for the mental workload included: operating the tractor/self-propelled unit; searching for guidance information; monitoring rate controllers and pressure gauges; controlling the sprayer pump, GPS system, and boom height; and talking on a cell phone. The analysis revealed that in a 60 min driving period, an operator spends a maximum of 49 min to guide the sprayer, which also accounted for the highest amount of relative mental workload (R.M.W.) (Table 2). The guidance information (34 min), sprayer pump (23 min), and rate controller and pressure gauge (33 min) tasks

Table 1. The percentage of driving time spent looking at different visual displays.

Visual displays	Time spent (%)			
	<25	26–50	51–75	>75
Tachometer	60	5	–	–
Temperature gauge	65	–	–	–
Oil pressure gauge	60	–	–	–
Mapping system	35	15	5	–
Rear-view mirrors	65	–	–	–
Right boom	40	40	10	5
Left boom	40	40	10	5
Field ahead	20	15	30	**35**
Lightbar	15	20	**35**	30

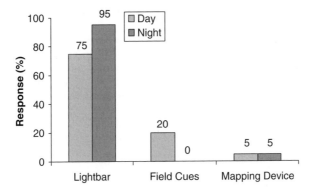

Figure 1. Source of guidance information during day and night driving.

Table 2. Average time spent for each task in a 60 min driving period and relative mental workload (R.M.W.) involved in each spraying task.

	Vehicle driving	GPS system	Boom height	Guidance information	Sprayer pump	Cell phone	Rate controller
Time spent (min)	49	22	11	34	23	4	33
R.M.W.	6	4	2	5	4	3	1

were less demanding. Limited-time activities, like controlling boom height (11 min) and talking on cell phone (3 min), were the second- and third-most demanding tasks. Watching the lightbar involved 22 min of the total driving time, but it caused the same amount of mental workload as talking on a cell phone.

Fatigue

The analysis of the questionnaire further revealed that the indicators of fatigue, in decreasing order of importance, included: yawning (50%), sore eyes and poor concentration (40%), sleeping at the wheel and moodiness (35%), restlessness (30%), and boredom (10%). Taking a break and opening a window (40%) were the common methods employed to cope with fatigue.

Discussion

Most subjects were experienced sprayer operators and thus, it was expected that the reported information would be reliable. Operators spent an average of 14.6 h per day in the sprayer, including day, dusk, and night driving, while carrying out different visual, physical, and mental workload tasks. The operators indicated a preference for the lightbar under both day and night driving conditions (Fig. 1), however, 35% of the subjects felt they spent more than 75% of their driving time looking at external cues (Table 1). These two results are contradictory.

When ditches or obstacles are encountered, two major tasks are done: the sprayer pump is shut off and boom height is adjusted. Therefore, solenoid on/off was the most demanding physical workload task followed by boom height control, hydrostatic control, and transmission control. Steering and speed correction needs constant attention, thus they were the most mentally-demanding tasks. Even a brief cell phone conversation triggers a huge mental workload while spraying.

Despite the presence of a promising guidance technology, the subjects felt the spraying job was demanding. Therefore, there is a need to investigate how these factors influence the overall workload and how to organize different spraying tasks to reduce the workload. It is clear that subjective self-reflection has limitations and it sometimes yields contradictory results. Our previous pilot study showed that, even during night driving, operators spent twice as much time looking outside as looking at the lightbar. Therefore, to supplement the survey questionnaire, future experimental studies should include eye glance behaviour and heart rate variability under day, dusk, and night driving conditions.

Conclusion

The research has established that the spraying tasks involve visual, physical, and mental workload. It was expected that the participants' driving experience (21 yr of average driving experience with an average of 2.7 yr of experience using GPS guidance systems) would provide satisfactory and reliable information. However, the participants provided contradictory information on the usage of the lightbar under day and night driving conditions. Therefore, subjective self-reflection has limitations. The important factors for the visual workload included looking at the horizon, booms, lightbar, mapping system, application display, tachometer, temperature gauge, and pressure gauge. Controlling the solenoid and boom height involved more physical workload while guiding the sprayer to maintain the speed and keeping the system on track contributed the highest amount of mental workload. Fatigue indicators identified were sore eyes, poor concentration, restlessness, yawning, moodiness, boredom, and sleeping at the wheel. The common methods (40%) for avoiding fatigue included opening the window and taking a break.

Acknowledgements

The authors acknowledge the financial support of the Manitoba Sustainable Development Innovations Fund, the Natural Sciences and Engineering Research Council of Canada, and University of Manitoba Graduate Fellowship.

References

Boer, E.R. 2001, Behavioral entropy as a measure of driving performance, *In Proc. Driver Assessment*, 225–229.

Ima, C.S. and Mann, D.D. 2004, Ergonomic concerns with lightbar guidance displays. *Journal of Agricultural Safety and Health*, **10**(2): 91–102.

Macaulay, N.S. 1988, Road lighting – the visual task. *The Lighting Journal*, 70–73.

Mann, D.D. and Dey, A.K. 2004, Eye-glance behaviour during simulated operation of an agricultural sprayer, *Annual Conference of the Ergonomics Society*, 14–16 April Swansea, Wales.

Palmer, R.J. and Matheson, S.K. 1988, The economic impact of precise field positioning and guidance systems on prairie agriculture, Paper 88–1602. St. Joseph, MI: ASAE.

Young, S.J. 2003, Design of an agricultural driving simulator for ergonomic evaluation of guidance displays, *M.Sc. thesis*, Winnipeg, MB: Department of Biosystems Engineering, University of Manitoba.

EFFECTS OF THE COMMON COLD ON SIMULATED DRIVING

Andy Smith

Centre for Occupational and Health Psychology, School of Psychology, Cardiff University, 63 Park Place, Cardiff CF10 3AS

Tye (1960) reported an association between the incidence of colds and influenza and the number of road accidents in the UK. The present study examined this issue by using a virtual reality driving simulator. Driving performance was assessed in terms of deviations from target speed, deviations from the centre line of the track (RMS error), kerb hits and reaction times to unexpected targets. The results showed that volunteers with colds hit the kerb more frequently and had slower response times to the unexpected targets. Further research on driver fatigue induced by minor illness is now clearly required.

Introduction

Upper respiratory tract illnesses (URTIs), such as the common cold and influenza, are widespread, frequent and a major cause of absenteeism from work and education. Initial evidence for effects of these illnesses on performance efficiency at work came from anecdotal reports and case histories (e.g. Tye, 1960). Studies of the effects of experimentally induced URTIs have confirmed that they can impair performance and influence mood (see Smith, 1992, for a review). These findings have been replicated in studies of the effects of naturally occurring illnesses (Hall and Smith, 1996; Smith et al, 1993b; Smith et al, 1995; Smith et al, 1998; Smith et al, 2000; Smith et al, 2004). It is now of major interest to determine whether these illnesses influence real-life activities and the present study examined whether having a cold impaired performance on a simulated driving task.

Method

This study was carried out with the approval of the ethics committee of the School of Psychology, Cardiff University, and the informed consent of the volunteers.

Design

Volunteers were tested on two occasions. On the first occasion some of the volunteers had a cold (N = 18) and others were healthy (N = 28). On the second occasion all volunteers were healthy. The scores from the second session were used as covariates to adjust for unwanted differences in performance.

Participants

The volunteers with colds did not differ from those who were healthy in terms of age (mean = 20.3 years s.d. = 1.49), gender or other demographic characteristics.

Similarly, there were no significant differences in personality or psychosocial scores for the two groups. In addition, the groups did not differ in terms of driving experience or frequency of playing computer games.

Nature of the driving task

Performance was assessed using a virtual reality driving simulator (Snowden et al, 1998). This involved simulated driving with a secondary reaction time task. Participants were seated in front of a large VDU. A joystick was used (with left hand) for acceleration (pushing joystick forward) and braking (pulling joystick back). Steering was controlled by using the mouse (with right hand). Volunteers were required to drive along a road following a dashed centre line at a constant speed (30 miles per hour in the slow condition and 60 miles per hour in the fast condition). The road consisted of a sequence of straight line segments, the texture mapped with a dashed line in the centre. The width of the road was equal to 7.3 metres. The view in the vehicle was from the centre of the vehicle at a height of 1.5 metres from the road. The width of the vehicle was 1.8 metres. The vehicle could travel at a maximum of 90 miles per hour. Actual acceleration or braking rate at any moment in time depended on how far the joystick was pushed forward or pulled back.

In the reaction time task the stimuli presented to the volunteers were white squares that were deleted as soon as the volunteer clicked the left mouse button. If the participant failed to respond to the stimulus the target disappeared after five seconds. The white squares were displayed whatever speed the volunteer travelled at. At each testing session volunteers were given the same instructions on how to use the driving simulator. They completed a five-minute familiarisation for both the slow and fast speeds at both sessions before the tests. Participants then completed two fifteen-minute test sessions (30 mph and 60 mph).

The variables derived from the task and used in subsequent statistical analyses were:
Kerb hits: analyses were performed using the mean number of kerb hits made by participants during their fifteen minute test sessions.
Reaction time to stimuli: analyses were performed using the ranked log reaction time taken by performing a log transformation ($\log(n + 1)$) on the ranked mean reaction times. Reaction times were ranked according to high, medium and low reaction times (e.g. high = the five longest reaction times, low = the five shortest reaction times).
Deviation from target speed: analyses were performed using participants' deviation scores from the target speed of 30 mph or 60 mph (the respective slow or fast condition).
Deviation from centre line: analyses were performed using participants' deviation from the centre line of the track as scores.

Results

Symptoms

Upper respiratory tract symptoms were measured using a standard symptom check-list (see Smith et al, 2004). Volunteers with colds at session 1 had a mean symptom score of 8.9 (s.d. = 3.3) whereas those who were healthy had a mean symptom score of 1.1 (s.d.01.2). This difference was highly significant ($p < 0.001$).Those with colds had significantly higher scores for the following symptoms: runny nose, blocked nose, sneezing, sore throat, cough, headache and fever. The colds group also reported significantly greater fatigue, a more negative mood state and more cognitive difficulties. There were no significant differences between the two groups on the second visit when all volunteers reported that they were healthy.

Table 1. Adjusted cell means for kerb hits.

	Colds	Healthy
Kerb hits (fast speed)	29.80	21.19
Kerb hits (slow speed)	11.17	7.90

Table 2. Mean speed (MPH) for cold and healthy participants.

	Colds	Healthy
Mean speed (fast speed)	53.35	54.16
Mean speed (slow speed)	31.39	30.47

Table 3. Mean deviation from the Centre line (RMS error) for cold and healthy participants.

	Colds	Healthy
Mean RMS (fast speed)	0.60	0.60
Mean RMS (slow speed)	0.53	0.52

Driving performance

Overall, participants in the colds group demonstrated significantly slower reaction times on the secondary task of responding to the stimuli (colds mean RT = −748 msec; healthy group mean RT = 681 msec; $p < 0.05$) at visit 1 (when they were symptomatic). This confirms results found using single simple reaction time tasks. Additionally, participants in the colds group demonstrated a greater number of kerb hits throughout the driving task when compared to the group of healthy subjects (see Table 1). ANCOVA analyses with visit 2 scores as the covariates revealed a significant effect of condition (cold or healthy) on the number of kerb hits ($p < 0.05$) and the log reaction times ($p < 0.05$) to the stimuli. No significant effects of condition were found on deviation from the target speed or deviation from the centre line of the road (see Tables 2 and 3).

Conclusions

The results from the present study have shown that those with colds are impaired on a simulation of driving. It should be noted that the simulation was far from realistic and it is now essential to examine this topic using a more sophisticated driving simulator. However, the study does represent one step towards realism from the simple performance tasks used in earlier research. It is important to note that the impairments observed here agree closely with those found in the early research using simple components of driving. In addition, laboratory research has shown that having a cold makes one more susceptible to the effects of other factors (e.g. noise, Smith et al, 1993a; alcohol, Smith et al, 1995; prolonged work, Smith et al, 2004) and safety limits based on studies of healthy individuals may not be appropriate for the person who is driving while ill or after recent illness.

Acknowledgement

Research reported in this paper was supported by the Economic and Social Science Research Council (grant RO22250143). Data collection was carried out by Anna Leach and Susan Williamson.

References

Hall, S. R. and Smith, A. P. 1996, Investigation of the effects and after- effects of naturally occurring upper respiratory tract illnesses on mood and performance. *Physiology and Behavior,* **59**, 569–577

Smith, A. P. 1992, Colds, influenza and performance. In A. P. Smith and D. M. Jones (Eds.), *Handbook of Human Performance. Vol. 2.* London: Academic Press. 196–218

Smith, A. P., Brice, C. F., Leach, A., Tiley, M. and Williamson, S. 2004, Effects of upper respiratory tract illnesses in a working population. *Ergonomics,* **47**, 363–369

Smith, A. P., Thomas, M. and Brockman, P. 1993a, Noise, respiratory virus infections and performance. *Actes Inrets,* **34**, 311–314

Smith, A. P., Thomas, M., Brockman, P., Kent, J. and Nicholson, K. G. 1993b, Effect of influenza B virus infection on human performance. *British Medical Journal,* **306**, 760–761

Smith, A. P., Thomas, M., Kent, J. and Nicholson, K. G. 1998, Effects of the common cold on mood and performance. *Psychoneuroendocrinology,* **23**, 733–739

Smith, A. P., Thomas, M. and Whitney, H. 2000, Effects of upper respiratory tract illnesses on mood and performance over the working day. *Ergonomics,* **43**, 752–763

Smith, A. P., Whitney, H. Thomas, M., Brockman, P. and Perry, K. 1995, A comparison of the acute effects of a low dose of alcohol on mood and performance of healthy volunteers and subjects with upper respiratory tract illnesses. *Journal of Psychopharmacology,* **9**, 267–272

Snowden, R. J., Stimpson, N., Ruddle, R.A. 1998, Speed perception fogs up as visibility drops. *Nature,* **392**, 450

Tye, J. 1960, The invisible factor – an inquiry into the relationship between influenza and accidents. *British Safety Council.* London

TOP GEAR ON CARS – EXPERTS' OPINIONS AND USERS' EXPERIENCES

Patrick W. Jordan

University of Leeds, Leeds LS2 9JT

A study compared journalists' opinions about new cars with people's subsequent experiences of driving and owning them. Car reviews in *Top Gear* Magazine were compared with the outcomes of a survey that asked owners to rate their cars on a number of dimensions. Results were mixed – sometimes journalists' reviews were a good predictor of people's experiences with the car, but often they weren't. The kinds of discrepancies that arose have implications for expert appraisals of user-experience. These implications are discussed and recommendations made about how to make such appraisals both accurate and relevant.

Expert appraisal and the new human factors

Within human factors, there is a long tradition of expert appraisal. Essentially this involves people with human factors training using and analysing products and services and making judgements about how usable they are likely to be based on their own experience of use and a number of ergonomic factors in their design.

In recent years, human factors has moved beyond looking simply at usability issues and has started to tackle some of the wider issues of user experience. These include, for example, the emotional aspects of using a product and other factors which relate to how pleasurable a product or service is to own or use.

Judging products in terms of the overall quality of experience that they bring to the user is not something new. Opinion leaders have been doing this in a variety of areas for quite some time.

Motoring journalism and product reviews

One of the most prominent contexts in which product reviewing is done is motoring journalism. Cars are a high-interest product for many people and a large and profitable industry has developed around the testing and reviewing of vehicles. Go to any news-stand in any industrialised country and you will see a whole bunch of magazines dedicated to cars and motoring. Newspapers nearly all have a weekly section dedicated to motoring – mainly new car reviews – and there are also a host of TV shows on the subject.

Top Gear

The UK motoring show with the biggest TV audience is called *Top Gear*, a show fronted by opinionated journalist Jeremy Clarkson, who seems to be loved and loathed in equal measure by the British public. The opinions expressed on the show and in the show's associated publication *Top Gear* Magazine can have a significant influence on people when it comes to making new-vehicle purchase decisions.

In this article, we compare the ratings and comments that *Top Gear* gave to vehicles with the opinions of those who have actually driven them over a significant period of

p.jordan@leeds.ac.uk

time. We did this by comparing the views expressed in the *Top Gear New Car Buyers Guide Winter 2003* (Top Gear 2003) with the JD Power Drivers Survey published in July 2005 (What Car 2005). The survey included many of the vehicles included in the buyers guide from 2 years before.

JD Power owners survey

J.D. Power received thousands of responses to their survey asking vehicle owners about their cars. The owners were asked to give an overall satisfaction score for each vehicle. These were converted this to a star-rating, with 1 being the lowest and 5 the highest. They were also asked to rate the cars on three scales, with respect to the following issues:

Quality – the build and reliability of the car; Appeal – performance, comfort and the driving experience; Running – costs and quality of dealer service.

Method

In order to compare journalists' ratings with those of owners we picked a selection of cars divided into categories according to the size and function of the vehicles. Within each category we looked at three vehicles – the one owners rated best in the category, the one they rated worst and one of the cars that they rated towards the middle. We then looked at the journalists' opinions of these cars and compared them with those of the owners.

Results

The journalists' comments about each vehicle are paraphrased and the star rating that they gave it is revealed. The ratings given by the owners are also listed and what the journalists said is compared to the comments of the owners.

Super-Minis

Honda Jazz (Top Gear Rating: 4). Engineering ingenuity allied to the way the car is packaged endear it to potential buyers. The car is practical and spacious, has a high tech feel and is fun to drive. *Owners Ratings – Overall: 5, Quality: 5, Appeal: 5, Running: 5*

Volkswagen Polo (4). A solidly built understated car which gives a smooth drive. However, the weight of the car limits the performance and handling and it feels a bit dated. *Owners Ratings – Overall: 3, Quality: 2, Appeal: 4, Running: 3*

Rover 25 (2). Dated car that has been re-branded many times, poor value for money compared with rivals. Suspension gives the car a sporty feel but experience is ruined by the driving position. *Owners Ratings – Overall: 1, Quality: 1, Appeal: 1, Running: 2.*

The views of the journalists generally reflected those of the owners who rated the Jazz as excellent in every respect and liked virtually nothing about the Rover. The owners were less positive about the Polo than the journalists were. Although they liked the interior of the vehicle and thought it performed well, they were disappointed by poor reliability.

Small cars

Toyota Corolla (3). Good quality design and decent materials make the car practical and well built but a little dull. Performance is reasonable and the car drives nicely. *Owners Ratings – Overall: 5, Quality: 4, Appeal: 4, Running: 5*

Ford Focus (4). The car is spacious and refined and the handling is little short of stunning for a family car. Great fun to drive with a well thought out interior. *Owners Ratings – Overall: 4, Quality: 4, Appeal: 3, Running: 4*

Alfa Romeo 147 (3). This car can be unreliable and the build quality is inconsistent. However, it is beautifully styled, is comfortable, handles well and is great fun to drive. *Owners Ratings – Overall: 1, Quality: 2, Appeal: 3, Running: 1*

The only thing that readers rated positively about the Alfa was its looks, with build quality, reliability, dealer service and running costs getting very poor ratings. The Corolla by contrast was loved by owners, getting excellent ratings right across the board. Owners were far more positive than the "faint praise" from the journalists. Owners agreed with journalists that the Focus had good ride and handling, but there were some quality issues that annoyed them. Overall, the journalists' comments were not a good predictor of user experiences with these cars.

Family cars

Skoda Octavia (3). The car is generally well built although the interior does include some cheap plastics. The car is spacious and offers a refined, if not always responsive, driving experience. *Owners Ratings – Overall: 5, Quality: 4, Appeal: 3, Running: 5*

Ford Mondeo (4). A superb all-round car with well made engines. It handles excellently as well as giving a smooth and comfortable ride. *Owners Ratings – Overall: 3, Quality: 3, Appeal: 3, Running: 3*

Renault Laguna (3). A well built car that is stylish, well-equipped and comfortable. The ride is OK, although it does not handle particularly well. *Owners Ratings – Overall: 3, Quality: 2, Appeal: 3, Running: 2*

Owners felt that the Laguna had some quality issues associated with it and were generally unhappy with the level of service that they got from dealers, but otherwise rated the car as OK in most respects. The Mondeo was also rated as pretty good across the board without being seen as outstanding in any respects. By contrast the Octavia scored very well in all categories, with low running costs being a feature that owners particularly liked. The owners' enthusiasm for the Octavia does not reflect the comments of the journalists, who rated the Mondeo more highly.

Executive cars

Lexus IS 200/300 (3). This is a well-equipped car although the use of materials in the interior is a little incoherent. Although the car handles well, its lack of power is disappointing. *Owners Ratings – Overall: 5, Quality: 4, Appeal: 5, Running: 4*

Mercedes C-Class (4). The car feels solid but in fact build quality can be inconsistent. It feels competent to drive, but some of the competition offers better driving experiences. *Owners Ratings – Overall: 3, Quality: 3, Appeal: 3, Running: 2*

Alfa Romeo 156 (4). A comfortable car that looks sublime. This is a wonderful car to drive with superb engines and great ride and handling. *Owners Ratings – Overall: 1, Quality: 1, Appeal: 2, Running: 1*

In this case owners views were largely at odds with those of the journalists. While they agreed that Alfa looked nice, they strongly disliked it in almost every other respect. The Lexus, on the other hand, was absolutely loved by owners, who gave it outstanding scores right across the range. Owners' views of the Mercedes were slightly less positive than those of the journalists with respect to the driving experience, but they agreed that quality was somewhat below par.

Multi-Person Vehicles (MPVs)

Nissan Almera Tino (3). Generally a well thought out family car, despite some glaring omissions such as wipe-clean trim. The driving experience is OK. *Owners Ratings – Overall: 4, Quality: 3, Appeal: 4, Running: 4*

Citroen Xsara Picasso (3). A well thought-out and spacious vehicle. A comfortable ride with a good driving position, although visibility can be poor. *Owners Ratings – Overall: 3, Quality: 2, Appeal: 4, Running: 3*

Fiat Multipla (4). A very original design which is fun and functional. Smooth and satisfying to drive. *Owners Ratings – Overall: 1, Quality: 2, Appeal: 3, Running: 1*

As far as owners were concerned this was not a sector of the market in which there were any truly outstanding vehicles, but the Almera's consistently good ratings across all categories gave it top place in the survey. The Picasso was consistently average. However, in complete contrast to the journalists, owners had very negative views about the Multipla, rating it as badly built, unreliable and not fun to drive. They were also unhappy with the dealer service and running costs.

Discussion

Taken overall, the results give a very mixed picture in terms of the extent to which the journalists' comments are predictive of the experiences of the users. In one of the categories – super-minis – the journalists comments are pretty much in line with user experience. Journalists and owners agree on what are the best and worst of the three cars looked at and their comments about the cars contain many similarities. In another category – family cars – there is some degree of agreement, but the journalists' comments do not reflect the very positive feelings that users have about their favourite model, the Skoda Octavia. In the executive and small cars categories the journalists comments are predictive of neither how positive owners were about the best-rated cars – the Lexus 200/300 and the Toyota Corolla – nor how negative they were about the two worst-rated cars, both Alfa-Romeos. Finally, in the MPV category journalists' and owners' comments seemed completely at odds. Journalists were most positive about the Multipla, yet this came bottom with the owners by a clear margin.

An obvious issue with respect to giving an expert analysis of a new product is that it can be easier to give meaningful insight into some aspects of it than others. For example, we might expect it to be easier to comment meaningfully on things such as the performance, comfort and appearance of the car than on issues such as reliability and running-costs, which are properties that are more likely to emerge over time. However, looking at the results, it seems that owners' views were hardly more likely to coincide with those of the journalists on the subject of appeal – driving performance, aesthetics and the interior, than they were with respect to quality, reliability, running costs and dealer experience.

A possible reason for this is that there could be a "halo effect" going on here. Owners who found their cars to be reliable and less costly to run may have become predisposed to think of their cars positively overall and therefore be more positive about their appeal. On the other hand, owners of unreliable cars may have had the effect in reverse, becoming critical of their cars overall and thus rating the appeal factors lower.

Another possible way of looking at it is that their could be a hierarchy of needs issue going on, with the reliability and build-quality issues at the bottom of the hierarchy and the appeal factors above. If this is the case it would suggest that the appeal factors become important only if users are positive with respect to the quality factors. If the car isn't reliable and well built then the owners will not take any pleasure in the vehicles performance and design.

Perhaps not surprisingly the journalists tended to be impressed by cars that were beautiful to look at or had novel and interesting designs – the two Alfa-Romeos in the survey scored far better with journalists than they did with owners as did the Fiat Multipla. The cars that the owners liked best sometimes seemed to be ones that the journalists seemed to think of as being a little bland – the Toyota Corolla, Lexus 200/300 and

Skoda Octavia being examples of this. A possible danger of this from the manufacturers' point of view is that because the strengths of these cars is in the long-term quality factors, they are not picked up by journalists and opinion leaders when the vehicles are launched. This means that the cars may not be "talked up" by the press as much as they deserve to be. On the other hand cars that look nice and are fun to drive around a track for a couple of hours can get rave reviews.

Conclusions

A problem faced by experts making an appraisal of a new product is that there are a number of factors, especially those relating to issues such as build quality and reliability that may only emerge after a significant period of use. It can be difficult, if not impossible, to pick these up in an expert appraisal session and yet they can often be – and in many cases are likely to be – the most crucial factors in determining the quality of experience that a person has with a product.

Over the last few years, human factors as a discipline has become adept at looking beyond usability to the wider factors which can affect the quality of a person's experience with a product. In practice, this has tended to mean looking at factors "above" usability – the things that give products or services a special emotional appeal. Perhaps a lesson from this study is that we also need to look at the factors "beneath" usability – at the basic factors, such as reliability and build-quality which determine whether or not the user can live happily with the product on a day to day basis.

References

Top Gear 2002. *Top Gear New Car Buyers Guide Winter 2003*
What Car 2005. *What Car*, July 2005

BIONIC – "EYES-FREE" DESIGN OF SECONDARY DRIVING CONTROLS

Steve Summerskill[1], J. Mark Porter[1], Gary Burnett[2] & Katharine Prynne[3]

[1]*Department of Design & Technology, Loughborough University, UK*
[2]*School of Computer Science & Information Technology, University of Nottingham, UK*
[3]*Honda Research & Development Europe, Swindon, UK*

The BIONIC project (Blind Operation of In-car Controls) was set-up to develop an "eyes-free" prototype interface, enabling drivers to access secondary and ancillary controls whilst minimising the visual demands within the car. This research was initiated out of concern for the increasing use of multi-function screen based interfaces that place an additional visual workload on the driver. BIONIC has created new guidelines for the design of highly tactile control interfaces, based upon a series of experimental studies and the development of prototype designs. Working prototypes were installed within a Honda Civic demonstrator vehicle and these novel controls were compared to the current interface in on-road trials. A reduction in total glance duration of 10% was stated as our target in the grant proposal; the BIONIC interface surpassed this benchmark.

Aims of the BIONIC project

We did not plan to enable people with severe visual impairments to drive cars – rather we hoped to demonstrate that accessing complex technology whilst driving could be made safer by being less demanding upon the driver, both visually and mentally. This was to be accomplished by developing an "eyes-free" prototype control interface that provided high levels of tactile and kinaesthetic information. BIONIC learnt from the experience and strategies adopted by people who are visually impaired when operating control interfaces on consumer products. The BIONIC consortium comprised HONDA Research & Development, Loughborough University, Visteon, ARRK Formation Ltd., SAMMIE CAD Ltd., RNIB and Nottingham University. The research was funded by the EPSRC, via the Loughborough Innovative Manufacturing and Construction Research Centre.

The research context

The basic need for this research emerged from four core issues:

1. With the predicted explosion in the application of Information Technology within future vehicles, many more functions will be available to drivers with their associated controls and displays. This is already evident in high end cars produced by BMW, Lexus and others. Research has shown that interfaces of this kind require large "eyes off road" time, and increase the likelihood of drivers wandering out of lane (Zwahlen et al, 1988 and Porter et al, 1999).
2. Although there is a wide range of manual controls in current vehicles, the tactile sense is commonly under-used. Tactile coding can be provided by the physical design of the control in terms of its location, size, shape, texture, orientation, and tactile and auditory feedback characteristics (Prynne, 1995).

3. Speech recognition has potential for the "eyes-free" operation of a restricted range of non-safety related functions, but given recognition rates of less than 100%, individual driver preferences, and the unpredictable noisy environment, it is considered that supplemental manual controls will always be necessary.
4. There is limited specific and up-to-date guidance available to vehicle designers on how to design physical controls to minimise the need for vision.

The thrust of the research was to support the design of novel control interfaces for future vehicles so that visual and mental demands are kept to a minimum. BIONIC has created new guidelines for the design of manual controls for secondary and ancillary functions in future cars. These guidelines are based upon a series of experimental studies and the development of prototype designs that are described in more detail in Porter et al (2005). Drawing from current data, a "pool" of control concepts for operating a sub-set of functions of vehicles was generated. Specific experimental studies subsequently identified the characteristics of in-vehicle controls that reduced the requirement for vision. The results of this research enabled a range of first iteration prototype controls to be built that were assessed using a driving simulator trial. Second iteration working prototypes were then built for evaluation within a Honda Civic demonstrator vehicle. In the final stage of the project, the usability of the novel controls in relation to current designs was evaluated in on-road trials. A strong emphasis was placed on measures that directly relate to safety, such as the number and duration of glances made to the control and/or display (Wierwille, 1995). A reduction in total glance duration of 10% was stated as our target in the grant proposal.

Project stages

Brief details of the various project stages are given below: A more detailed description can be found in Porter et al (2005).

Knowledge gathering

This stage involved a literature review and a review of current control interface design was conducted by visiting showrooms for all major car manufacturers.

Study 1: To determine the tactile cues that visually impaired participants find most useful in determining control location and function

Individual interviews were held during which the visually impaired participants were asked to use an unfamiliar portable hifi. Their techniques for exploration of the device were recorded along with their comments, providing a number of design recommendations for tactile control coding.

Study 2: To identify which secondary controls can be used by drivers without looking

The participants were asked to keep looking at a point through the windscreen of their own car, some distance away, whilst attempting to operate a selection of secondary controls. This baseline study provided clear evidence that using existing in-car secondary controls non-visually can be particularly difficult for some cars/designs/tasks.

Development of functionality specification and scenarios for critical use

A complete list of functionality for current cars was compiled with reference to high specification cars such as the BMW 7 series. A series of focus groups with experienced drivers reduced this list of functionality down to that which would need to be easily accessed before, during and after normal driving and in the event of an emergency.

SAMMIE evaluation to identify control and display locations in a Honda Civic

The SAMMIE CAD system (Porter et al, 2004) was used to model the interior of the Honda Civic and identify alternative areas which were suitable for positioning the pods and the screen for easy reach and vision by drivers of all shapes and sizes.

Generation of concept designs for an "eyes-free" interface

Brainstorming sessions were held which resulted in a large number of concepts being generated. A strong modal design was produced by providing 3 separate "pods", one each for SAT NAV (satellite navigation), HVAC (heating, ventilation and air conditioning) and ICE (in-car entertainment). Each pod has an integrated hand control reference point where the palm of the hand can rest.

Study 3: To identify the relative strengths of coding using shape, size and location

Ten participants were blindfolded and asked to find specific cardboard cut-out numbers (1–9) on 6 different boards. Separate boards were prepared for different coding methods, singularly and in combination. The results showed conclusively that the most important coding was by location.

Study 4: Evaluation of BIONIC pods using a driving simulator

Twelve participants (6 male, 6 female, aged 19–52 years) evaluated the non-operational pods using a driving simulator. The results showed that high success rates, in the range 80–100% across the participants and conditions, were achieved with the prototype BIONIC design. Improvements to the design were identified from the errors observed.

Study 5: Simulated use of a touchpad for destination entry using the left or right hand

Sixteen participants "drove" in the driving simulator for 10 minutes. They were asked to draw letters and numbers on a simulated touchpad. Participants were asked to use their left and right hand, in a balanced order. There was some evidence that using the non-preferred (left) hand for data entry was more demanding (increased glances, perceived workload) and the quality of handwriting was poorer, as compared with the right hand.

Construction of a working prototype for BIONIC pods and screen interface

The design was finalised in Pro-Engineer. The electronics were developed and hand-built and fitted to the next generation of rapid prototypes. The labels and graphics were produced and applied to the pods. The structure and graphics of the screen interface were designed and passed to a Visteon partner for software development.

Preparation of an instrumented demonstrator Honda Civic

The Honda Civic was fitted with the working prototype BIONIC interface (pods and screen) as well as "lipstick" cameras for monitoring the drivers' visual behaviour, physical actions and the road and traffic environment.

Road trial evaluation of the BIONIC interface and results

Sixteen drivers (8 male, 8 female, with a 50:50 split between a "younger" group aged 21–33, and an "older" group aged 55–70) were selected for the road trials. Each

Figure 1. The BIONIC controls and the view from the lipstick cameras during experimental trials.

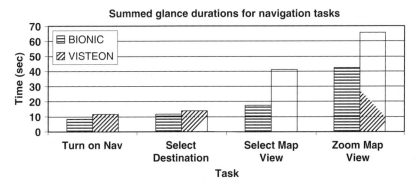

Figure 2. Comparing the summed glance durations for the BIONIC and Visteon satellite navigation conditions.

participant drove the instrumented car, fitted with either the BIONIC or the standard interface, in a balanced presentation order. They were then asked to perform 13 tasks with the SAT NAV, HVAC and ICE interfaces whilst driving down an "A" road e.g.: turn on the radio and select classic FM; turn on navigation; select the shortest route to Thistle Street Edinburgh. The mean total glance duration for all tasks was 51.7 seconds for the BIONIC interface compared to 62.3 seconds for the standard interface. This represents a 17% reduction in the "eyes off road" time, and exceeds the target that was set of 10%. The BIONIC interface achieved an overall reduction in total glance duration of 20% and 32% for the HVAC and SAT NAV tasks, respectively. The graph below shows a comparison of the summed glance durations for stages of the navigation interaction.

The graph above shows that the BIONIC system reduces the "eyes off road" time for both multi-stage and single stage operations. This is attributed to the combination of tactile coding, functional grouping, and simplified interface when compared to the baseline system. The BIONIC ICE tasks required a 7% increase in total glance duration. This was due to the relatively poor location of the ICE pod. This constraint was imposed by the user of the Honda Civic interior as it stood. If the BIONIC concept is used in future models, then it would be conceived as an integrated part of the interior and would not be so vulnerable to such constraints.

Development of guidelines for "eyes-free" interfaces

This series of studies and evaluations has led to the compilation of generic guidelines for designing "eyes-free" interfaces. These guidelines can be found in Porter et al (2005).

The future

The BIONIC project has shown tangible safety benefits for "eyes-free" interface design. The generic guidelines arising from this research could be used to design more effective and efficient interfaces for people who do not have the freedom to stop whatever they are doing every time they need to interact with technology, whether on-the-job or on-the-move. Examples include driving, examining screen displays from remote camera systems, operating machinery, sail-boarding, cycling and commuting. Any additional visual distraction during these activities could have a serious safety issue. Other potential benefits of "eyes-free" interfaces would include improved productivity and pleasure of use. We are also keen to be involved in developing innovative interfaces for elderly users and people who are visually impaired.

References

Porter, J.M., Summerskill, S., Burnett, G. and Prynne, K., 2005, BIONIC – "eyes-free" design of secondary controls. British Computer Society Workshops in Computing (eWIC) Series: Accessible Design in the Digital World Conference, Dundee.

Prynne, K., 1995, Tactile Controls. Automotive Interiors International, summer edition, pp. 30–36.

Zwahlen, H.T., Adams, C.C. and Debald, D.P., 1988, Safety aspects of CRT touch panel controls in automobiles. In: A.G. Gale et al (Eds.), Vision in vehicles II (pp. 335–344). Amsterdam: Elsevier Science Publishers B.V.

Porter, J.M., 1999, Evaluations of CIMS interfaces in the Mercedes S Class, Jaguar S Type, BMW 7 Series and the Lexus LS500. Confidential report, Vehicle Ergonomics Group, Loughborough University.

Wierwille, W.W., 1995, Development of an initial model relating driver in-vehicle visual demands to accident rate. Third Annual Mid-Atlantic Human Factors Conference Proceedings, Blacksburg, VA: Virginia Polytechnic Institute and State University.

Porter, J.M., Marshall, R., Freer, M. and Case, K., 2004, SAMMIE: A Computer Aided Ergonomics Design Tool. In: Working Postures & Movements, tools for evaluation and engineering, N. Delleman, C. Haslegrave and D. Chaffin (Eds.), pp. 454–470 LLC.

Lomas, S.M., Burnett, G.E. Porter, J.M. and Summerskill, S.J., 2003, The use of haptic cues within a control interface, In: Proceedings of HCI International conference, Vol 3 (Human-Centred Computing), pp. 502–506, Crete, Greece, June 23–27, 2003.

MIND THE GAP? – WHAT GAP!

G. Hayward[1] & S. Bower[2]

[1]*Director, Consumer Risk Ltd, 83 Station Road, London N3 2SH*
[2]*Engineering Manager (Rail Infrastructure) Heathrow Express Operating Company*

Heathrow Express (HEx) is a railway designed to 21st century airline travellers' expectations including easy accessibility Could it be possible that it was seeing a higher incidence of accidents to passengers getting on and off trains than railways built during the 19th century with much worse gaps between the carriage and platform? In the peak year of 2000, HEx was recording on average one "stepboard accident" (a fall or trip at the interface between the carriage floor and station platforms) every 3 weeks – one per three hundred thousand passenger journeys.

Following studies of the human factors issues interventions were implemented both to reduce the susceptibility to human error and to physically design-out the risk as far as possible. Within four years, the overall incident rate halved. However while rates were reduced by an order of magnitude at the modern stations (where initially incidence had been highest) the trend reversed (ie incidents increased) at the one 19th century legacy station used by HEx trains.

This paper describes the investigation of possible causes, analyses the effects of the remedial measures implemented and examines connections with theories of risk-homeostasis and habit-intrusion.

Introduction

Heathrow Express (HEx) is a state-of-the-art railway opened in 1998 by the company that operates London's Heathrow airport. Its trains run from underground stations serving Heathrow's terminals non-stop to Paddington Station in the centre of London.

Although the trains were built to the standard external dimensions for British main line railways, 21st century technology and styling are apparent in the passenger environment, operational controls and signalling. HEx trains offer full wheelchair-accessibility, achieved through matching the height of the platforms and the carriage floor. In principle, there should be less risk of accidents when boarding Heathrow Express trains because there is less of a step (and less of a gap) than on most UK railway lines.

Accident records and site observations

Brief free-text descriptions had been recorded of each incident. Accidents while boarding outnumbered those while alighting (by at least 2:1). A few passengers admitted not having been paying sufficient attention or not having noticed the gap. Other contributory factors were recorded in about half the cases, but no particular factors or sequence of events was associated with more than a small percentage of cases. Those most frequently noted were: (1) handling luggage – including pulling wheeled luggage;

(2) stepping backwards on or off the train; (3) hurrying to catch a train or flight; (4) engaging in conversation or a task with one foot on the train and one on the platform; and (5) heavy flows of other passengers.

Female passengers outnumbered males in these accident records by a factor of 2:1. Generally the range of their estimated ages was between 20 and 70 years, with just one child (aged 11) and one or two more elderly. It appears that these accidents most commonly occur to fit and healthy adults.

The horizontal gap between the platform edge and the carriage stepboard was 90 to 100 mm at a straight platform (increasing to 210 mm at the maximum platform curve). The stepboards were practically level with the platform edge, and 240 mm deep, but there was then a small step up (60 mm) from the step board to the edge of the carriage floor.

The platforms at the airport stations are light in colour and brightly lit with no strong visual contrasts. Paddington platforms are less brightly lit than the train doorways, but at all platforms, there was more contrast between the edge of the carriage flooring and the step board than between the step board and the platform. However the feature of the train that really stands out is the black band at window height running the full length of each set of four carriages, which appears to be unbroken because the windows are themselves smoke-black. The only visual relief around the door area in the original design was provided by two small no-smoking pictograms and a small yellow triangle beside the door operating button.

Most passengers alighting were observed to step right over the stepboard. Passengers boarding did not exhibit such a consistent pattern with regard to where they placed their feet, but all displayed an awareness of the small step from the step board to the carriage floor. Many passengers were travelling with a considerable amount of luggage (which the train interiors are designed to accommodate), but the only ill-advised behaviours observed were (1) dragging wheeled luggage straight across the interface with the expectation that the wheels would not catch either in the gap or on the step up to the floor, and (2) standing for a short while with one foot on the step board and the other on the platform (eg while waiting for the train's departure to be announced).

Applicability of ergonomic theories and analysis to the accidents

If the unexpected frequency of stepboard injuries being recorded on Heathrow Express actually represents higher risk of such incidents (rather than being explained by a greater likelihood of minor accidents being recorded on HEx than on other lines), then this could be seen as an example of risk compensation predicted by "risk homeostasis theory" (Wilde 1994). The hypothesis would be that when a person has to make an abnormally high or wide step to get onto a train, they do it with more attention and thus less risk of making an error, but when the step and gap appear to be small, the same person might not interrupt other tasks (eg talking on a mobile phone or looking at their watch).

In contradiction, it is arguable that risk homeostasis theory can really only be applied to continuous activities where participants have time to adjust the level of risk to their "target" (eg by driving faster) and is not appropriate to short discrete tasks where the "gain" in taking less care is negligible and the amount of extra mental attention required to negotiate the obstacle safely is minimal.

A more practical insight into accidents on Heathrow Express stepboards may be gained by analysing them through the ergonomics theories of human error (Reason 1984 & 1990). All the immediate causes ascribed to the recorded accidents would (in Reason's structure) be classified as lapses or "omitted checks" by people who were all quite capable of crossing the interface safely but who failed at a crucial moment make the appropriate observational checks of the surface under their feet. Reason describes one frequently observed cause of such lapses as "habit intrusion".

However, for this to happen repeatedly to different people just around the doors of Heathrow Express trains there must be something about the environment encouraged (at a subconscious level in some passengers) that encourages association with types of activity less hazardous than boarding and alighting from railway trains.

Arguably many of the visual cues (or reminders) people might (sub consciously) associate with being on a railway platform and boarding a train are absent and there is a strong similarity to the environment (and the activity) associated with the doorway of a lift in a hotel or office building. The long flat vertical black band dominating the carriage sides is somewhat reminiscent of the appearance (close-up) of some styles of plate glass office architecture, since when standing close to the train, this band fills the field of vision. A contributory factor may also be that right next to the doors there is no view into the carriage other than through the doorway, so the opportunity for hazard intrusion would be strongest when passengers paused for a short time at the interface or just before crossing it (eg waiting for the doors to open).

Unfortunately behavioural habits (or safety expectations) associated with entering or leaving a lift would not trigger precautions adequate for the HEx platform interface. Apart from the change of level between the carriage floor and the stepboard, the 100 mm gap between the stepboard and the platform presents far more of a residual hazard for feet is encountered in any modern lift (where the gap between the floor of the lift car and the landing is limited to a maximum of 20 mm).

Interventions

This analysis suggested that the risks could be reduced by making the carriage entrances appear less like a building or lift entrance and by generally increasing the use of strong colour and contrast to capture attention. Following HSE guidelines for highlighting floor level hazards in workplaces, the appearance of HEx train entrances was modified by the addition of (a) pictograms and text warnings clearly visible on both sides of each door and (b) black/yellow chevron markings on the step board.

These visual changes were applied to the whole fleet of carriages over a period of time. This was completed by June 2002, but the accident records note that chevrons had

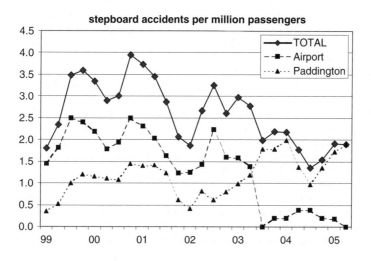

appeared on the stepboards of some carriages as much as a year earlier. There was also at least one major awareness raising campaign directed at staff and passengers.

These measures were followed by the more fundamental intervention of reducing the width of the gap by adding strips to the edges of the platforms. This was completed at the two airport stations by mid 2003, but was not possible at Paddington due to the curvature of part of the platform and the potential use of these platforms by other companies' trains.

Observed trends in accident rates

The effect of these interventions was evaluated by analysis of the annual incident records and passenger figures *on a rolling quarterly basis* (to provide some smoothing) with the use of some interpolation in order to present rates over a 7.5 years time span.

The steady decline in stepboard accident rates from the peak of 4 per million at the end of 2000 through to the beginning of 2002 (which was reflected at all stations on the line) coincided with the first major intervention programme, which began with 7-week awareness – campaign, continued with trials of painted stepboards and ended when all trains had received yellow hatch markings on their stepboards and warning labels on their doors. By this time the overall rate of stepboard accidents was halved.

Subsequently – between the first trials platform edge strips in the second half of 2002, through to completion of their installation at both airport stations – a 90% decline was observed (but only at those stations).

However, the data also show a counter-trend of increasing accidents at Paddington over a two-year period from the beginning of 2002. The most likely explanations of this trend are events outside the control of the railway. When HEx opened, British Airways and other major airlines offered check-in facilities at Paddington so that passengers did not have to load their own hold luggage onto trains (though many of them chose to do so). Use of this facility by airline passengers declined markedly as all airlines increased their scrutiny of passengers luggage following the plane hijackings on September 11 2001, and the Paddington check-in was closed permanently in July 2003. The period between these two dates – over which increasing numbers of HEx passengers would have been taking their "hold" luggage onto the train at Paddington – corresponds fairly closely with the period over which stepboard accidents were rising again at Paddington. (A simultaneous rising trend was also initially observable at the airport stations (where the passengers had to unload this luggage) – until the gap reduction work started to have an dominant effect.)

Conclusions

Seemingly benign design improvements may produce unpredictable human factors side effects. The improved accessibility of Heathrow Express trains may have led – through risk compensation – to passengers putting themselves at greater risk of their feet slipping between a station platform and a train's stepboard. However, the more significant contributory factors were probably increased manual handling of luggage and habit intrusion lapses prompted by the visual styling of the carriages.

In order to achieve continuing reductions in risk of injury there are roles for the routine recording and investigating of minor incidents, raising awareness of staff on the ground, involving outside experts, and adopting good practice from non-transport sectors.

Accident statistics are subject to many influences that can make it difficult to determine quantitatively the true long-term value of implementing prevention measures.

References

HSE Guidance for employers on identifying hazards and controlling risks – Slips and trips, 1996

HSE Guidance on the Safety Signs and Signals Regulations 1996

HSE Guidance: reducing error and influencing behaviour (2nd edn) 1999

Reason, JT 1984 "Lapses of attention" in Parasuraman, R & Davies, R (eds) "Varieties of attention" – Academic Press, New York

Reason, JT 1990 "Human error" – Cambridge University Press

Wilde, GJS 1994 "Target risk" – PDE Publications

DESIGNING A SYSTEM FOR EUROPEAN ROAD ACCIDENT INVESTIGATION

C.L. Brace[1], H. Jahi[2], L.K. Rackliff[1] & M.E. Page[1]

[1]*Vehicle Safety Research Centre, Ergonomics and Safety Research Institute, Holywell Building, Holywell Way, Loughborough, Leicestershire LE11 3UZ*
[2]*Institut national de recherche sur les transports et leur sécurité, 25, avenue François Mitterrand Case 24, 69675 Bron cedex, France*

The toll of accidents and injuries on Europe's roads is unacceptably high. In order to reduce the shocking figures, independent and reliable accident investigation needs to be conducted so that effective countermeasures can be determined. Despite systems being in place to examine aviation, rail and maritime accidents, there is currently no standard approach that addresses road safety. This paper describes the needs for such a system and the approach that is being adopted to design best practice guidelines for Europe for implementation in the development and management of road accident data capture processes, and in the subsequent use of any resulting data. A requirement of the guidelines is that they are designed for optimum independence and transparency. This work is being undertaken as a work package within the SafetyNet integrated project (http://safetynet.swov.nl/).

Introduction

Each year within the European Union (EU-15), there are approximately 40,000 people killed on the roads and over 1.7 million people injured (European Commission, 2005a). Such accidents cost the Community over 180 billion Euros annually, equal to 2% of the EU's GNP. With the growth in the number of EU member states (to EU-25), the European road death toll is set to increase to even more dramatic heights. To put these figures into perspective, "road crashes are the second most serious cause of death and hospital admission for EU citizens, preceded by cancer and followed by coronary heart disease" and for Europeans under 45 years of age, road crashes are the largest single cause of death (ETSC, 1999).

The number of people killed and injured on the roads started to decrease considerably from 2002 onwards (European Commission, 2005b), with improvements year on year for 2003 and 2004. However, there has not been such steep decline in the overall number of accidents; accidents are still occurring frequently although improvements in vehicle design have helped to reduce the severity of injuries to the people involved in accidents (especially car occupants), but the number of slight injuries has not decreased.

Despite these improvements in injury outcomes, it is estimated that 97% of all socio-economic costs for transport accidents within the EU are as a result of road accidents, and that 97% of the transport related fatalities occur in the road sector (ETSC, 1997).

The EU target of a 50% reduction in fatalities on the roads by 2010 (European Commission, 2005a) will only be achieved by the introduction of the most effective

---

On behalf of the SafetyNet Consortium.

Table 1. Number of people killed within the different modes of transport during 2002*.

Transport Mode	Road	Rail	Maritime**	Aviation
Number of people killed (in EU-15)	38604	121	1273	101

*Figure from European Commission, 2004. **This is a Worldwide figure. The year depicted was above average with the mean number of fatalities between 1998–2003 standing at 526.

countermeasures. It relies on the existence of basic knowledge of crashes and their causation and the availability of road safety data to monitor and assess performance.

Regarding the legal framework of accident investigation at European level, the issue is fairly well addressed in the more public transport modes, and it is interesting to examine the accident statistics across the main transport modes, Table 1. For civil aviation there is a comprehensive legal structure with two definitive European Directives addressing the issue of accident and incident investigation and reporting. In the field of maritime transport, the European Directive is not specific to the accident investigation and does not require the Member States to establish an independent investigation body. However, while Member States have no formal obligation to establish an independent investigation body for the investigation of maritime casualties, this remains an objective. In the mode of rail transport, a recent European Directive requires the Member States to establish an independent accident investigation body.

For road transport, there is a significant difference in traffic accident investigation from the "public" transport modes. While road transport safety has more recently emerged as an issue on all political decision-making levels, it has for a long time been neglected compared to the issue of safety in the rail, aviation and maritime sectors. There is no international or European legal framework that currently exists within the EU for investigating and learning from road accidents (SafetyNet, 2005; ETSC, 2001).

Brief of the work

This work is being undertaken to design best practice guidelines for Europe to be implemented in the development and management of road accident data capture processes (input), and in the subsequent use of any data resulting (output), e.g. accident investigation protocols, data collection, data storage (databases), data use. A requirement of the guidelines is that they are designed for optimum independence and transparency. This work is being undertaken as a work package within the SafetyNet integrated project (http://safetynet.swov.nl/), funded under the European Commission's 6th Framework Programme. The "inputs" refer to the type of response an institution should have, e.g. data capture processes, who investigates, how many investigators should be involved dependent on experience etc., and what level this experience should be, how the data is protected etc., in order to design and implement an independent and transparent road accident investigation process. It can be emphasised that the parts of the best practice guidelines which give suggestions for the way an accident investigation procedure should be carried out focus on the areas to investigate rather than "how" to investigate. For example, it is not in the project brief to provide forms for accident investigators to complete at the scene of an accident. Instead the guidelines should suggest areas that they should investigate, i.e. the driver's behaviour, the car's structure.

The "outputs" refer to what happens to the resulting data from an independent accident investigation process (i.e. what happens to the data once it has been collected), e.g. how data is stored (e.g. by whom – security, confidentiality), how data is accessed (e.g. database of raw data, aggregated data or e.g. www, book (annual data release)) and use of data (e.g. by whom, for what).

A key objective of this work is to determine what level of investigation (if any) should take place for any given accident type. At one end of the spectrum, recommendations will be made for standard EC procedures that should be followed in the event of a major road incident or where there is a strong public interest in the reasons for the crash. At the other end of the spectrum, guidelines for "best-practice" for investigations of road accidents of a more routine nature will be proposed. The task will consider the nature of the response of institutions to the need to investigate major road incidents and the roles of road safety stakeholders in identifying new opportunities for casualty reduction.

Methodological approach

A field research approach was developed, designed to optimise research efficiency and expertise, whilst capitalising on the allocated resources.

Step 1: Review of best practice

This exercise focuses on gathering information from actual users of systems currently in place for accident investigation, both intrinsic to road transportation, and extrinsic to other transport modes. It can be suggested that much of the information will relate to the "inputs" with some information also collected with respect to the "outputs". This process will examine how to deal with the different levels of response required to be incorporated into the guidelines. For example, the different modules of information required and what should be required of different countries.

(A) Organisations who have a "driving" workforce will be contacted, e.g. postal/delivery companies etc. in order to understand strategy protocols regarding road accident investigation so that an understanding can be gained of the methods/approaches followed when a road accident occurs amongst a workforce.
(B) A representative proportion of police forces across Europe will be contacted, in order to understand the approaches used in road accident investigation, and the legislation/background to these methods.
(C) A cross-section of independent accident investigation bodies (including Government related) will be contacted in order to understand strategy protocols regarding road accident investigation so that an understanding can be gained of the methods/approaches followed when a road accident occurs within their remit.
(D) The accident investigation processes within the other main modes of transport (air, maritime, rail) will be examined.

Within activities [A]–[D], structured interviews will be used to understand the methods employed. Additional to these activities, the wider research literature will be reviewed to capture any additional processes that can be learnt from. All the methods ascertained will be critiqued and the potential techniques that can be considered for incorporation into the independent European guidelines for road accident investigation will be extracted and documented.

Step 2: Identification of users

A list of potential users of the best practice guidelines will be compiled for each country represented by the project partnership. These users will include:

- National public administrations, e.g. ministries and departments
- Bodies directly involved in accidents, e.g. police, dedicated investigation bodies (where they exist)

- Insurance companies
- Industry, e.g. vehicle manufactures, road constructors
- Research and scientific institutions, e.g. public/private institutions, universities
- Professional associations, e.g. freight transport associations, unions
- Other, e.g. road users associations, charities, AA, RAC

This matrix of potential users and other interested parties is required for Step 4: Period of Consultation.

Step 3: Preparation of draft procedure

The critique of the techniques uncovered in Step 1 will assist in the development of a set of independent road accident investigation best practice (draft) guidelines. It is likely that these guidelines will include techniques regarding: development of data capture processes, management of data capture processes, storage of collected data, and use of collected data. The guidelines will also outline the financial implications of adopting the different elements of approach. It is likely that these guidelines will be separated into sections, including guidelines for how to manage the investigation of different types of accidents, e.g. fatal, serious, etc.

Step 4: Assembly of "top level" opinion

Although the sponsors of the work will be given opportunity prior to this exercise to inform the direction of the project during routine project monitoring, this exercise will focus on gathering other "top level" opinion. Therefore, (senior) practitioners and policy makers (stakeholders), across Europe, including the European Commission (EC), will be interviewed. Generally, these individuals do not themselves conduct accident investigation but rather use the results of such systems to plan policy etc. It is anticipated that most of the information will relate to the "outputs" with some information also collected with respect to the "inputs". This process will also examine how to deal with the different levels of response required to be incorporated into the guidelines. For example, the different modules of information required and what should be required of different countries. The key themes to be examined include: general feedback on draft guidelines, anticipated difficulties with proposed methods in guidelines and indications of who else the project consortium should be targeting for feedback.

Step 5: Period of consultation

This process will continue to examine how to deal with the different levels of response required to be incorporated into the guidelines. For example, the different modules of information required and what should be required of different countries. The draft procedure will be circulated to the identified contacts in Step 2 and feedback and comments will be requested using a pro forma that addresses the different aspects of the guidelines. A workshop will be conducted using the knowledge and experience of the EC's National Experts panel. A second workshop will be held with other types of individuals identified in Step 2 and where suitable, with individuals identified during Step 4. In each workshop, the ideas in the draft guidelines will be presented and expert feedback from these policy makers and practitioners will be obtained, enabling feedback to be obtained using top down and bottom up approaches. Additionally, interviews will be conducted with other nominated contacts from Steps 2 and 4 (non-workshop attendees) where the ideas in the draft guidelines will be presented and additional expert feedback from practitioners and policy makers will be obtained.

Step 6: Period of iteration

The results of the Steps 4 and 5 consultations will be complied and the draft best practice guidelines adjusted accordingly. The revised guidelines will be re-circulated to the consultation group (Steps 2 and 4) and feedback requested.

Step 7: Preparation of final guidelines

Upon receipt of final comments and feedback in Step 6, a set of full guidelines will be produced. This will be in the form of a report in November 2007.

References

European Commission (2004). *Energy and transport in figures 2004*. Accessed online [9 December 2005] from: http://europa.eu.int/comm/dgs/energy_transport/figures/pocketbook/2004_en.htm

European Commission (2005a). Accessed online [9 December 2005] from: http://europa.eu.int/comm/transport/road/roadsafety/index_en.htm

European Commission (2005b). Accessed online [9 December 2005] from: http://europa.eu.int/comm/transport/road/figures/accidents/quickindicator/index_en.htm

European Transport Safety Council (ETSC) (1997). Transport Accident Costs and the Value of Safety. ETSC, Brussels

European Transport Safety Council (ETSC) (1999). Risk of Death and Injury of Travel in the European Union. Accessed online [9 December 2005] from: http://www.etsc.be/documents/pre_8jun99.htm

European Transport Safety Council (ETSC) (2001). Transport Accident and Incident Investigation in the European Union. ETSC, Brussels. Accessed online [9 December 2005] from: http://www.etsc.be/documents/accinv.pdf

SafetyNet (2005). Deliverable D4.1 "Bibliographical Analysis". Accessed online [9 December 2005] from: http://safetynet.swov.nl/

CHALLENGES IN THE USABILITY EVALUATION OF AGRICULTURAL MOBILE MACHINERY

Piia Nurkka

*MTT Agricultural Engineering Research, Vakolantie 55,
03400 Vihti, Finland*

Agricultural work will change when converting to Precision Agriculture. More site specific information about the crop and the field is needed and supplied. The functionality and the degree of automation in the machinery increase. New control system for this machinery needs to be developed. To prevent the human operator from becoming overloaded with information, adequate forethought has to be given during system development to what information the system will display and what information it will withhold. The interaction between the human and the machine should be evaluated to confirm the suitability of the predicted interactions. This paper introduces the concept of situation awareness as a possible means of a usability evaluation methodology of agricultural mobile machinery control system on the base of literature and experimental research.

Introduction

Agricultural work routines change due to the introduction of automation in machinery. The driver is faced with increased monitoring demands imposed by automation that stem from the component proliferation that automation in field machinery invites. This means that the human and the machine subsystems should be structured and designed to work in mutually cooperating ways. In the conventional design of human-machine systems the design of interaction between them is based on the technology-centered idea (e.g. Sawaragi and Murasawa, 2001). That has also been the case in agricultural engineering. However, the allocation of tasks between the human operator and automation should be done carefully to make it possible for the human and machine to collaborate to achieve the operating objectives (e.g. Hollnagel and Bye, 2000). That is why the interaction between the human and machine as well as the work context needs to be studied. Though much work has already been done in improving human centered design processes, some fields are still to be researched. New context of use require adapted usability tests (e.g. Zhang and Adipat, 2003) and new measures of usability to adequately capture what is considered important in the particular context.

Usability evaluation in agricultural engineering

Agricultural engineering is a new field in human centred design so there are no established usability techniques and measures. Examples in other domains have shown that an easy-to-use interface is critical for successful adoption and use of application. An important research issue therefore is how to conduct an appropriate usability test. Usability testing is an evaluation method used to measure how well a specific user can use a specific product or a system in a specific situation, aiming at to achieve the best possible interaction and relation between the user, the product and the task.

Definition of usability

There are several definitions of usability. E.g. ISO 9241-11 (1998) defines usability as "effectiveness, efficiency and satisfaction with which specified users can achieve goals in particular environment". According to e.g. Nielsen (1993) usability as well as utility is subcategory of usefulness and they all influence the system acceptability. Further, usability can be broken down to five attributes of use: easy to learn, efficient to use, easy to remember, few errors and subjectively pleasing. As Hornbæk (2005) points out, measures of usability are various, not established, as they depend on the context. To conclude, the term usability is to a large extent determined by how we measure it.

Challenges of usability evaluation in agricultural engineering

In agriculture it is difficult to foresee the exact situations of the product use (especially interface and control systems of machinery) as the user needs to perform diverse tasks in varied situations and environments. More over the user may be performing many demanding tasks simultaneously (e.g. driving a tractor while monitoring and controlling implement) that can be safety critical. The method needs therefore be fit to purpose.

Situation awareness

The definition of situation awareness lies in the separation of the human understanding of the system status and the actual system status. Basically, situation awareness is being aware of what is happening around you and understanding what that information means to you now and in the future (e.g. Endsley et al, 2003). The concept has been studied particularly in aviation industry as it is usually applied to operational situations, where people must have good situation awareness as poor situation awareness may have negative and safety critical consequences. Therefore it has important application for design as it is relevant when designing displays. The information in displays has to be usable both physically and cognitive. The key benefit of examining situation awareness is that it helps to understand how the data needs to be combined and understood.

The development process of systems to support situation awareness should according to Ensley (2001) begin with situation requirements analysis as it guides the development process. The design concepts should thereafter be evaluated by directly and objectively measuring operator situation awareness while using the product.

The experimental research

The experimental research explored the assessment of situation awareness as a means of usability evaluation in the field of agriculture. In focus of the evaluation was the information and control system of the implement, to be exact crop sprayer.

Five operators carried out a predetermined crop spraying task in a given test field. Operators' situation awareness was measured with both subjective and objective measures. Usability was measured by observing the system use and by user comments.

The results were compared to find out whether there is congruency with good operator situation awareness during the task and good usability of information and control system of the implement.

Conclusions and future work

The increased monitoring demands in agricultural field work changes the normal work routines. To prevent the human operator from becoming overloaded with information, the information processing capabilities of the operator should be analyzed, recognized and taken into consideration in the design of a new control system. The evaluation methodology of new design needs to be fit to purpose.

Literary research on situation awareness and the experimental research found shared characteristics of the two operational environments, namely: 1) simultaneous multiple goals 2) multiple tasks competing for the operator's attention; and 3) time stress and negative, safety critical consequences associated with poor performance. To conclude, although the roots of the concept situation awareness are in aviation it can be applicable in land based domains, like agriculture as well as.

The experiment proved that the assessment of situation awareness provides data of the actual use of the information provided to the driver as well as the sources of the information. Also the experienced situation awareness of the operators has some analogy with the usability of the system. This data can be used when making requirement for design of new control systems for mobile machinery. However more research has to be done to establish and distinguish the concepts of situation awareness and usability and to be able to compare the measures of them in the field of agriculture.

References

Endsley, M.R. 2001, Designing for situation awareness in complex systems. *Proceedings of the Second international workshop on symbiosis of humans, artifacts and environment*, Kyoto, Japan

Endsley, M.R., Bolte, B. and Jones, D.G. 2003, What is situation awareness? In *Designing for Situation Awareness. An Approach to User-Centered Design,* (CRC Press), 13

Hollnagel, E. and Bye, A. 2000, Principles for modelling function allocation. *International Journal of Human-Computer Studies,* **52**, 253–265

Hornbæk, K. 2006, Current practices in measuring usability: Challenges to usability studies and research. *International Journal of Human-Computer Studies,* **64**, 79–102

ISO 1998. ISO 9241-11:1998. Ergonomic requirements for office work with visual display terminals (VDTs) – Part11: Guidance on usability

Nielsen, J. 1993, *Usability Engineering*, (Academic Press Inc.), 362

Sawaragi, T. and Murasawa, K. 2001, Simulating behaviors of human situation awareness under high workloads. *Artificial Intelligence in Engineering,* **15**, 365–381

Zhang, D. and Adipat, B. 2003, Challenges, methodologies and issues in the usability testing of mobile applications. *International Journal of Human-Computer Interaction,* **18**, 293–308

RESEARCH ON THE INFLUENCE OF DESIGN ELEMENTS ON DRIVING POSTURE IN CHINA

Ning Zou[1,2], Shou-Qian Sun[1], Ming-Xi Tang[2] & Chun-Lei Chai[1]

[1]*Industrial Design Institute, College of computer science and technology, ZheJiang University, Hang Zhou, China 310027*
[2]*Design Technology Research Center, School of Design, The Hong Kong Polytechnic University*

> Research on driving comfort for Chinese people has just started. This paper presents the results of our experiments on driving posture from which a digital model of driving comfort has been developed. The experiment was concerned with identifying the relations between human body joint comfort and driving posture. It aimed at deciding the key design elements influencing car drive comfort in order to determine which elements have the most important influence on the joints of a human body in the Chinese populations. Based on the data obtained from the experiments, we have developed a digital driving posture model with design rules and calculation methods, to be used as the guidance on the design for driving comfort.

Introduction

With car sales in China reach more than five millions per year, research into car safety and driving comfort for Chinese drivers is becoming more and more important. Based on a review on international research on driving posture and comfort, we have found that it is necessary to study the characteristics of Chinese drivers in order to provide ergonomics information for the design of cars in China market. Our long term research objective in this area is to develop a computer aided ergonomics system, using the specialist knowledge and modules abstracted from experiments and analysis, to guide car design for China market with improved driving comfort and safety.

The ergonomic design for cars has mostly concentrated on the interior of car. Increasingly this is becoming problematic for several reasons: (1) the working posture of a driver may be incorrectly predicted, resulting in drivers having to twist their bodies when in the driving condition because of different pressures coming from the seats; (2) due to the narrow space, a driver's body may easily collide with the instruments; (3) the workplace may be improperly configured, with unnecessary increased length of the operating path.

Building on our previous projects, this research focused on the height of the steering wheel, the angle of the seat, the distance from the centre of the wheel to the panel, and their influences on diving postures. After a platform of a mock-up environment was built, 30 people were involved in the experiments. We obtained 18 groups of data for each person. When they adjusted their postures, we captured the data of the postures by using a motion capturing equipment. The data and the result are to be used in the implementation of driving posture forecast model in a computer aided ergonomics design system called ZJU-ERGO, which is a major research initiative being undertaken in ZheJiang University.

Design of the experiments

The experiments were conducted in an environment in which the seat height, the steering wheel position and the cushion angel were adjustable. The measuring of driving posture was done by a motion capture machine produced by Motion Analysis Co.

Three important variables, the steering wheel height, the seat angel and the distance from the center of the wheel to the pedal were considered independently in the experiments. Each variable stood for a kind of typical assemblage. The heights of the steering wheel were set to 57 mm, 97 mm, and 137 mm, respectively, which represented several types of vehicles ranging from sports car to truck. The seat angle (SAEL27) was defined as 5 degree and 20 degree respectively. The distance from the wheel center to point BOF (SW-BOFX) was adjusted from 300 mm to 400 mm, including 18 kinds of assemblage.

We marked fifteen joint points on the bodies of the subjects with the mark balls (Figures 1 and 2), which were distributed to two ankle joints, two knee joints, two sciatic joints, two wrist joints, two elbow joints, two shoulder joints, neck, thorax, and the center point between the eyes.

Result analysis

After the experiments, we gathered the data of the driving postures stored in the motion capture equipments, and we analyzed them in order to establish the influence of each group of data on the driving postures, and on every posture variable.

Figure 1. The forepart orientation.

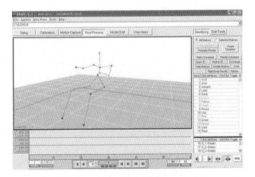

Figure 2. Driving posture sketch.

The influence of test data on the driving posture

We identified the important and independent influence of steering wheel height, wheel position and seat angle on the driving posture through ANOVA analysis. We found that the wheel position had a linear influence on driving posture. The method of regression analysis was used to evaluate the nonlinear characteristics of the data obtained, and to determine the relations between body measuring variables and posture variables. In terms of the relation between the body measuring variable and posture variable, we were particularly interested in knowing, for example, when the steering wheel position was changed, what different influence it would have on drivers who were tall or short.

In order to avoid the influence of pertinence on the data, we choose the ratio of height to weight (mm/kg) to analyze the effect of the posture. It was more easily to use less pertinence regression analysis than directly use height and weight. This reduced the mistake of statistics caused by pertinence. We performed regression analysis on all the captured data, and then identified the influence of the range of each design parameter on the posture variable, as shown in Table 1.

From the analysis, we observed the following relations:

1. The influence of the steering wheel
   The horizontal position of the steering wheel has a strong impact on the Z axis direction of center eye point. Moving back the wheel for 50 mm, the center eye point moved back 39.7 mm along Z axis. The back movement of the steer wheel caused a 19-degree reduction of the elbow angle, and a 6-degree increase of the angle between the thigh and rachis.
2. The influence of the seat angle
   Obvious influence on driving posture could be seen as the seat angle changed from 5 degree to 20 degree. When the angle increased, the hip point moved forward for about 26.275 mm, and the elbow joint angle increased by about 12 degree.
3. The influence of body dimension
   Under the same condition as in above (2), the H point of taller people moved back correspondingly. The height had a distinct impact on the eye center point.
4. The influence of the steering wheel
   The steering height had a strong influence on elbow, and as the steering wheel was hoisted by 80 mm, the elbow angle decreased for 13 degree.

Table 1. Regression analysis of the results.

Posture variable	Steering wheel distance (mm)	Seat angle	Seat height (mm)
Range	100	15	80
Left Knee Angle	0	0	0
Right Knee Angle	0	0	0
Left Elbow Angle	−19.7	12.435	−14.24
Right Elbow Angle	−19	11.385	−12.48
Cervical Flexion	−3.5	0	0
Thorax Angle	0	0	0
Center Eye Point x	0	0	0
Center Eye Point y	0	26.76	0
Center Eye Point z	39.7	0	0
Hip Z	0	22.275	0
Left Hip Angle	5.9	0	0
Right Hip Angle	6.9	0	0

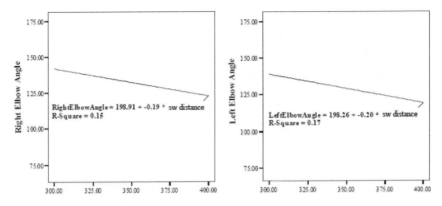

Figure 3. The influence of the steering wheel distance on the elbow joint.

Table 2. Coefficients used in regression analysis of the data

Posture variable	Steering wheel distance (mm)	Seat angle	Seat height (mm)	Ratio of H to W (mm/kg)
L-Knee Angle	0	0	0	0.21
R-Knee Angle	0	0	0	0.15
L-Elbow Angle	−0.197	0.829	−0.178	−1.50
R-Elbow Angle	−0.19	0.759	−0.156	−0.94
Cervical Flexion	−0.035	0	0	0.18
Thorax Angle	0	0	0	1.95
Center Eye Point x	0	0	0	2.81
Center Eye Point y	0	−1.784	0	−6.96
Center Eye Point z	0.397	0	0	0.7699
Hip Z	0	1.485	0	−8.92
Left Hip Angle	0.059	0	0	1.10
Right Hip Angle	0.069	0	0	0.91

The analysis of driving posture variables

By carrying out regression analysis of each driving posture variable and design variable with the captured data, the influence of driving posture on different joints was identified. The elbow angle can be calculated easily through regression analysis. Both of steering wheel's height and distance had distinct impacts on two elbows and the angle of two elbows would decrease as the two variables increased.

Figure 3 illustrates the influence of the steering wheel distance on the elbow joint with regression equation of the right elbow angle and steering wheel distance, $y = -0.19x + 198.91$, and regression equation of left elbow angle and steering wheel distance, $y = -0.197x + 198.26$ respectively.

Performing regression analysis of each driving posture variable and design variable with the captured data, we obtained the following data as shown in Table 2.

Conclusions

In this research, we obtained initial data from which the relations among driving posture, seat, pedal, steering wheel, and the body parameters can be analyzed. Using the method of regression analysis, we obtained the functional expressions representing these relations. These can be highlighted as:

1. The four variables have different influences on each part of our body, among which the steering wheel distance represents the most intensely influence on the elbow joint and hip point angle. It also has certain impact on the cervix angle and eye height, but not very obviously.
2. The seat angle influences the eye horizontal position, the horizontal position of hip joint and the elbow joint angle intensely, especially the eye and hip point.
3. The height of steering wheel only has some influence on elbow joint, but hardly on any other parts of the body.
4. The ratio of height to weight is the most active factor among four variables. It influences all the other factors except for the hip point angle and the eye position. It also has a distinct impact on the knee joint, the elbow joint and the horizontal position of the hip point.

After analyzing the influence of the design factors on driving posture, a rational model identifying the relation between the driving posture and design parameters was then developed. With this model and the data, the ergonomics levels of our car design can then be evaluated and simulated at real time.

Acknowledgements

This research has been supported by a key national research project funded by 973 State Science and Technology Initiative. We acknowledge the collaboration of CAD lab in ZheJiang University as well as all volunteers involved in this project.

References

Chunlei Chai, 2005, Research on the Technology of Ergonomics Design based on Driving Posture Prediction Model, *Dissertation Submitted to Zhejiang University in partial fulfillment of the requirement for the degree of Doctor of Philosophy.*

Byun, S.N., 1991, A computer simulation using a multivariate biomechanical posture prediction model for manual material handling tasks, *Ph.D. Dissertation, The University of Michigan, Ann Arbor, MI.*

Hou Honglun, Sun, S.Q. and Pan, Y.H., 2000, Research on virtual human in ergonomics simulation, *Chinese Journal of Mechanical Engineering*, 112–117.

Hsiang, M.S. and Ayoub, M.M., 1994, Development of methodology in biomechanical simulation of manual lifting, *International Journal of Industrial Ergonomics,* 13, 271–288.

DEVELOPING SYSTEMS TO UNDERSTAND CAUSAL FACTORS IN ROAD ACCIDENTS

C.L. Brace

*Vehicle Safety Research Centre, Ergonomics and
Safety Research Institute, Holywell Building, Holywell Way,
Loughborough, Leicestershire LE11 3UZ*

The persistent lack of accident and causation data to help inform and monitor road and vehicle safety policy is a major requirement at EU level. A recent analysis conducted by the European Transport Safety Council identified that no single accident database could meet all of the needs and that there were major gaps including in-depth accident causation. Specific policy questions at EU level involve the role of infrastructure in accident causation, the monitoring of progress towards the 2010 targets and the improvement of vehicle performance in accident and injury causation. This paper will give an overview of the situation and describe the development process of two databases designed to fill these gaps. The first is a broad ranging, intermediate level, fatal accident database. The second is an in-depth accident causation database. The two databases have divergent character and together will contribute a major advance of the knowledge of accidents and injuries at EU level. This work is being undertaken as a work package within the SafetyNet integrated project (http://safetynet.swov.nl/).

Introduction

Each year within the European Union (EUR-15), there are over 40,000 people killed on the roads and over 1.7 million people injured (European Commission, 2005). Such accidents cost the Community over 160 billion Euros annually, equal to 2% of the EU's GNP. If the additional road toll of approximately 23,000 persons killed each year in the EU's Candidate States were to be taken into account, the annual socio-economic cost would be around 250 billion Euros. To put these figures into perspective, "road crashes are the second most serious cause of death and hospital admission for EU citizens, preceded by cancer and followed by coronary heart disease" and for Europeans under 45 years of age, road crashes are the largest single cause of death (ETSC, 1999).

A core element of the EC road safety strategy includes a reduction of fatalities by 50% by the year 2010 (European Commission, 2005). Central to this strategy is the requirement for good quality in-depth accident data. Such data are seen as a fundamental pre-requisite for the formulation and monitoring of road safety policy in the EU. Data are needed to assess the performance of road and vehicle safety stakeholders and are needed to support the development of further actions. A recent analysis conducted by the European Transport Safety Council (ETSC, 2001) identified that no single accident database could meet all of the needs and that there were major gaps including in-depth accident causation. Specific policy questions at EU level involve the role of infrastructure in accident causation, the monitoring of progress towards the 2010 targets and the improvement of vehicle active and passive safety performance in accident and injury causation.

On behalf of the SafetyNet Consortium.

Brief of the work

This work is being undertaken as a work package within the SafetyNet integrated project (http://safetynet.swov.nl/) funded under the European Commission's 6th Framework Programme. This work uses an existing accident investigation network to develop two main databases;

(1) A broad ranging, intermediate level, fatal accident database; and
(2) An in-depth accident causation database.

This work addresses the need for a range of in-depth accident data and will provide two road accident databases which deal specifically with the causation of accidents. The information provided in the Fatal Accident Database will contribute a major advance of the knowledge of fatal accidents at EU level and tie in with the EU targets for fatal accident reduction. In 2001, the European Transport Safety Council (ETSC, 2001) stated that there is a lack of systematically collected data regarding representative samples from in-depth accident investigation which could be integrated into new safety policies. These issues are being addressed by the Accident Causation Database.

Methodolosssgical approach

The two databases will have divergent characteristics but will both contribute a major advance of the knowledge of accident causation factors at EU level. The data collection areas for the two Tasks will be from the countries with the largest fatality populations in Europe (Italy, France and Germany) as well as northern (Sweden, Finland) and middle European (UK, Netherlands) countries. Independent groups with no interest in commercial attributes of the study outcomes will conduct the accident investigation activities using a common protocol specifically developed for the study.

(1) Independent fatal accident database

The data will be derived from the police records of fatal accident investigations but will record strictly factual data only. The data recorded will describe the environmental factors, vehicle and driver factors and casualty factors to provide a description of the whole crash. The level of detail recorded will be considerably greater than is obtainable from databases of current specification; approximately 150 variables with 500+ items of data will typically be gathered. Specific areas of data will describe the overall accident circumstances, driver and vehicle characteristics, specific infrastructure features, and descriptions of other crash participants. A pilot and review activity will take place before the main data collection phase commences. The main data collection period will involve investigation of a representative sample of between 2% and 10% of the fatal crashes in each country covered, depending on the magnitude of the fatal population. It is anticipated that around 1,300 sets of fatal accident data will be gathered over one year and entered onto a database. Subsequently, a variety of data analysis techniques will be used to establish key trends.

(2) Independent in-depth accident causation database

The independent in-depth accident causation database will have major applications in the areas of new technology development and active safety systems as well as the more traditional area of infrastructure and road safety. It will be a new accident investigation activity and there is no direct model available internationally. Specialist teams will investigate the causes of accidents in six countries to give a representative survey of those Member States. It is expected that on-scene methods will generally be used to investigate the crashes. The main data collection will investigate 1,000 cases over a two year period. The purpose of the independent in-depth accident causation data will be to put together a crash investigation process that identifies the main risk factors leading to a crash. The data gathering methodology will record factual data on the circumstances of the crash that will be interpreted to provide descriptive information on the causation factors. It is expected that a multitude of variables, the numbers dependent on the type of scenario, will be recorded for each crash. The data will primarily address the development of new technology based countermeasures but it is inevitable that it will also be applicable to policy issues at national and EU level. The act of gathering the data and conducting in-depth crash investigations can also potentially initiate further awareness and develop insight into crash causation at national and local levels. The method used for the causation case analyses which will be the SafetyNet Accident Causation System. This is based on the existing Driving Reliability and Error Analysis Method (DREAM), which, in turn, is an adaptation for the area of vehicle traffic of the Cognitive Reliability and Error Analysis Method (CREAM), developed by Hollnagel (1998). The DREAM method has a Human-Technology-Organisation perspective, which implies that accidents happen when the dynamic interaction between people, technologies and organizations fails in one way or another, and that there are a variety of interacting causes creating the accident (Ljung, 2002). Subsequently, a variety of data analysis techniques will be used to establish key trends. The data will be of a detail that will enable clear understanding of the factors involved in a representative proportion of road accidents within the participating countries.

References

European Commission (2005). Accessed online [9 December 2005] from: http://europa.eu.int/comm/transport/road/roadsafety/index_en.htm

European Transport Safety Council (ETSC) (1999). Risk of Death and Injury of Travel in the European Union. Accessed online [9 December 2005] from: http://www.etsc.be/documents/pre_8jun99.htm

European Transport Safety Council (ETSC) (2001). Transport Accident and Incident Investigation in the European Union. ETSC, Brussels. Accessed online [9 December 2005] from: http://www.etsc.be/documents/accinv.pdf

Hollnagel, E. (1998). Cognitive Reliability and Error Analysis Method – CREAM. Oxford, Elsevier.

Ljung, M. (2002). DREAM – Driving Reliability and Error Analysis Method. Master's thesis, Linköping University.

AUTHOR INDEX

Author Index

Akbaba, M. 38
Alhemood, A. 50
Allatt, T. 378
Allen, P. 546
Amati, C. 551
Anderson, M. 570
Anokhin, A. 98
Ashdown, I. 285
Atkinson, P. 475
Axelrod, L. 208

Bagnall, P. 358
Ball, E. 399
Banks, A.P. 73
Barnes, C. 174
Barone, R. 45, 67
Baxter, G. 236
Beevor, A. 392
Benedyk, R. 231
Beswick, J. 570
Beucheler, I. 263
Bichard, J.-A. 431
Birchley, L. 395
Blackwell, A. 224
Bobjer, O. 338
Bouchard, C. 164
Bower, S. 635
Brace, C.L. 640, 653
Brown, S. 423
Burnett, G. 219, 630
Bust, P.D. 541

Case, K. 200, 465, 470, 499
Cauvain, S.P 45
Chai, C.-L. 648
Chandler, E. 448
Chandrashekar, S. 231
Cheng, P.C.-H. 45, 67
Chesters, R. 382
Chhibber, S. 199
Childs, T.H.C. 157, 174
Chirivella, C. 159
Choi, Y.S. 404
Choo, D. 174
Chui, Y.P. 144
Clarkson, P.J. 258, 338, 368, 373
Clift, M. 423
Coates, M. 122
Cockshell, S. 103
Cole, J. 321
Cook, S. 70

Cross, S. 555
Cullen, L. 575

Dadashi, Y. 280
David, H. 79, 83, 276, 299
Davies, G. 395
Davis, S.J. 139, 290, 293, 296
de Souza, M. 184
De, M. 263
Deighton, C. 54
Dejean, P.-H. 184
Dejoz, R. 159
Derrer, N.M. 139, 290, 293, 296
Dewsbury, G. 358
Dey, A.K. 616
Ditchen, D.M. 590
Dixon, E. 448
Dong, H. 338, 368
Dowie, M. 521
Drakopoulos, D. 611
Dubuc, L. 224

Edge, D. 224
Egan, E. 516
Eklund, J. 194
Ellegast, R.P. 590, 593
Erbug, Ç. 149
Eriksson, P. 179
Evyapan, N. 189

Fisher, J. 560
Flynn, F. 122
Fullwood, C. 139, 290, 293, 296

Gale, A.G. 17, 215, 318, 321, 348
Gamage, S. 395
Garaj, V. 419
García, A.C. 164
Genaidy, A. 50, 509, 603
George, K. 513, 516
Georgiou, D. 453
Gibb, A.G.F. 409, 436, 541
González, J.C. 164
Goodman, J. 368
Gould, V. 313

Greed, C. 426, 431
Gyi, D.E. 200, 443, 465, 470, 541
Gültekin, P. 149

Haines, V 461, 489
Haley, P. 521
Hallas, K. 531, 536
Hamilton, W.I. 560
Hanna, S. 103
Hanson, J. 431
Harrison, C.M. 241
Hasdoğan, G. 189
Haslam, R.A. 117, 409, 436
Haynes, F. 122
Hayward, G. 635
Healey, L. 199
Heaton, N. 521
Helander, M.G. 144
Henderson, J. 555
Henson, B. 174
Hermanns, I. 593
Hignett, S. 303, 308
Hill, B. 203
Hill, T. 333
Hilton, B. 122
Hitchcock, D. 122, 461, 489
Hodgkinson, C. 395
Hoehne-Hückstädt, U.M. 590
Hone, K. 208
Högberg, D. 499

Jacko, J.A. 404
Jackson, A. 570
Jackson, J. 453
Jackson, M. 122
Jahi, H. 640
Jarrell, J. 509
John, J. 333
Jones, A. 308
Jordan, P.W. 414, 625

Karbassioun, D. 219
Karlsson, I.C.M. 169
Keith, S. 246
Kessler, E. 587
Kletz, T.A. 3
Koh, C.W. 28
Kokkinaki, A. 448
Korkut, F. 189

Laberg, T. 343
Langdon, P.M. 258, 368, 373
Lansley, P. 363
Lardner, R. 580
Law, C.M. 404
Lee, I. 504
Lewis, C. 333
Lewis, T. 258
Lim, K.Y. 28, 88, 127
Lindberg, A. 194
Liu, X. 17
Livingstone, H. 54
Long, J. 203
Luther, R. 54

Mack, Z. 575
Maeda, S. 598
Makola, M. 603
Mann, D.D. 611, 616
Mansfield, N. 598
Marlow, P. 313
Marshall, E. 98
Marshall, R. 200, 465, 470
Martino, O.I. 139, 290, 293, 296
Mason, R. 521
Mateo, B. 159
May, M. 93
McBride, P. 338
McDermott, H.J. 409, 436
McKeran, W.J. 73
McNeill, M. 494
McTavish, A.-M. 263
Middleton, J. 382
Miller, M. 536
Mills, S. 587
Milnes, E. 526
Mitchell, J. 378, 382, 392, 395
Moes, N.C.C.M. 483
Monk, A. 236
Montero, J. 159
Morris, N. 139, 290, 293, 296
Mortby, P. 392
Mortimer, R. 122
Mourthe, C. 184
Mª Gutierrez, J. 164

Nendick, J.V. 62
Newell, G. 598

Newton, R. 329, 419
Nicolle, C. 399
Nurkka, P. 645

Olaso, J. 159
Onditi, V. 358
Ormerod, M. 329, 419
Ovenden, S. 378
Özdener, N. 38
Özdener, O.E. 38

Page, M.E. 640
Pappa, N. 45
Pasquire, C.L. 541
Pearce, K. 164
Pereira, L.M. 448
Persad, U. 373
Petersen, J. 93
Petrie, H.L. 241
Pettitt, M. 219
Pinder, A.D.J. 313
Pinnington, L. 423
Porter, J.M. 199, 200, 465, 470, 630
Porter, S. 199
Powell, S. 570
Prat, J. 164
Prynne, K. 630
Purdy, K. 17, 215

Qi, W. 271
Qiu, Y.F. 144

Rackliff, L.K. 640
Reeves, G. 580
Reilly, T. 513, 516
Roe, P. 448
Rouncefield, M. 358
Russell, E. 122

Sanders, M. 392
Scaife, R. 551
Schütte, S. 194
Scott, A. 531, 536
Scott, H.J. 318
Sener, B. 149
Sharples, S. 280, 285
Shi, F. 215, 348
Siemieniuch, C.E. 58, 62, 70
Simpson, L. 443
Sims, R. 200, 465, 470
Sinclair, M.A. 58, 62, 70
Slater, L. 263
Smethurst, J. 378, 395

Smith, A. 546, 565, 621
Smith, I.H. 58
Solves, C. 159, 164
Sommerville, I. 358
Song, T. 17
Spencer, D. 521
Sperling, L. 179
Stammers, R.B. 132
Stanton, N.A. 253
Such, M.-J. 159, 164
Summerskill, S. 465, 630
Sun, S.-Q. 648
Sutherland, S. 103
Swain, S. 489
Sütoluk, Z. 38

Tainsh, M.A. 13
Tang, M.-X. 648
Taylor, K. 303
Teasdale, D. 122
Thomas, P. 329
Thorley, P. 23
Tito, M. 157
Townsend, C. 392
Turner, M. 378

Vera, P. 159

Wadsworth, E. 546
Wallace, J.A. 513, 516
Waller, R. 392
Walsh, J. 461
Weinmann, S. 587
Wells, S. 23
Whitney, G. 246
Wikström, L. 169
Williams, C. 117
Wilson, J.R. 45
Wilson, T. 607
Wong, W. 246
Wood, J. 109
Woodcock, A. 33, 263, 453
Woolner, A. 453
Wright, A. 353
Wright, C. 387
Wu, J. 28

Yi, J.S. 404
Young, L.S. 45
Young, M.S. 253

Zou, N. 648